STUDIES IN INTERFACE SCIENCE

Dynamics of Adsorption at Liquid Interfaces

Theory, Experiment, Application

STUDIES IN INTERFACE SCIENCE

SERIES EDITORS
D. Möbius and R. Miller

Vol. 1

Dynamics of Adsorption at Liquid Interfaces
Theory, Experiment, Application
by S.S. Dukhin, G. Kretzschmar and R. Miller

Dynamics of Adsorption at Liquid Interfaces

Theory, Experiment, Application

STANISLAW S. DUKHIN

Ukrainian Academy of Sciences
Kiev, Ukraine

GÜNTER KRETZSCHMAR

Max-Planck-Institut für Kolloid- und Grenzflächenforschung
Berlin-Adlershof, Germany

REINHARD MILLER

Max-Planck-Institut für Kolloid- und Grenzflächenforschung
Berlin-Adlershof, Germany

ELSEVIER

Amsterdam – Boston – London – New York – Oxford – Paris
San Diego – San Francisco – Singapore – Sydney - Tokyo

ELSEVIER SCIENCE B.V.
Sara Burgerhartstraat 25
P.O. Box 211, 1000 AE Amsterdam, The Netherlands

First edition 1995
Second impression 2003

Library of Congress Cataloging-in-Publication Data

Dukhin, S. S. (Stanislav Samuilovich)
    Dynamics of adsorption at liquid interfaces : theory, experiment,
application / Stanislav S.Dukhin, Günter Kretzschmar, Reinhard
Miller.
        p.   cm. -- (Studies in interface science ; v. 1)
    Includes bibliographical references and index.
    ISBN 0-444-88117-4 (alk. paper)
    1. Adsorption. 2. Interfaces (Physical science)   I. Kretzschmar,
Günter. II. Miller, Reinhard. III. Title. IV. Series.
QD547.D84 1995
541.3'3--dc20   95-7868
                                                              CIP

ISBN: 0 444 88117 4

Transferred to digital printing 2005

Printed and bound by Antony Rowe Ltd, Eastbourne

STUDIES IN INTERFACE SCIENCE

INTRODUCTION TO THE SERIES

Interfaces and processes involving interfaces have gained growing interest beyond the classical and continuing studies of phenomena like surface activity, adhesion, and wetting. In this domain, catalysis involving inorganic active materials and well defined solid surfaces is a vast field of fundamental investigations as well as industrial applications. On the other side, biomembranes represent dynamic structures built of organic molecules with interfaces capable of molecular recognition and signal processing (cybernetics) or catalysis and charge transport (dynamics and energetics). A particular aspect here is the ordering of organic molecules at interfaces with the consequence of formation of functional units of molecular dimensions, i.e. supramolecular structures. A different approach to the construction of molecular machines in modern chemistry is the synthesis of complex supermolecules mimicking such functions by linking active subunits in a well defined manner and geometry.

The goal of interface science is a description of interfacial phenomena on the basis of supramolecular structure and intermolecular interactions. This series is devoted to contributions describing and reviewing recent theoretical and experimental developments. We intend to cover progress in the following fields: *interfacial phenomena* like statics and dynamics of contact angle and wetting, the surface free energy, adhesion, adsorption, electrochemical phenomena, mechanical and optical properties of interfaces; *monolayers and organized systems* including insoluble monolayers, monolayer phases, long range order, phase diagrams, multicomponent monolayers, formation and functions of designed organized monolayer assemblies; *colloidal systems* with formation and stability of emulsions, dispersions, foams, liquid crystalline phases; *thin films between solids* involving surface forces, the influence of water structure on the interaction between solid surfaces, and the effect of dipole layers near the surface; *impact of interfacial phenomena on environmental problems*, addressing phenomena such as adsorption and penetration of pollutants in model membranes and cells, synergetic effects of surfactants and pollutants, interfacial enrichment and separation of pollutants from aqueous solutions; *new technologies based on interfacial phenomena*, mentioning here only sensors and biosensors based on organized systems, designed organized assemblies for nonlinear optical phenomena, new lubricants, and coatings for controlled adhesion.

Emphasis will be on the scientific content, although new experimental methods should also be described in this series. There will be no restriction with respect to the materials, i.e. biopolymers may be included as well as polymeric and low molecular weight tensides, dyes, and solid materials of any kind. However, we do not want to consider UHV techniques for the investigation of adsorption of gases on solids and the catalysis of gas reactions on solid surfaces as subjects of this series. Reviews or monographs in rapidly developing areas are preferred, but new aspects in classical fields are also of particular interest. Each volume will be devoted to a single subject. The books are no textbooks, and the addressed readership will be specialists and graduate students involved in studies in interface science. We hope that this series may contribute to the development of this fascinating field.

Dietmar Möbius and Reinhard Miller, Series Editors

x

PREFACE

The important aspect of adsorption processes at a liquid interface is lateral mobility which can lead to lateral excess transport of adsorbed molecules. Lateral transport disturbs the equilibrium state of an adsorption layer. In many important systems, such as emulsions, foams, and bubbly liquids, the properties of a non-equilibrium adsorption layer can be essential. This has been demonstrated in the systematic work of the Russian and Bulgarian schools summarised in monographs like "Thin Liquid Films" by Ivanov, "Coagulation and Dynamics of Thin Films" by Dukhin, Rulyov and Dimitrov, and "Foams and Foam Films" by Krugljakov and Exerowa. These books pay most attention to thick film drainage and stabilisation/destabilisation of thin liquid films. This book is focused on other dynamic processes at liquid interfaces in general or connected with phenomena of emulsions and foams.

A topic closely linked to the developments of the dynamics of foams and emulsions is the experimental verification of existing theoretical models of adsorption dynamics and the derivation of new theories for more complex experiments. Existing models fail for example when the relaxation time of an adsorption layer exceeds the characteristic time of surfactant transport. This is a situation which quite often occurs because both parameters change in a wide range of time. Systematic experimental investigations are necessary at least to cover the application range of the adsorption dynamics models. Further progress towards understanding the physical mechanisms of so-called kinetic-controlled adsorption dynamics also requires special experimental studies. The elaboration of the theoretical and experimental foundation of adsorption dynamics at liquid/fluid interfaces is therefore one of the priorities of this book.

A second major topic is the understanding of adsorption dynamics as a transport phenomenon. Despite 30 years of research the theory of the dynamic adsorption layer on rising bubbles in surfactant solutions for the significant regime of large Reynolds numbers has been developed only for the extreme cases of strong or weak retardation of the bubble surface as a whole. The most interesting phenomenon of a rear stagnant cap formation is quantified only for small Reynolds numbers. This regime is insignificant as even thoroughly purified water contains sufficiently large amounts of surface active matter to provide a complete retardation of the bubble surface and the formation of a rear stagnant cap is impossible.

So far and with the exception of too-complex cases only the close link between adsorption dynamics and surface rheology has been satisfactorily described. The textbook by Edwards, Brenner and Wasan, "Interfacial Transport Processes and Rheology", fortunately deals with this and we need to touch this topic only briefly. We do this by outlining the role of surface rheology in dynamic interfacial studies and its interconnection with relaxation processes of adsorption layers, two strongly-interrelated topics.

Dynamic adsorption layers (DAL) influence practically all sub-processes which manifest themselves in particle attachment to bubble surfaces by collision or sliding. Surface retardation by DAL affects the bubble velocity and the hydrodynamic field and consequently the bubble-particle inertial hydrodynamic interaction. It also affects the drainage and thereby the minimum thickness of the liquid interlayer achieved during a first or second collision or sliding. Thus elementary acts of microflotation and flotation is systematically considered in this book for the first time with account of the role of DAL. Extreme cases of weakly and strongly retarded bubble surfaces are discussed which assists to clarify the influence of bubble and particles sizes on flotation processes.

Many of these topics were originally developed or inspired by B.V. Derjaguin and his school. For this reason this book is dedicated to B.V. Derjaguin to honour his outstanding contribution to the dynamics of adsorption layers and its application to other fields of colloid and surface science.

The authors have not managed to prevent this book growing overlong, although it is only a very subjective selection of the huge variety of material. A first reading to learn about the contents and future ideas and expected developments could be made by just running through the summaries of the twelve chapters.

Finally we want to use the chance to thank colleagues who supported us in the preparation of the book. We thank J. Eastoe / Bristol, A. Howe / London, L. Simister / London, and N. van Os / Amsterdam, for many hints and discussions to improve the text. Discussions with V.B. Fainerman /Donetsk, H. Linde /Berlin, S. Karaborni /Amsterdam, C.J. Radke / Berkeley, and J. Ralston / Canberra on recent developments allowed us to incorporate very new results and ideas. We thank also B. Buchmann / Berlin, D. Knight / London, S. Kozubovsky / Kiev, and S. Siegmund / Berlin for their technical support during the preparation of the manuscript.

CHAPTER 1

1. INTRODUCTION

This monograph deals with the dynamics of adsorption layers of amphiphiles at liquid/fluid interfaces. Its main characteristic is that many phenomena at such interfaces are caused by a non-equilibrium state of the adsorption layer. Effects based on non-equilibrium adsorption layers are widespread in industry and in nature, for example in foam and emulsion science, flotation and microflotation, enhanced oil recovery, wetting, fibre production, coating processes, rising bubbles in surfactant solutions, surface waves, dynamic contact angles, flow properties and thinning of liquid films, mass transfer by Marangoni instabilities at liquid interfaces, sedimentation and filtration, cleaning processes.

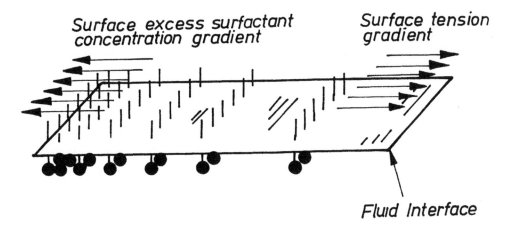

Fig. 1.1.   Schematic of the coupling between hydrodynamic flow with compression and dilatation of the surfactant adsorption layer; the adsorption/desorption processes are marked by arrows

A major issue in this book is the physical and mathematical modelling of the coupling of hydrodynamic to surface forces. This coupling is inherent in processes such as thinning of liquid films, rising of bubbles in surfactant solutions, and propagation and damping of surface waves. Fig. 1.1 represents a simple demonstration of this coupling. The two forces have opposite direction and cancel each other out.

We learn from Fig. 1.1 that adsorbed surfactant molecules, presented graphically by tails (lines) and head groups (circles) are more densely packed in the direction of the hydrodynamic flow at the interface. As the thermodynamic potential of the adsorbed molecules must be the same at any point of the interface, the surfactant molecules at closer packed areas desorb into the bulk phase or spread in direction of lower surfactant surface concentration, while at places at the interface less packed by surfactants adsorption takes place.

1.1. SURFACES, SURFACE TENSION AND SURFACE PHENOMENA

A liquid interface is the first requirement for all the many forms of capillarity. At least one phase must be sufficiently fluid. The shape of a liquid with an interface to air or another liquid is determined by the surface or interface tension. The shape of a liquid surface or interface at rest changes only when the surface tension, forces of gravity, and in some cases the electric field forces (Lorentz forces) are altered, provided the solid boundaries have constant dimensions and certain angles. The shape of pendant drops, sessile drops on a solid substrate, a meniscus against a solid wall and the length of so-called surface waves are well-known examples of capillarity. These phenomena need separate examination of liquid interfaces, as the surface state between two phases cannot be deduced from their bulk properties.

A general survey of surface tension is restricted to measurement of mechanical forces. There is a variety of mechanical methods to determine this surface force. In recent years the measurement of surface or interfacial tension has been determined direct from the surface shape. Using a computer program to fit the fundamental Gauss-Laplace equation (Gauss 1830, Laplace 1806) to experimental drop shape coordinates yields exact surface tension data. This procedure can be applied to hanging as well as sessile drops. The most efficient fitting procedure, the axisymmetric drop shape analysis, was introduced by Neumann and co-workers (Rotenberg et al 1983, Cheng et al. 1990)

From the thermodynamic point of view surface or interfacial tensions are intensive parameters and their value does not depend on the extension of the interface. Originating in the physics and chemistry of surfaces, surface thermodynamics is now an independent field.

In this introduction it may be useful to give a brief definition of surface tension and surface free energy. The dimension of the surface tension is related to unit length. Lenard's classical experiment (1924) is one of the best demonstrations of the surface force of a liquid acting on an extended wire in contact with a liquid surface. By carefully lifting the wire from the level of the surface, a force can be measured for as long as the pendent lamella remains in contact with the liquid bulk. The force measured in this way, divided by the length of the wire, leads to a well-defined surface tension. It is obvious that the contribution of both free lamella surfaces has to be taken into account in the calculation of surface tension. Instead of an extended wire a ring or a thin plate, known as Wilhelmy plate, can be used. Although their work needed subsequent correction, the problem of the contact zone of the tensiometer ring and the liquid was first thoroughly investigated by Harkins & Jordan (1930). Their excellent work led to correction factors as a function of the geometry of the ring being adopted for the calculation of interfacial tensions. Lunkenheimer & Wantke (1981), and Lunkenheimer (1982, 1989) only very recently performed systematic experimental and theoretical studies on unknown phenomena in ring tensiometry and found substantial errors in this technique not taken into account by the Harkins/Jordan correction factors caused by the stretching of the liquid film on the walls of a hydrophilic vessels as well as the liquid surface. The contact angle effect of the liquid against the ring has also been examined by Lunkenheimer (1989). In principle, the Wilhelmy method is as accurate as the ring method, provided the errors discussed for the ring are negligible. Fig. 1.2 shows an illustration of the basic principles of direct mechanical measurement of interfacial tension. There are also other well-established methods, such as the drop shape, drop volume, and the maximal bubble pressure technique. Liquid meniscii and problems of exact surface tension measurements by different methods were intensively discussed by Padday (1976).

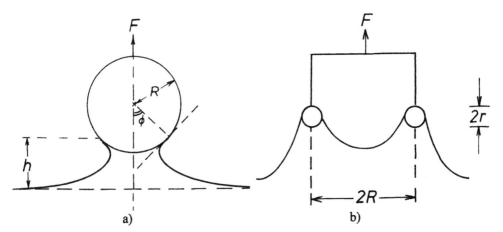

Fig. 1.2. Determining the surface tension of a liquid by direct force measurements a: extended wire of diameter 2R, b: ring method according to Lecomte du NoÜy (1919)

In contrast to the definition of the surface tension of a liquid by the acting force along a unit length, the surface free energy is defined by the reversible work necessary to extend the surface area by a square unit, e.g. 1 cm^2. The basic form of this definition is given by

$$\gamma = \left(\frac{d\overline{G}}{dA}\right)_{P,T,i}. \tag{1.1}$$

Here γ is the surface or interfacial tension, $\overline{G}$ is the Gibbs surface free energy, A is a unit surface area, and i is a component forming the interface.

The surface tension is the parameter representing the uncompensated intermolecular forces acting in the bulk phase (cf. Fig. 1.3). In the bulk each molecule is surrounded by identical neighbours, and the sum of all short and long range forces, acting at any point in the bulk is the same. Molecules close to the interface have different neighbours and these forces are not compensated for. The transition from a water phase to an air phase is illustrated in Fig. 1.3, using spheres as simple models for the molecules (Dynarowicz 1993).

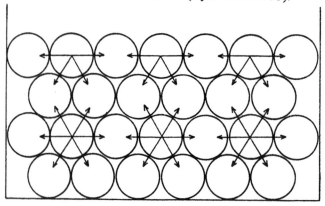

Fig. 1.3.   Simple representation of the asymmetric forces acting at a molecule close to the interface, according to Dynarowicz (1993)

Here, we need to say something about the behaviour of solid surfaces, such as metals, polymers, semiconductors, glasses, ceramics and woods. All these solid substrates have a surface tension like a liquid. Unlike liquids, the intermolecular forces of solids are strong enough (e.g. by lattice forces), to mean that the shape of their surface is not determined by the corresponding surface tension. Above the melting point, provided the solid can melt, such systems have the properties of a typical liquid. So for example, under the action of the surface tension a curved interface appears.

The equilibrium state of a surface is characterised by certain conditions which include the validity of equilibrium thermodynamics. Further essential prerequisites are the absence of

temperature gradients,

tangential and normal diffusion layers,

transport of mass, heat and impulse,

polarisation of an electric double layer,

hydrodynamic stresses in the adsorption layer,

external electric field forces.

Practically all surface properties can be modified by suitable surface active substances. Such detergents represent an important class of organic compounds and the volume of their world production is second only to polymers.

Since this book is dedicated to the dynamic properties of surfactant adsorption layers it would be useful to give a overview of their typical properties. Subsequent chapters will give a more detailed description of the structure of a surfactant adsorption layer and its formation, models and experiments of adsorption kinetics, the composition of the electrical double layer, and the effect of dynamic adsorption layers on different flow processes. We will show that the kinetics of adsorption/desorption is not only determined by the diffusion law, but in selected cases also by other mechanisms, electrostatic repulsion for example. This mechanism has been studied intensively by Dukhin (1980). Moreover, electrostatic retardation can effect hydrodynamic retardation of systems with moving bubbles and droplets carrying adsorption layers (Dukhin 1993). Before starting with the theoretical foundation of the complicated relationships of non-equilibrium adsorption layers, this introduction presents only the basic principles of the chemistry of surfactants and their actions on the properties of adsorption layers.

1.2.   SURFACE CHEMISTRY OF SURFACTANTS AND BASIC ADSORPTION PHENOMENA

The production of surfactants has increased enormously throughout the world over the last few decades due to the rise in applications of this class of substances. Although a considerable part of surfactants produced is still used for conventional purposes (washing and cleaning), the variety of applications can scarcely be underestimated. One application is as stabilisers of emulsions, foams, liquid films, suspensions and as wetting agents. Surfactants such as alkyl sulphate, alkyl sulphonates and ethylene oxide adducts, account for the bulk of this production. Since the number of unconventional structures of surfactants is continuously increasing the establishment of relevant structure/effect relationships is becoming increasingly complex. Depending on the particular effect we talk for example about emulsifiers, demulsifiers, wetting agents, foaming agents, or flotation agents.

Within this introduction we attempt to elucidate a number of colloid and interfacial chemical aspects of importance for the non-equilibrium surface states. New mechanisms of action will be discussed and, for clarity, the influence of surfactants on the dynamics of adsorption will take up most of our time.

1.2.1.   SURFACE CHEMISTRY OF SURFACTANTS

To begin with, the chemical structure of two selected representatives of surfactants is shown in Fig. 1.4. The surface activity of a chemical substance is based on its so-called amphiphilic character, comprising a non-polar hydrocarbon part, in most cases a hydrocarbon chain and/or aromatic or cycloalkane ring, and a polar part, e.g. a carboxyl, sulphate, sulphonate, phosphate, amino or polyoxyethylene group. The non-polar part, in general, determines the detergent activity, characterised by a functional connection between the number of adsorbed molecules at the air/water or air/oil interface and the relevant volume concentration of the surfactant. The polar group is responsible for the water solubility and, to a certain extent, counteracts the surface activity. Increasing the number of C-atoms in the non-polar part of the molecule increases amount of adsorption at these interfaces.

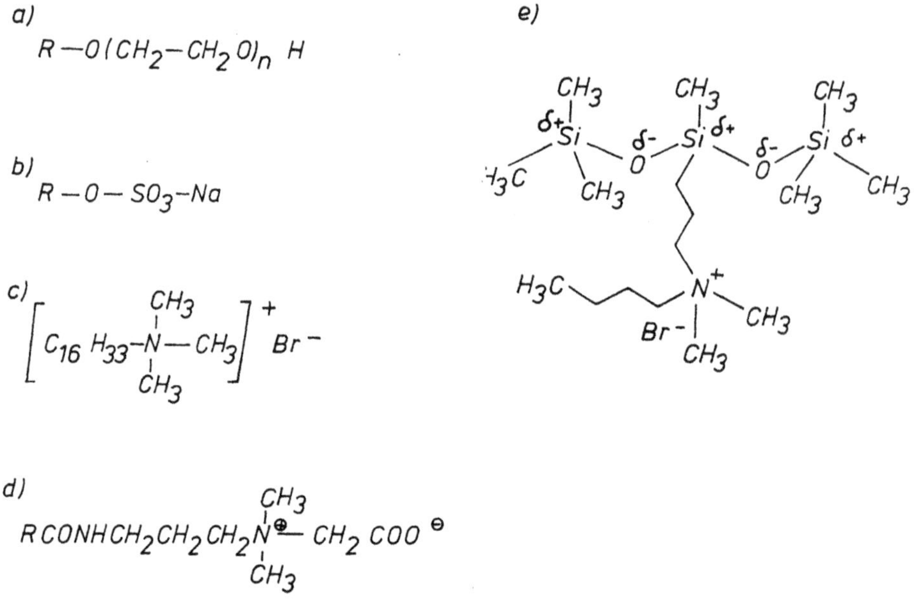

a)
$$R-O(CH_2-CH_2 O)_n H$$

b)
$$R-O-SO_3-Na$$

c)
$$\left[C_{16} H_{33}-\overset{\underset{\displaystyle CH_3}{|}}{\underset{\underset{\displaystyle CH_3}{|}}{N}}-CH_3 \right]^+ Br^-$$

d)
$$R\,CONH\,CH_2CH_2\,CH_2\overset{\underset{\displaystyle CH_3}{|}}{\underset{\underset{\displaystyle CH_3}{|}}{N}}^{\oplus}\!\!-CH_2\,COO^{\ominus}$$

e)

Fig. 1.4.   Main types of surfactants; a- nonionic on the basis of long-chain alcohol and ethylene oxide, b - anionic represented by sodium dodecyl sulphate, cationic represented by cetyl trimethyl ammonium bromide, d - betain surfactant represented by fatty acid amidopropyl dimethyl ammonio acetate, e - siloxane surfactant represented by N,N,N-trimethyl-3-(1.1.1.3.5.5.5-heptamethyl trisiloxane-3-yl) propylammonium bromide

Recently there have been remarkable developments in new classes of surfactants, for example of silicon-containing surfactants (Schmauks 1993).

The adsorption of the anionic surfactant sodium dodecyl sulphate (SDS), probably the most frequently studied surfactant and often used as model substance at the air/water and at the decane /interface is given in Fig. 1.5. The surface and interfacial tension have been plotted as a function of SDS concentration in the aqueous phase. From the slope of the tangents to the curves in Fig. 1.5 the interfacial excess concentration (adsorption density) Γ at different interfacial tensions can be calculated directly using Gibbs' fundamental adsorption isotherm (see section 2.4.1),

$$\Gamma = -\frac{1}{RT}\frac{d\gamma}{d\ln c}.$$ (1.2)

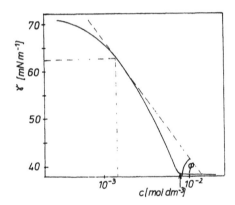

Fig. 1.5.   Graphic determination of surface concentration from adsorption isotherms, SDS adsorbed at the water/air interface, according to Vollhardt & Wittig (1990)

The Gibbs adsorption isotherm is the logical consequence of Gibbs thermodynamics, briefly described in Section 2.3. From this data the area occupied by a single adsorbed molecule can be calculated. In the limiting case, for a closely packed layer, the minimal molecular area is obtained. For SDS this value has been found to be 4 nm²/molecule, whereas for oxyethylene oxide adducts the values are substantially higher because they need greater space. While the Gibbs' isotherm is derived from fundamental thermodynamics (see Chapter 2), there are a number of other isotherms which are derived empirically or from a couple of assumptions. The following isotherms are the most important for the description of surfactant adsorption:

von Szyszkowski isotherm (1908),

$$\Delta\gamma = \gamma_0 - \gamma = \alpha \log(1 + c/\beta) \tag{1.3}$$

Langmuir isotherm (1918),

$$\Gamma = \frac{\Gamma_\infty x}{x + \exp(\Delta\mu_0^s / RT)}, \tag{1.4}$$

where x is the molar fraction of the surfactant in the bulk.

Frumkin isotherm (1925),

$$\Delta\gamma = -\Gamma_\infty RT \cdot \ln\left(1 - \frac{\Gamma}{\Gamma_\infty}\right) - \alpha'\left(\frac{\Gamma}{\Gamma_\infty}\right)^2, \tag{1.5}$$

in which γ_0 is the surface tension of the aqueous phase, γ the surface tension in the presence of a surfactant of bulk concentration c. An adsorption isotherm which also takes into consideration electrical effects and is applicable to systems close to the equilibrium state has recently been described by Borwanker & Wasan (1983).

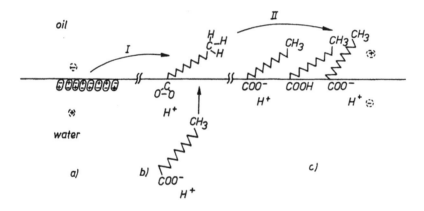

Fig. 1.6.   Energetic aspects of surfactants a liquid interfaces, I - restructuring by adsorption of single surfactant molecules, II - restructuring by molecular interactions in the adsorption layer, according to Kretzschmar & Voigt (1989)

The von Szyszkowski isotherm establishes the connection between the change in surface tension γ and the surfactant bulk concentration. Stauff (1957) has evaluated the parameters of this semi-empirical adsorption isotherm and has shown that it is in agreement with interfacial thermodynamics. Frumkin's isotherm has often recently been used to describe the adsorption of different types of surfactants, for example by Lunkenheimer (1983), Miller (1986), Wüsteneck et al. (1993), and others. One of the main aims of this book is to show that in the many applications of surfactant systems an full understanding of the equilibrium adsorption behaviour is vital to understanding the effects of a disturbance of the equilibrium state of adsorption. A deviation from the equilibrium adsorption, by which is meant the adsorption density at the interface as well as all parts of the electrical double layer, can produce several different effects. Non-equilibrium adsorption layers are related to phenomena such as surface elasticity, polarisation of the electrical double layer, non-linearity of electrokinetic effects, relaxations of the shape of a liquid surface and dynamic contact angles (cf. Fig. 1.6).

1.2.2. BASIC ADSORPTION PHENOMENA

Adsorption isotherms can be applied to any surface. In the following we focus our attention on surfaces covered with adsorption layers under dynamic conditions, the kinetics of adsorption and desorption of surfactants to and from soluble adsorption layers for example. Another phenomenon is the spread of surfactant molecules tangential to the surface that effect takes place if the adsorption layer is inhomogeneous (cf. Fig. 1.1).

At this point we want to note that adsorption layers, monolayers, and bilayers (thin foam and emulsion films, liposomes) can be formed not only by soluble amphiphilic molecules but also by those which are almost insoluble in the adjacent bulk phases. Such compounds, mainly phospholipids, form biological structures, such as biological membranes. Even foam films can be formed by insoluble amphiphiles as Exerowa et al. (1987) have shown. One of the parameters controlling the stability of biological membranes and thin liquid films against mechanical actions are the elasticity forces. Fig. 1.7 shows typical deformations, such as stretching and bending forces at the membrane. Starting from a flat membrane bending produces areas of higher and lower packing density. A similar effect of elastic forces can be seen in the adsorption layers of soluble surfactants.

The effect of surface dilational elasticity was first formulated by Gibbs (1957) as

$$E_o = \frac{d\gamma}{d \ln A} \qquad (1.6)$$

In Chapter 3 we will show that the modulus of surface elasticity is really quite a complex quantity. Whereas the flow properties of liquid/air and liquid/liquid interfaces are determined strongly by elastic forces, the solid/liquid interface is mainly related to wetting processes.

Other effects of significance for the application of surfactants include forced and spontaneous emulsification, stabilisation of emulsions and dispersions, the wetting of hydrophobic surfaces by surfactants solutions, and many more.

We will now discuss surfactant adsorption at a liquid interface. When the surface of a surfactant solution has been formed the adsorption equilibrium takes time to be established. This process is determined by transport from the bulk to the surface, mainly by diffusion processes (undisturbed diffusion, laminar or convective diffusion).

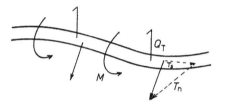

Fig. 1.7.   Model of a curved biological membrane with area elements of compression and dilatation, M - momentum resultant, Q_T - transverse resultant shear

In the so-called "sublayer"-model the jump of surface active molecules from the sublayer to the interface can play a decisive role. Such adsorption mechanisms are called kinetic-controlled. The complicated field of electrostatic or steric controlled mechanisms still has to be developed. Some preliminary ideas will be discussed in Chapters 4 and 7. The main principles of studying the dynamics of adsorption/desorption is the measurement of dynamic surface tension. A number of means of determining dynamic surface tensions in different time windows are given in Chapter 5. Short time studies can be performed by oscillating jet, maximum bubble pressure and drop volume methods. For longer time intervals of say 30 seconds up to hours, static methods can be applied, such as direct force measurements by ring tensiometry, Wilhelmy plate or hanging drop methods. Very short adsorption times can be studied by the oscillating jet method first used by Lord Rayleigh (1879). A complex formula developed by Bohr (1909) enables us to calculate the surface tension from the wave length of the jet, which refers to a dynamic surface tension.

A dynamic method for the measurement of dynamic surface potentials in the range 0.005 to 1 seconds has been described by Kretzschmar (1965). This device consist of a rotating rod dipping into the surfactant solution. The rotating rod transports a film of the surfactant solution below a vibrating plate condenser to determine the surface potential. As the age of the liquid surface is a direct function of the rotational speed, the time dependence of the surface potential is obtained. Fig. 1.8. demonstrates of this principle.

The device yields a time dependence of electrical surface potential. Such studies allow us to construct relationships between surface tension and surface potential as seen in Fig. 1.9. It becomes evident that a low surface tension drop correlates to a high surface potential change.

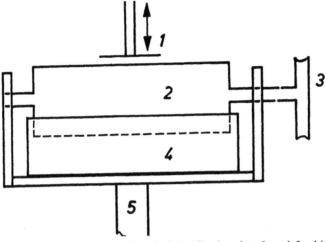

Fig. 1.8.   Schematic of the rotating rod method, 1 - vibrating plate, 2 - rod, 3 - drive, 4 - solution, 5 - lift, according to Kretzschmar (1975)

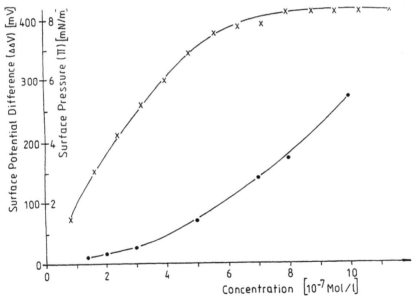

Fig. 1.9.   Surface potential (x) and surface pressure (●) as functions of concentration of heptaoxyethylene dodecylether, according to Kretzschmar (1976)

1.3.   ADSORPTION DYNAMICS AND DYNAMIC ADSORPTION LAYERS. QUALITATIVE APPROACH

We have already shown how surface active substances are able to modify significantly the properties of interfaces by adsorption. This is a law used in many processes and many new technologies are based on such adsorption effects. In general, these technologies work under dynamic conditions and their efficiency can be improved by a controlled use of interfacially active material. To optimise the use of surfactants, polymers and mixtures of the two, specific knowledge of their dynamic adsorption behaviour rather than of equilibrium properties is of great interest (Kretzschmar & Miller 1991). The importance of dynamics of adsorption in different applications has been recently discussed: froth flotation (Malysa 1992), foam generation (Fainerman et al. 1991), demulsification (Krawczyk et al. 1991), or emulsification (Lucassen-Reynders & Kuijpers 1992).

Two problems are inherent in many technologies. The interfaces involved are freshly formed and have an age of only seconds, sometimes even less than a millisecond. On the other hand the movement of interfaces, for example of drops in emulsions or bubbles in foams, disturbs the equilibrium of their adsorption layers.

After the creation of a fresh surface (or an almost bare surface) the adsorption increases until equilibrium has been reached. Liquid flow lateral to the surface creates concentration and surface tension gradients which induce additional surface flow. This effect is called the Marangoni-Gibbs effect.

In distinction to the equilibrium state the dynamics of adsorption is characterised by time dependence and inhomogeneous distributions of surface and bulk concentration. This is accompanied by hydrodynamic convection and convective diffusion as transport process for the molecules between the bulk phase and the adsorption layer.

While "kinetics" means only time-dependent, the terminology "adsorption dynamics" includes the coupling of transport by diffusion and hydrodynamic fields. It comprises surface concentration changes, movement in the adsorption layer and correlation between the distribution of surface concentration and velocities along the surface. The adjacent liquid bulk is involved in the diffusion and hydrodynamic flows which exhibit mutual interrelation. The term "dynamic adsorption layer" refers only to the non-equilibrium state of the adsorption layer.

Current activities in modelling adsorption dynamics are restricted to transport phenomena involving only simple flow geometries. More complicated problems of dynamic adsorption layers (DAL) remain to be solved.

In the simplest theories any liquid movement is neglected. The main feature of these theories is the description of the adsorption/desorption exchange as a two-stage process: transport of surfactant by diffusion in the bulk and exchange of molecules between the adsorption layer and the sublayer (or subsurface) adjacent to it, yielding two independent equations. The two unknown functions, surface and sublayer concentrations, can be determined from these equations.

In more complicated models both equations have to be generalised by coupling surface and bulk convective diffusion and hydrodynamics. The situation is finely balanced since the motion of the surface has an effect on the formation of the dynamic adsorption layer, and vice versa. Adsorption increases in the direction of the liquid motion while the surface tension decreases. This results in the appearance of forces directed against the flow and retards the surface motion. Thus, the dynamic layer theory should be based on the common solution of the diffusion equation, which takes into account the effect of surface motion on adsorption-desorption processes and of hydrodynamics equations combined with the effect of adsorption layers on the liquid interfacial motion (Levich 1962).

Quantitative models of adsorption dynamics for single rising bubbles are described in Chapter 8. The problem is mathematically so difficult that groups (such as Dukhin, Harper, Maldarelli, Saville) have produced only approximate solutions for some extreme cases, i.e. small Reynolds numbers, large Reynolds numbers at strong and weak surface retardation.

In conclusion adsorption dynamics at liquid interfaces is a very specific transport phenomenon and a challenge to mathematical physics. For the most important case of large Reynolds numbers and any degree of surface retardation no solution yet exists and we hope that this book will stimulate the solution to this problem.

Many theories is the restriction to low surface coverage, although larger coverages are of more technological relevance. The higher the adsorption layer is packed to stronger is the manifestation of specific mechanisms of rheology (surface viscosity and elasticity). Existing models are based on the universal Marangoni mechanism of surface retardation.

1.4.   SOME SURFACE PHENOMENA

This section is dedicated to some selected examples of the very broad spectrum of surface phenomena directly influenced by the dynamics of adsorption of surfactants. There are books and monographs which show these and other examples in much more detail (Adamson 1990, Dörfler 1994, Hunter 1992, Ivanov 1988, Krugljakov & Exerowa 1990, Lyklema 1991, Schulze 1984). However, to make the reader acquainted with some of the huge variety of applications, we inserted this section into the book.

1.4.1.   DYNAMIC CONTACT ANGLE AND WETTING

The last mentioned method is particularly connected with the wetting behaviour of a solid surface by a solution. If a hydrophobic solid surface is not wetted in such a way that a continuous stable liquid film is formed but disintegrates instead into isolated droplets. If the surfactants possesses adequate wetting effects the liquids spreads over the solid surface even when the pure solvent itself does not. The spreading coefficient, the difference between the adhesion work and the cohesion work of the liquid, controls the spreading. If this coefficient is positive the liquid spreads spontaneously over the solid surface. This is a typical dynamic effect stimulated by the hydrodynamics of the thin liquid film and the kinetics of adsorption of the surfactant at the film surface and the three-phase contact zone. This process is highly dynamic because kinetics of adsorption takes place during surface area expansion in a short time frame.

The terms "cohesion work" and "adhesion work" can be defined as follows. When the liquid and the solid are brought into contact with each other a new interface is formed (index 1 corresponds to the liquid, index 2 to the solid phase, index 12 to the interface). The total decrease of Free Energy of this process is given by

$$\Delta F = A(\gamma_1 + \gamma_2 - \gamma_{12}).$$
(1.7)

In the last equation $\gamma_2 - \gamma_{12}$ is the so-called wetting tension b. The cohesion work of the liquid is twice the surface tension value, equal to $2\gamma_1$. The difference between adhesion and cohesion work is the spreading coefficient, which now reads

$$S = \gamma_1 + \gamma_{12} - 2\gamma_1 = \gamma_2 - \gamma_1 - \gamma_{12}.$$
(1.8)

Thus the spreading coefficient is equal to $b - \gamma_1$, the difference between wetting an surface tension of the liquid. Since the spreading of a liquid presumes a contact angle $\theta = 0$ between the liquid and the solid phase, a positive value of S is needed. Positive S-values are obtained only if the surface tension of the liquid is less than of the solid. The surface tension of liquids can be determined very easily, while for solid surfaces it is impossible to determine directly. If we assume that the minimal surface tension of an aqueous solution of a conventional surfactant is in the range of 25 - 30 mN/m, then paraffin can be wetted only if its surface tensions is higher.

There is often a problem of determining the wettability of an aqueous dispersions as a model substrates with a constant surface energy. Assuming a contact angle $\theta > 0$ of the disperse phase the basic equation for the force balance is the equation of Young (1855) and Dupré (1869). The physical significance of this can be seen in Fig. 1.10a. Thus, we obtain

$$\gamma_2 = \gamma_{12} + \gamma_1 \cos\theta.$$
(1.9)

For those instances in which the contact angle is $\theta > 0$ the work of adhesion can be written as

$$A_A = \gamma_1(1 + \cos\theta). \tag{1.10}$$

An experiment suitable for determining the wetting tension is shown in Fig. 1.11. A glass capillary with a modified surface is immersed in the liquid under investigation. With a contact angle $\theta < 90°$ the height h of the liquid in the capillary above the liquid level of outside the tube is a measure of the wetting tension. The weight of the liquid column in the capillary tube $\pi r^2 h \rho$ is equal to the wetting tension $2\pi r\gamma_1 \cos\theta$ acting along the inner circumference of the capillary (cf. Fig. 1.10b).

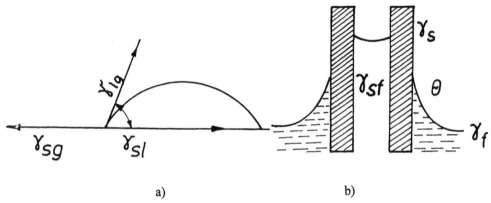

a)   b)

Fig. 1.10.   Force balance in a three phase contact, a) forces acting in the three-phase contact , b) liquid meniscus in a a capillary

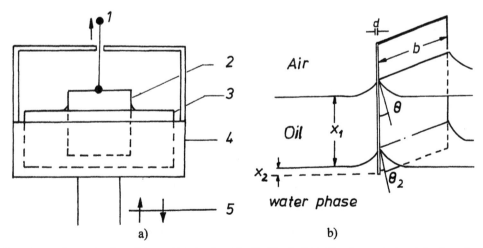

a)   b)

Fig. 1.11.   Determination of the wetting tension of a liquid, a) schematic of an experimental set-up, 1 - force balance, 2 - plate, 3 - level of the liquid surface, 4 - temperature control jacket, 5 - lift; b) rewetting at a liquid/liquid interface, after Richter (1994)

16

The wetting tension then is

$$b = \gamma_1 \cos\theta = \frac{rh\rho}{2}.$$ (1.11)

Two extreme cases can be distinguished when a hydrophobic surface is wetted. The aqueous solution of the surfactant hydrophilises the surface of the solid without removing the hydrophobic layer from the surface. This mechanism is seen whenever the solid itself consists of a hydrophobic material such as organic polymers or when the hydrophobic layer adheres sufficiently firmly. The latter may be the case with wax fat layers above their melting points. When the solid is covered with a hydrophobic liquid layer there is another possible mechanism, depicted in Fig. 1.12. The aqueous solution of the surfactant possesses a higher affinity to the surface of the solid than the hydrophobic oil. This case gives rise to the phenomenon of so-called re-wetting, where the thin oil layer is deformed into droplets with a contact angle of $\theta = 180°$ and in place of the oil film a coherent aqueous layer is formed on the solid surface. This is shown in Fig. 1.12 in which γ_{so}, γ_{sw} and γ_{ow} stand for the interfacial tensions between solid/oil, solid/water and oil/water, respectively.

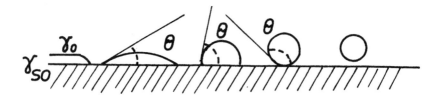

Fig. 1.12.   Re-wetting equilibrium of interfaces solid/oil/water, according to Richter & Kretzschmar (1982)

Wetting and contact angle, described by the Young equation (1.9), is an essential part of the surface energetics. Gibbs was the first to focus scientific attention on the case where droplets in contact with another fluid or solid phase become so small that the energetic properties of the 3-phase contact line is no longer negligible. The four terms in the Young equation are given by the energetics of the three-phase contact line. In analogy with extended phases this line is characterised by a two-dimensional pressure, the line tension κ. The line tension becomes important, if the diameter r of the contact zone is small enough and the Young equation is modified to

$$\gamma_{sl} - \gamma_1 \cos\theta - \gamma_s - \frac{r}{\kappa} = 0. \tag{1.12}$$

After Scheludko et al. (1981) the line tension is defined as demonstrated in Fig. 1.13. the force $F = 2\kappa \sin(\delta\alpha/2)$ acts parallel to N so that the distance ab tends to zero. After Gibbs (1957), papers by Gershfeld & Good (1967), Torza & Mason (1971), Good & Koo (1979), Platikanov et al. (1980), Kralchevsky et al. (1986) were dedicated to experimental and theoretical studies of the line tension.

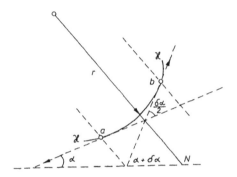

Fig. 1.13. Acting forces in the three phase contact zone including the line tension κ, r - radius of curvature, according to Scheludko et al. (1981)

Wetting phenomena are not only restricted to extended solid/liquid bulk phases. Droplets of a liquid in contact with a solid surface for instance can transform into a thin liquid film, called wetting film. Its stability on a solid surface, for example on polished quartz, is determined by all components acting in thin liquid films (cf. Appendix 2D); long-range forces, van der Waals forces, and double-layer forces.

The equilibrium thickness h of a wetting film was obtained by Derjaguin as early as 1940 as the result of the minimum of van der Waals and gravitational forces. It was determined from the height H of a horizontal plate held over the liquid the vapour of which forms the wetting film. The basic equation for the relationship of disjoining pressure Π and film thickness h has the form

$$\Pi(h) = \rho g(h + H). \tag{1.13}$$

De Gennes (1985) analysed the static and dynamic problems of different wetting phenomena. In this book only the dynamics of spreading and the wetting transition are of interest. Wetting transition takes place at a characteristic temperature where the contact angle diminishes and a wetting film is formed.

Churaev et al. (1994) recently published actual results on the relationship between wetting film thickness h and disjoining pressure Π, from which the macroscopic contact angle of the liquid on the substrate can be calculated. Today a compact monolayer is known as a French pancake, a bilayer a Swedish pancake and thick films as an American pancake. To sum up, wetting films and their transitions are combined with dynamic processes which elucidate rather complex and yet unsolved issues.

Conditions can be deduced from the energy balance mentioned above under which the re-wetting tension is positive, a precondition for the mechanism shown in Fig. 1.12. It should be said that re-wetting is the first step in the complex process of removal of hydrophobic layers from a solid . The oil droplets formed must also be sufficiently stabilised by the surfactant to prevent them coalescing. This takes place in many cleaning processes. When a water-insoluble substance is to be dispersed in water we can distinguish between thermodynamically stable and unstable dispersions. Thermodynamically unstable dispersions are the usual emulsions or dispersions of solids. Solubilisation systems and optically transparent emulsions, so-called micro-emulsions, are in a metastable state where drop growing by collision and coalescence cannot be completely suppressed. These systems are frequently called thermodynamically stable.

Essential criteria for the preparation of emulsions, solubilisates and micro-emulsions will be dealt with in more detail below. On the subject of stabilisation and flocculation we will restrict ourselves to emulsions rather than dispersions in general as the basic laws are transferable. However, dispergates of solid particles are more difficult to treat than emulsions because of their often rough and inhomogeneous surface structure. With emulsions experimental complications arise over drop size distribution caused by the mechanical work input during emulsification, the nature of the emulsifier and by time, which is of considerable importance too.

1.4.2.   STABILITY CRITERIA FOR EMULSIONS

The two extreme cases of stabilisation of dispersed systems by electrostatic repulsion and steric hindrance are shown schematically in Fig. 1.14. When two droplets possessing electric charges of the same sign approach each other the electrostatic repulsion forces between them are responsible for the stability of the system. The droplets' electric charge, which is expressed by the so-called ψ_0-potential of the interface, is given by the adsorption of ionic emulsifiers. A certain part of the counterions is present in the adjacent diffuse layer with an effective thickness of κ^{-1}. For a 1:1 electrolyte the Debye length κ^{-1} is defined by

$$\kappa = \left(\frac{2e^2c_{el}}{\varepsilon kT}\right)^{1/2},$$
(1.14)

where e is the electric unit charge, c_{el} is the bulk electrolyte concentration, ε is the dielectric constant of the medium. According to the DLVO-theory the stability of colloidal particles in a liquid is characterised by a maximum of energy of particle interaction. Two types of forces determine the position of the energy maximum, electrostatic repulsion and van der Waals attraction. The Free Energy of electrostatic repulsion is always positive and for larger particles, for which $\kappa a_p \gg 1$ (a_p is the particle radius) it can be calculated according to Groves (1978) by the following approximation,

$$V_E(h) = 2\pi\varepsilon a_p \Psi_\delta^2 \ln(1 + \exp(-\kappa h)).$$
(1.15)

The total force balance in the narrow gap between two particles is governed by the van der Waals attraction forces. Two plates made from the same material and separated by another medium develop attraction forces acting over small distances.

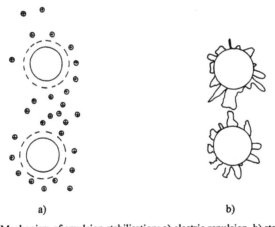

a)   b)

Fig. 1.14.   Mechanism of emulsion stabilisation; a) electric repulsion, b) steric hindrance

According to Hamaker (1937) the Free Energy of the attraction forces is approximately

$$V_A(h) \approx -\frac{a_p\left(\sqrt{A_1} - \sqrt{A_2}\right)^2}{12h}.$$
(1.16)

20

A_1 and A_2 are the so-called Hamaker constants of the particles and the solution respectively. The Hamaker constants of most materials are in the range of 5-100 kT, e.g. 10.6 kT for water.

Fig. 1.15 shows the free Energy of electrostatic repulsion (dotted curve) and van der Waals attraction (dashed curve). The solid curve shows the overlap of the van der Waals attraction and electrostatic repulsion.

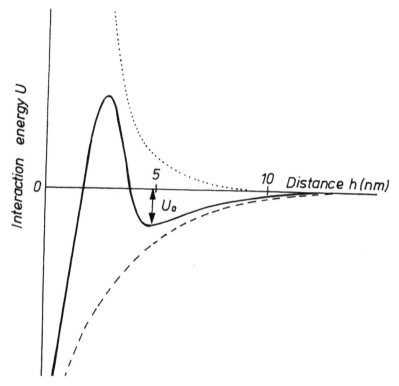

Fig. 1.15.   Energy-distance function between two particles (solid) as summation of functions for the electric repulsion (dotted) and the van der Waals attraction (dashed)

From the brief sketch of the Derjaguin, Landau, Verwey, Overbeek (DLVO)-theory the following three important practical conclusions, suitable for foams and emulsions, can be drawn:

- By increasing the interfacial charge-density by strongly adsorbing ionic surfactants a high Ψ_o-potential can be established, resulting in larger distances between stabilised particles. The formation of a lower energy minimum, equivalent to a more stable position of the equilibrium is important here.

- The electrostatic repulsion between dispersed particles can be diminished by increasing the concentration of background electrolyte (e.g. NaCl, $CaCl_2$). Polyvalent ions are more effective than monovalent. There is a critical electrolyte concentration for every system at which flocculation or coalescence takes place. These principles must be taken into account when emulsions have to form in very hard water.

- The energy minimum corresponds to the most stable equilibrium distance between the particles and can be lowered by addition of electrolyte.

This minimum can be exceeded by increasing the kinetic energy of the particles for example. This can be done by raising the temperature or by applying a centrifugal field. The kinetic energy of particles depends on mass and velocity. Thus dispersions (emulsions) are the more stable the smaller the particles are, with all other conditions (emulsifier concentration, temperature, salt content), constant.

A class of substances, which has gained widespread use in emulsion manufacture is the ethylene oxide adducts. These surfactants exist as non-ionics or, in the presence of sulphate, carboxyl or phosphate groups, as ionic compounds. For non-ionics the emulsion stabilisation mechanism of Fig. 1.14b prevails. Although a change in the electrical interface potential Ψ_0 is possible through the adsorption of protons or other ions, the influence of the steric stabilisation predominates. The effect may be explained by the strength of the adsorption layer itself or a steric hindrance due to chain segments (loops) mutually repelling each other. Investigation of the equilibrium distances between individual oil droplets, which showed a shift of the energy minimum towards larger distances with increasing degree of ethoxylation, support the idea of a steric hindrance mechanism.

More important for the application of nonionics in practice is knowledge of the relationship between the stabilising effect and the average ethylene oxide number in the molecule. The surfactant concentration c_{st} necessary to increase the shelf-life of a standard emulsion is taken as a measure of the stabilising effect. Fig. 1.16 shows such a relationship for ethoxylated dodecanol (Müller & Kretzschmar 1982). The ratio of the hydrophilic and hydrophobic parts in the molecule has a direct bearing on the concentration needed to stabilise the toluene-in-water system. The ethylene oxide adducts possesses particularly high variability and are readily available.

According to Bancroft's rule the best type of emulsion raises the solubility of the emulsifier in its homogeneous phase. Hence, nonionic ethylene oxide adducts with a low degree of ethoxylation, for example the tetra glycol dodecyl ether, stabilise a water/oil type and higher ethoxylated products, such as EO 10, stabilise oil/water emulsions. Given a sufficient solubility surfactants form micelles both in water and non-aqueous solvents. This leads to another property called solubilisation. The simplest case is the solubilisation of water in oil or oil in water. In all cases the presence of micelles in the respective homogenous phase is a necessary condition for this.

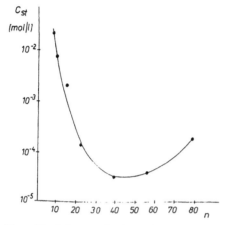

Fig. 1.16.   The influence of polyoxyethylene chain length on the stabilisation of toluene droplets in water, according to Müller & Kretzschmar 1982

1.4.3.   SOLUBILISATION AND MICROEMULSION

Optically transparent micellar solutions are often able to dissolve considerable quantities of a substance molecular insoluble in the pure solvent. The transparency of such a solution is not affected by this. The solubilisation is controlled by the shape and size of the micelles which determine the amount of components incorporated into the interior of the micelles (Schulman et al. 1959).

In Fig. 1.17a an example of the solubilisation of oil in water and of water in oil as a function of temperature is illustrated. The system consists of 5 wt% polyoxyethylene-8,6-nonylphenyl ether, 47.5 wt% water and 47.5 wt% cyclohexene (Shinoda & Friberg 1975).

With rising temperature the volume of the aqueous phase grows, the micelles swell until suddenly, at the so-called phase inversion temperature, the oil phase has the larger volume. This effect is explained by the polyoxyethylene chains dehydrating as the temperature rises. The hydrophilic/hydrophobic balance of the molecule is thereby altered and the solubility in the oil phase grows. When concentration is great enough, micelles are formed in the oil phase and water is solubilised. If the two phases do not exist as stratified layers but as emulsion a phase inversion is obtained at higher temperature. At the phase inversion temperature PIT a minimum in the interfacial tension is observed (Fig. 1.17b).

A similar phenomenon was observed by Müller & Kretzschmar (1982) with ethylene oxide adducts at significantly increased concentration of the background electrolytes connected with clouding of the emulsion.

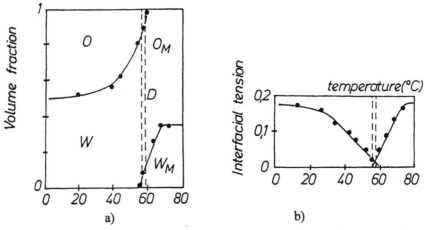

Fig. 1.17. The solubilisation of oil or water in a micellar solution of non-ionic surfactant, a) two-phase diagram (O - oil, W - water, O_M - oil in micellar solution, W_M - water in inverse micellar solution, D - phase separation temperature region), b) interfacial tension as a function of T, according to Shinoda & Friberg 1975

The quantity of insoluble substance which can be solubilised in micelles depends to a considerable extent on the chemical structure of the surfactant and is influenced by the presence of other components, which may influence either the micelle formation concentration (CMC) or the micelle geometry (aggregation number, shape). The transition from solubilisation to another important phenomenon, the formation of a micro-emulsions, is continuous. Micro-emulsions form spontaneously, whereas typical solubilisation systems attain their equilibrium state often only after extreme long periods of intensive mixing of both phases.

As common emulsions described above, micro-emulsions also exist as oil/water and water/oil emulsions. Unlike macro-emulsions, micro-emulsions are optically transparent and usually thermodynamically stable; they do not tend to separate into phases. The diameter of micro-emulsion droplets is of the order of 10-100 nm, less than the wave length of visible light.

Typical micro-emulsions are formed in the presence of a nonionic surfactant without a so-called co-surfactant, or with an ionic surfactant in presence of a co-surfactant. As "co-surfactants" lower alcohols, such as butanol, pentanol or amino-alcohols, e. g. 2-amino-1-methylpropanol, can be used along with ethylene oxide adducts. It is always the degree of solubility of the surfactant in the homogeneous phases which determines the type of emulsion formed. Potassium oleate, for example, is soluble in water and pentanol is soluble in benzene, so that a micro-emulsion of the water/benzene type is formed. Polyoxyethylene-3-nonylphenyl ether is soluble in benzene and polyoxyethylene-20-nonylphenyl ether is soluble in water. With such a surfactant/co-surfactant system micro-emulsions of the benzene/water type are formed. It is important to note that micro-emulsion systems are formed only at a certain concentration ratio of surfactant and co-surfactant and at higher concentrations. The phase behaviour of a quaternary system water/oil/surfactant/co-surfactant is shown in Fig. 1.18 (Friberg 1983).

Solubilisates and micro-emulsions are of interest in special applications where the use of small quantities of water and a relatively high surfactant/co-surfactant concentration do not destroy the system or produce synergistic effects. One advantage of the formation of typical micro-emulsions is that, in contrast to macro emulsions, high proportions of oil-in-water or water-in-oil dispergates can be prepared which do not coalesce and form spontaneously. The sequence of admixed components is also insignificant. The spontaneous formation of micro-emulsions is a process of great importance because their preparation requires a minimum of mechanical energy.

The formation of new phases is accompanied by the formation of new interfaces and for a complete understanding of the phase transitions kinetics, the dynamic properties of the related adsorption layers stabilising the system have to be known as well.

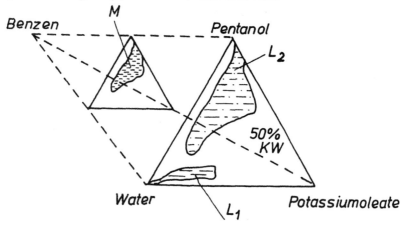

Fig. 1.18.   Quaternary phase diagram, L_1 - micellar solution, L_2 - inverse micellar solution, M - microemulsion, after Friberg (1983)

According to Langevin (1986) the surfactant adsorption layers between oil and water domains in microemulsions can be considered as lyotropic liquid crystals. In contrast to typical lyotropic systems a high flexibility exists which allows for a fast kinetic exchange of components between the domains.

1.4.4. SPONTANEOUS EMULSIFICATION

When an emulsion (toluene in water for example), is to be prepared there has to be a certain amount of work in form of stirring or shaking to overcome the energy of cohesion of the phases to be dispersed. If the toluene phase contains 10 wt% methanol the emulsification is spontaneous, because the methanol transfers from the toluene into the water phase. The mechanism is called "kicking" and it is linked to a mass transfer of the alcohol across the interface. This leads to a non-uniform distribution of the alcohol at the toluene/water interface and thus to interfacial tension gradients. Areas of different interfacial tensions at a liquid interface are unstable and flow processes set in, the well-known Marangoni instabilities. This process is characterised by the formation of vortices, which have their origin in the motive processes at the interface and extend far out into the phases. Fig. 1.19 shows a Schlieren picture of the formation of vortices over time in the system amyl alcohol/water with the material transfer of a surfactant. The strong influence of time on the formation of cellular structures can also be seen here.

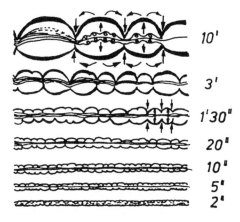

Fig. 1.19.   Stream lines of first order roll-cells during transfer of SDS across the iso-amylol/water interface; scaled-up picture of a section of Fig. 3C.1a; after Linde (1978)

This process with its vortex formation initiates the break-off of one continuous phase into drops and finally into a dispersed phase.

Another example is the spontaneous emulsification of xylene/cetyl alcohol mixtures in an aqueous solution of sodium dodecyl sulphate. Here the extreme reduction of the interfacial tension by mixed adsorption of alkyl sulphate and cetyl alcohol causes the spontaneous emulsion formation. The spontaneous dispersion of the phases leads to a reduction in the Free Energy of the system and proceeds freely.

The observation that mixed adsorption layers often provide an extreme reduction of the interfacial tension and thereby enable easy or even spontaneous emulsification was studied systematically by Shinoda & Friberg (1975) using the two surfactants $CH_3(CH_2)_8 - (OCH_2CH_2)_{8.5} OP(OH_2)$ and $CH_3(CH_2)_8 (OCH_2CH_2)_3 OP(OH)_2$. By changing the composition of the mixed components in the system water/heptane conditions for the formation of liquid/crystalline phases and spontaneous emulsification were found. The results support the idea that the use of two suitably compatible emulsifiers is often much more efficient than single surfactant systems.

The formation of emulsions or microemulsions is connected with several dynamic processes; the time dependence of surface tensions due to the kinetics of adsorption, the dynamic contact angle, the elasticity of adsorption layers as a mechanic surface property influencing the thinning of the liquid films between oil droplets, the mass transfer across interfaces and so on. Kahlweit et al. (1990) have recently extended Widom's (1987) concept of wetting or non-wetting of an oil-water interface of the middle phase of weakly-structured mixtures and micro-emulsions. They pointed out that the phase behaviour of microemulsions does not differ from that of other ternary mixtures, in particular of mixtures of short-chain amphiphiles (cf for example Bourrell & Schechter (1988).

Chapters 10 and 11 are dedicated to nex concepts of flotation and microflotation. As a main act of these processes the formation and stability of thin liquid films are discussed in Appendix 2D.

1.5. ORGANISATION OF THE BOOK

The present book is organised in the form of main text, summaries, and appendices. To preserve the main idea of the individual chapters, additional material to the topics of secondary importance is presented in the appendices. There are also appendices to introduce fields which play a marginal role for this book. Some tables of correction factors for experimental methods, special functions, and data on surfactants and solvents have also been included.

The first three chapters should be regarded as an introduction to the broad area of the dynamics of adsorption, while subsequent chapters contain original results, some of them not yet published.

Before starting with dynamic effects at a liquid interface, the equilibrium state of adsorption is described and adsorption isotherms as basic requirements for theories of adsorption dynamics are reviewed. Chapter 2 presents the transfer from thermodynamics to macro-kinetics of adsorption. As Chapter 7 deals with the peculiarities of ionic surfactant adsorption and introduces some properties of electric double layers.

The aim of Chapter 3 is to demonstrate the complexity of the coupling of surface rheology and bulk transport processes. It also demonstrates the close link of surface rheology and interfacial relaxation processes.

An overview of the separate chapters of the book can be easily gained by reading the chapter summaries. These contain analysis of unsolved problems and ideas for further developments of specific areas.

1.6.   REFERENCES

Adamson, A.W., "Physical Chemistry of Surfaces", John Wiley & Sons, Inc., New York, Chichester, Brisbane, Toronto, Singapore, (1990)

Bohr, N., Phil. Trans. Roy. Soc., London Ser. A, 209(1909)281

Borwankar, R.P. and Wasan D.T., Chem. Eng. Sci, 43(1983)1323

Bourrell, M. and Schechter, R.S, "Microemulsions and Related Systems", Marcel Dekker, Inc., New York and Basel, (1988)

Cheng, P., Li, D., Boruvka, L., Rotenberg Y. and Neumann, A.W., Colloids & Surfaces, 43(1990)151

Churaev, N.V. and Zorin , Z.M., Colloids Surfaces A, (1994), submitted

De Gennes, P.G., Rev. Modern Physcis, 57(1985)827

Derjaguin, B.V., Zh. Fiz. Khim., 14(1940)137

Derjaguin, B.V. and Dukhin, S.S., Trans. Amer. Inst. Mining Met., 70(1960)221

Dörfler, H.-D., Grenzflächen- und Kolloidchemie, VCH, 1994

Dukhin, S.S., Croatica Chemica Acta, 53(1980)167

Dukhin, S.S., Adv. Colloid Interface Sci., 44(1993)1

Dupré, A., Theorie Mechanique de la Chaleur, Paris, 1869

Dynarowicz, P., Adv. Colloid Interface Sci, 45(1993)215

Exerowa, D., Cohen, R. and Nikolova, A., Colloids and Surfaces, 24(1987)43

Fainerman, V.B., Khodos, S.R., and Pomazova, L.N., Kolloidny Zh. (Russ), 53(1991)702

Friberg, S., Progr. Colloid Polymer Sci., 68(1983)41

28

Frumkin, A., Z. Phys. Chem. (Leipzig), 116(1925)466

Gauss, C.F., "Commentationes societaties regiae scientiarium Gottingensis recentiores" Translation "Allgemeine Grundlagen einer Theorie der Gestalt von Flüssigkeiten im Zustand des Gleichgewichts" (1903), Verlag von Wilhelm Engelmann, Leipzig, (1830)

Gershfeld, N.L. and Good, R.J., J. Theor. Biol., 17(1967)246

Gibbs, J.W., "The Collected Works of J. Willard Gibbs", Vol. 1, New Haven, Yale University Press, 1957

Good, R.J. and Koo, M.N., J. Colloid Interface Sci., 71(1979)283

Groves, M.J., Chemistry & Industry, (1978)417

Hamaker, H.C., Physica (Utrecht), 4(1937)1058

Harkins, W.D. and Jordan, H.F., J. Amer. Chem. Soc., 52(1930)1751

Hunter, R.J., Foundations of Colloid Science, Volumes I and II, Clarendon Press, Oxford, (1992)

Ivanov, I.B., (Ed.), "Thin Liquid Films", Marcel Dekker, Inc. New York, Basel (Surfactant Science Series), Vol. 29, (1988)

Kahlweit, M., Strey, R. and Busse, G., J. Phys. Chem., 94(1990)3891

Kralchevsky, P.A., Ivanov, I.B. and Nikolov, A.D., J. Colloid Interface Sci., 112(1986)108

Krawczyk, M.A., Wasan, D.T., and Shetty, C.S., I&EC Research 30(1991)367

Kretzschmar, G., Kolloid Z., 206(1965)60

Kretzschmar, G., Proc. Intern. Confer. Colloid Surface Sci., Budapest, 1975

Kretzschmar, G., VII Intern. Congr. Surface Active Substances, Moscow, (1976)

Kretzschmar, G. and Voigt, A., Proceeding of "Electrokinetic Phenomena", Dresden, 1989

Kretzschmar, G. and Miller, R., Adv. Colloid Interface Sci. 36(1991)65

Krugljakov, P.M. and Exerowa, D.R., "Foam and Foam Films", Khimija, Moscow, 1990

Langevin, D., Mol. Cryst. Liqu. Cryst., 138(1986)259

Langmuir, I., J. Amer. Chem. Soc., 15(1918)75

Laplace, P.S. de, "Mechanique Celeste", (1806)

Lecomte du Noüy, P., J. Gen. Physiol., 1(1919)521

Lenard, P., von Dallwitz-Wegener and Zachmann, E., Ann. Phys., 74(1924)381

Levich, V.G., Physicochemical Hydrodynamics, Prentice-Hall, Englewood Cliffs, N.Y., 1962

Linde, H., Sitzungsber. AdW der DDR 18N(1978)20

Lord Rayleigh, Proc. Roy. Soc. (London), 29(1879)71

Lucassen-Reynders, E.H. and Kuijpers, K.A., Colloids & Surfaces, 65(1992)175

Lunkenheimer, K., Tenside Detergents, 19(1982)272

Lunkenheimer, K., Thesis B, Berlin, (1983)

Lunkenheimer, K., J. Colloid Interf. Sci., 131(1989)580

Lunkenheimer, K. and Wantke, D., Colloid & Polymer Sci., 259(1981)354

Lyklema, J., "Fundamental of Interface and Colloid Science", Vol. I - Fundamentals, London, Academic Press, 1991

Malysa, K., Adv. Colloid Interface Sci., 40(1992)37

Miller, R., Thesis B, Berlin, (1986)

Müller, H.J. and Kretzschmar, G., Colloid Polymer Sci., 260(1982)226

Padday, J., Pure & Applied Chem., 48(1976)485

Platikanov, D., Nedyalkov, M. and Scheludko, A., J. Colloid Interface Sci., 75(1980)612

Richter, L. and Kretzschmar, G., Tenside Detergents, 19(1982)347

Richter, L., Tenside Detergents, 31(1994)189

Rotenberg , Y., Boruvka, L. and Neumann, A.W., J. Colloid Interface Sci., 93(1983)169

Scheludko, A., Tchaljovska, S. and Fabrikant, A., Disc. Faraday Soc., (1970)1

Scheludko, A., Schulze, H.J. and Tchaljovska, S., Freiberger Forschungshefte, A484(1971)85

Scheludko, A., Toshev, B.V. and Platikanov, D., "On the mechanism and thermodynamics of three-phase contact line systems", in "The Modern Theory of Capillarity", Akademie-Verlag, Berlin, 1981

Schmauks, G., PhD Thesis, Merseburg, 1993

Schulman, J.H., Stoeckenius, W. and Prince, L.M., J. Phys. Chem., 53(1959)1677

Schulze, H.J., "Physio-chemical Elementary Processes in Flotation", Elsevier, Amsterdam, Oxford, New York, Tokyo, (1984)

Shinoda, K. and Friberg, S., Adv. Colloid Interface Sci., 4(1975)285

Stauff, J., Z. Phys. Chem., N.F., 10(1957)24

Szyskowski, B. von, Z. Phys. Chem., 64(1908)385

Torza, S. and Mason, S.G., Colloid Polymer Sci., 253(1971)558

Vollhardt, D. and Wittich, M., Colloids and Surfaces, 47(1990)233

Widom, B., Langmuir, 3(1987)12

Wüstneck, R., Miller, R. and Kriwanek, J., Colloids and Surfaces, 81(1993)1

Young, T., Miscellaneous Work, Vol. 1, G. Peacock (Ed.), Murray, London, 1855

CHAPTER 2

2.   THERMODYNAMICS AND MACRO-KINETICS OF ADSORPTION

2.1.   DEFINITION OF LIQUID INTERFACES

In Chapter 1 we briefly described an interface as a layer with uncompensated intermolecular forces. The thermodynamics of a liquid interfaces covered with a soluble or insoluble monolayer layer has been describe in detail by many other competent authors and we want to present only the thermodynamic basis needed for the subsequent chapters of this book. Let us consider the interface between water and air. The specific properties of the bulk water, e.g. the freezing point, boiling point, vapour pressure, viscosity, cluster formation and hydrophobic bonds, are well described by long and short-range intermolecular forces and strong and weak intramolecular forces. Israelachvili recently (1992) remarked in a short note on the usefulness of this classification, although it is not clear whether the same interaction is counted twice or two normally distinct interactions are strongly coupled.

Nevertheless, we have to work with the same intermolecular forces as in chemistry. Additionally, in surface and colloid chemistry interactions of macroscopic bodies exist and the summation of all intermolecular forces leads to another force due to a specific force-distance function. First there are the well-known van der Waals forces which are attractive and decay due the power of sixth. Other attractive forces are:

forces between polar molecules,
forces between permanent and induced dipoles,
forces between nonpolar molecules.
Another class of interacting forces are the so-called chemical forces (Prausnitz 1969). In contrast to the physical forces these forces are counterbalanced. Typical examples are the covalent bonds, electron donor-acceptor interactions, acidic solute - basic solvent interactions. Association and solvatation are effects well-known to every chemist.

Following the concept of Tanford (1980), we find interactions of surfactant molecules in aqueous solution encouraging the formation of micelles and adsorption layers at a liquid interfaces. These hydrophobic interactions rely on the degrees of freedom of the water molecules. For example, the interaction between the hydrophobic parts of amphiphilic molecules is well established by van der Waals forces. During adsorption or the formation of micelles the area of contact of the hydrophobic part of the amphiphilic molecule with water molecules decreases and the non-ordered part of water molecules increases. In the term of thermodynamics the increase in entropy by adsorption or association can be interpreted as the motive force of these amphiphile effects.

In the case of interacting molecules with distances short enough for the electronic shells to overlap, Mie (1903) suggested a relationship with an empirical constant and a certain decay law of high power against distance. Later, on the basis of Mie's equation and the van der Waals law h^{-6} Lennard-Jones (1936) derived an intermolecular pair potential

$$u(r) = -\frac{A}{r^6} + \frac{B}{r^{12}}.$$ (2.1)

The interaction forces between single molecules of amphiphilic character with solvent molecules at a liquid interface were clearly demonstrated by Bourrell & Schechter (1988) (cf. Fig. 2.1.)

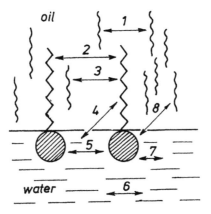

Fig. 2.1.   Interaction energies of amphiphiles at a water/oil interface, 1 - hydrophobic oil/oil, 2 - tail/tail, 3 - oil/tail, 4 - water/oil, 5 - head/head, 6 - hydrophilic water/water, 7 - head/water, 8 - head/oil, after Bourrell & Schechter (1988)

Beside the forces mentioned above, we are also dealing with Coulomb interacting forces as the strongest attraction or repulsion forces. Such forces are relevant to the adsorption layers of charged surfactants. The formation of an electrical double layer by the adsorption of charged amphiphiles produces a number of surface effects on the properties of the corresponding adsorption layers, and on the kinetics of adsorption itself. Before we move to problems connected with an electrical double layer at a liquid interface it is important to know something more about a pure liquid interface. The most frequently used solvent, water, offers two main properties significant for interfacial phenomena: the thickness of the interfacial layer, and the orientation of the water dipoles at the interface, their dynamics and fluctuations. Neither problem has yet been adequately explained. The range of surface forces into the water bulk was detected experimentally for the first time by Derjaguin and his school.

This first attempt, made by Derjaguin & Obuchov (1936) and Derjaguin & Kussakov (1939), suggests the existence of a surface layer thicker than a single monolayer. From investigations on thin liquid films, Derjaguin & Zorin (1955) postulated a structural component by investigation of the disjoining pressure, measured as vapour pressure, of a thin film at a quartz surface. This problem has recently been theoretically and experimentally discussed by Rabinovich et al. (1982). The classic version of the common blowing off-method is based on the principle shown in Fig. 2.2. In a chamber a liquid layer is placed on a solid substrate. A stream of nitrogen is blown along the surface to form a liquid wedge. The tangential blowing-off acting with the stress τ and the Newton equation yields

$$\tau = \eta \frac{dv}{dh},$$ (2.2).

with η, v and h are viscosity, liquid velocity, and thickness of the liquid film respectively.

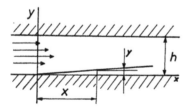

Fig. 2.2.   Principle of the blowing-off method, according to Derjaguin & Karasev (1957)

To learn about the degree of ordering of the water layer at the water interface in which the jump of the dielectric constant takes place, e.g. from water to air, there must be suitable experimental conditions. One of the best and simplest methods is the measurement of the surface potential jump at the interface. As we will show, absolute electrical surface potentials, like single electrode potentials, are not measurable. In every case we need a reference system.

The hydrogen electrode well known to electrochemistry has a potential which is set to zero by convention. To solve this problem, computer simulations using the Monte-Carlo or molecular dynamics models have been made (Stillinger & Ben-Naim 1967, Aloisi et al. 1986, Matsumoto & Kataoka 1988, Wilson et al. 1988, Barraclough et al. 1992). The obtained results differ in magnitude (by more than one decade) and sign. Thus, the surface properties discussed in the present book are relative to the surface state of the pure solvent, in most cases water. In the presence of an adsorption layer, say water/air interface, the displacement of water molecules from the interfacial layer by surfactant molecules, a reorientation of water molecules forming the interface, as well as a change of the intermolecular forces between the solvent molecules at the interface have to be calculated. It is understandable that related assumptions are speculative if not enough is known about the molecular picture of the pure liquid interfacial layer.

Dynarowicz (1993) has recently given an overview of these problems. One model for a liquid interface is based on the assumption of an oriented dipole layer, equivalent to the classical Helmholtz capacitor (cf. Fig. 2.3.). Both the water molecules and the adsorbed molecules carry dipoles. The different kinds of dipole orientations and their distances to each other and to a defined interface have resulted in various models being suggested.

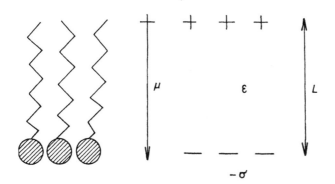

Fig. 2.3.   Dipole orientation at an interface based on the Helmholtz capacitor model, μ - dipole moment, σ- charge, according to Dynarowicz (1993)

As we have shown above, there are two issues in the description of the molecular state of liquid surfaces, firstly the presence and the range of a structured interface and secondly the kind of surface structure. Since Derjaguin and other's work, no radically different models about the existence of one or more oriented water layers at an interface have been published.

This makes the early papers of Irvin Langmuir interesting. Langmuir (1918) wrote: "The theory of surface tension which I developed in 1916 and which was later elaborated by Harkins, also furnishes striking proof that the surface layers in pure liquids are normally of the thickness of a single molecule, and that molecules are oriented in definite ways. Although with liquids, direct proof of the existence of these monomolecular films was obtained, the evidence in the case of adsorption by solid bodies was indirect. Only in one instance, namely the adsorption of atomic hydrogen on glass surfaces, was direct evidence of this kind found. "

2.2.   MODELS OF THE SURFACE OF WATER

The evidence for the orientation of water molecules which form the interface is particularly complex. Apart from the thermodynamic treatments given by Gibbs (1906) and Bakker (1928), thermodynamics does not lead to the molecular architecture of the interfacial layers. The orientation of water dipoles at a liquid interface is determined by the electrostatics of the permanent as well as induced dipoles. The water molecules are polarised, as shown in Fig. 2.4.

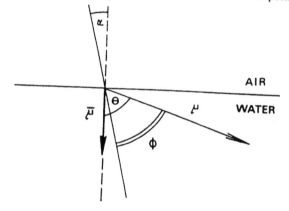

Fig. 2.4.   Polarisation of the O···H-bonds in the water molecule and the net dipole moment of the molecule

The net dipoles of the water layer forming the interface fluctuate primarily around an axis with a tilt angle to the interface. Without any assumptions there are two opposite orientations of the water molecules possible; directed to the air (or oil) phase are either the oxygen or hydrogen atoms directed to the air (or oil) phase. Frumkin (1956) suggests an orientation of the oxygen atoms to the gas phase. By evaluations of experimental data from the second harmonic generation Goh et al. (1988) have recently confirmed Frumkin's idea.

Relating to the dipole-dipole, dipole-induced dipole and dipole-ion interaction forces which exist in practically all surface layers covered by an amphiphile adsorption layer, the dipole orientation in the pure interphase is already important.

In order to understand the terminology of electric surface properties we have to distinguish between inner and outer electrical potentials or Galvani and Volta potential differences. The Section 2.8. will deal with electrostatic properties of interfaces, so we will give a definition of these potentials here, which is based on the nomenclature given by Lange & Koenig (1933). In order to describe the dipole orientations of all components forming the interfacial layer, we assume that each molecule at the interface has a net dipole moment as the vector sum of all bonding dipole moments of the "surface" molecules μ_s and its orientation to the surface, given by the angle α. Generally we have to discriminate between the orientation along the main axis of the adsorbed or spread molecule at an interface and the axis of the net dipole moment of such molecules.

Further, we assume that the surface forming molecules, represented by their net dipole moments are oriented in two separated layers like a condenser, which is identical to the Helmholtz condenser model (Fig. 2.3). From elementary physics it is well known that the potential difference ΔV of this condenser is given by the charge density σ, the distance d between the separated plates and the dielectric constant ε of the medium inside the condenser,

$$\Delta V = \frac{4\pi\sigma d}{\varepsilon}, \tag{2.3}$$

where ε is the dielectric constant and d is the thickness of the surface condenser. We have to transfer this formula in order to solve for the oriented water dipoles forming an interfacial layer or for the dipoles of adsorbed amphiphiles. We introduce the adsorption density Γ or the number of the surface forming water molecules n, as the dipole density per unit area, $\sigma d = n\mu_s$, where e is the electric unit charge and $n(\Gamma)$ is the number of charges.

Taking into account the orientations of the surface dipoles with the angle of inclination of the net molecular dipoles μ_s we obtain

$$\Delta V = \frac{4\pi\mu_s n(\Gamma)}{\varepsilon} = 4\pi n\mu \cos\alpha, \tag{2.4}$$

where α is the tilt angle, $\mu_s = \mu\cos\alpha$, μ is the actual dipole moment, and ε is assumed to be unity (see Fig. 2.3). In this equation the charge density σ is given in electrostatic cgs-units. Now ΔV can be obtained in Volt, while 1 Volt in electrostatic units equals 300 ordinary Volts. In the SI-system we have to introduce $\varepsilon_0\varepsilon$ and obtain

$$\Delta V = \frac{n\mu_s \cos\alpha}{\varepsilon_0\varepsilon}, \tag{2.5}$$

with $\varepsilon_0 = 10^7/4\pi c^2 = 8.85\cdot10^{-12}$. The charge density σ is given in Coulomb/m^2 and ΔV is obtained in ordinary Volts.

For practical purposes we are measuring ΔV in ordinary Volts and μ_s is given in Debye, where 10^{-18} e.s.u. cm is 1 Debye.

As mention above, the Helmholtz model of an interface enables us to calculate the effective dipole moment $\tilde{\mu}$ due to the polarisation of the electric field $\Delta V/d$,

$$\tilde{\mu} = P_N A d,$$

(2.6)

where P_N, A, and d are the polarisation normal to the plates of the condenser, the area occupied by one molecule at the interface, and d the distance between the plates of the "surface condenser", respectively. The effective dipole moment is described in Fig. 2.3 and equals the magnitude of the net dipole moment of a molecule at the interface in the normal direction.

The model of an interface, presented in the monograph of Davies & Rideal (1963) was often used. They suggested that the resulting (effective) surface dipole moment of an adsorbed molecule is the sum of μ_1, μ_2 and μ_3, as μ_1 indicates the polarisation of the surface water molecules and μ_2 and μ_3 represent the contributions of the hydrophilic head groups and the hydrophobic tails.

Demchak & Fort (1974) developed a three layer capacitor model, shown in Fig. 2.5. The main features are the introduction of three components to the effective dipole moment

$$\tilde{\mu} = \tilde{\mu}_1 / \varepsilon_1 + \tilde{\mu}_2 / \varepsilon_2 + \tilde{\mu}_3 / \varepsilon_3.$$

(2.7)

The three terms of Eq. (2.7) are related to the water phase, the layer of the hydrophilic head groups and to the hydrophobic tails in the air phase, respectively.

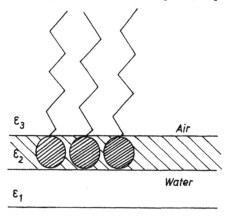

Fig. 2.5.   Three layer capacitor model, according to Demchak & Fort (1974)

By introducing a local dielectric permittivity for the water part and the air part of a real interface Vogel & Möbius (1988a, b) suggested a two layer capacitor model shown in Fig. 2.6, which is in agreement with their experimental findings. This model considered the polarisation effects of ionic head groups as well as the influence of an electrical double layer.

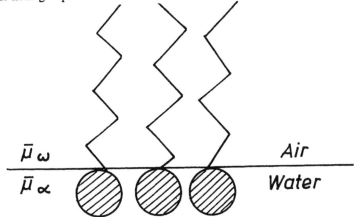

Fig. 2.6.   Double layer model of a water surface, after Vogel & Möbius (1988a)

2.3.   GENERAL PRINCIPLES OF EQUILIBRIUM SURFACE THERMODYNAMICS

The thermodynamic description of the interfacial state of liquid systems is the basis for the development of relationships between the adsorption density Γ at a liquid interface, the surface tension and the surfactant bulk concentrations. Beside this the interfacial tension is an intrinsic parameter which determines the shape of a curved interface as well as different other types of capillary phenomena.

There are three treatments for the state of an interface. Guggenheim (1959) pointed out that the treatment of van der Waals jr. & Bakker (1928) is less abstract than Gibbs' dividing surface concept. Good (1976), Defay et al. (1966) and Rusanov (1981) presented actual issues of surface thermodynamics. A very clear overview of this topic was published recently by Hunter (1992). The Gibbs phase rule, applied to the wetting process, was recently derived by Li et al. (1989). More focused on solid/liquid interfaces are the treatments by Jaycock & Parfitt (1981) and Widom (1990), whereas the import field of thermodynamics of thin liquid films was developed by the schools of Derjaguin (1993), de Feijter (1988) and Ivanov (1988). For ease of comprehension we define the surface tension on the basis of a real physical model of an interface using the pressures of the bulk and surface phases. In contrast to Gibbs and agreement with Guggenheim (1959) we assume a three-phase system, as shown in Fig. 2.7. α and β are fluid bulk phases; β can be for example air. Between the phases α and β the surface phase s is placed comprising all molecules with asymmetric interacting forces.

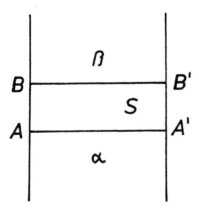

Fig. 2.7.   Three phase surface model, according to Guggenheim (1959)

The thickness of the phase s determined by the extension of the molecules forming the interface is given by the boundaries A-A´ and B-B´. Beyond these planes the interaction forces between the molecules become identical to those in the bulk phases α and β. The phase s can therefore be described as a zone of continuously changing properties. The change of properties is related to intrinsic parameters because the surface model is determined by the molecular state and does not depend on the surface area. In other words, all properties of s are constant in the direction parallel to A-A´ but not in the direction normal to A-A´.

The volume of the s-phase depends on the size and shape of the molecules. There are three extreme possibilities for the molecular composition of s:

i)   The phase β consist of air and s contains the same kind of molecules like the phase α.
ii) The phases α and β are partially miscible. In practice that happens in all liquid/liquid systems. Therefore, the phase s contains molecules of α and β. The different molecules offer a specific tendency to adsorb at the interface (phase s). This results in the concept of discontinuity for the phase s. Thus the interface is not formed by the interaction between molecules of the same kind but is structured by the interaction forces between different molecules.
iii) At the interface typical surfactants are adsorbed. There is a huge variety of surfactants with molecular weights from about 200 to 2000, not counting surface active macromolecules, such as proteins.

According to Gibbs we obtain surface excess quantities as the basic for adsorption isotherms and surface rheology. For flat or only slightly curved surfaces one can define the surface excess quantity convenient for discussion of a particular problem. It is assumed that this freedom exists. Additional simplification is possible for case i) and iii) because the solute can be assumed to be in one bulk phase only if the vapour pressure of the solute is negligible (i) or the surfactant is practically soluble only in one bulk phase (iii).

The surface excess amount or Gibbs adsorption of component i denoted by n_i^s, may be positive or negative. This amount is given by

$$n_i^s = n_i - V^\alpha c_i^\alpha, \qquad (2.8)$$

where n_i is the total amount of component i in the system, c_i^α and V^α are the bulk concentration of component i in phase α and the volume of the bulk phase, respectively. This can also be expressed also by the following equation,

$$n_i^s = \int_{\text{phase } \alpha} \left(c_i - c_i^\alpha \right) dV. \qquad (2.9)$$

The concentration difference in the integrand deviates from zero only inside phase s. So Eq. (2.9) transforms into the well-known definition of the surface concentration,

$$\Gamma_i = \int_0^\infty \left(c_i(x) - c_i(\infty) \right) dx, \qquad (2.10)$$

when the surface is located at $x = 0$. This simple and convenient definition fails if case ii) applies or solubility of the surfactant has to be considered in both liquid bulk phases and the question arises about the exact localisation of the interface between the two bulks.

Gibbs' main idea was to introduce a dividing (mathematical) surface: its position (inside of the s-phase of Bakker's model) is defined in such a way that the adsorption of one of the components is zero and with a surface energy independent of surface curvature. This surface is noted as the surface of tension. In essence we can restrict Gibbs' thermodynamics to the IUPAC recommendation for colloid and surface chemistry, prepared by Everett (1971). Gibbs' idea of a dividing surface agrees with the definition by Rusanov (1981) mention above.

The excess amount of a component i actually present in the system compared to a reference system of the same volume as the real system in which the bulk concentrations in the two phases remain uniform up to the Gibbs dividing surface. This situation is shown in Fig. 2.8.

This amount is given by $n_i^s = n_i - V^\alpha c_i^\alpha - V^\beta c_i^\beta$, where n_i is the total amount of the component i in the system, c_i^α and c_i^β are the concentration in the two bulk phases α and β, V^α and V^β are the volumes of the two phases defined by the Gibbs dividing surface.

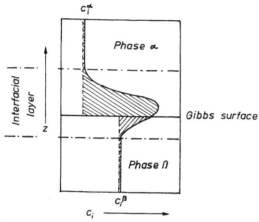

Fig. 2.8.   Schematic presentation of Gibbs' dividing surface

If c_i is the concentration of component i in a volume element dV, then

$$n_i^s = \int (c_i - c_i^\alpha) dV + \int (c_i - c_i^\beta) dV . \tag{2.11}$$

Phase α up to   Phase β up to
Gibbs' surface   Gibbs' surface

By introducing the surface area A^s the Gibbs surface concentration or surface excess concentration Γ_i^s is given by

$$\Gamma_i^s = n_i^s / A^s . \tag{2.12}$$

Corresponding definitions can be given for the surface excess mass of component i, n_i^s, and for the related surface excess molecular concentration and surface excess mass concentration.

In general, the choice of the position of the Gibbs dividing surface is arbitrary. It is possible to define quantities which are invariant with respect to the choice of its location. If $\Gamma_i^{(1)}$ or $\Gamma_{i,1}$ are the relative adsorption and Γ_i^s and Γ_1^s the Gibbs surface excess concentrations of components i and 1, respectively, then the relative adsorption of component i with respect to component 1 is defined by

$$\Gamma_i^{(1)} = \Gamma_i^s - \Gamma_1^s \left\{ \frac{c_i^\alpha - c_i^\beta}{c_1^\alpha - c_1^\beta} \right\} , \tag{2.13}$$

and is invariant to the location of the Gibbs dividing surface.

Rusanov (1981) commented that Gibbs' surface thermodynamic is simple and complicated at a time. Gibbs was the first to work out a detailed theory of interfacial phenomena independently of the work of Laplace and Gauss. Guggenheim (1959) wrote "it is much less to use Gibbs' formulas than to understand them".

In a homogeneous bulk phase the force across any unit area is equal in all directions, except in the surface phase s. Here this force depends on the direction and is called pressure which is assumed to be a tensor (in the bulk phase the pressure is a scalar). The normal and tangential components of the pressure tensor determine the interfacial tension γ. According to Bakker (1928) we obtain

$$\gamma = \int_{-\infty}^{\infty} (P_N - P_T) dz, \tag{2.14}$$

where P_N and P_T are the normal and the tangential component of the pressure tensor (cf. Fig. 2.9).

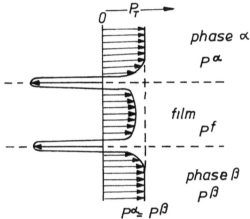

Fig. 2.9. Pressure vectors at an interface after Bakker (1928) and modified by de Feijter (1988) for a thin liquid Film

Sanfeld (1968) generalised Bakker's equation to a plane layer with an electric field normal to the interface

$$\gamma = \int_{-\infty}^{\infty} \left(P_N - P_T - \frac{\varepsilon E^2}{4\pi} \right) dz. \tag{2.15}$$

The definition of surface tension γ for a spherical interface can be defined by the Bakker equation (after Ono & Kondo 1960 and Buff 1955)

$$\gamma = \frac{1}{r_\gamma^2} \int_{r_i}^{r_e} \left(P_N^{tot} - P_T^{tot} \right) r^2 dr. \tag{2.16}$$

The parameter r_e, r_i and r_γ, respectively are explained by the schematic of a curved interface in Fig. 2.10.

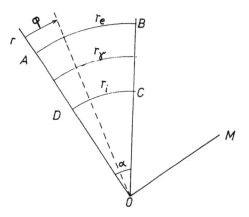

Fig. 2.10.   Demonstration of a curved interface at the distance r_γ

The surface or interfacial tension causes the pressure drop across a liquid interface. For spherical bubbles we obtain the well-known Thomson equation (1871)

$$\Delta p = \frac{2\gamma}{R}.$$

(2.17)

For liquid surfaces apart from spherical geometry the pressure drop is expressed through the main radii of curvature of the surface R_1 and R_2,

$$p^\alpha - p^\beta = \gamma \left(\frac{1}{R_1} + \frac{1}{R_2} \right).$$

(2.18)

This equation was first developed by Laplace and Gauss on the basis of mechanics.

Eq. (2.18) is the exact definition of the experimental relationship for the determination of surface tension by measuring the corresponding pressure differences and radii of curvature. This relationship is the basis of many experimental surface and interfacial tension methods measuring for example the volume of detaching drops (Section 5.2), the pressure inside bubbles (Section 5.3) or drops (Section 5.5), and the shape of sessile or pendent drops (Section 5.4).

2.4. ADSORPTION AT LIQUID/FLUID INTERFACES

Alternatively, $\Gamma_i^{(1)}$ may be regarded as the Gibbs surface concentration of i when the Gibbs surface is chosen so that Γ_1^s is zero, i.e. the Gibbs surface is chosen so that the reference system contains the same amount of component as the real system, hence $\Gamma_1^{(1)} \equiv 0$.

2.4.1. GIBBS ADSORPTION ISOTHERM

Gibbs thermodynamics leads to the following expressions and to the most important adsorption isotherm related to the excess quantities of a fluid interface:

$$U^s = U - U^1 - U^2 \qquad \text{(energy)}, \qquad (2.19)$$

$$S^s = S - S^1 - S^2 \qquad \text{(entropy)}. \qquad (2.20)$$

For the excess (Helmholtz) energy it follows

$$F^s = U^s - TS^s, \qquad (2.21)$$

and for the excess enthalpy

$$H^s = U^s - \gamma A. \qquad (2.22)$$

The Gibbs excess energy of the surface is therefore

$$G^s = H^s - TS^s = F^s - \gamma A. \qquad (2.23)$$

The chemical potential of the component i at the surface is consequently

$$\mu_i^s = \left(\frac{\partial F^s}{\partial n_i^s}\right)_{T,A,n_j^s} = \left(\frac{\partial G^s}{\partial n_i^s}\right)_{T,p,\gamma,n_j^s}. \qquad (2.24)$$

Equilibrium of the component i in the entire system exists, if $\mu_i^s = \mu_i^1 = \mu_i^2$. The Gibbs energy of the surface is given by

$$G^s = \sum n_i^s \mu_i^s. \qquad (2.25)$$

and the surface tension can be written

$$\gamma = \left(\frac{\partial G^s}{\partial A}\right)_{T,p^s,n_i^s} = \left(\frac{\partial G}{\partial A}\right)_{T,p,n_i^s,n_i^1,n_i^2}. \qquad (2.26)$$

The relationship between the surface tension, the chemical potential and the Gibbs surface energy is given by the general relation

$$\gamma A = F^s - \sum n_i^s \mu_i^s = F^s - G^s,$$
(2.27)

or for a unit surface area

$$\gamma = f^s - \sum \Gamma_i \mu_i^s.$$
(2.28)

One can write

$$dG_i^s = \mu_i^s d\Gamma_i + \Gamma_i d\mu_i^s$$
(2.29)

and for the equilibrium adsorption we have

$$d\Gamma_i = 0 \text{ and } d\mu_i^s = d\mu_i^{1,2}.$$
(2.30)

From the fundamental principles of thermodynamics it follows

$$d\mu_i^s = RT\, d\ln a_i^{1,2}$$
(2.31)

and

$$\Gamma_i = -\frac{1}{RT}\frac{d\gamma}{d\ln a_i}.$$
(2.32)

a_i is the activity of the component i in the bulk phase. For a system containing only one surface active component, the adsorption isotherm of Gibbs reads

$$\Gamma = -\frac{1}{RT}\frac{d\gamma}{d\ln c},$$
(2.33)

where we replaced the activity by the bulk concentration of surfactants, $a_i \cong c_i$.

So far the Gibbs adsorption isotherm represents the best founded theoretical background for the calculation of the adsorption excess densities of surfactants. Statistical thermodynamics may enable us in future to calculate adsorption densities by accounting for the chemical structure of a surfactant. Beside the direct calculation of excess adsorption densities Γ with the help of Γ-log c-plots, relationships of Γ and the interfacial tension γ as functions of the surfactant bulk concentration are very helpful.

These adsorption isotherms are known as Henry (1801), von Szyszkowski (1908), Langmuir (1916), Frumkin (1925), Volmer (1925) or Hückel-Cassel-isotherm (Hückel 1932, Cassel 1944), respectively. The constants in these isotherms refer to kinetic models of adsorption/desorption, interactions between adsorbed molecules and/or to the minimum area of adsorbed species.

Below in Section 2.4.4. and Appendix 2B, the physical background and derivation of the most frequently used adsorption isotherms will be discussed in more detail. Examples of the application of the Langmuir and Frumkin isotherms to experimental data are summarised in Appendix 5D.

2.4.2. ADSORPTION OF CHARGED SURFACTANTS

Another quite important type of adsorption isotherms is given by a generalisation due to the electrical repulsion in adsorption layers of ionic surfactants. This was explained in the monographs of Davies & Rideal (1961) and Lucassen-Reynders (1981).

Narrowly defined, the main contributions to film pressure or interfacial tension decrease come from the osmotic term and the repulsion of the electrical double layers of ionic surfactants including the effects of counterions. Interactions in mixed adsorption layers are of broad interest for the description of the state of surfactant adsorption layers. For the clarification of the adsorption mechanism at liquid interfaces the replacement of solvent molecules, mainly water, has been intensively studied by Lucassen-Reynders(1981).

In this book we are only concerned with the electrostatic repulsion aspect of the surface behaviour of charged surfactants, and then, as shown later, their electrostatic retardations effects. Such kinetic effects are in some cases decisive dynamic properties of a liquid interface and therefore significant for applications in colloid science and technology.

Lucassen-Reynders (1981) derived the electrostatic term of the film pressure. Her model is that of a surface charge by long chain ions, without taking into consideration inorganic counterions. Starting from the equation of Davies (1951) for the electrostatic repulsion in surface films

$$\pi_r = \int_0^{\Psi_0} \sigma \, d\Psi, \tag{2.34}$$

(σ represents the surface charge density and Ψ the Volta potential of the charged surface) and using the Gouy-Chapman model of the electric double layer (see Section 2.8. on electric double layers) the electrostatic repulsion acting in a surface films reads

$$\pi_r = \frac{2\pi kt}{\beta} \sqrt{c_c} \left\{ \cosh \sinh^{-1} \left(\frac{\beta}{A\sqrt{c_c}} \right)^{-1} \right\}. \tag{2.35}$$

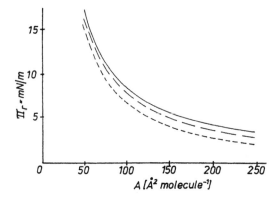

Fig. 2.11.　Electrostatic repulsion of the adsorption layer of an ionic surfactant as function of the occupied area per adsorbed molecule at different electrolyte concentration: $c_{el} = 0.0001M$ (solid), 0.01M (———), 0.1M (- -), according to Lucassen-Reynders (1981)

The coefficient β depends on temperature and the dielectric constant of the substrate. c_c is the concentration of counterions in the electrical double layer. According to Lucassen-Reynders (1981) we obtain a dependence of π_{el} on A (Fig. 2.11).

Kretzschmar & Voigt (1989) have recently examined the contribution of interacting forces in surfactant adsorption layers to the film pressure. A detailed knowledge of the geometry of the electrical double layer with respect to the plane of the interface is an essential item in the theoretical description of charged monolayers, thin liquid films and membranes. Fig. 2.12. shows an illustration of structural and energetic aspects of the surfactant monolayer formation.

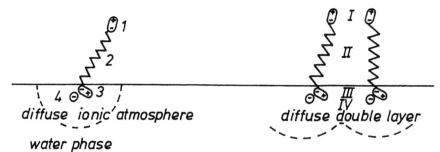

Fig. 2.12.　Adsorption and interaction energy contributions in surfactant monolayers; phase transfer: 1 - dipole in apolar phase, 2 - hydrocarbon chain, 3 - polar group in polar phase, 4 - charged groups; interaction energies: I - dipole in apolar phase, II - hydrocarbon chain, III - polar group in polar phase, IV - charged groups, according to Kretzschmar & Voigt (1989)

2.5. TREATMENT OF THE LANGMUIR ADSORPTION ISOTHERM AS INTRODUCTION TO ADSORPTION DYNAMICS

Langmuir's basic treatment for an adsorption isotherm was set out in his paper "The condensation and evaporation of gas molecules" (1917). Langmuir (1917, 1918) and Volmer (1925) derived a similar formula for the interpretation of the kinetics of adsorption/desorption. In these papers the process of adsorption was proposed as a dynamic situation for the first time. The basic relationship describing the adsorption as a kinetic mechanism results from the balance of the adsorption flux j_{ad} and desorption flux j_{des}.

$$\frac{d\Gamma}{dt} = j_{ad} - j_{des}.$$ (2.36)

The simplest case results when a non-localised adsorption is assumed (Baret 1968a, b), so that $j_{ad} \sim c_o$ and $j_{des} \sim \Gamma$. As the result we obtain Eq. (4.31), where are k_{ad} and k_{des} are the rate constants of adsorption and desorption, and c_o is the bulk concentration of the adsorbing species. On the basis of a localised adsorption the Langmuir mechanism Eq. (4.32) results. Further transfer mechanisms used to describe the kinetics of adsorption are given in Section 4.4, Eqs (4.31) - (4.34). To use these so-called transfer mechanisms for model of dynamic adsorption layers they have to be coupled with the transport process in the bulk. Baret (1969) suggested replacing c_o by the so-called subsurface or sublayer concentration. This is per definition the bulk concentration adjacent to the adsorption layer $c(0,t)$ localised at $x = 0$. The following two flux balance equations for the molecular transfer results,

$$\frac{d\Gamma}{dt} = k_{ad}\, c(0,t)(1 - \frac{\Gamma}{\Gamma_\infty}) - k_{des}\, \frac{\Gamma}{\Gamma_\infty},$$ (2.36a)

$$\frac{d\Gamma}{dt} = k_{ad} c(0,t) - k_{des}\Gamma.$$ (2.36b)

In adsorption equilibrium, when the adsorption flux equals the desorption flux, the Langmuir and Henry isotherms results.

In this book, adsorption and desorption processes play a key role. Thus it is important to discuss the physical background of adsorption isotherms and their use in describing the mechanism of surfactants in relaxation processes of adsorption layers at liquid interfaces.

The examination of the adsorption of gases at crystal surfaces by Langmuir (1917, 1918) and the derivation of isotherms through a cycle process by Volmer (1925) leads to almost the same structure of an equation describing the relationship between the surface concentration Γ and the bulk concentration c_o.

$$\Gamma = \Gamma_\infty \frac{c_o}{a_L + c_o}. \tag{2.37}$$

From comparison with Eq. (2.36a) we get $a_L = k_{des} / k_{ad}$. Behind the thermodynamically based adsorption equation of Gibbs (2.33) the Langmuir equation is the next important equation for describing the state of adsorbed surfactants. In comparison to the Gibbs equation the Langmuir equation is derived for a definite system, namely for a plane crystal surface and for gases well below their saturation.

Volmer (1925) was the first who mentioned the equivalence of the Langmuir and the von Szyszkowski isotherm, given by von Szyszkowski in the following from,

$$\gamma_0 - \gamma = B \ln\left(\frac{c_o}{a_L} + 1\right), \tag{2.38}$$

with $B = RT\Gamma_\infty$. Eq. (2.37) transforms into Eq. (2.38) via Gibbs fundamental equation. Another relation can be obtained in this way (Traube equation 1891)

$$\gamma = \gamma_0 + RT\Gamma_\infty\left(1 - \frac{\Gamma_0}{\Gamma_\infty}\right), \tag{2.39}$$

which is equivalent to (2.37) and (2.38) and a suitable relationship for theoretical calculations of dynamic surface tensions from adsorption kinetics models (cf. Chapter 4).

Many adsorption experiments on long chain fatty acids and other amphiphiles at the liquid/air interface and the close agreement with the von Szyszkowski equation is logically one proof of the validity of Langmuir's adsorption isotherm for the interpretation of $\gamma - \log c$ -plots of typical surfactants in aqueous solutions (cf. Appendix 5D). This evidence is also justification for use of the kinetic adsorption/desorption mechanism based on the Langmuir model for interpreting the kinetics and dynamics of surface active molecules.

This result also involves some fundamental issues of the kinetics and dynamics of processes of large molecules in the bulk layer adjacent to the interfaces.

2.6.   ADSORPTION ISOTHERM FOR SINGLE AND MIXED SURFACTANT SYSTEMS

The adsorption equation (2.38) of von Szyszkowski (1908) is originally an empirical relationship. Stauff (1957) later discussed the physical background of the constants a_L and B.

For very low surfactant concentrations of long chain fatty acids with surface tensions close to that of the pure solvent Traube (1891) found a linear relationship between interface tension and surfactant bulk concentration

$$\pi = \Delta\gamma = \gamma_0 - \gamma = RTk\, c.$$ (2.40)

The parameter k has the dimension of a length and results from Eq. (2.36b) as the ratio of adsorption over desorption rate constants $k = k_{ad} / k_{des}$. The description of an adsorption layer as a two dimensional gas is more thermodynamically based,

$$\pi\, A^s = RT,$$ (2.41)

where π is the so-called film pressure. Relating to the area of a molecule at an interface we obtain the Henry isotherm

$$\pi = \Gamma\, RT,$$ (2.42)

which is equivalent to the equation derived by Traube and can be transferred directly into each other via the Gibbs adsorption equation (2.33). Eq. (2.40) results from the Langmuir-Szyszkowski isotherm (2.38) at low concentrations $c_0 \ll a_L$ with $k = RT\Gamma_\infty / a_L$.

There are two different ways of deriving the Langmuir adsorption isotherm. One of the derivations is based on kinetic transfer mechanisms of adsorption/desorption of adsorbing molecules. The second, thermodynamic derivation starts from the equivalence of the chemical potentials of adsorbing molecules in the bulk phase and in the adsorbed state. Frumkin (1925) introduced additional interaction forces between adsorbed molecules into the Langmuir adsorption isotherm.

This idea is a consequent transfer of the three-dimensional van der Waals equation into the interfacial model developed by Cassel and Hückel (cf. Appendix 2B.1). The advantages of Frumkin's position is a more realistic consideration of the real properties of a two-dimensional surface state of the adsorption layer of soluble surfactants. This equation is comparable to a real gas isotherm. This means that the surface molecular area of the adsorbed molecules are taken into consideration. Frumkin (1925) additionally introduced, on the basis of the van der Waals equation, the intermolecular interacting force of adsorbed molecules represented by a',

$$\gamma = \gamma_0 - RT\Gamma_\infty \left(\ln(1 - \frac{\Gamma}{\Gamma_\infty}) + a'(\frac{\Gamma}{\Gamma_\infty})^2 \right).$$ (2.43)

Via Gibbs' equation (2.33) we can obtain a relationship between c and Γ,

$$c = a_F \frac{\Gamma}{\Gamma_\infty - \Gamma} \exp\left(-2a' \frac{\Gamma}{\Gamma_\infty}\right),$$ (2.44)

while a relationship between c and γ does not exist. Thus a comparison of experimental $\gamma - \log(c)$ data with the Frumkin isotherm is not trivial. However, this adsorption isotherm has for decades been the most frequently used equation of state and describes many surfactant systems very well.

Lucassen-Reynders (1976, 1981) derived more general thermodynamic relationships which take into consideration different molecular self-areas of solvent ω_1 and solute molecules ω_2. Her relationship reduces to a Frumkin-type isotherm by setting $\omega_1 = \omega_2 = 1/\Gamma_\infty$,

$$\pi = -RT\Gamma_\infty \left[\ln\left(1 - \frac{\Gamma}{\Gamma_\infty}\right) + \frac{H^s}{RT}\left(\frac{\Gamma}{\Gamma_\infty}\right) \right].$$ (2.45)

where H^s is the partial heat of mixing at the interface.

Lucassen-Reynders (1981) also derived isotherms for surfactant mixtures, which were used for example by Pomianowski & Rodakiewicz-Nowak (1980) or more recently by Wüstneck et al. (1993). As the most simple form of an isotherm for surfactant mixtures, a generalised Henry isotherm can be used,

$$\Gamma_i = k_i c_{oi}, \ i=1,2,...r.$$ (2.46)

This isotherm does not consider any interaction between the adsorbing molecules. The generalised Langmuir isotherm regards for a competitive adsorption of all surface active component r at the interface,

$$\Gamma_i = \Gamma_\infty \frac{c_{oi}/a_{Li}}{1 + \sum_{j=1}^{r} c_{oj}/a_{Lj}}, \ i=1,2,...r.$$ (2.47)

However, this isotherm does not consider specific molecular interfacial areas of the different components, as it was proposed by Damaskin (1969) and Damaskin et al. (1972), which can be relevant for realistic surfactant mixtures (Wüstneck et al. 1993).

2.7. MACRO-KINETIC ASPECTS OF LANGMUIR'S THEORY AND ITS APPLICATION TO ADSORPTION DYNAMICS

As we have seen in Sections 2.5. and 2.6. there are different isotherms for the description of the adsorption density Γ or the interfacial tension γ as functions of the known surfactant concentrations c_o in the bulk. From the thermodynamic point of view it is difficult to ask whether or not an equilibrium adsorption isotherm is valid under non-equilibrium conditions.

The basic model for the calculation of the adsorption flux of amphiphiles to an interface is shown in Fig. 2.13. This model consists of a so-called "sublayer", the coordinate x is oriented normal to the interface. The second assumption is an equilibrium between the sublayer and the interface at any time. The weakest point in this physical model is the problem of the validity of an equilibrium adsorption isotherm in a non-equilibrium state. In any case close to the equilibrium state the adsorption isotherms provides a good approximation.

The location of the sublayer at x=0 is an arbitrary but necessary part of the boundary conditions for the diffusion model.

Eq. (2.36) in its general form and Eqs (4.31) to (4.34) as particular cases have to be modified essentially in order to use them for the description of adsorption kinetics processes. This modification is the replacement of the bulk concentration c_o by the sublayer concentration $c(0,t)$, which was first suggested by Baret (1969), which leads to Eq. (4.35) used in Chapter 4 as the basis for the so-called kinetic-controlled adsorption model.

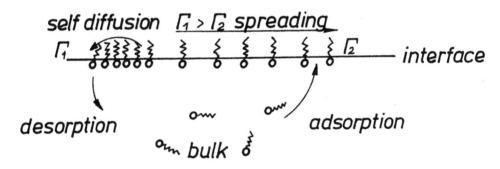

Fig. 2.13.   Schematic of adsorption/desorption fluxes to and from an interface

From surface thermodynamics itself one cannot know whether a sublayer model is correct. To discover if the sublayer model is a physically acceptable boundary condition, Fordham (1954) studied the use of equilibrium adsorption isotherms for systems in non-equilibrium. He concluded that the gradient of the diffusion profile has to be considered in the balance of the free energy of the system.

Later Hansen (1961) pointed out that due to a concentration gradient near an interface the Gibbs definition of the surface excess is not correct. As a reason he gave

$$\Gamma = \int_0^\infty (c(x,t) - c_o)dx + n(t) = 0, \qquad (2.48)$$

which is contradictory, but on first glance only. The contradiction arises from the integration in Eq. (2.48) over the whole diffusion layer. However the adsorbed amount defined by Gibbs thermodynamic suggests an integration over the surface phase s (cf. Eq. 2.11). Thus, in Eq. (2.48) c_o has to be replaced by the subsurface concentration while the integration has to be performed over the surface layer thickness.

Kretzschmar (1974, 1975, 1976) tried to find a correction term for the application of the adsorption isotherm of Hückel and Cassel, Eq. (2B.7), under dynamic conditions. The semi-quantitative relationship is based on partition functions in the system solvent/solute and its influence on the free energy in a non-equilibrium system.

Vollhardt (1966) and Kretzschmar & Vollhardt (1967) carried out experiments on dynamic desorption systems. The difference between the calculated desorption fluxes and the experimental results were explained on the basis of Fordham's treatment.

We must emphasise that the combination of the boundary condition of type (2.36) with the solution of the transport equation (diffusion equation) needed to calculate the subsurface concentration, is not trivial. In Section 4.4. the complexity of this type of models is demonstrated in detail.

Another difficulty is caused by the application of Gibbs thermodynamic concept to non-equilibrium conditions of an adsorption layer, as it is done in many theoretical models (cf. Chapter 4). From non-equilibrium thermodynamics we learn (cf. Eq. (2C.8) derived by Defay et al. (1966), Appendix 2C) that the diffusional transport causes an additional term to the Gibbs equation. However, this term seems to be negligible in many experiments. The experimental data discussed in Chapter 5, obtained from measurements in very different time windows, support the validity of Gibbs' fundamental equation (2.33) also under non-equilibrium conditions.

2.8.   CHARGED LIQUID INTERFACES

Before starting with some specific models of a charged interface and the adjacent bulk phase we want to discuss the terminology used in this book. Unfortunately for electric phenomena at interfaces, in contrast to the general thermodynamics, no unambiguous rules for the potential drop and charges at liquid interfaces exist. For example the magnitude and position of surface charges, the diffuse layer and Stern or Helmholtz layer charges are based on a large number of different theories. The situation is additionally complicated by the need to split the electrical potential drop into those parts caused by dipoles and those caused by free charges.

2.8.1. GENERAL REMARKS ON THE NOMENCLATURE OF CHARGED INTERFACES

All thermodynamic properties of liquid interfaces can be projected onto the Gibbs' dividing surface model and there is no reason to make assumptions about the dimension of the adsorption layer. The extension of the compact and diffuse electrical double layers is however essential for all electrical properties at interfaces like electrokinetics and film pressure contributions. Further, the adsorption of counterions is a sensitive influence on the general behaviour of an adsorption layer and depends strongly on the amount of charges as well as on its position in the interfacial phase. Surface charges and electrokinetic charges are different parameters and should not be mixed up. Charges are generally the primary parameters, rather than the potential. The IUPAC recommendation made a distinction for interfacial electrochemistry between indifferent and specifically adsorbing ions. Indifferent ions are adsorbed by Coulomb forces, whereas specifically adsorbing ions possess a chemical affinity to the surface in addition to the Coulomb interaction. Here "chemical" is a collective adjective for Van der Waals, hydrophobic, π-electron exchange and complex formation forces. Specifically adsorbing ions can adsorb on an initially uncharged surface, giving it a charge.

The origin of charges is well defined. The distribution of free charges and that of partial charges in oriented dipole layers at interfaces, which also generate potentials, is more difficult. With respect to surfaces we speak about (after Lange 1951, 1952) inner, Galvani and Volta potentials. Their definitions are given below.

2.8.2. THE ELECTROSTATIC DOUBLE LAYER

If a unit charge e approaches the surface of an ideally isolated phase (to a distance of approximately 10^{-4} cm) from infinity, the work of $e\psi$ must be done. ψ represents the Volta potential. In other words the Volta potential denotes the electrical potential difference inside the phase.

54

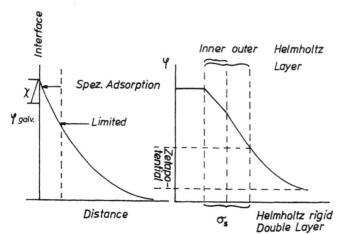

Fig. 2.14.   Model of the electric double layer and the specific adsorption of different types of charges at interfaces

To transfer charge carriers (electrons, ions) from one phase to another, the surface must be crossed. The surface itself as well as the adsorption layer of surfactants consist of carrying dipoles. That is the reason for an additional jump of the electrical surface potential. This potential may be (Lange 1951) denoted as the dipole part χ of the surface potential.

The simplest model for the electrical double layer is the Helmholtz condenser. A distribution of counterions in the bulk phase described by a Boltzmann distribution agree with the Gouy-Chapman theory. On the basis of a Langmuir isotherm Stern (1924) derived a generalisation of the double layer models given by Helmholtz and Gouy. Grahame (1955) extended this model with the possibility of adsorption of hydrated and dehydrated ions. This leads to a built-up of an inner and an outer Helmholtz double layer. Fig. 2.14. shows schematically the model of specific adsorption of ions and dipoles.

2.8.3.   EXAMPLE FOR THE CALCULATION OF $\Delta\chi$ POTENTIAL FROM KNOWN DIPOLE MOMENT OF AN ADSORBED MOLECULE IN THE NORMAL DIRECTION TO THE INTERFACE

Let us assume the adsorbed surfactant molecule occupy an area of 50Å^2 at the interface which corresponds to $2\cdot10^{14}$ dipoles per cm². We also use the Helmholtz equation

$$\Delta V = 4\pi n\mu, \tag{2.49}$$

where n is the number of dipoles per unit area and μ the resulting dipole moment in the direction normal to the interface. If we take a reasonable value of a dipole moment of 500 mDebye   we   obtain $\mu = 0.5\cdot10^{-18}$ electrostatic   units   and   consequently   we   get $\Delta V (= 12\cdot2\cdot10^{14}\cdot0.5\cdot10^{-18}) \cong 1.2\cdot10^{-3}$ electrostatic units/cm ². If we consider that 1 electrost.

cgs units is about 300 Volts, it follows for $\Delta V = 1.2 \cdot 10^{-3} / 3.3356 \cdot 10^{-3} = 0.360 \text{V}$. This value is typical for experimentally measured surface potentials.

2.8.4. ION ADSORPTION AND ELECTRIC DOUBLE LAYER

The adsorption of ionic surfactants as carrier of electrical charges leads to the built-up of a surface charge. The kinetics of adsorption is coupled with the formation of an electrical double layer at the interface. There is evidence that the electrical double layer can retard the adsorption flux of the surface active ions with an electrostatic barrier.

Modern aspects of the structure of the electrochemical double layer at interfaces were recently summarised by Watanabe (1994) in line with IUPAC recommendation for the description of the different parts of an electrical double layer forming at an interface as well as the conditions of electroneutrality and the behaviour of different ions at interfaces.

The essential factor in the formation of the surface charge is the specificity of ion interaction with the interface, the general mechanism of the compensating layer, the Stern-Helmholtz as well as the diffuse layer, depends on the forces of electrostatic attraction and the screening happens without any affinity of the counterions to the interface. These are ions which form the screening charge. On the contrary, ions which built up the surface charge are called potential or charge determining ions. The introduction of two different concepts of charge determining ions and specifically adsorbed ions emphasise their different role in the structure of the double layer. Both types of ions are specifically adsorbing. However, the adsorption of only one type predominates and controls the surface charges. Consequently, the sign of surface charge coincidences with that of the charge building ions so that we call them charge determining ions.

2.8.5. THE DISTRIBUTION OF IONS IN AN ELECTRIC FIELD NEAR A CHARGED SURFACE

The distribution of ions in the diffuse layer in the form of a concentration profile is the result of electrostatic attraction forces and the thermal equilibrium.

Thus the counterions migrate toward the surface under the action of the electric field. Since their concentration increases in this direction, this electromigration is fully compensated for by the diffusion current. The conditions for compensation of electromigration in diffusion can be mathematically represented in the following form

$$j_x^\pm = -D^\pm \frac{dC_{el}^\pm(x)}{dx} \mp \frac{F}{RT} D^\pm z^+ z^- C_{el}(x)\frac{d\Psi}{dx} = 0. \tag{2.50}$$

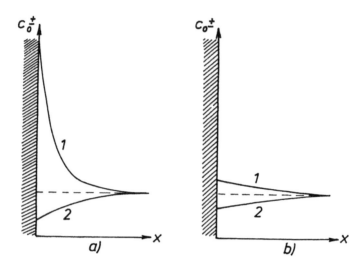

Fig. 2.15. Distribution of ions in an electric field near a charged surface; (a) high surface potential, (b) low surface potential, 1 Distribution of counterions; 2 distribution of coions, according to Dukhin (1974)

We consider two equations for the three functions $\Psi_{el}(x)$, $C_{el}^\pm(x)$ and can express the ion concentration distribution through the potential distribution. By separation of the variables one can carry out the integration which yields

$$C_{el}^\pm(x) = C_{o,el}^\pm \exp\left[\frac{\mp z^\pm \Psi(x)}{kt}\right]. \tag{2.51}$$

The relevant boundary conditions are

$$C_{el}^\pm\big|_{x\to\infty} = C_0^\pm,\ \Psi\big|_{x\to\infty} = 0. \tag{2.52}$$

As expected, the ions are distributed across the double layer according to Boltzmann's formula. As a result, the general relationship between the electric potential $\Psi_{eq}(x)$ and the density of the charge in point x, can be expressed by Poisson's equation

$$\mathrm{div}\big[\varepsilon(x)\mathrm{grad}\Psi(x)\big] = 4\pi\rho(x). \tag{2.53}$$

The transformation into a one-dimensional space we obtain

$$\frac{d}{dx}\left[\varepsilon(x)\frac{d\Psi}{dx}\right] = 4\pi\rho(x). \tag{2.54}$$

Substitution of Eqs (2.51) and (2.54) leads to

$$\frac{d}{dx}\left[\varepsilon(x)\frac{d\Psi}{dx}\right] = -4\pi e \sum_i z_i c_{i\infty} \exp\left[-\frac{z_i e\Psi(x)}{kT}\right]. \tag{2.55}$$

Eq. (2.55) can be integrated directly to

$$\frac{d\Psi}{dx} = \pm\left(\frac{8\pi RT}{\varepsilon}\sum C_{i\infty}\left\{\exp\left[\frac{\mp z_i e\Psi_{eq}(x)}{kT}\right]-1\right\}\right)^{1/2}, \tag{2.56}$$

if a possible change in ε is neglected. An approximation results when using the Debye-Hückel theory for strong electrolytes (linear equation)

$$\Psi(x) = \Psi_0 \exp(-\kappa x), \tag{2.57}$$

with

$$\kappa^2 = \frac{4\pi F^2 \sum_i C_{i\infty} z_i^2}{\varepsilon RT}. \tag{2.58}$$

2.8.6. THE STERN-GOUY-CHAPMAN MODEL

In the introduction to the electrochemistry of interfaces the splitting of the charges compensating the surface charge in a compact part near the interface (Stern-Helmholtz layer) and in a diffuse part (Gouy-Chapman theory) was shown. The sum of surface charges and the charges with the opposite sign, denoted generally as counterions, must be zero. Therefore the equation

$$\sigma_s + \sigma_{St} + \sigma_d = 0 \tag{2.59}$$

must be valid. Here the subscripts s, St, and d refer to the surface, the Stern layer and the diffuse layer, respectively. The compact double layer of opposite electric charges adjacent to the interface is one of the most complicated issues to describe. There are some major problems which can be partially solved only by theoretical treatments and experiments:

-the distance of the counterions from the plane of surface charge,

-the theoretical considerations in connection with the presence of point charges, dehydrated charges and hydrated charges.

In this case the notation of the Stern potential is convenient. The Stern potential characterises the potential at the boundary of the Stern and the diffuse layer. If the potential drop across the Stern layer can be neglected the state of the counterions within it can be characterised by the Stern potential too. Thus the charges of Stern and diffuse layers can be expressed by one potential. At low electrolyte concentration the diffuse layer thickness exceeds the thickness of the Stern layer which can be estimated as a multiple of the ion dimension. This means that the potential drop across the Stern layer is very small in comparison to the Stern potential and can be neglected. However the neglecting of the potential drop across the Stern layer at high electrolyte concentration, when the diffuse layer thickness is small, is questionable.

2.8.7. THE MODEL OF THE INNER PART OF THE ELECTRIC DOUBLE LAYER

In the Stern-Gouy-Chapman (SGC) theory the double layer is divided into a Stern layer, adjacent to the surface with a thickness d_1 and a diffuse layer of point charges. The diffuse layer begins at the Stern plane in a distance d_1 from the surface. In the simplest case the Stern layer is free of charges. In real cases the Stern layer is formed by specifically adsorbed ions. The condition of electroneutrality was given by Eq. (2.59) In addition to σ_{St} and σ_d, the surface charge can be represented by the Stern potential. It transforms the conditions of electroneutrality into the equation for the determination of the Stern potential.

SGC-type models have been criticised by Cooper & Harrison (1977) but experiments have shown that the model of dividing the double layer into a diffuse and a non-diffuse part works very well in practice. A review on the use of modern statistical methods for the present topic has been published for example by Carnie & Torrie (1984).

2.8.8. THE MODELS OF ADSORPTION AND SURFACE CHARGE

As mentioned above, the adsorption of ions at interface can be divided into different cases. The solid can be an extended surface or can consist of disperse particle, such as latices and specific and non-specific adsorption is possible. At the solid surface active sites are distributed. This active sites are the geometrical spots for localised adsorption. Two different planes of adsorption are possible. The first one is the solid/liquid interface itself. We could take the adsorption of negatively charged phosphate ions of amino groups as part of a solid surface as an example of this. The phosphate groups are a classical case of non-point charges which can approach the interface only to a distance dependent on the geometric extension of the hydrophilic head group. This adsorption is called localised adsorption and can be described by an isotherm analogous to the Langmuir isotherm. The layer of adsorbed counterions and the layer of the interface form a Helmholtz condenser. This adsorption layer of counterions is called Stern layer. The specifically adsorbing ions are described in a limited range by the charge density relationship of Stern (1924),

$$\frac{\Theta_i}{1-\Theta_i} = \frac{c_i}{55.5}\exp\left(-\frac{W_i}{RT} - \frac{z_i F\Psi_{St}}{RT}\right). \tag{2.60}$$

Here, Θ_i is the fraction of surface sites covered by specifically adsorbed ions i, W_i is the specific Gibbs energy of adsorption, $c_i/55.5$ is the mole fraction of i and $z_i F\tilde{\Psi}_{St}$ the coulomb contribution to the adsorption energy.

In contrast, the liquid interface can be considered homogeneous which is a precondition for a non-localised adsorption. Any free space at the liquid interface is available for ion adsorption. At the transition from a localised to a non-localised adsorption the exponential term of the Stern-Langmuir equation can be preserved and the pre-exponential multiplier must be changed. This was done by Martynov (1979). The basis of this approach is the notation of a homogeneous potential well along the liquid interface as a whole.

The homogeneity of a liquid surface leads to identical conditions for the adsorption at any place of the surface which is reflected by the notion of a homogeneous potential well in a layer of thickness Δ. Its depth is equal to the adsorption energy, its thickness represents the decay of interaction between ions and the interface as the distance increases. Martynov considered this decay and initially introduced the potential well of a complicated configuration. In the framework of this model the adsorption is characterised as the total quantity of the ions within the potential well, i.e. as the product of the well thickness and the ion concentration within it.

The recombination of Martynov's theory, which is based on a Boltzmann distribution and a Lennard-Jones potential, the isotherm for a "smeared adsorption" can simplified to

$$\Gamma = \Delta e^{-W/kt - \tilde{\Psi}_{St}}. \tag{2.61}$$

There are some uncertainties about the choice of Δ. It is significant that one unknown parameter $\Delta e^{-W/kT}$ has been used in the Martynov theory instead of the two of the Stern theory. The existence of many stable states of adsorbed organic ions with different adsorption energies corresponds to some not fully understood surface effects. It is obvious that Martynov's model needs generalisations in that respect.

Recently from the electrochemical point of view the adsorption of weak as well as strong organic electrolytes has been examined by Koopal (1993) who also considered specific electrochemical aspects of adsorbed ionic surfactants.

The structure of a surfactant adsorption layer get more complicated as its bulk concentration increases. This book is devoted to dynamic adsorption properties of liquid interfaces. The dynamic problems at interfaces are so complicated that their solution is only possible for the simplest cases of adsorption layers formed by dilute surfactant solutions. For this reason we have not considered any special problems relating to surfactant adsorption layer structures.

2.8.9. *EXTENSION OF MODELS OF THE INNER PART OF THE ELECTROCHEMICAL DOUBLE LAYER*

The inner part of the DL consists of the adsorption layer and specifically adsorbed counterions. The state of charge-determining ions and specifically adsorbed counterions depend on potentials which are determined by their localisation.

A distinction is often made between the plane where the centres of charge of the partially dehydrated specifically adsorbed ions reside, the inner Helmholtz plane, and Stern plane at distance d_1, which is also called the outer Helmholtz plane. The double layer model consists of an inner and outer Helmholtz layer and a diffuse layer. This is often called the triple layer model.

In the present treatment we will use the simplified Stern layer model in which the specifically adsorbed ions are located in the Stern plane, i.e., we let the inner and outer Helmholtz plane coincide. Here, however, is a peculiarity caused by the indefiniteness of the location of the charged head of adsorbed surfactant molecules which can be some distance from the interface in direction to the water phase.

As the electrolyte concentration is low and the Debye radius many times exceeds this distance an identification of the potential in this zone with the potential of the adsorbing ions is reasonable. At high electrolyte concentration the diffuse layer thickness can be comparable to this distance. Even if the counterions are indifferent their distribution in this layer cannot be neglected because it decreases the electrostatic component of surfactant ion adsorption., i.e. enhance its adsorption. In this case we have to consider a discreteness of charges must, the formation of a counter ion atmosphere around the adsorbed ions and their overlap with the neighbour adsorbed ions.

Two extremes of specifically adsorbed counterions behaviour can be distinguished. In general the counterions can interact specifically with the interface and with the charged heads of the adsorbed ions. The extreme case is of interest when only one type of interaction strongly predominates. If the interaction of ions with the interface dominates the adsorption is non-localised and the Stern-Martynov theory must be used. In contrast, if the interaction of adsorbed ions with the counterions predominates, or with respect to the definition for a localised adsorption, the Stern theory must be applied.

For liquid/liquid systems we consider counterion adsorption from water because the solubility of inorganic ions in an oil phase is negligible. The electrostatic interaction of the inorganic ions with the uncharged section of the water/oil interface is repulsive because the dielectric pemittivity of water exceeds that of oil. This effect is called the Onsager force of mirror image caused by the polarisation of the interface (Dukhin et. al. 1980) Cations are usually unable to specifically adsorb at an uncharged water interface. The specific interaction of counterions with the charged head of the adsorbed charge-determining ions is the most frequent and significant outcome. So only the Stern isotherm is applicable.

2.9.   SUMMARY

FROM THERMODYNAMICS TO MACRO-KINETICS

Langmuir was the first to provide of a theory of the interrelation between thermodynamics and macro-kinetics. It introduces the balance of adsorption and desorption fluxes into the adsorption theory and defines the equality as the equilibrium state of the adsorption layer. A disturbance of the adsorption equilibrium leads to a net adsorption or net desorption flux. This idea serves as a bridge from thermodynamics to macro-kinetics and allows a deeper understanding of the equilibrium state of adsorption as a dynamic process as demonstrated by de Boer in his monograph "Dynamic Character of Adsorption".

This dynamic character of the adsorption equilibrium has contributed significantly to developments in non-equilibrium thermodynamics. The balance of adsorption and desorption fluxes as the first step in the description of the dynamics of adsorption is a key point in this book. The second step is the introduction of a sublayer concentration and the diffusion layer to describe the non-equilibrium state in the bulk phase. While the system surface-bulk is in non-equilibrium the presence of local equilibrium is assumed between the adsorption layer and the sublayer as the third important step. This allows us to generalise Eq. (2.36) to Eqs (2.36a) and (2.36b). The first examples of dynamic adsorption layers of rising bubbles were given already by Frumkin & Levich (1947) and Levich's book (1962) on "Physico-Chemical Hydrodynamics" (cf. Chapter 8) offered the first theories. Simultaneously, Frumkin & Levich

(1947) emphasised that local equilibrium is only an extreme case and disturbance of the adsorption equilibrium leads to relaxation processes (subsequent Chapter 3).

In the transition from thermodynamics to dynamics of adsorption transport and energetic aspects have to be distinguished. The main question on energetic aspects belongs to the validity of Gibbs' equation under non-equilibrium conditions. Defay et al. (1966) demonstrated the outcome of non-equilibrium thermodynamics for this topic. This direction deserves further attention. Comparison of experimental data with theory enables us to determine the range of applicability of Gibbs' theory to non-equilibrium systems.

ADSORPTION ISOTHERMS

Studies of dynamic adsorption layers require the use of equilibrium adsorption isotherms, as will be demonstrated in detail in Chapter 4. The use of Langmuir's isotherm is restricted to large molecular areas of adsorbing molecules because specific interaction is left out of his account. This isotherm is however the most frequently-used relationship because of its simple form and comparatively good agreement with many experimental data.

A comprehensive analysis of the family of adsorption isotherms is given and their use discussed for investigations of non-equilibrium adsorption layers.

IONIC SURFACTANTS AT THE WATER SURFACE

Two main aspects are considered here. First is the coupling of the potential drop caused by the preferential orientation of dipoles and by the inhomogeneity of ion distribution. Second is the specificity of the structure of the inner part of the double layer formed by ionic surfactants at the water-air interface.

At comparable thicknesses of the dipole and diffuse layers their coupling becomes important and has so far been underestimated. However, the Debey length increases with decreasing electrolyte concentration and the diffuse layer becomes independent. The counter-ion distribution in the inner part of the double layer can be described as localised or non-localised adsorption dependent on some conditions.

2.10.   REFERENCES

Adamson, A.W., "Physical Chemistry of Surfaces", John Wiley & Sons, Inc., New York, Chichester, Brisbane, Toronto, Singapore, (1990)

Aloisi, G., Guidelli, R., Jackson, R.A., Clark, S.M. and Barnes, P., J. Electroanal. Chem., 206(1986)131

Bakker, G., in "Handbuch der Experimentalphysik", Akademische Verlagsgesellsch., Leipzig, VI(1928)

Baret, J.F., J. Phys. Chem., 72(1968a)2755

Baret, J.F., J. Chim. Phys. , 65(1968b)895

Baret, J.F., J Colloid Interface Sci., 30(1969)1

Barraclough, C.G., McTigue, P.T. and Leung Ng, Y., J. Electroanal. Chem., 329(1992)50

Bockris, J.O.M. and Reddy, A.K.N., "Modern Electrochemistry" Plenum Press, New York, (1973)

Boruvka, L., Rotenberg, Y. and Neumann, A.W., J. Physical Chemistry 90(1986)125

Bourrell, M. and Schechter, R.S, "Microemulsions and Related Systems", Marcel Dekker, Inc., New York and Basel, (1988)

Buff, F.P., J. Chem. Phys., 23(1955)419

Butler, J.A.V., Trans. Faraday Soc., 19(1924)734

Butler, J.A.V., Trans. Faraday Soc., 28(1932)379

Carnie, G.C. and Torrie, G.M., Adv. Chem. Phys., 56(1984)141

Cassel, H., J. Phys. Chem., 48(1944)195

Chapman, D.D., Phil. Mag., 25(1913)475

Corkill, J.M., Goodman, S.P., Harold, P. and Tate, J.R., Trans. Faraday Soc., 63(1967)247

Cooper, I.L. and Harrison, J.A., Electrochim. Acta, 22(1977)519

Damaskin, B.B., Frumkin, A.N., Borovaja, N.A., Elektrochimija, 6(1972)807

Damaskin,B.B., Elektrochimija, 5(1969)346

Davies, J.T., Proc. Roy. Soc., Ser. A, 208(1951)244

Davies, J.T. and Rideal, E.K., "Interfacial Phenomena", Academic Press, New York, (1963)

de Boer, J.H., "The Dynamical Character of Adsorption", Oxford Univ. Press, London, (1953)

Debye, P. and Hückel E., Physik. Z., 24(1923)185

Defay, R., Prigogine, I., Bellemans, A. and Everett, D.H., Surface Tension and Adsorption, Longmans, Green and Co. Ltd., London, (1966)

Defay, R. and Petré, G., in "Surface and Colloid Science", Matijevic´, E., (Ed.), John Wiley, New York, 3(1971)

Defay, R., Prigogine, I. and Sanfeld, A., J. Colloid Interface Sci., 58(1977)498

de Feijter, J.A., "Thermodynamics of Thin Liquid Films" in "Thin Liquid Films", Ivanov, I.B., (Ed.), Marcel Dekker, Inc. New York, Basel (Surfactant Science Series), 29(1988)1

64

de Groot, S.R. and Mazur, P., Non-Equilibrium Thermodynamics, Dover Publications, Inc., New York, (1984)

Demchak, R.J. and Fort, T., J. Colloid Interface Sci., 46(1974)191

Derjaguin, B.V. and Obuchov, E., Acta Physicochim. USSR, 5(1936)1

Derjaguin, B.V. and Kussakov,M., Acta Physicochim. USSR, 10(1939)153

Derjaguin, B.V. and Landau, L., Acta Physicochim. USSR, 14(1941)633

Derjaguin, B.V. and Zorin, Z., Zh. Fiz. Khim., 29(1955)1010

Derjaguin B.V. and Karasev, V.V., Proc. 2nd Int. Congr. Surface Activity, Butterworth, London, Vol. 2, (1957), pp 531

Derjaguin, B.V. and Churaev, N.V., J. Colloid Interf. Sci., 49(1974)249

Derjaguin, B.V., Colloids & Surfaces A, 79(1993)1

Dukhin, S.S., Glasman, J.M. and Michailovskij, V.N., Koll. Zh., 35(1973)1013

Dukhin, S.S., Croatica Chemica Acta, 53(1980)167

Dukhin, S.S. and Derjaguin, B.V., in "Surface and Colloid Science", Matijevic´, E., (Ed.), John Wiley, New York, 7(1974)

Dynarowicz, P., Adv. Colloid Interface Sci, 45(1993)215

Everett, D.H., "Manual of Symbols and Terminology for Physocichemical Quantities", Appendix II, Butterworth, London, (1971)

Everett, D.H., Pure & Appl. Chem., 52,(1980)1279

Exerowa, D., Nikolov, A.D. and Zachariova, M., J. Colloid Interface Sci., 81(1981)491

Exerowa, D., Kachiev, D. and Balinov, B., in "Microscopic Aspects of Adhesion and Lubrication", Elsevier, Amsterdam, (1982)107

Exerowa, D., Balinov, B., Nikolova, A. and Kachiev, D., J. Colloid Interface Sci., 95(1983)289

Exerowa, D. and Kachiev, D., Contemp. Phys., 27(1986)429

Fijnaut, H.M. and Joosten, J.G.H., J. Chem. Phys., 69(1978)1022

Frumkin, A., Z. Phys. Chem., 116(1925)466

Frumkin, A.N. and Levich, V.G., Zh. Phys. Chim., 21(1947)1183

Frumkin, A.N., Zh. Phys. Chim., 30(1956)1455

Gibbs, J.W., Scientific Papers, Vol. 1, 1906

Goh, M.C., Hicks, J.M., Kemnitz, K., Pinto, G.R., Bhattacharaya, K. and Eisenthal, K.B., J. Phys. Chem., 92(1988)5074

Good, R.J., Pure & Applied Chem., 48(1976)427

Gouy, G., J. Phys., 9(1910)457

Gouy, G., Ann. Phys., 7(1917)129

Grahame, D.C., Chem. Rev., 41(1947)44

Grahame, D.C., Z. Elektrochem., 59(1955)773

Grimson, M.J., Richmond, P. and Vassiliev, C.S., in "Thin Liquid Films", Surfactant Science Ser., I.B. Ivanov, Ed., Vol. 29, Marcel Dekker, 1988

Guggenheim, E.A., Thermodynamics, North-Holland Publishing Company, Amsterdam, (1959)

Hamaker, H.C., Physics, 4,(1937)1053

Hansen, R.S., J. Colloid Sci., 16(1961)549

Hansen, R.S. and Beikerkar, K.G., Pure & Appl. Chem., 48(1976)435

Henry, W., Nicholson's J., 4(1901)224

Hückel, E., Trans Faraday Soc., 28(1932)442

Hunter, R.J., Foundations of Colloid Science, Volumes I and II, Clarendon Press, Oxford, (1992)

Israelachvili, J., "Intermolecular and Surface Forces", Academic Press, London, San Diego, New York, Boston, Sydney, Tokyo, Toronto, 1992

Ivanov, I.B. and Toshev B.V., Colloid & Polymer Sci. 253(1975)593

Ivanov, I.B., Ed., Surfactant Science Ser., "Thin Liquid Films", Vol. 29, Marcel Dekker, 1988

Jaycock M.J. and Parfitt, S.D., Ellis Horwood Ltd., John Wiley & Sons, New York, Chichester, Brisbane, Toronto, (1981)

Klevens, H.B., J. Am. Oil Chem. Soc., 30(1953)74

Kohler, H.H., "Surface Charge and Surface Potential" in: "Coagulation and Flocculation" Dobias, B., (Ed.), Marcel Dekker, New York, 37(1993)

Kohler, H.H., "Thermodynamics of Adsorption from Solution" in "Coagulation and Flocculation", B. Dobias, (Ed.), Marcel Dekker (Surfactant Science Series), New York, (1993)

Koopal, L.K., Wilkinson, G.T. and Ralston, J., J. Colloid Interf. Sci., 126,(1988)493

Koopal, L.K., "Adsorption of Ions and Surfactants" in: Coagulation and Flocculation, B.Dobias, (Ed.), Marcel Dekker, New York, 101(1993)

Kretzschmar G and Vollhardt D, Berichte d. Bunsengesellschaft, 71(1967)410

Kretzschmar, G. Z. Chemie, 14(1974)261

Kretzschmar, G., Proc. Intern. Confer. Colloid Surface Sci., Budapest, 1975

Kretzschmar, G., VII Intern. Congr. Surface Active Substances, Moscow, (1976)

Kretzschmar, G. and Voigt, A., Proceeding of "Electrokinetic Phenomena", Dresden, 1989

Lange, E. and Koenig, F.O., Handbuch der Experimentalphysik, Vol. 12, Leipzig, (1933)

Lange, E., Z. Elektrochem., 55(1951)76

Lange, E., Z. Elektrochem., 56(1952)94

Langmuir, I., Physical Review, 8(1916)2

Langmuir, I., Proc. Nat. Academy of Sciences, 3(1917)141

Langmuir, I., J. Amer. Chem. Soc., 15(1918)75

Langmuir, I., J. Chem. Physics, 1(1933)187

Lennard-Jones, J.E., in "Fowler's Statistical Mechanics", Cambridge, 1936

Levich, V.G., Physico-Chemical Hydrodynamics, Prentice-Hall, Englewood Cliffs, N.Y., 1962

Li, D., Gaydos, J. and Neumann, A.W., Langmuir, 5(1989)1133

Lucassen-Reynders, E.H., Progr. Surface Membrane Sic., 3(1976)253

Lucassen-Reynders, E.H., Adsorption at Fluid Interfaces, in "Surfactant Science Series",
     Marcel Dekker, New York, 11(1981)

Matsumoto, M. and Kataoka, Y., J. Chem. Phys., 88(1988)3233

Martynov, G.A., Elektrokhimiya, 15(1979)494

Mie, G., Ann. Physik, 11(1903)657

Milner, S.R., Phil. Mag., 13(1907)96

Motomura, K., J. Colloid Interf. Sci., 48(1974)307

Ono, S. and Kondo, S., in "Handbuch der Physik", Flügge, S., (Ed.), Springer, Berlin, (1960)

Verwey, E.J.W. and Overbeek, Th.G., Theory of Stability of Lyophobic Colloids, Elsevier,
     Amsterdam, (1948)

Platikanov, D., Nedyalkov, M. and Scheludko, A., J. Colloid Interf. Sci., 75(1980)612

Platikanov, D., Nedyalkov, M. and Nusteva, V., J. Colloid Interf. Sci., 75(1980)620

Pomianowski, A. and Rodakiewicz-Nowak, J., Polish J. Chem., 54(1980)267

Rabinovich, Ya.I., Derjaguin, B.V. and Churaev, N.V., Adv. Colloid Interface Sci., 16(1982)63

Richter, L., Platikanov, D. and Kretzschmar, G., Proc. Intern. Conf. Surface Active Agents,
     Akademie-Verlag, Berlin, (1987)

Rotenberg, Y., Boruvka, L. and Neumann, A.W., J. Colloid Interface Sci., 93(1983)169

Rusanov, A.I., "The Modern Theory of Capillarity", Goodrich, F.C. and Rusanov, A.I., (Ed.),
     Akademie-Verlag, Berlin, (1981)

Sanfeld, A., Thermodynamics of Charged and Polarized Layers, Wiley Interscience, London,
     New York, Sydney, Toronto, (1968)

Scheludko, A., "Colloid Chemistry", Elsevier Amsterdam, London, New York, (1966)

Scheludko, A., Radoev, B. and Kolarov, T., Faraday Soc., 69(1968)2213

Scheludko, A., Exerowa, D. and Platikanov, D., Bulgarian Academy of Sciences,
     Communication of the Chemistry Dept., 11(1969)499

Sonntag, H. and Strenge, K., "Koagulation und Stabilität disperser Systeme", VEB Deutscher
     Verlag der Wissenschaften, Berlin, (1970)

Stauff, J. and Rasper, J., Kolloid Z., 151(1957)148

Stern, O., Z. Elektrochem., 30(1924)508

Stillinger, F.H. and Ben-Naim, A., J. Phys. Chem., 47(1967)4431

Szleifer, I. and Widom, B., Molecular Physics 75(1992)925

Szyskowski, B. von, Z. Phys. Chem., 64(1908)385

Tanford, C., "The Hydrophobic Effect", Wiley Interscience, New York, 1980

Thomson, W., Phil. Mag., 42(1971)448

Toshev, B.V. and Ivanov, I.B., Colloid & Polymer Sci 253(1975)558

Traube, I., Liebigs Ann., 265(1891)27

Usui, S.,Vol. 7 "Electrical Double Layer" in: "Electrical Phenomena at Interfaces", Vol. 15,
     Marcel Dekker, New York, (1984)

Vogel, V. and Möbius, D., Thin Solid Films, 159(1988a)73

Vogel, V. and Möbius, D., J.Colloid Interface Sci., 126(1988b)408

Vollhardt, D., PhD Thesis, Berlin, (1966)

Volmer, M., Z. Phys. Chem., 115(1925)253

Vrij, A., Disc. Faraday Soc., 42(1966)23

Ward, A.F.H. and L.Tordai, J. Phys. Chem. 14(1946)453

Watanabe, A., "Electrochemistry of Oil-Water Interfaces", Matijevic′, E. and Good, R.J., (Ed.)
     Plenum Press, New York, London, Surface and Colloid Sci, 1(1994)13

Widom, B. and Clark, A.S., Physica A, 168(1990)149

Wilson, M.A., Pohorille, A., and Pratt, L.R., J. Chem. Phys., 88(1988)3281

Wüstneck, R., Miller, R. and Kriwanek, J., Colloids and Surfaces, 81(1993)1

Yao, Y.L., Irreversible Thermodynamics, Van Nostrand Reinhold Company, New York,
     Cincinatti, Toronto, London, Melbourne, (1981)

Zimmels, Y., J. Colloid Interf. Sci., 130(1988)320

CHAPTER 3

3. **SURFACE PHENOMENA, SURFACE RHEOLOGY AND RELAXATIONS PROCESSES AT LIQUID INTERFACES**

The important peculiarity of adsorption processes at liquid interfaces is its lateral mobility which leads to lateral excess transport of adsorbed molecules. Lateral transport causes non-equilibrium adsorption layers which can initiate relaxation processes and hydrodynamic flow. And vice versa, liquid flow can produce non-equilibrium adsorption layers which causes lateral transport. This is the reason why non-equilibrium adsorption layers are of distinguished importance for liquid interfaces.

Surface rheology has to be taken into account when describing surface movement and adsorption/desorption kinetics. Some brief information on surface rheology is given in Section 3.2. The rate of normal and lateral surfactant transport and dynamic surface tension response on external disturbances depends on relaxation properties, into which Section 3.1. introduces.

The coupling of surface rheology and adsorption/desorption kinetics causes great difficulties in quantitative models of dynamic adsorption layers. Emulsions and foams are the most important systems in which the non-equilibrium state of adsorption is essential. Although there are even additional complications met in the elaboration of dynamic models of foams and emulsions, substantial results were achieved by the American scientists Edwards, Brenner and Wasan and described in their book (1991) "Interfacial Transport Processes and Rheology" and by the group of Ivanov in Bulgaria put down in the monograph "Thin Liquid Films" (1988).

This book deals mainly with dynamic properties of amphiphiles at liquid/air and liquid/liquid interfaces rather than at solid/liquid interfaces. However static and dynamic contact angles are discussed in Appendix 3B as these phenomena are determined by the kinetics of adsorption of surfactants also at the fluid interface. Some specific aspects of lateral transport phenomena studied by many authors are briefly review in Appendix 3D.

Probably the most striking phenomenon which is caused by adsorption dynamics is the Marangoni instability. A short introduction into this topic and few demonstrations of observed features are given in Appendix 3C.

3.1.   RELAXATIONS AND CHEMICAL REACTIONS

As we have already seen, the state of soluble as well as insoluble monolayers can deviate from a equilibrium state defined at constant temperature, pressure, bulk and surface concentrations. A deviation from the equilibrium state of the corresponding adsorption layer can be triggered by vertical and lateral concentration gradients due to adsorption/desorption processes or by hydrodynamic or aerodynamic shear stresses, as shown in Fig. 3.1.

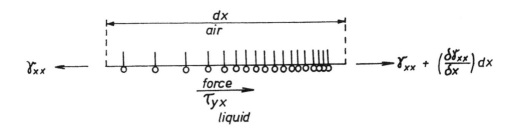

Fig. 3.1.   Action of hydrodynamic or aerodynamic shear stresses at adsorption layers of insoluble surfactants

Another familiar experiment is the compression or dilation of insoluble monolayers on a Langmuir trough. By this operation the film passes different states, such as mesophases. The transition of the film from one state into another needs time, which is a characteristic parameter for such processes starting from a non-equilibrium state and directed to the re-establishment of equilibrium. The principle of "relaxation" coordinates for any process was first introduced by Maxwell (1868) in his work on relaxations of tensions. After Maxwell, a liquid body under deformation can be described by the shear stress

$$\tau = e^{-\frac{G}{\eta}t}\left[\tau_o + G\int \dot{\xi}e^{\frac{G}{\eta}t}\ dt\right], \tag{3.1}$$

where G/η is equivalent to the relaxation time τ, G is the elasticity module and η is the viscosity. A relaxation curve is obtained after the displacement ξ becomes constant, it means $\dot{\xi}=0$, and can be described by

$$\tau = \tau_o\ e^{-\frac{G}{\eta}t}. \tag{3.2}$$

The time η/G is denoted as retardation time, applicable to bodies other than a Maxwell body. The relaxation curve according to Eq. (3.2) or its spectrum (the process may be complex and can have more than one relaxation time), was applied for fitting the pressure decay obtained after a sudden stop of the compressing of a monolayer at a certain surface pressure. A typical experimental result is shown in Fig. 3.2 (Tabak et al. 1977).

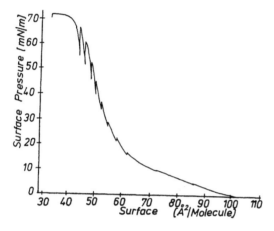

Fig. 3.2.   Typical pressure relaxation after the continuous compression of a monolayer has been stopped suddenly at a certain molecular area; according to Tabak et al. (1977)

In considering the dynamic behaviour of amphiphiles at interfaces, we have to include several dynamic processes. There are not only "simple" adsorption/desorption processes but also time-dependent orientations and lateral transport phenomena. Each of these processes is connected to a characteristic time, denoted as relaxation time. Two extreme cases exist,

$$\tau_1 \ll \tau_{exp} \tag{3.3}$$

and

$$\tau_2 \gg \tau_{exp}, \tag{3.4}$$

where τ_1 and τ_2 are the relaxation times connected with a process, such as diffusion or orientation and τ_{exp} is the time of resolution by the used experiment. For example, the time of orientation of small amphiphiles may be very short, say nanoseconds, while a reorientation of surface active proteins at interfaces needs minutes or hours. In both cases, these relaxation experiments fail to find out the relaxation mechanism which is the main target of such studies. Only if the real relaxation time is of about the same order of magnitude as the experimentally available time resolution, will the experiment yield reasonable results.

Lucassen-Reynders (1981) and van den Tempel & Lucassen-Reynders (1983) have analysed different relaxation processes at interfaces and classified them into:

- relaxation by diffusional surface-bulk exchange of matter,
- relaxation due to micellar formation or breakdown,
- relaxation due to exchange with multilayer particles,
- relaxation due to adsorption-barrier processes.

The adsorption kinetics in micellar systems was studied first by Lucassen (1975). He defined a micelle relaxation time

$$\tau^{-1} = \frac{-\partial \ln \Delta_M}{\partial t} = \frac{\partial \ln \Delta c_0}{\partial t}, \tag{3.5}$$

where Δc_M and Δc_0 are the respective derivations of micelle and monomer concentrations from their equilibrium values.

Before going on to the main principles of surface rheology (2D-rheology), we have to deal with the problem of visco-elasticity studied by relaxation (or retardation) phenomena in bulk phases of fluids.

Each chemical reaction has its own relaxation time which can be studied in a number of ways. Moelwyn Hughes (1971) gave a systematic introduction into the problem of relaxation processes in chemical reactions. We learn that a lot of principles of relaxation phenomena in solutions, in theory as well as in experiments, are applicable to studies of related subjects at interfaces. We can also transfer the main principles of relaxation theories from the bulk phase to the interface. This step has a strong parallel in the foundation of the surface rheology by the principles of the rheology of bulk phases (3D-rheology). However, a significant difference between bulk and surface rheology remains. Parts of the surface can appear or disappear while in the bulk this cannot happen. In 3D-rheology the tension is determined by the change in the volume shape of a body while the surface tension is strongly determined by the composition of the surface phase.

The change of the composition of a reacting system and its coupling with the displacement of particles (atoms, molecules) due to mass transfer effects is typical of chemical reactions in solutions. Beside this, relaxations in a fluid can be generated by wave propagation, e.g. ultrasonic waves, shock waves, alternating electric fields, or in general by pressure jumps. Pressure jump experiments at interfaces are described for example by Loglio (1986a, b, 1988, 1991) and Miller et al. (1992). Loglio et al. (1986a) introduced the Fourier transformation for the analysis of an area jump disturbance. Similar experiments were published by Hachiya et al. (1987).

3.2. STRESS-STRAIN RELATIONSHIPS - GENERAL DESCRIPTION

The classical form of describing relaxations, as mention above, was founded by Maxwell (1868). Following Moelwyn-Hughes (1971) the relaxation time τ is connected with the displacement of a particle by a simple equation. It is interesting, that surface rheological properties can be studied by the displacement of hydrophobic particles at viscoelastic interfaces (Maru & Wasan 1979), using an equation like

$$-\frac{d(x-x_e)}{dt} = \frac{1}{\tau}(x-x_e) \tag{3.6}$$

or

$$-\frac{d\Delta x}{dt} = \frac{\Delta x}{x}\frac{1}{\tau} \tag{3.7}$$

as the basic law for a relaxation process. On the basis of a generalised strain-stress relationship $s(\xi)$ acting in a body under the displacement

$$\frac{d\xi}{dt} = \frac{1}{\beta}\frac{ds}{dt} - \frac{\xi}{\tau}. \tag{3.8}$$

In the case of a solid, τ is large enough to make the last term approximately zero. If the body is viscous, ξ does not remain constant, it means the body relaxes. In a simple compressions experiment ξ is proportional to the outer pressure Δp and the strain is $d\ln V$. The relaxation equation becomes

$$\frac{dp}{dt} = -\frac{1}{\beta}\frac{d\ln V}{dt} - \frac{\Delta p}{\tau}. \tag{3.9}$$

With $dp/dt = \text{const.}$, Eq. (3.9) can be written as

$$B = -\frac{1}{\beta}\frac{d\ln V}{dt} \tag{3.10}$$

By integration

$$\frac{dp}{dt} = B - \frac{\Delta p}{\tau} \text{ of} \tag{3.11}$$

we obtain

$$\Delta p = B\tau\left[1 - \exp(-t/\tau)\right]. \tag{3.12}$$

A relaxation process described by Eq. (3.12) is shown in Fig. 3.3.

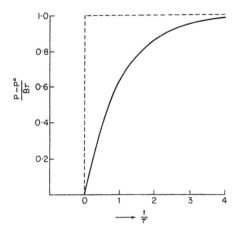

Fig. 3.3.   Demonstration of the relaxation process of a liquid with real and hypothetical establishment of the stationary state, according to Moelwyn-Hughes (1971)

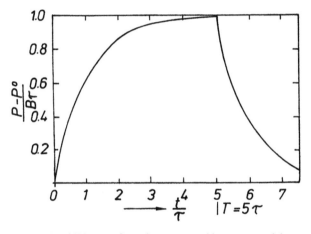

Fig. 3.4.   Establishment of a stationary state with an exponential stress decay from $B\tau$ to zero

The relaxation time is the reciprocal first-order constant which governs the process of equilibrium establishment. On the other hand there are experiments in which the strain is set constant at a time $t = \tau$. Such processes can be characterised by pressure-time relationships with $B = 0$. Thus, the solution of

$$\frac{dp}{dt} = -\frac{\Delta p}{\tau} \qquad (3.13)$$

has the form

$$\ln \Delta p = -t/\tau + \text{const.} \qquad (3.14)$$

At $t = \tau$ it follows

$$\ln B\tau = -t_o/\tau + \text{const.} \qquad (3.15)$$

and

$$\Delta p = B\tau \exp\left[-(t-t_o)/\tau\right].$$

(3.16)

where t_o is the time during which the steady state has been reached. This process is demonstrated in Fig. 3.4.

From a phenomenological point of view, relaxations are simple fluxes of energy (de Groot 1960). The fluxes of energy between two parts of a system are stimulated by the different temperatures T' and T'' that exist as long as the system as a whole is not in a thermodynamic equilibrium. Therefore, for the change of entropy as "driving force" for the establishment of the thermodynamic equilibrium one can write

$$dS = \frac{dU'}{T'} + \frac{dU''}{T''}.$$

(3.17)

For a thermal isolated system, containing any number of subsystems, here only denoted by $'$ and $''$, it is valid

$$dU = dU' + dU'' = 0$$

(3.18)

Together with (3.17) it leads to

$$\frac{dS}{dt} = \frac{dU'}{dt}\left(\frac{1}{T'} - \frac{1}{T''}\right) = \frac{dU'}{dt}\left(\frac{T''-T'}{T^2}\right).$$

(3.19)

If dU'/dt is the flux, than $T''-T'$ is the definition of a force acting between the two subsystems.

From the phenomenological relationship one obtains

$$\frac{dU'}{dt} \leq \frac{T''-T'}{T^2}.$$

(3.20)

de Groot (1960) mentioned that in theories of relaxations phenomena given by Eq. (3.20) are usually described by

$$\frac{dT}{dt} = \frac{1}{\tau}(T''-T')$$

(3.21)

This follows from $dU' = C' dT''$ with C' being the specific heat of one of the systems and $C'T^2/L = \tau$ the so called "relaxation time".

Propagation of sound is an established method of studying irreversible thermodynamics. Sound propagation is accompanied by heat production, viscous flow, relaxation phenomena and chemical reactions, each of which is determined by a particular relaxation time.

To demonstrate the "overlapping" conditions between chemical reactions and surface processes controlled by "specific" relaxation times we can compare processes in both areas. Such areas are: diffusion of components, formation of new phases, e.g. crystallisation, evaporation, freezing, the transformation from an aggregate to a monomer state or vice versa. For example the component A produces the component B in dependence of temperature, pressure or other forces, such as outer electric fields.

To describe the surface state we have to take into consideration the vertical and lateral diffusion needed to establish the same equilibrium chemical potential of single adsorbed and associated molecules (clusters) at liquid interfaces. We can therefore describe chemical reactions and surface chemical processes by the same thermodynamic formalism, including entropy production, fluxes, forces and affinities and their deviation to the coordinate of displacement.

3.3.   BRIEF INTRODUCTION INTO SURFACE RHEOLOGY

This section is dedicated to some general definitions and a brief introduction into the topic of rheological properties of interfaces. An adsorption layer can change the flow properties of interfaces drastically. Today surface rheology or 2D-rheology is an independent area of surface science. The notations and definitions used in surface rheology are taken from the well-established bulk- or 3D-rheology. The main difference between these two rheologies consists in the fact that surface rheology deals with open systems, because during compression and dilatation of a soluble adsorption layer adsorption and desorption can take place. Direct measurements of the complete rheological state is more difficult than in bulk rheology. So surface rheology is more restricted to simplest rheological models, such as the Maxwell-Oldroyd liquid, and to a linear superposition of elastic and viscous parts of strain-stress relationships including the corresponding phase angles.

The main principles of surface rheology were set up by Landau & Lifschitz (1953), Stuke (1961), Levich (1962), Lucassen (1968), Wasan et al. (1971), van den Tempel (1977), Goodrich (1981). Goodrich(1969), Izmailova (1979), Lucassen-Reynders (1981), Wüstneck & Kretzschmar (1982) and, more recently, Edwards et al. (1991) and Miller et al. (1994) have provided overviews.

In the notation of tensor components we are using some symbols taken from the classical work of Landau & Lifschitz (1953), especially

u_{ii} - relative volume change

γ_{ik} - surface tension tensor

u_{ik} - components of the deformation tensor.

In Fig. 3.5. the tensor components for shear and compression/dilatation at a small cube is shown.

From this figure we learn that pure deformation without any change in size (pure shear) and change in size without any change in shape of the body (dilatation or compression) have to be considered. For the latter case the tensor of strain becomes $u_{ik} \sim \delta_{ik}$ (Landau & Lifschitz 1953). Any deformation can be given as the sum of pure shear and dilatation deformations. Therefore, we get

$$u_{ik} = \left(u_{ik} - \frac{1}{3} \delta_{ik} u_{ll} \right) + \frac{1}{3} \left(\delta_{ik} u_{ll} \right), \tag{3.22}$$

where the first term corresponds to pure shear and the second to pure dilational deformation. For a dilational deformation all components $u_{ik, i \ne k}$ diminish and the strain tensor is constant for the whole volume, for examples for a hydrostatic compression or ordinary stretch of a rod.

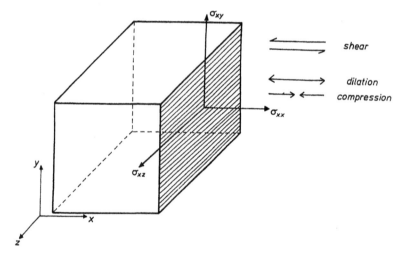

Fig. 3.5.   Description of shear and stress tensor components acting on a cube, after Landau & Lifschitz (1953)

We will now examine the simplest rheological model, the Hooke law for elasticity and Newton law for viscosity. In Hooke's law the tensor of strain u_{ik} is a linear function of the stress tensor of s_{ik}, i.e. the deformation is proportional the acting forces. If the inertial stress, the elastic or Hooke's stress and the viscous stress are additive, we can write

$$\left(\frac{m}{g_c f} \right) \frac{d^2 \xi}{dt^2} + \eta_k \frac{d\xi}{dt} + G_k \xi = s(t). \tag{3.23}$$

Here are ξ - strain, s - stress, t - time, G - shear modulus, f - a form factor for an undefined function A / L_2, g_c - dimensional constant relating force, m - mass, α - angle.

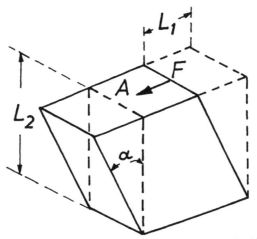

Fig. 3.6.   Schematic picture of acceleration relationships in a liquid element

Newton's second law of motion has the form $F = m \cdot (d^2L_1 / dt^2) / g_c$, so that the acceleration can be written as (cf. Fig. 3.6.),

$$dL_1 = L_2 \partial \xi. \tag{3.24}$$

For the step from the 3D-rheology to the 2D-state, to the surface rheology, it is best to use the vector treatment for describing the complex variables of strain s^*, stress ξ^*, complex viscosity η^*, complex shear modulus G^*, respectively. η^* and G^* are viscoelastic vectors. The relating vector treatment for strain in a shear deformation is shown in Fig. 3.7.

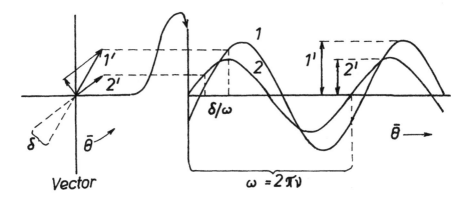

Fig. 3.7.   Vector treatment for a strain in shear, $\omega = 2\pi\nu$ corresponds to a full revolution of the vector s_o, Θ - time, δ - phase angle, s - stress, σ - shear rate

The complex variables for strain and stress can be written as

$$\xi^* = \xi_o e^{i\omega\Theta}, \tag{3.25}$$

$$s^* = s_0 e^{i(\omega\Theta+\delta)}. \tag{3.26}$$

Differentiation of these expressions is equivalent to a counter clockwise 90 degrees rotation of the vector

$$d\dot{\xi}/d\Theta = i\omega\xi_0 e^{i(\omega\Theta+\pi/2)}. \tag{3.27}$$

The subscript "o" refers to the amplitude of the vector. Therefore, we have two complex quantities:

$$G^* = s^*/\xi^*, \tag{3.28}$$

and

$$\eta^* = s^*/d\xi^* d\Theta, \tag{3.29}$$

or

$$G^* = G' + iG'' \tag{3.30}$$

and

$$\eta^* = \eta' - i\eta''. \tag{3.31}$$

The relations between G', G'', η' and η'' are

$$G' = \omega\eta'' \tag{3.32}$$

and

$$G'' = \omega\eta. \tag{3.33}$$

Hence, $G''/G' = \eta'/\eta''$ and $G''/G' = \tan\Theta$, where $\tan\Theta = \eta_k/(G_k/\omega) = I\omega/g_c f$.

In analogy with electrical systems we can formulate the amplitude of the mechanical impedance Z as the ratio of an effective stress over the effective rate of strain. The mechanical impedance directed against the flow, just as in an inelastic system, is the apparent viscosity. In general we have

$$Z = |\eta^*| = \sqrt{(\eta^*)^2 + (\eta'')^2}, \tag{3.34}$$

$$\eta' = \eta_k, \tag{3.35}$$

$$\eta'' = \left[(G_k/\omega) - (I\omega/g_c f) \right]. \tag{3.36}$$

In reality we find a superposition of elastic and viscous effects in the strain-stress relationship. In 3D-rheology properties of a given system turned out to be described by the combination of mechanical springs and damping elements, dash-pots. The rheological models of Maxwell, Kelvin, Burgers and Voigt are familiar combinations of these mechanical elements.

Maxwell's and Kelvin's models, as shown in Fig. 3.8., give the most important analogue to mechanical elements

The Maxwell model for shear is given by

$$d\xi / dt = (1/\eta_M)_s + (1/G_M)ds/dt \tag{3.37}$$

and the Kelvin model

$$S = \eta_k / dt + G_k \xi. \tag{3.38}$$

The ratio shear viscosity to shear modulus is often symbolised by the time $\tau = \eta / G$. For the Maxwell model, τ is called the stress relaxation time. In the Kelvin model τ is a measure of the time required for the extension of the spring to its equilibrium length under a constant stress. τ is called the retardation time.

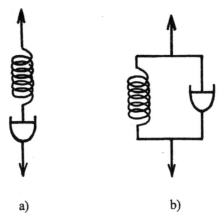

a)   b)

Fig. 3.8.   Maxwell (a) and Kelvin (b) model as linear or parallel combination of spring and dash-pot

As is well known, a lot of effects of surfactants, like damping of surface waves, the rate of thinning of liquid films, foaming and stabilisation of foams and emulsions, cannot just be described by a decrease in interfacial tension or by van der Waals and electrostatic interaction forces between two interfaces. The hydrodynamic shear stress at an interface covered by a surfactant adsorption layer is a typical example for the stimulation of an important surface effect. This effect, shown schematically in Fig. 3.9., is called the Marangoni effect.

If we consider the simplest equation of state $\pi = RT\Gamma$ each value Γ corresponds to a film pressure value π. In other words, the gradient of surface density Γ, shown in Fig. 3.9. is equivalent to a surface tension gradient. Such gradient at a surface or interface leads to a

80

deformation of a flat surface, a consequence of the Gauss-Laplace equation. This effect is shown in Fig. 3.10.

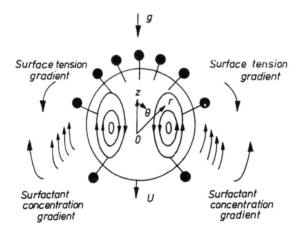

Fig. 3.9.   Inhomogeneous density of adsorbed species due to the action of a hydrodynamic force initiates a Marangoni flow in the opposite direction; after Edwards et al. (1992)

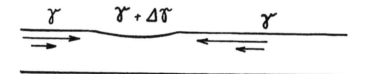

Fig. 3.10.   Deformation of a flat surface under the influence of a surface tension gradient

In practical systems, the motion of bubbles or droplets in surfactant solutions is strongly retarded by gradients of the surface tension. The schematic in Fig. 3.11. clearly demonstrates the retardation effect on rising bubbles or sinking drops. In surfactant solutions gravitation and the Marangoni effect move in the opposite direction.

This subject is one of the important themes of this book. In Chapters 8, 9, 10, 11 and 12 this effect is used for the qualitative and in some cases quantitative explanation of many phenomena connected to rising bubbles in surfactant solutions.

There is evidence that surface rheology follows the same principles and terminologies developed intensively for "bulk rheology". Focusing at typical applications, the flow properties of liquid interfaces are often drastically influenced by the hydrodynamic retardation of surface rheology, such as surface elasticity, surface viscosity and the related loss angle. This parameters are acting during compression and dilatation as well as in shear.

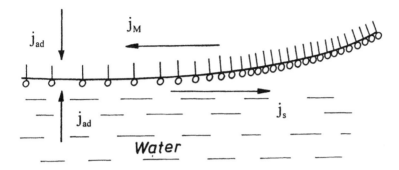

Fig. 3.11.   Opposite direction of hydrodynamic shear stress j_s and Marangoni flow j_M, j_{ad} - adsorption
flux, after Edwards et al. (1992)

In the cube shown in Fig. 3.5. the tensor components for the strain-stress relationship of a 3D-body can be seen. Neglecting the z-coordinate, the tensor reduces from a 3x4 to a 2x2 matrix. The use of the 3D-rheology for related surface problems is only valid if a 3D-analogue for the relaxation is introduced. This is the only way to learn about the surface state in the absence of ideally elastic behaviour of the adsorption layer.

There are several observable or possible effects. They can be classified into molecular orientations, molecular structure formations (associations or domain formation), lateral and vertical transport processes. In real non-equilibrium systems we have to consider the overlap of two or more contributions.

For soluble surfactant adsorption layers the vertical mass transfer occurs under two different conditions, after the formation of a fresh surface of a surfactant solution and during periodic or aperiodic changes of the surface area. From the thermodynamic point of view the "surface phase" is an open system. The theoretical and practical aspects of this issues have been outlined in many classical papers, published by Milner (1907), Doss (1939), Addison (1944, 1945), Ward & Tordai (1946), Hansen (1960, 1961), Lange (1965). New technique for measuring the time dependence of surface tension and a lot of theoretical work on surfactant adsorption kinetics under modern aspects have recently been published by Kretzschmar & Miller (1991), Loglio et al. (1991), Fainerman (1992), Joos & Van Uffelen (1993), MacLeod & Radke (1993), Miller et al. (1994). This topic will be discussed intensively in Chapters 4 and 5. The relevance of normal mass exchange as a surface relaxation process is discussed in Chapter 6.

82

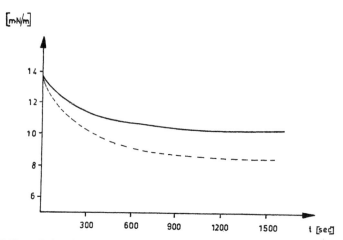

Fig. 3.12.   Relaxation of a surface film of dipalmitoyl lecithin after halting the compression, compression rates: 2 cm/min - solid line, 10 cm/min - dashed line

A typical effect related to surface relaxations is obtained in measurements of π-A isotherms of insoluble monolayers. In most of the measurements with spread amphiphiles there are differences between the curve for compression and expansion of the surface films. Usually this characteristic behaviour is described as hysteresis. One experimental example of a spread dipalmitoyl lecithin is shown in Figs 3.12. This phenomenon corresponds to one or more of these surface relaxations.

All these phenomena are contributions to the surface loss angle in the strain-stress relationship. 3D-rheology deals with a closed system. There is no mass transfer across the limits of this system. In contrast, the so-called surface phase with an adsorption layers can exchange matter with the bulk phase depending on the boundary conditions, such as adsorption after formation of a fresh surface or periodic adsorption and desorption due to periodic changes of the surface area.

First of all, surface rheology is completely described by four rheological parameters: elasticity and viscosity of compression/dilatation and of shear. In every case surface flow is coupled with the hydrodynamics of the adherent liquid bulk phase. From interfacial thermodynamics we know that the integration over the deviation of the tangential stress tensor from the bulk pressure represents the interfacial tension γ (after Bakker 1928).

Starting from the basic equation for the coupling of movement of an interfacial element and the adherent bulk liquid we can write, after Goodrich (1969) and van den Tempel (1977),

$$\frac{\partial \gamma_{xx}}{\partial x} + \frac{\partial \gamma_{yx}}{\partial x} = \left(\xi_{zx}^a - \xi_{zx}^b \right)_{z=0}. \tag{3.39}$$

Here $\xi^{a,b}$ are the shear stresses in the phases a and b, respectively, with the interface of tension γ. By introducing the shear stress s through the Newton's equation of liquid motion and by considering a uniform dilatation $\gamma_{yx} = \gamma_{xy} = 0$, the surface stress tensor becomes scalar and Eq. (3.39) can be rewritten as

$$\frac{\partial \gamma}{\partial x} = -\eta^a \left(\frac{\partial v_x}{\partial z} + \frac{\partial v_z}{\partial x} \right)_{z=-o} + \eta^b \left(\frac{\partial v_x}{\partial z} + \frac{\partial v_z}{\partial x} \right)_{z=+0} . \tag{3.40}$$

For a gas/liquid interface $\eta^a \gg \eta^b$ and the boundary condition simplifies to

$$\frac{d\gamma}{dx} = \eta^a \frac{dv_x}{dz} \bigg|_{z=0} . \tag{3.41}$$

The surface dilational elasticity was firstly defined by Gibbs (1906)

$$E = \frac{d\gamma}{d \ln A}, \tag{3.42}$$

where A is the surface area covered by adsorbed or spread molecules. Combined with Eq. (3.41) one obtains

$$\frac{d\gamma}{dx} = E \frac{d^2\xi}{dx^2} . \tag{3.43}$$

Here, ξ is the displacement of a surface element in x-direction.

In respect of the classical mechanics, E is an "ideal" coefficient, like the elasticity modulus in Hooke's model. Most of the practical compressions/dilatation experiments carried out with adsorption layers are comparable to the screening of elastic properties in material science. In analogy to the coefficients of the 3D-elasticity theory, we have to consider complex coefficients in surface rheology. The surface elasticity coefficient written as a complex modulus therefore has the form

$$E = E' + E''. \tag{3.44}$$

Here, E' is the real part and E'' the imaginary part of the complex modulus.

3.4.   SURFACE RHEOLOGY AND ADSORPTION DYNAMICS IN DRAINAGE PROCESSES OF THIN LIQUID FILMS

The rate of thinning (drainage) of liquid films is drastically influenced by the rheological properties of the related adsorption layer. We will restrict ourselves to just a few examples. A detailed description of various sites of thin film problems is given for example by Ivanov (1988) and Hunter (1993). The immobilisation of a cylindrical plane film is a precondition for the application of Reynolds equation to determine the disjoining pressure isotherm by the dynamic method of Scheludko et al. (1961). This film state was denoted by Ivanov & Dimitrov (1988) as a lubrication approximation. The authors pointed out the basic principles which influence the rate of drainage of thin liquid films by coupling the conservation of momentum in the bulk phase with the conservation of surfactants (developed by Levich 1962) and the surface momentum conservation (Scriven 1960, Aris 1962, and Maru 1977).

The application of the lubrication approximation demands that the radius of the liquid film exceeds its thickness 10^6 times. In thin film research the thickness of a film is usually denoted by h, the radius R corresponds to the flat part of the thin film as shown by Manev et al. (1976). For a symmetric geometry of the Scheludko cell the equation of Reynolds (1886) for the rate of liquid drainage is valid. The main condition for the determining of the disjoining pressure of a thin liquid film is the complete immobility of the two surfaces of the liquid film. The "Scheludko cell" is shown in Fig. 3.13. The driving force for the liquid flow-out of the film is, in the simplest case, given by the capillary pressure of the curved borders between the flat part of the film and the wall of the glass tube with a diameter of approximately 0.5 cm.

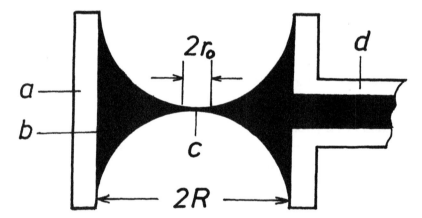

Fig. 3.13.   Sketch of an axisymmetric liquid film in a "Scheludko cell"; a - glass ring of diameter 2R, b - surfactant solution, c - thin film of diameter $2r_0$, d - tube for the sucking off of the liquid of a relatively thick film

For the rate of thinning for rigid surfaces Reynolds' equation can be written in the following form (Scheludko 1967),

$$V_R = \frac{dh}{dt} = \frac{2}{3} \frac{h^3}{\mu R^2} \Delta P .$$ (3.45)

This equation is the basis for Scheludko's dynamic method (1961) for determining the disjoining pressure in thin films. An additional condition is

$$\Delta P = P_c - \pi, \tag{3.46}$$

where P_c is the capillary pressure of the curved interface with radius $R_c - R$, as demonstrated in Fig. 3.14. The rate constant of the drainage of a thin liquid film is coupled with the rheological properties of the film interfaces. These rheological properties include elastic as well as viscous components. In real systems one has to consider viscoelastic properties, like in the 3D-rheology.

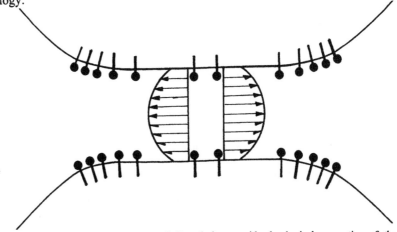

Fig. 3.14.   Coupling of the rate constant of film drainage with rheological properties of the film interfaces, according to Scheludko (1966)

Thin liquid films in foam and emulsion systems are usually stabilised by soluble surfactants. During the formation of such films the flow-out process of liquid disturbs the surfactant equilibrium state in the bulk and film surfaces. The situation of drainage of a surfactant containing liquid film between two oil droplets is shown in Fig. 3.15. (after Ivanov & Dimitrov 1988). Here j^d and j^f are the bulk fluxes in the drops and the film, respectively, j^s and j^c are the fluxes due to surface diffusion or spreading caused by the Marangoni effect, respectively.

When the trajectories of the liquid flow are comparable, different types of adsorption and desorption mechanisms set in, which leads to the respective boundary conditions. The main principle for the surface momentum conservation is that the surface tension as a scalar parameter is effected by surface viscous stresses described by a tensor $\underline{\underline{s}}'$. From the related surface stress tensor $s\underline{\underline{I}}_{=s}$ (after Landau & Lifschitz 1953), where $\underline{\underline{I}}_{=s}$ is the unit surface tensor, and the balance of surface stresses it follows

$$\nabla^s \bullet \left(s\underline{\underline{I}}_{=s} + \underline{\underline{s}}' \right) = \underline{n} \bullet \left[\left|\underline{\underline{P}}\right|\right]. \tag{3.47}$$

Here $\left[\left|\underline{\underline{P}}\right|\right] = P^f - P^d$ is the difference of the bulk stresses $\underline{\underline{P}}$ on both sides of the film.

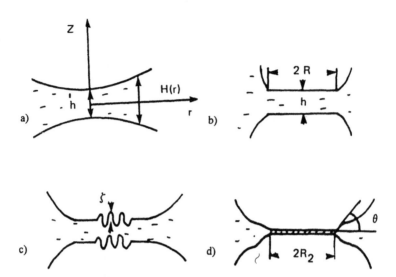

Fig. 3.15. Drainage of a surfactant containing liquid between two oil droplets or air bubbles, a) - approach, b - planar film after dimple flow out, c - thermal fluctuations, d - equilibrium state, 2R - radius of the film, h -film thickness, according to Ivanov & Dimitrov (1988)

Ivanov & Dimitrov (1988) wrote the components of the surface viscous stress tensor in surface Cartesian coordinates

$$s'_{ik} = \eta_s \left[\frac{\partial v_i}{\partial x_k} + \frac{\partial v_k}{\partial x_i} \right] - \delta_{ik} \left(\frac{\partial v_l}{\partial x_l} \right) + E_d \delta_{ik} \left(\frac{\partial v_l}{\partial x_l} \right) \tag{3.48}$$

Here, η_s, E_d, δ_{ik} are the surface shear viscosity, the surface dilational viscosity and the Kronecker symbol, respectively. By introducing a foam film mobility parameter ε^f the surface velocity U for mobile surfaces and for the case of slow diffusion in the lubrication approximation yields

$$U = \frac{V_r}{2h(1+\varepsilon^f)} \tag{3.49}$$

with

$$\varepsilon^f = -\frac{(\partial \gamma_o / \partial c_o) \Gamma_o}{3\mu D} \left[1 + \frac{2D_s (\partial \Gamma_o / \partial c_o)}{Dh} \right] \tag{3.50}$$

and V_r is the rate of film expansion. The same is possible for emulsions by defining ε^e as an emulsion film mobility parameter. Radoev et al. (1974) and Ivanov & Dimitrov (1974) introduced additional surface effects to describe the rate of thinning V relative to Reynolds' rate V_{Re} for rigid surfaces

$$\frac{V}{V_{Re}} = 1 + b + \frac{h_s}{h} = 1 + \frac{1}{\varepsilon^f}, \tag{3.51}$$

with $b = 3\beta_d$, $h_s = 6\beta_s h$, β_d, β_s are bulk and surface diffusion parameters, respectively.

Ivanov & Dimitrov (1974) mentioned that Eq. (3.51) is not consistent with the lubrication approximation. By combining the Eq. (3.74) and γ_{ik} Aris (1962) obtained

$$\nabla_s \bullet \left(s \underset{=s}{I} \right) = \underset{=s}{I} \bullet \nabla_s s + s \nabla_s \bullet \underset{=s}{I} \nabla_s s + 2s K_m \tag{3.52}$$

K_m represents the main curvature. The general tangential stress balance is given by

$$\mu_{sh} \Delta_s^2 \underline{V}_s + \mu_d \nabla_s (\nabla_s \bullet \underline{V}_s) = \nabla_s s + \underline{n} \bullet \left[\lVert P \rVert \right] \bullet \underline{I}_s. \tag{3.53}$$

3.5.   SURFACE RHEOLOGY AND STABILITY OF FOAMS AND EMULSIONS

As shown in Section 3.4. the rate of drainage of liquid films depends on surface rheological properties. This is valid for foam and emulsion films. However, the drainage of the films between fluid particles is only the first common step in stability. the DLVO theory controls films and films stabilised in the second minimum of the π/A- isotherm.

In order not to extend this topic too fat we will restrict ourselves to general mechanisms responsible for the lifetime and stability of foams. The mechanisms in emulsions are similar, although complicated by the distribution of diameter and shape of the oil droplets and its Brownian motion. Foams are regular systems as demonstrated by Kruglyakov et al. (1991) and shown in Fig. 3.16.

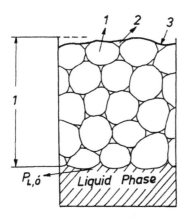

Fig. 3.16.   Foam structure on a liquid surface, according to Kruglyakov et al. (1991)

Khristov et al. (1984) noted that in foams stabilising forces (elasticity forces) are high enough to counteract the destructive forces caused by foam formation. Such liquid foam films can therefore reach thicknesses at which specific thermodynamic properties appear, as described in Section 2.10.

On the basis of a differential equation Ivanov (1977) described all stages of thin liquid film evolution. He distinguished the effects of Marangoni-Gibbs and of surface viscosity. Additionally, the substantial effects of surface diffusion and slow adsorption (barrier or kinetic controlled mechanisms) are taken into consideration. A selection of basic equations can be find in Chapter 4.

In general we can discriminate between

-the rate of thinning of thick liquid films to established equilibrium of all acting forces and

-the life-time of the common black films.

After the theory of Vrij (1966) surface waves play an important role. The critical thickness for the rupture of thin liquid films derived from the behaviour of surface waves is much smaller than the equilibrium thickness. Fig. 3.17. shows the thinning of a film due to surface waves generated by disturbances with squeesing modes.

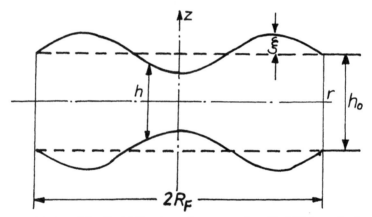

Fig. 3.17.   Thinning of free liquid films by surface waves; the critical thickness for the film rupture is reached by this disturbance, ξ - displacement of liquid elements, $2R_F$ - film diameter, h_o - equilibrium film thickness, according to Vrij (1966) and Fijnaut & Joosten (1978)

An extended work, experimental as well as theoretical, on mechanisms in thin microscopic foam lamellas and macroscopic foams has been published by Krugljakov & Exerowa (1990). In the theory of Exerowa & Kashiev (1986) thin-film stability is explained in terms of lateral diffusion of vacancies in a lattice like adsorption layers. As an example, selected experimental results of Exerowa et al. (1983) are shown in Fig. 3.18.

An actual overview on interfacial rheological properties of adsorbed surfactant layers stabilising foams and emulsions has been given by Malhotra & Wasan (1988). The results of

Khristov et al. (1984) on the dependence of lifetime of foams on the pressure difference in the Plateau-Gibbs borders are shown in Fig. 3.19.

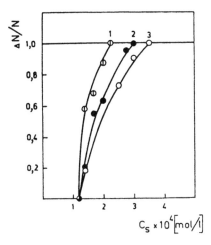

Fig. 3.18.   Probability $\Delta N / N$ of spot formation in a Newton black film as a function of SDS concentration in 0.5M NaCl, film radius R= 0.005 cm (1), 0.01 cm (2), 0.05 cm (3), according to Exerowa et al. (1983)

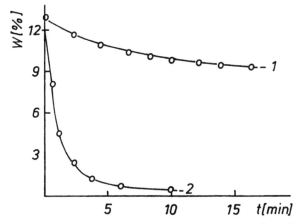

Fig. 3.19.   Pressures acting on the water content W in BSA foam films, 1 - without additional pressure, 2 - additional pressure of $\Delta P = 10^4$ N / m^2, according to Lalchev et al. (1979)

Another demonstration of a critical phenomenon, the rate of coalescence of emulsions in dependence of surface elasticity and viscosity, based on the work of Boyd et al. (1972), can also be found in the review of Malhotra & Wasan (1988) shown in Fig. 3.20.

Malysa et al. (1985) measured the retention time of foams in aqueous n-alcohol solutions. The surface activity of the surfactants and the adsorption kinetics determine the elasticity of the related adsorption layers. The dilational elasticity of the stabilisers is additionally influenced by the frequencies of the outer disturbances. In Fig. 3.21. the retention time as a function of reduced fatty acid concentrations is shown. The relating efficient dilational elasticity as a function of reduced fatty acid concentrations is shown is given in Fig. 3.22.

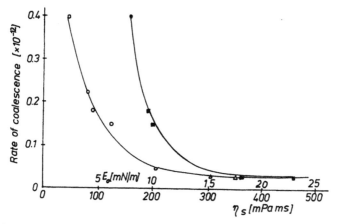

Fig. 3.20.   Rate of coalescence of emulsions in dependence of surface elasticity (o) and viscosity (■), according to Boyd (1972)

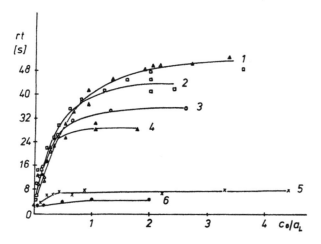

Fig. 3.21.   Retention time of froths as a function of concentration of n-alcanoic acids of different alkyl chain lengths, 1 - octanoic acid, 2 - heptanoic acid, 3 - hexanoic acid, 4 - pentanoic acid, 5 - nonanoic acid, 6 - decanoic acid, according to Malysa et al. (1985)

In a later paper Malysa et al. (1991) interpreted their results more theoretically. The highly dynamic system leads to a sum of interesting issues to the subject "dynamic properties at interfaces". Recently Wantke et al. (1994) gave an improved description of the experimental data in form of a semi-quantitative theory.

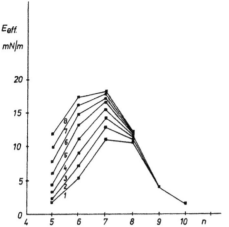

Fig. 3.22.   Effective surface elasticity as a function of chain lengths n of n-alcanoic acid at different frequencies of external disturbance, $c_o / a_L = 1$, $t_{ad} = 0.15s$, 1 - 200 Hz, 2 - 400 Hz, 3 - 800 Hz, 4 - 1600 Hz, 5 - 3200 Hz, 6 - 6400 Hz, 7 - 12800 Hz, 8 - 25600 Hz, according to Malysa et al. (1991)

3.6.   SURFACE WAVES

One of the most remarkable and most frequently studied effects of surface rheology is the damping of surface waves. Surface waves, transversal and longitudinal waves, are described by dispersion equations. Mann (1984) gave recently an overview on modern aspects of dynamic surface tension and capillary waves.

After Levich (1941) a transversal surface waves is described by

$$\rho\omega^2 = \gamma k^3,$$ (3.54)

where $k = 2\pi / \lambda$ is the wave number. The solution of Eq. (3.54) for the damping of transversal waves was derived by Hansen & Mann (1964) and van den Tempel & van de Riet (1965). From the measurement of the damping coefficient of these waves in dependence of surfactant bulk concentration, good agreement with the theory is obtained. A pronounced maximum in the damping coefficient at a certain surfactant concentration is observed, as shown in Fig. 3.23.

Lucassen (1968) and Lucassen-Reynders (1969) worked out the theory for longitudinal surface waves, which appear at elasticity modules higher than 30 mN/m and behaves like a stretched membrane. The related dispersion equation has the form

$$\eta\rho\omega^3 = i\varepsilon^2 k^4.$$ (3.55)

92

Transversal surface wave measurements are related to small surface area changes and work with frequencies in the range of a hundred to some thousand Hertz, while longitudinal waves work at much lower frequencies.

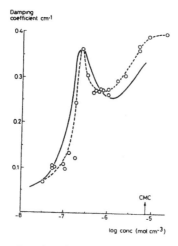

Fig. 3.23.   Damping of transversal capillary waves, after van den Tempel & van de Riet (1965)

While the propagation of surface waves is characterised by small surface area changes the movable barrier method mounted at a Langmuir trough works at much higher amplitudes. The principles and difficulties are described by van Voorst Vader et al. (1964), Lucassen & Barnes (1972), Lucassen & Giles (1975), and Kretzschmar & König (1984). Fig. 3.24. shows the schematic set-up of the oscillating barrier method as an example.

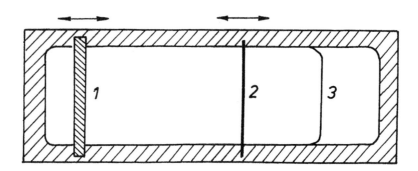

Fig. 3.24.   Method of the oscillating barrier, 1 - compression barrier, 2 - oscillating barrier (0.05 to 1 Hz), 3 - force balnce

Another recently developed method for determining surface rheological properties is the damping of a radial oscillating bubble, firstly described by Lunkenheimer & Kretzschmar (1975) and established theoretically by Wantke et al. (1980). This technique is described in more detail in Chapter 6. It is based on damping effects and yields dilational rheological parameters through the damping coefficient, while the oscillating barrier method leads directly to complex elasticity through measurements of surface pressure oscillation and phase angle between generation of the oscillating and the resulting pressure oscillations. The complex modulus of the monolayer elasticity is given by the formula

$$E = \frac{-d\pi}{d\ln A} = E' + iE'' = |E|\cos\alpha + i|E|\sin\alpha \qquad (3.56)$$

Fig. 3.25. gives an example of $\Delta\pi$ versus time for the determination the phase angle α. ε can be calculated by using the approximation

$$\frac{-d\pi L}{dL} \approx \frac{\Delta\pi L}{\Delta L}. \qquad (3.57)$$

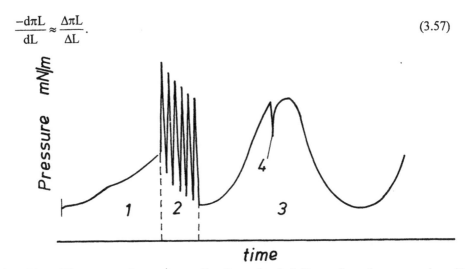

Fig. 3.25. Film pressure change $\Delta\pi$ vs. time for a phospholipid monolayer, 1 - compression with 10cm/min, 2 - oscillation with f = 0.05 Hz, 3 - one oscillation in higher time resolution, 4 - phase lag tick, according to Kretzschmar (1988)

After Lucassen (1968) the situation of a pure elastic and a viscoelastic surface can be clearly described by the picture in Fig. 3.26.

Using the classical equation of Boussinesq (1913) for the surface dilational viscosity, a relationship between surface tension change, surface elasticity and surface dilatation viscosity is obtained,

$$\Delta\gamma = E\Delta\ln A + \frac{d\ln A}{dt}. \qquad (3.58)$$

For the description of surface rheological properties as we have said we need to consider the surface stress caused by dilatation/compression as well as by shear. As in the case of area changes two coefficients of surface shear flow exist. Using the same symbols for the shear and stress tensors, as given in Fig. 3.5., we obtain

$$\gamma_{yx} = \eta_s \frac{\partial v_x}{\partial y}$$
(3.59)

as the basic definition of the surface shear viscosity η_s.

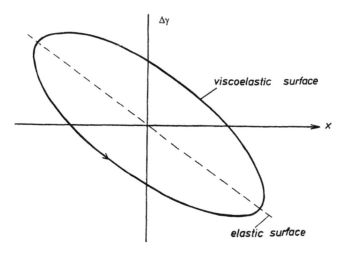

Fig. 3.26.   Schematic of a purely elastic and visco-elastic surface, after Lucassen (1968)

3.7.   SUMMARY

SURFACE RHEOLOGY AND ITS COUPLING WITH ADSORPTION DYNAMICS

In liquid/fluid disperse systems (emulsions, foams for example) the liquid interface is usually covered by an adsorption layer and often under lateral movement. This movement causes lateral transport along the interface and brings the adsorption layer out of its equilibrium state so that an adsorption/desorption exchange of matter sets in.

The lateral liquid und surface layer movement is influenced by surface rheology which is coupled with the exchange of matter process (surface relaxations). This coupling creates great difficulties in the quantitative description of dynamic adsorption layers (DAL). A brief introduction into surface rheology and surface relaxations is given and results are specified for use in Chapters 8, 9, 10 and 11.

MANIFESTATION OF ADSORPTION DYNAMICS AND SURFACE RELAXATION AT LIQUID INTERFACES AND IN DISPERSE SYSTEMS

Chapter 3 is devoted to processes strongly influenced by hydrodynamics, adsorption kinetics and surface rheology. There are at the moment several weak points in the description of surface flow properties because of relaxations stimulated by chemical reactions, lateral and vertical diffusion, adsorption kinetics and interfacial instabilities.

Sanfeld et al. (1990) have theoretically analysed the competition between chemical reactions at the surface and in the volume. They define a surface elasticity which is determined by the kinetics of all these processes and conclude that the effect of capillary forces may be considered as an external field acting on the reactive system. Their conception started from the principle of De Donder, as did our general description of relaxation in chemical reactions. To generalise relaxation phenomena at interfaces can be described as ordinary chemical reactions. In principle there is no distinction between the application of the laws to chemical kinetics in bulk phases and at interfaces.

Accepting this convention opens new trends in chemical technology and in environment protection. Differences in the rate constants of chemical reactions in a bulk phase and at an interface can for example be exploited to increase the selectivity of separation processes.

Beside "classical" applications of surface flow properties, like increased stability of disperse systems, the damping of surface disturbances and influence of small particles flows near gas bubbles, the complex actions of surface forces modified by surfactants appear to be responsible for the presence and the enrichment of micro-flotable components in marine aerosols (Loglio et al. 1985). Loglio et al. (1986a) also define the action of rising bubbles on heavy-metal enrichment at the sea surface.

Adsorption of vanadium and chromium are of direct significance in geo-chemical cycles. Hydroxide surfaces largely determine their transformation and complex formation. Such surfaces are additionally built up by a thin adherent water film. The adsorption kinetics of vanadyl (IV) and chromium (III) depend on the surface OH ligands (Wehrli & Stumm 1988). For the interpretation of the results the relation between the adsorption rate of different ions and the rate of water exchange is based on the data obtained by Hachiya et al. (1984).

For the transport of heavy ions to a solid surface coated with an adherent water film, like aluminium oxide, the visco-elastic properties of electric field forces and the concentration of heavy ions may be important for the rate of adsorption. For this reason we need information not only on relaxations restricted to a surface of an extended liquid, but also on the adherent water layer at the adsorbents. The last issue may be a bridge to the thermodynamics and flow properties of thin liquid films have been studied by some excellent research groups.

Unresolved problems include overall theory and related experiments on the contribution of chemical reactions at interfaces to the relaxation spectrum for a better theoretical description of surface rheological parameters. As discussed above each chemical reaction is characterised by at least one relaxation time. On the other hand surface rheological properties, for example dilational elasticity, are connected to relaxations taking place at interfaces.

Along with the basic research on modelling of flow properties at liquid interfaces, the development of related devices, computer simulations, the description of transport phenomena in connection with surface rheology, we should also look out for new relaxations arising from chemical reactions at interfaces. Such chemical reactions can be the formation of associates of single molecules at liquid interfaces, complex formation of heavy metal ions, geometrical effects of reductions in interfacial reactions (Astumian & Schelly 1984), reaction of liquid metal droplets with gases (Sanfeld et al. 1990), polymerisation, polycondensation, stereospecific reactions, leaching of adsorbed dyes, and enzymatic reactions. We have to conclude that the surface rheology of melted polymers, vital to many technologies, is not yet sufficiently developed.

The problem of modelling dynamic contact angles under the influence of high shear stresses and in the presence of wetting agents is rather complicated. In coating processes under high speed for example intensive mathematical research for solving the problems connected with the complex action of hydrodynamic stresses which effect the dynamic contact angle, adsorption kinetics of wetting agents in a flow field and the counterbalance of capillary forces.

3.8. REFERENCES

Addison, C.C., J. Chem. Soc., (1944)252

Addison, C.C., J. Chem. Soc., (1945)98

Aris, R., "Vectors, Tensors and the Basic Equations of Fluid Mechanics", Prentice-Hall, Englewood Cliffs, New Jersey (1962)

Astumian, R.D. and Schelly, Z.A., J. Am. Chem. Soc., 106(1984)304

Bakker, G., "Handbuch der Experimentalphysik", Vol. VI, Akademie-Verlagsgesellschaft, Leipzig, 1928

Bénard, H., Ann. Chim. Phys., 23(1901)62

Blake, T.D. and Haynes, I.M., J. Colloid Interface Sci., 30(1969)421

Blake, T.D., Ver. Deut. Ing., Berlin, 182(1973)117

Blake, T.D., AIChE Internat. Symp. Mechanics of Thin-Film Coating, New Orleans, 1988

Boussinesq, J., Ann. Chim. Phys., 29(1913)349

Boussinesq, J., Acad. Sci. Paris, 156(1913)1124

Boyd, J., Parkinson, C. and Sherman, F., J. Colloid Interface Sci., 41(1972)359

Cerro, R.L., and Whitaker, St., J. Colloid Interface Sci., 37(1971)33

Cherry, B.W. and Holmes, C.M., J. Colloid Interface Sci., 29(1969)174

Clark, D.C., Dann, R., Mackie, A.R., Mingins, J., Pinder, A.C., Purdy, P.W., Russel, E.J., Smith, L.J. and Wilson, D.R., J. Colloid Interface Sci., 138(1990a)195

Clark, D.C., Coke, M., Mackie, A.R., Pinder, A.C. and Wilson, D.R., J. Colloid Interface Sci., 138(1990b)207

de Boer, J.H., "The Dynamical Character of Adsorption", Oxford Univ. Press, London, (1953)

de Groot, S.R., Thermodynamics of Irreversible Processes, North-Holland Publ. Comp., Amsterdam, (1960)

de Donder, Th., "L' affinité", Gauthier-Villars, Paris, (1927)

Dimitrov, D.S., Panaiotov, I., Richmond, P. and Ter-Minassian-Saraga, L., J. Colloid Interface Sci., 65(1978)483

Doss, K.S.G., Koll. Z., 86(1939)205

Edwards, D.A., Brenner, H. and Wasan, D.T., Interfacial Transport Processes and Rheology, Butterworth-Heineman Publishers, 1991

Einstein, A., Sitz.ber. Akad. Wiss., Math.-Phys., Berlin, K1(1920)380

Fainerman, V.B., Colloids & Surfaces, 62 (1992) 333

Fijnaut, H.M. and Joosten, J.G.H., J. Chem. Phys., 69(1978)1022

Frenkel, J. and Obraztsov, J., Zh. Fiz. (USSR), 3(1940)131

Fritz, G., Z. angew. Physik, 19(1965)374

Gibbs, J.W., Scientific Papers, Vol. 1, 1906

Giordano, R.M. and Slattery, J.C., AIChE J., 25(1983)483

Goodrich, F.C., "Rheological properties of fluid Interfaces" in: Solution Chemistry of Surfactants, Mittal, K.L., (Ed.), Plenum Press, New York, Vol. 1, (1969)738

Goodrich, F.C., "Surface viscosity as a capillary excess transport property" in "The Modern Theory of Capillarity", Akademie-Verlag, Berlin, 1981

Grader, L., PhD Thesis, Leuna-Merseburg, 1985

Hachiya, K., Sasaki, M., Ikeda, T., Mikami, N. and Yasunaga, T., J. Phys. Chem., 88(1984)27

Hachiya, K., Takeda, K. and Yasunaga, T., Adsorption Science & Technology, 4(1987)25

Hansen, R.S., J. Phys. Chem., 64(1960)637

Hansen, R.S. and Mann, J.A., J. Appl. Phys., 35(1964)152

Hansen, R.J. and Toong, T.Y., J. Colloid Interface Sci., 37(1971)196

Hard, S. and Neumann, R.D., J. Collolid Interface Sci., 120(1987)15

Heckl, W.M., Miller, A. and Möhwald, H., Thin Solid Films, 159(1988)125

Hoffmann, R.L., J. Colloid Interface Sci., 94(1983)470

Huh, C. and Scriven, L.E., J. Colloid Interface Sci., 35(1971)85

Huh, C. and Mason, S.G., J. Fluid Mech., 81a(1977)401

Ismailova, V.N., "Structure Formation and Rheological Properties of Proteins and Surface-Active Polymers of Interfacial Adsorption Layers" in: Progress in Surface and Membrane Science 13(1979)

Ivanov, I.B., Thesis, University of Sofia, (1977)

Ivanov, I.B., Ed., Surfactant Science Ser., "Thin Liquid Films", Vol. 29, Marcel Dekker, 1988

Ivanov, I.B. and Dimitrov, D.S., in "Thin Liquid Films", Surfactant Science Ser., Vol. 29, I.B. Ivanov, Ed., Marcel Dekker, 1988

Joos, P., Bracke, M. and van Remoortere, P., AIChE Meeting, Orlando, (1990)

Joos, P. and Van Uffelen, M., J. Colloid Interface Sci., 155(1993)271

Joosten, J.G.H., Vrij, A. and Fijnaut, H.M., Int. Conf. Phys. Chem. Hydrodyn., Levich Conference 1977, Advances Publications, Guernscy, 2(1978)639

Kretzschmar, G. and König, J. SAM, 9(1981)203

Kretzschmar, G., Progr. Colloid Polymer Sci., 77(1988)72

Kretzschmar, G., J. Inf. Rec. Mat. 21(1993)439

Kretzschmar, G., Lunkenheimer, K. and Miller, R., Material Science Forum, 25-26(1988)211

Khristov, Khr., Malysa, K. and Exerowa, D., Colloids & Surfaces, 11(1984)39

Krugljakov, P.M. and Exerowa, D.R., "Foam and Foam Films", Khimija, Moscow, 1990

Kruglyakov, P.M., Exerowa, D. Khristov, Khr.I., Langmuir, 7(1991)1846

Laddha, S., Machie, A.R. and Clark, D.C., Membrane Biology, 142(1994), in press

Landau, L. and Lifschitz, E., "Theoretical Physics", Moscow, 1953

Lange, H., Koll. Z. Z. Polymere, 201(1965)131

98

Langevin, D., J. Colloid Interface Sci., 80(1980)412

Levich, V.G., Acta Physicoxchimica U.R.S.S., XIV(1941)307

Levich, V.G., Physicochemical Hydrodynamics, Prentice Hall, Inc., New York, 1962

Liebermann, L., Phys. Rev., 76(1949)1520

Linde, H., Mber. Detsch. Akad. Wiss., 1(1959)586, 699

Linde, H. and Kretzschmar, G., J. prakt. Chemie, 15(1962)288

Linde, H., Chu, X.L., Velarde, M.G. and Waldheim, W., Phys. Fluids, A5(1993)3162

Linde, H. and Friese, P., Z. phys. Chem. (Leipzig), 247(1971)225

Linde, H. and Shuleva, N., Monatsberichte DAW Berlin, 12(1970)883

Linde, H. and Schwartz, P., Nova acta Leopoldina, NF61(1989)268

Linde, H. and Engel, H., Physica D, 49(1991)13

Linde, H. and Zirkel, Ch., Z. phys. Chem. (München), 174(1991)145

Linde, H., Schwartz, P. and Wilke, H., "Dynamics and Instability of Fluid Interfaces", Lecture
      Notes in Physics, 105, Springer, Berlin, (1979)

Loglio, G., Tesei, U. and Cini, R., J. Colloid Interface Sci., 71(1979)316

Loglio, G., Tesei, U. and Cini, R., Colloid Polymer Sci., 264(1986a)712

Loglio, G., Tesei, U. and Cini, R., Bolletino Di Oceanologica Theorien ed Applicata,
      IV(1986b)91

Loglio, G., Tesei, U. and Cini, R., Rev. Sci. Instrum., 59(1988)2045

Loglio, G., Tesei, U., Miller, R. and Cini, R., Colloids & Surfaces, 61(1991)219

Lucassen, J., Trans. Faraday Soc., 64(1968a)2221

Lucassen, J., Trans. Faraday Soc., 64(1968b)2230

Lucassen, J. and Barnes, G.T., J. Chem. Soc. Faraday Trans. 1, 68(1972)2129

Lucassen, J. and Giles, D., J. Chem. Soc. Faraday Trans. 1, 71(1975)217

Lucassen, J., Faraday Disc. Chem. Soc., 59(1975)76

Lucassen, J., "Dynamic Properties of Free Liquid Films and Foams", in Surface Science
      Series, Vol. 11, Marcel Dekker, New York, (1981)217

Lucassen-Reynders, E.H. and Lucassen, J., Adv. Colloid Interface Sci., 2(1969)347

Lucassen-Reynders, E.H., J. Colloid Interface Sci., 42(1973)573

Lucassen-Reynders, E.H., "Surface Elasticity and Viscosity in Compression/Dilatation", in
      Surfactant Science Series, Vol. 11, (1981)173

Lucassen-Reynders, E.H., Marcel Dekker, Surface Science Series, 11(1986)1

Lucassen-Reynders, E.H., J. Colloid Interface Sci., 117(1987)589

Lucassen-Reynders, E.H., Colloids & Surfaces, 25(1987)231

Ludviksson, E.N. and Lightfoot, J., AIChE J., 14(1968)674

Lunkenheimer, K. and Kretzschmar, G., Z. Phys. Chem. (Leipzig), 256(1975)593

MacLeod, C.A. and Radke, C.J., J. Colloid Interface Sci., (1993)

Malhotra, A.K. and Wasan, D.T., Surfactant Science Ser., Vol. 29, (1988)767

Malysa K, Lunkenheimer K, Miller R, and Hempt C., Colloids & Surfaces, 16(1985)9

Malysa K, Miller R, and Lunkenheimer K., Colloids & Surfaces, 53(1991)47

Manev, E.D., Vassilieff, Chr.St. and Ivanov, I.B., Colloid Polymer Sci., 254(1976)99

Mann, J.A., in "Surface and Colloid Scince", E.Matievic and R.J. Good (Eds), Vol. 13,
      Plenum Press, New York, London, 1984

Maru, H.C., Wasan, D.T. and Kintner, R.C., Chem. Eng. Sci., 26(1971)1615

Maru, H.C. and Wasan, D.T., Chemical Eng. Sci., 34(1979)1295

Maxwell, J., Phil. Mag., 4(1968), 35, 129, 185

Miller, R. and Kretzschmar, G., Adv. Colloid Interface Sci. 37(1991)97

Miller, R., Loglio, G. and Tesei, U., Colloid Polymer Sci., 270(1992)598

Miller, R., Kretzschmar, G. and Dukhin, S.S., Colloid Polymer Sci., 272(1994)548

Moelwyn-Hughes, E.A., The Chemical Statics and Kinetics of Solutions, Academic Press, London, New York, (1971)

Moffat, H.K., J. Fluid Mechan., 18(1964)1

Neumann, A.W. and Good, R.J., J. Colloid Interface Sci., 38(1972)341

Petrov, J.G., Kuhn, H. and Möbius, D., J. Colloid Interface Sci., 73(1980)66

Petrov, I.G. and Radoev, B.P., Colloid Polymer Sci., 259(1981)753

Prigogine, I. and Glansdorff, P., "Thermodynamic Theory of Structure, Stability and Fluctuations, Whiley-Interscience, London, 1971

Princen, H.M., Surfaces and Colloid Sci., 2(1969)1

Radoev, B.P., Dimitrov, D.D. and Ivanov, I.B., Colloid Polymer Sci., 252(1974)50

Reynolds, O., Phil. Trans. Roy. Soc., London, A177(1886)157

Sanfeld, A., Passerone, A., Ricci, E. and Joud., J.C., Il Nuovov Cimento, 12(1990)353

Scheludko, A. and Platikanov, D., Koll. Z., 175(1961)150

Scheludko, A., "Colloid Chemistry", Elsevier Amsterdam, London, New York, (1966)

Scheludko, A., Adv. Colloid Interface Sci., 1(1967)391

Schwartz, P., Bielecki, J. and Linde, H., Z. phys. Chem. (Leipzig), 266(1985)731

Scriven, L.E., Chem. Eng. Sci., 12(1960)98

Sternling, C.R. and Scriven, L.E., AIChE J., 5(1959)514

Stuke, B., Chemie Ing. Techn., 33(1961)173

Tabak, G.A., Notter, R.H. and Ultman, J.S., J. Colloid Interface Sci., 60(1977)117

van den Tempel, M. and van de Riet, R.P., J. Chem. Phys., 42(1965)2769

van den Tempel, M., J. of Non-Newtonian Fluid Mechanics, 2(1977)205

van den Tempel, M. and Lucassen-Reynders, E.H., Adv. Colloid Interface Sci., 281(1983)

van der Waals, jr., J.D. and Bakker, G., Handbuch der Experimentalphysik, Vol. 6, Leipzig, 1928

van Voorst Vader, F., Erkelens, Th.F. and Van den Tempel, M., Trans. Faraday Soc., 60(1964)1170

Velarde, M.G. and Chu, X.L., "Waves and Turbulence at Interfaces", Phys. Ser., T25(1989)231

Velarde, M.G. and Normand, C., Convection, Ser. A, 243(1980)92

Voinov, O.V., Dokl. Akad. Nauk SSSR, 243(1976)1422

Vollhardt, D., Zastrow, L. and Schwartz, P., Colloid Polymer Sci., 258(1980)1174

Vollhardt, D., Zastrow, L., Heybey, J. and Schwartz, P., Colloid Polymer Sci., 258(1980)1289

Wantke, K.D., Miller, R. and Lunkenheimer, K., Z. Phys. Chem. (Leipzig), 261(1980)1177

Wantke, K.D., Malysa, K. and Lunkenheimer, K., Colloids Surfaces

Wasan, D.T., Gupta, L. and Vora, M.K., AIChE J., 17(1971)1287

Wassmuth, F., Laidlaw, W.G. and Coombe, D.A., Chem. Eng. Sci., 45(1990)3483

Wehrli, B. and Stumm, W., Langmuir, 4(1988)753

Weidman, P.D., Linde, H. and Velarde, M.G., Phys. Fluids, A4(1992)921

Wilde, P.J. and Clark, D.C., J. Colloid Interface Sci., 155(1993)48

Wüstneck, R. and Kretzschmar, G., Z. Chemie, 22(1982)202

Yao, Y.L., "Irreversible Thermodynamics", Van Nostrand Reinhold Comp., New York, Cincinnati, Toronto, London, Melbourne, (1981)

Yin, T.P., J. Phys. Chem., 73(1969)2413

CHAPTER 4

4. THE DYNAMICS OF ADSORPTION AT LIQUID INTERFACES

Surface active substances are able to modify significantly the properties of interfaces by adsorption. This fact is used in many processes and many new technologies are based on adsorption effects. In general, these technologies work under dynamic conditions and improvement of their efficiency is possible by a controlled use of interfacially active material. The interfaces involved are freshly formed and have only a small effective age of some seconds or sometimes even less than a millisecond.

To optimise the use of surfactants, polymers and mixtures of them, specific knowledge of their dynamic adsorption behaviour rather than of equilibrium properties is of great interest (Kretzschmar & Miller 1991). The importance of dynamics of adsorption in different applications has been recently discussed: froth flotation (Małysa 1992), foam generation (Fainerman et al. 1991), demulsification (Krawczyk et al. 1991), or emulsification (Lucassen-Reynders & Kuijpers 1992).

The most frequently used parameter to characterise the dynamic properties of liquid adsorption layers is the dynamic interfacial tension. Many techniques exist to measure dynamic tensions of liquid interfaces having different time windows from milliseconds to hours and days.

The aim of this chapter is to present the fundamentals of adsorption at liquid interfaces and a selection of techniques, for their experimental investigation. The chapter will summarise the theoretical models that describe the dynamics of adsorption of surfactants, surfactant mixtures, polymers and polymer/surfactant mixtures. Besides analytical solutions, which are in part very complex and difficult to apply, approximate and asymptotic solutions are given and their range of application is demonstrated. For methods like the dynamic drop volume method, the maximum bubble pressure method, and harmonic or transient relaxation methods, specific initial and boundary conditions have to be considered in the theories. The chapter will end with the description of the background of several experimental technique and the discussion of data obtained with different methods.

4.1. GENERAL IDEA OF ADSORPTION KINETICS

The adsorption kinetics of interfacial active molecules at liquid interfaces, for example surfactants at the aqueous solution/air or solution/organic solvent interface, can be described by quantitative models. The first physically founded model for interfaces with time invariant area was derived by Ward & Tordai (1946). It is based on the assumption that the time dependence of interfacial tension, which is directly correlated to the interfacial concentration Γ of the adsorbing molecules, is caused by a transport of molecules to the interface. In the absence of any external influences this transport is controlled by diffusion and the result, the so-called diffusion controlled adsorption kinetics model, has the following form

$$\Gamma(t) = 2\sqrt{\frac{D}{\pi}}\left(c_o\sqrt{t} - \int_0^{\sqrt{t}} c(0, t-\tau)d\sqrt{\tau}\right), \tag{4.1}$$

where D is the diffusion coefficient and c_o is the surfactant bulk concentration. The integral equation describes the change of $\Gamma(t)$ with time t. The application of the very complex Eq. (4.1) to dynamic surface tensions $\gamma(t)$ is not trivial and many attempts were undertaken to further develop the theory and to make it applicable to a large variety of liquid systems.

There are two general ideas to describe the dynamics of adsorption at liquid interfaces. The diffusion controlled model assumes the diffusional transport of interfacially active molecules from the bulk to the interface to be the rate-controlling process, while the so-called kinetic controlled model is based on transfer mechanisms of molecules from the solution to the adsorbed state and vice versa. A schematic picture of the interfacial region is shown in Fig. 4.1. showing the different contributions, transport in the bulk and the transfer process.

Transport in the solution bulk is controlled by diffusion of adsorbing molecules if any liquid flow is absent. The transfer of molecules from the liquid layer adjacent to the interface, the so-called subsurface, to the interface itself is assumed to happen without transport. This process is determined by molecular movements, such as rotations or flip-flops. As pointed out in Chapter 2 adsorption of surface active molecules at an interface is a dynamic process. In equilibrium the two fluxes, the adsorption and desorption fluxes, are in balance. If the actual surface concentration is smaller than the equilibrium one, $\Gamma < \Gamma_o$, the adsorption flux to the interface predominates, if $\Gamma > \Gamma_o$, the actual amount adsorbed at the interface is higher than the equilibrium value Γ_o, and the desorption flux prevails.

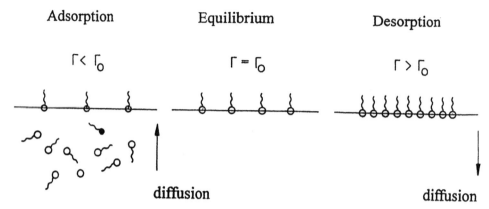

Fig. 4.1.   Schematic diagram of the fluxes of adsorbing molecules near a liquid interface in absence of
any forced liquid flow

Milner (1907) was the first to discuss diffusion as a process responsible for the time-
dependence of surface tension of soap solutions. Later, several models were developed taking
into account transfer mechanisms having the form of rate equations (Doss 1939, Ross 1945,
Blair 1948, Hansen & Wallace 1959). More complicated models take into account diffusion
and transfer mechanisms simultaneously (Baret 1968a, b, 1969, Miller & Kretzschmar 1980,
Adamczyk 1987). If surface tension gradients are created at the interface, the Marangoni-Gibbs
effect leads to a flow along the interface. Temperature or density gradients in the bulk phase
also initiate flows in a direction which is controlled by the direction of the gradients.

Models which consider diffusion in the bulk as the only rate-controlling process are called
diffusion controlled. If the diffusion is assumed to be fast in comparison to the transfer of
molecules between the subsurface and the interface the model is called kinetic-controlled or
barrier-controlled. Both steps are taken into account in mixed diffusion kinetic controlled
models.

In technological applications as well as in scientific experiments specific boundary conditions
are often given, such as definite changes of the interfacial area. A schematic representation is
given in Fig. 4.2. which shows various bulk and interfacial transport processes of surface active
molecules: diffusion in the bulk, interfacial diffusion, bulk flow of different origin, interfacial
compression and dilation.

Theoretical models for the interpretation of experimental data and quantitative descriptions of
processes in particular technologies have to consider these specific conditions which usually
lead to very complex theoretical models.

If there is, for example, a liquid flow in a layer adjacent to the interface, a surface concentration
gradient is created. Any concentration gradient at an interface results in an interfacial tension

gradient, which initiates a flow of mass along the interface in direction to the higher interfacial tension. This effect is the so-called Marangoni effect. This situation can happen, for example, if an adsorption layer is compressed or stretched. Close to the barrier used in compression experiments, an increase in surface concentration results and desorption starts. If a barrier is used for stretching the adsorption layer an adsorption process is initiated. In both cases a Marangoni flow is initiated simultaneously.

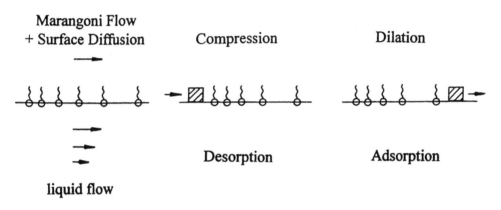

Fig. 4.2.   Schematic of surfactant transport at the surface and in the bulk of the liquid

4.2.   THEORETICAL MODELS OF DIFFUSION-CONTROLLED ADSORPTION KINETICS

The adsorption kinetics of surfactant molecules to a liquid interface is controlled by transport processes in the bulk and the transfer of molecules from a solution state into an adsorbed state or vice versa. In this paragraph qualitative and quantitative models are discussed.

4.2.1. QUALITATIVE MODELS

First models can be derived under very simple boundary conditions. To describe the adsorption of a surfactant at an interface by a simple model, one can assume the situation schematically shown in Fig. 4.3.

The surfactant concentration distribution in the solution bulk at time t=0 is assumed to be equal to c_0 for x<0 and zero at x>0,

$$c(x,0) = \begin{cases} c_0 & \text{at } x < 0 \\ 0 & \text{at } x > 0 \end{cases}, t = 0. \tag{4.2}$$

If a diffusion process starts at t>0, the concentration distribution is given by

$$c(x,t) = \frac{c_o}{2}\left[1 - \mathrm{erf}\left(\frac{x}{2\sqrt{Dt}}\right)\right].$$

(4.3)

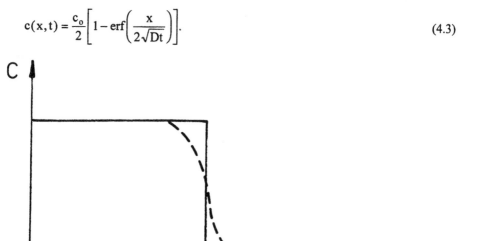

Fig. 4.3. Concentration distribution c(x,t) near a hypothetical interface located at x = 0 at different times; —— t=0, --- t > 0

The adsorbed amount can now be calculated for a model in which the change of surface concentration with time is assumed to be proportional to the concentration gradient at x=0, the location of the interface. This model is in accordance with the 1st diffusion law,

$$\frac{d\Gamma}{dt} = -D\frac{\partial c}{\partial x}, \, x = 0.$$

(4.4)

From Eqs (4.3) and (4.4) the first relation (4.5) is obtained, which describes in a very simple way the change of adsorption with time,

$$\Gamma(t) = c_o\sqrt{\frac{Dt}{\pi}}.$$

(4.5)

This relation is very often used as a rough estimate and results from Eq. (4.1) when the second term on the right hand side is neglected.

The easiest model for a desorption process can be derived when the adsorbed amount is assumed to be located in an ideal two-dimensional layer Γ_d (Fig. 4.4) and the change in concentration in this layer with time over the initial concentration is $\Gamma(t) - \Gamma_o$.

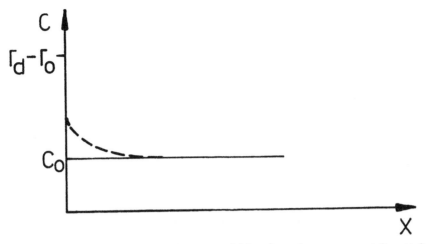

Fig. 4.4.   Point source at the interface as a model for a desorption process; ——— t=0, --- t > 0

The change of the concentration in the layer adjacent to the interface, the subsurface concentration, is then given by

$$c(0,t) = c_o + \frac{\Gamma_d - \Gamma_o}{\sqrt{\pi Dt}}.$$ (4.6)

If one assumes the adsorbed amount Γ_d is contained in a layer of thickness δ adjacent to the interface in addition to the equilibrium bulk concentration of the surfactant c_o, the change of the concentration at $c(0,t)$ is obtained in form of the relation

$$c(0,t) = c_o + (c_\delta - c_o) \text{erf}(\frac{\delta}{2\sqrt{Dt}}).$$ (4.7)

The above relations are qualitative and must be used with circumspection. If applied under the wrong conditions, the estimates made can be erroneous even by up to several orders of magnitude.

4.2.2. QUANTITATIVE MODELS OF DIFFUSION CONTROLLED ADSORPTION

The quantitative description of adsorption kinetics processes is much more complicated than the use of the simplified models mentioned above. An introduction into the variety of theoretical models and appropriate boundary conditions is given in a recent review (Miller et al. 1994a). The diffusion-controlled model assumes that the step of transfer from the subsurface to the interface is fast compared to the transport from the bulk to the subsurface. It is based on the following general equation,

$$\frac{\partial c}{\partial t} + (v \cdot grad) \, c = D \; div \; grad \; c. \tag{4.8}$$

If any flow in the bulk phase is neglected the diffusion equation (4.8) reads

$$\frac{\partial c}{\partial t} = D \frac{\partial^2 c}{\partial x^2} \quad \text{at } x > 0, \, t > 0. \tag{4.9}$$

The transport problem can be solved only after definition of the corresponding initial and boundary conditions. Fick's first law is used as the boundary condition at the surface located at x=0. It has the following general form

$$\frac{\partial \Gamma}{\partial t} + (v_s \cdot grad_s) \, \Gamma = D_s \; div_s grad_s \Gamma - \Gamma div_s v_s + D \frac{\partial c}{\partial x}. \tag{4.10}$$

This condition describes the surfactant flux from the bulk to the interface during adsorption. If the surface concentration Γ is higher than the equilibrium value Γ_o a desorption process creates a surfactant flux in the opposite direction. Neglecting again any flow and fluxes other than bulk diffusion, for example surface diffusion, the boundary condition becomes very simple,

$$\frac{\partial \Gamma}{\partial t} = j = D \frac{\partial c}{\partial x} \quad \text{at } x = 0; \, t > 0. \tag{4.11}$$

The diffusion problem of a surfactant mixture can be described in an analogous way. The generalised diffusion equation reads

$$\frac{\partial c_i}{\partial t} + (v \cdot grad) \, c_i = \sum_{j=1}^{r} D_{ij} div \; grad \; c_i. \tag{4.12}$$

To a first approximation, at low surfactant concentrations (Haase 1973) the mixed diffusion coefficients D_{ij} can be replaced by the individual ones D_i,

$$D_{ij} = \begin{pmatrix} D_i \text{ at } i = j \\ 0 \text{ at } i \neq j \end{pmatrix}. \tag{4.13}$$

Neglecting any flow in the bulk liquid leads to

$$D_i \frac{\partial^2 c_i}{\partial x^2} = \frac{\partial c_i}{\partial t} \text{ at } x > 0, t > 0, i=1,...,r. \tag{4.14}$$

Here r is the number of surfactant species in the system. In the same way as above the boundary conditions for each component is obtained, which all have the form of Eq. (4.11). To complete the transport problem one additional boundary condition and an initial condition are necessary. In most cases the assumption of an infinite bulk phase is advantageous

$$\lim_{x \to \infty} c(x,t) = c_o \text{ at } t > 0. \tag{4.15}$$

For systems having a bulk phases of limited depth h, the assumption of a diminishing concentration gradient at the border is possible

$$\frac{\partial c}{\partial x} = 0 \text{ at } x = h; t > 0. \tag{4.16}$$

The usual initial condition is a homogenous concentration distribution and a freshly formed interface with no surfactant adsorption

$$c(x,t) = c_o \text{ at } t = 0 \tag{4.17a}$$

$$\Gamma(0) = 0. \tag{4.17b}$$

In a surfactant mixture the initial conditions, for each component, are equivalent to those for a single surfactant system. When solving the given initial and boundary condition problem the result is Eq. (4.1). The derivation of the solution was performed using Green's functions (Ward & Tordai 1946, Petrov & Miller 1977) or by the Laplace operator method (Hansen 1961). Appendix 4E demonstrates the application of the operator method for solving such types of transport problems.

An equivalent relation to Eq. (4.1), derived by Panaiotov & Petrov (1968/69) via Abel's integral equation, has the following form,

$$c(0,t) = c_o - \frac{2}{\sqrt{D\pi}} \int_0^{\sqrt{t}} \frac{d\Gamma(t-\tau)}{dt} d\sqrt{\tau}. \tag{4.18}$$

By Hansen (1960) and Miller & Lunkenheimer (1978) numerical algorithms have been developed to solve the integral equation Eq. (4.1) or Eq. (4.18). The numerical procedure of using a trapezium rule approximation of the integral in Eq. (4.1) is described in detail in Appendix 4A. Some values of $\Gamma(t)$ calculated from Eq. (4.1) using a Langmuir adsorption isotherm Eq. (2.37) in the form,

$$\Gamma(t) = \Gamma_\infty \frac{c(0,t)}{a_L + c(0,t)}, \qquad (4.19)$$

are shown in Fig. 4.5 in terms of dynamic surface tension calculated from the corresponding Langmuir-Szyszkowski equation, which is obtained from the relation (2.39) by replacing Γ_0 by time-dependent adsorption $\Gamma(t)$:

$$\gamma = \gamma_0 + RT\Gamma_\infty \ln(1 - \frac{\Gamma(t)}{\Gamma_\infty}) \qquad (4.20)$$

The effects of the diffusion coefficient D as well as the surfactant concentration become evident from this graph.

The application of Eqs (4.1) or (4.18) to experimental data requires very complex numerical calculation, as shown in Appendix 4A. Therefore, other relations were derived in order to simplify the interpretation of experimental results.

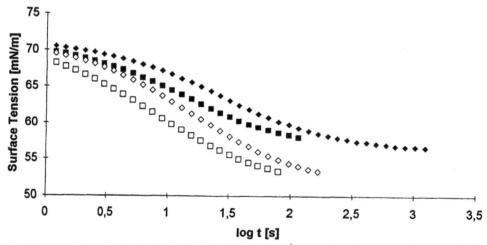

Fig. 4.5   Values calculated using Eq. (4.18) and different c_0 / a_L - values of a Langmuir isotherm (4.19) with $\Gamma_\infty = 4 \cdot 10^{-10}$ mol/cm², $a_L = 5 \cdot 10^{-9}$ mol/cm³; $c_0 = 2 \cdot 10^{-8}(\blacklozenge\blacksquare)$, $3 \cdot 10^{-8}(\square\diamondsuit)$ mol/cm³; $D = 1 \cdot 10^{-5}(\blacklozenge\diamondsuit)$, $2 \cdot 10^{-5}(\blacksquare\square)$ cm²/s

4.2.3. ANALYTICAL SOLUTION FOR A LINEAR ADSORPTION ISOTHERM

A simple equation derived by Sutherland (1952) can be used as a first approximation but its range of application is very limited (Miller 1990). The derivation of Sutherland is based on a linear adsorption equation of the form

$$\Gamma(t) = Kc(0,t), \tag{4.21}$$

resulting in the following analytical relation

$$\Gamma(t) = \Gamma_0 \left(1 - \exp(Dt/K^2) \, \text{erfc}(\sqrt{Dt}/K)\right). \tag{4.22}$$

The parameter Γ_0 denotes the interfacial concentration in the equilibrium state of adsorption. The determination of the function $\exp(\xi^2)\text{erfc}(\xi)$ is very complicated and values are obtained either from tables or from calculations via special numerical procedures (described in Appendix 4B).

Via Gibbs' equation (2.33) in the form

$$\Gamma(t) = -\frac{1}{RT}\frac{d\sigma(t)}{d\ln c(0,t)}, \tag{4.23}$$

which is valid under the condition of local equilibrium between the interface and the adjacent bulk phase (the subsurface), a relation for the dynamic interfacial tension results,

$$\gamma(t) = \gamma_0 - RT\Gamma_0\left(1 - \exp(Dt/K^2)\text{erfc}(\sqrt{Dt}/K)\right). \tag{4.24}$$

As mentioned before, this relation is valid only at very small coverages of the interface. Therefore, either $\Gamma(t)$ [and consequently $\gamma(t)$], has to be calculated numerically from Eq. (4.1) or other solutions to the problem have to be found when the interfacial coverage becomes too large. To show the range of application of Eq. (4.24) model calculations were performed, again using the Langmuir isotherm (Miller 1990). Under the condition $c_0/a_L \ll 1$ the Henry isotherm Eq. (4.21) results as a special case. It is visible that for $c_0/a_L > 0.5$ remarkable effects on the $\Gamma(t)$-dependencies appear and the use of approximation Eq.(4.24) leads to large errors.

4.2.4. COLLOCATION SOLUTION FOR A LANGMUIR-TYPE ADSORPTION ISOTHERM

Recently, the model of diffusion controlled adsorption was solved by a so-called collocation method using the Langmuir isotherm (2.37) (Ziller & Miller 1986). If we define a dimensionless time by $\Theta = Dt / (\Gamma_o / c_o)^2$, the solution of the diffusion controlled adsorption problem can be given in the following polynomial form

$$\frac{\Gamma(\Theta)}{\Gamma_o} = \sum_{i=1}^{N} \xi_i \tau^i, \qquad (4.25)$$

where the coefficients ξ_i are functions of the reduced concentration c_o/a_L and τ is defined by:

$$\tau = 1 - \frac{1}{1 + \sqrt{\Theta + \Theta^2 c_o / a_L}}. \qquad (4.26)$$

For the region of practical interest $0 \leq c_o / a_L \leq 100$ the coefficients ξ_i are tabulated by Miller & Kretzschmar (1991). In Appendix F, these coefficients are given for the interval $0 \leq c_o / a_L \leq 10$. Another polynomial solution was derived by McCoy (1983) which is more complicated to use.

The practical use of Eq. (4.25) is performed in the following way. First, from the measured $\gamma(t)$ - value the corresponding $\Gamma(t)$ -values are calculated via Eq. (4.27), which is equivalent to Eq. (4.20):

$$\Gamma(t) = \Gamma_\infty \left(1 - \exp\left[\frac{\gamma(t) - \gamma_o}{RT\Gamma_\infty} \right] \right). \qquad (4.27)$$

By variating D, all values of $\Gamma(\Theta)$ from Eqs (4.25) and (4.26) are fitted to the corresponding $\Gamma(t)$ - values, i.e., to the experimental $\gamma(t)$ - values. For these calculations the values of the specific parameters Γ_∞ and a_L of the studied surfactant are needed. These are to be determined beforehand from the corresponding equilibrium interfacial tension isotherm. If the fitted value of D is reasonable the studied surfactant adsorbs by a diffusion-controlled mechanism. Systematic studies of sufficiently pure surfactant systems have shown that surfactants usually adsorb by a diffusion-controlled mechanism (Kretzschmar & Miller 1991, Miller & Lunkenheimer 1986). Experimental results with different surfactant systems will be discussed in Section 4.3.

The analytical solutions presented above are most of all derived on the basis of the very simple Henry isotherm or the more physically sensible Langmuir isotherm. Beside these analytical solutions a direct integration of the initial and boundary value problem of the diffusion-controlled model is possible. To do so differentials are replaced by differences. This approximation leads to linear equation systems for each time step which have to be solved. As a result all time functions, $\Gamma(t)$, $c(0,t)$ and $\gamma(t)$ are obtained. The advantage of a direct integration is that it can be performed on the basis of any adsorption isotherm. A difference scheme was proposed by Miller (1981) and numerical results were produced using various adsorption isotherms. An example of such numerically calculated data is given in Fig. 4.6.

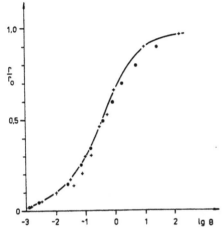

Fig. 4.6.   Documentation of a numerical integration of the diffusion problem by a finite difference scheme; effect of step size of the numerical procedure; $\Delta\Theta_o = 10^{-4}$, $\Delta\Theta_{i+1} = \Delta\Theta_i * 1.01$, $\Delta X = 0.1 (+), \Delta X = 0.01 (\bullet)$, variable $\Delta X (*)$; solid line refers to the solution via the integral equation (4.1)

The influence of the step size used in the procedure is significant if the steps size is not large. A very good agreement is obtained with sufficiently small step sizes, where the time and spatial step sizes are coupled with each other (see Appendix 4C). The difference scheme is discussed in more detail in Appendix 4C.

4.3.   DIFFUSION CONTROLLED ADSORPTION OF SURFACTANT MIXTURES

The description of the adsorption kinetics of a surfactant mixture is made in an analogous way as for a single surfactant solution. Instead of Eq. (4.9) a set of transport equations (4.14) has to be used, one for each of the r different surfactants. The initial and boundary conditions are defined for each component, in analogy to Eqs (4.11), (4.15) and (4.17). A system of r integral equations result, either in the form

$$\Gamma_i(t) = 2\sqrt{\frac{D_i}{\pi}} \left(c_{oi}\sqrt{t} - \int_0^{\sqrt{t}} c_i(0, t-\tau)d\sqrt{\tau} \right), \; i = 1, ..., r \tag{4.28}$$

or

$$c_i(0,t) = c_{oi} - \frac{2}{\sqrt{D_i\pi}} \int_0^{\sqrt{t}} \frac{d\Gamma_i(t-\tau)}{dt} d\sqrt{\tau}, i = 1, \ldots, r. \tag{4.29}$$

The set of r integral equations is linked together, for example via a multi-component adsorption isotherm. In the simplest case a set of r independent relations of the type of Eq. (4.22) results when linear adsorption isotherms Eq. (2.40) are assumed for each of the surfactants (Miller et al. 1979):

$$\Gamma_i(t) = \Gamma_{oi}\left(1 - \exp(D_i t / K_i^2)\text{erfc}(\sqrt{D_i t}/K_i)\right), i=1,\ldots,r. \tag{4.30}$$

The r equations are independent of each other and give only a reasonable result for very dilute surfactant mixtures. Except in this very simple case, no analytical solution exists. In Fig. 4.7. results of model calculations for a solution containing two surfactants are given using a generalized Langmuir isotherm Eq. (2.47). The data were calculated using a finite difference method described in Appendix 4D.

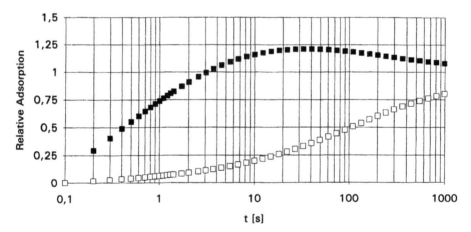

Fig. 4.7   Model calculations of a diffusion-controlled adsorption kinetics of a system containing two surfactants: $\Gamma_\infty = 3\cdot10^{-10}$mol/cm², $c_{01} = 10^{-6}$ mol/cm³, $c_{02} = 10^{-7}$ mol/cm³, $c_{01}/a_1 = 0.5$, $c_{02}/a_2 = 0.8$, $D_1 = D_2 = 5\cdot10^{-6}$ cm²/s, component 1 - ■, component 2 - □, according to Miller et al. (1979)

The graph shows the competitive adsorption of the two surfactants. In the beginning of the adsorption process, the less surface active component 1 adsorbs preferentially while component 2 starts to adsorb later. When the adsorbed amount of component 2 becomes larger molecules of component 1 have to leave the interface and the adsorbed amount of this compound passes a maximum. At longer times the adsorptions of both components level off and reach equilibrium values.

4.4. KINETIC-CONTROLLED MODELS

Further models of adsorption kinetics were discussed in the literature by many authors. These models consider a specific mechanism of molecule transfer from the subsurface to the interface, and in the case of desorption in the opposite direction ((Doss 1939, Ross 1945, Blair 1948, Hansen & Wallace 1959, Baret 1968a, b, 1969, Miller & Kretzschmar 1980, Adamczyk 1987, Ravera et al. 1994). If only the transfer mechanism is assumed to be the rate limiting process these models are called kinetic-controlled. More advanced models consider the transport by diffusion in the bulk and the transfer of molecules from the solute to the adsorbed state and vice versa. Such mixed adsorption models are called diffusion-kinetic-controlled The mostly advanced transfer models, combined with a diffusional transport in the bulk, were derived by Baret (1969). These diffusion-kinetic controlled adsorption models combine Eq. (4.1) with a transfer mechanism of any kind. Probably the most frequently used transfer mechanism is the rate equation of the Langmuir mechanism, which reads in its general form (cf. Section 2.5.),

$$\frac{d\Gamma}{dt} = k_{ad} c_o (1 - \frac{\Gamma}{\Gamma_\infty}) - k_{des} \frac{\Gamma}{\Gamma_\infty}. \tag{4.31}$$

Under the condition $\frac{\Gamma_o}{\Gamma_\infty} \ll 1$, i.e. at small interfacial coverage, the so-called Henry mechanism results,

$$\frac{d\Gamma}{dt} = k_{ad} c_o - k_{des} \Gamma. \tag{4.32}$$

The definition of k_{ad} and k_{des}, the rate constants of adsorption and desorption, respectively, results from the condition $\frac{d\Gamma}{dt} = 0$, the equilibrium state of adsorption. The solution of these adsorption models is easy and results in exponential expressions. For more complicated transfer mechanisms an analytical solution becomes complicated. For example, using a Frumkin mechanism (MacLeod & Radke, 1994)

$$\frac{d\Gamma}{dt} = k_{ad} c_o (1 - \frac{\Gamma}{\Gamma_\infty}) - k_{des} \frac{\Gamma}{\Gamma_\infty} \exp(a' \frac{\Gamma}{\Gamma_\infty}) \tag{4.33}$$

or a Langmuir-Hinshelwood mechanism (Chang & Franses, 1992), regarding for an additional adsorption barrier B,

$$\frac{d\Gamma}{dt} = k_{ad}\, c_o\, (1 - \frac{\Gamma}{\Gamma_\infty})\, \exp\left(-B\frac{\Gamma}{\Gamma_\infty}\right) - k_{des}\frac{\Gamma}{\Gamma_\infty}\exp\left(-B\frac{\Gamma}{\Gamma_\infty}\right), \qquad (4.34)$$

complex relations are obtained, which have again to be evaluated by using numerical procedures.

The pure diffusion-controlled and kinetic-controlled models are only limiting cases. Many real surfactant systems will need to be described by a combination of the two types of mechanisms. Therefore, a coupling between them is necessary. Following Baret (1969) the coupling of transfer mechanisms with the diffusional exchange of matter is arranged by replacing the bulk concentration c_0 by the subsurface concentration $c(0,t)$. For the Langmuir mechanism this leads to

$$\frac{d\Gamma}{dt} = k_{ad}\, c(0,t)(1 - \frac{\Gamma}{\Gamma_\infty}) - k_{des}\frac{\Gamma}{\Gamma_\infty}. \qquad (4.35)$$

Through the subsurface concentration $c(0,t)$ the Eqs (4.33) and (4.1) are coupled and an integro-differential equation system results. Such systems can be solved only numerically by different means (Miller & Kretzschmar 1980, Chang et al. 1992, Chang & Franses 1992).

Another attempt to combine the diffusion theory with transfer mechanisms was made by Fainerman et al. (1987). He derived approximate solutions by averaging the rate of the two different processes in the following way

$$\frac{d\Gamma}{dt} = \frac{(d\Gamma/dt)_{diff}(d\Gamma/dt)_{kin}}{(d\Gamma/dt)_{diff} + (d\Gamma/dt)_{kin}}. \qquad (4.36)$$

The indices "diff" and "kin" refer to the diffusional and transfer steps, respectively. Recently, theoretical models of surfactant adsorption kinetics were developed mainly to take into account specific experimental conditions or surfactant properties, such as molecular charge (Dukhin et al. 1983, 1991, Borwankar & Wasan 1986, Chang & Franses, 1992, Miller et al. 1994a), micelle formation (Rakita & Fainerman 1989, Dushkin & Ivanov 1991, Fainerman 1992, Serrien et al 1992) or other specific effects (Lin et al. 1991, Ravera et al. 1993, 1994, Jiang et al. 1993). Some of these models will be discussed in sections or chapters below.

4.5. MODELS WITH TIME DEPENDENT INTERFACIAL AREA

All models of adsorption kinetics discussed so far are based on interfaces with constant interfacial area. In the drop volume method this condition is fulfilled only in the so-called quasi-static mode after the stage of fast drop formation. In the "classical" dynamic mode in which the drops are growing continuously during the whole drop formation process due to a constant liquid flow, this condition is not fulfilled. Therefore, the change of the drop surface area in alliance with the flow inside the drop must be taken into consideration. At moderate growing rates this flow turns out to be radial. In the case of liquid-liquid systems flow is radial inside and outside the drop. A similar situation exists in bubble pressure experiments, where the bubble growth from its initial state up to the hemisphere has to be considered in the interpretation of experimental data. Again a flow normal to the surface has to be considered together with the interfacial area change. This paragraph is dedicated to adsorption kinetics models at interfaces with time dependent areas. Such area changes are usually coupled with flow patterns in the adjacent bulk phase which must be considered simultaneously.

4.5.1. GENERAL CONSIDERATION OF INTERFACIAL AREA CHANGES WITH TIME

As pointed out before, interfacial area changes are usually connected with a liquid flow in the adjacent bulk phase. Under certain conditions this flow becomes small and can be neglected. The diffusion-controlled adsorption kinetics are then described by the same diffusional transport in the bulk, given by Eq. (4.9). All initial and boundary conditions hold except the one at the surface x=0. This has to be modified to take into consideration the area change A(t) during the adsorption process. The modification can be made again on the basis of Fick's First law by setting the flux through the area A(t) equal to the change of adsorption at this area $\Gamma(t)\,A(t)$:

$$\frac{d(\Gamma A)}{dt} = A(t)\,D\frac{\partial c}{\partial x}; x = 0; t > 0 \qquad (4.37)$$

This adsorption model can be solved analytically using Laplace Transforms (Hansen 1961, Miller 1983) but the result is a non-linear Volterra integral equation similar to the Ward & Tordai equation (4.1):

$$\Gamma(t) = \Gamma_d - \int_0^t \Gamma\frac{d\ln A}{dt}dt + 2\sqrt{\frac{D}{\pi}}\left(c_o\sqrt{t} - \int_0^{\sqrt{t}}\frac{d\Gamma(t-\tau)}{dt}d\sqrt{\tau}\right). \qquad (4.38)$$

This integral equation cannot be used directly for the interpretation of experimental data, again a numerical procedure has to be applied. Another possibility consists in the application of a numerical method directly to the initial and boundary value problem. In Appendix 4C an algorithm is also given for the case of the boundary condition (4.37).

A more general model for a diffusion-controlled adsorption with a time dependent interfacial area was derived by Joos & van Uffelen (1993a). In their model the flow resulting from an area dilation or compression $\Omega = \dfrac{d \ln A}{dt}$ has been considered. For this case the transport equation (4.9) has to be modified to

$$\frac{\partial c}{\partial t} \pm \Omega x \frac{\partial c}{\partial x} = D \frac{\partial^2 c}{\partial x^2}, x > 0, t > 0. \tag{4.39}$$

The different signs stand for the compression (+) and dilation (-) case.

For constant dilation rate Ω Joos & van Uffelen discussed the dynamic interfacial behaviour (1993a, b). They compared the theory of Loglio, summarised by Miller et al. (1991), with the one presented by van Voorst Vader et al. (1964). For interfaces with constant dilation rate the relation

$$\Delta\gamma(t) = \frac{\varepsilon_o \Omega}{\omega_o} \int_0^{\omega_o t} \exp(x) \mathrm{erfc}(\sqrt{x}) dx \tag{4.40}$$

results. For $t \to \infty$ the interfacial tension difference tends to infinity which shows the limit of the approximate theory. Van Voorst Vader et al. (1964) obtained a steady state for small constant dilation rates,

$$\Delta\gamma_{ss} = \frac{\varepsilon_o}{\sqrt{\omega_o}} \sqrt{\frac{\pi\Omega}{2}} \tag{4.41}$$

which is exactly the expression resulting from the interfacial tension difference $\Delta\gamma(t)$ for $t \to \infty$ calculated in (1993c),

$$\Delta\gamma = \frac{RT\Gamma^2}{c_o} \sqrt{\frac{\pi\Omega}{2D}} \sqrt{\tanh\frac{\Omega t}{2}}. \tag{4.42}$$

These relations are useful for interpretations of experiments with, for example, constant area expansion, proposed by van Uffelen & Joos (1993c). These experiments are called "Peak

Tensiometry" as the surface tension passes a maximum due to the superposition of area expansion and adsorption rate of surfactants. Similar behaviour is observed at the surface of growing drops, discussed in the next paragraph.

4.5.2. CONSIDERATION OF INTERFACIAL AREA CHANGES AND RADIAL FLOW FOR GROWING DROPS

The kinetics of the adsorption process taking place at the surface of a growing drop or bubble is important for the interpretation of data from drop volume or maximum bubble pressure experiments. The same problem has to be solved in any other experiment based on growing drops or bubbles, such as bubble and drop pressure measurements with continuous, harmonic or transient area changes (for example Passerone et al. 1991, Liggieri et al. 1991, Horozov et al. 1993, Miller at al. 1993, MacLeod & Radke 1993, Ravera et al. 1993, Nagarajan & Wasan 1993).

Ilkovic (1934, 1938) was the first who discussed the adsorption at a growing drop surface. A complete diffusion-controlled adsorption model, considering the radial flow inside a growing drop, was derived much later by Pierson & Whittaker (1976). The diffusion equation for describing the transport inside or outside a spherical drop or bubble has the following form:

$$\frac{\partial c}{\partial t} + v_r \frac{\partial c}{\partial r} = D(\frac{\partial^2 c}{\partial r^2} + \frac{2}{r}\frac{\partial c}{\partial r}). \qquad (4.43)$$

Liquid/Gas Interface

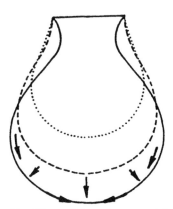

Fig. 4.8   Schematic representation of the diffusional transport inside a drop

For a drop of a surfactant solution this diffusion equation applies in the interval $0 < r < R$, where R is the radius of the drop. For a solvent drop or a bubble in a surfactant solution the diffusion process happens outside the drop or bubble, respectively, and Eq. (4.43) holds in the interval $R < r < \infty$. A schematic of the situation is shown in Fig. 4.8.

The drop growth is connected with a flow inside the drop and an area enlargement. The flow inside the drop can be assumed to be radial. Although the area stretching is not totally isotropic no significant deviations are expected and, therefore, no flow normal to the interface is initiated. Nevertheless, the stretching of the interface simultaneously leads to a stretching of the adjacent liquid layers. As a consequence, the concentration profile caused by diffusion is compressed. This situation is schematically shown in Fig. 4.9.

As a consequence, two processes overlap and are directed opposite to each other: diffusion layer compression due to the enlargement of the drop, and diffusion layer dilation due to the growth of the drop area and dilation of the adsorption layer coverage. Both counteracting processes have been taken into consideration in current theories.

diffusion layer compression

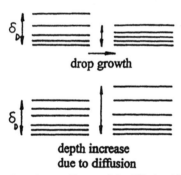

drop growth

depth increase
due to diffusion

Fig. 4.9   Change of the diffusional layer thickness adjacent to the surface of a growing drop

The radial flow, which is a good approximation of the real flow pattern, is given by (Levich 1962)

$$v_r = \frac{R^2}{r^2}\frac{dR}{dt}. \tag{4.44}$$

In analogy to the Ward & Tordai (1946) equation (4.1) the following non-linear integral equation was derived on the basis of these relations (Miller, 1980):

$$\Gamma(t) = 2c_o\sqrt{\frac{3Dt}{7\pi}} - \sqrt{\frac{D}{\pi}}t^{-2/3}\int_0^{\frac{3}{7}t^{7/3}}\frac{c(0,\frac{7}{3}\lambda^{3/7})}{\left(\frac{3}{7}t^{7/3}-\lambda\right)^{1/2}}d\lambda. \tag{4.45}$$

Eq. (4.45) can be used to describe the adsorption process at the surface of growing drops. The analysis of this rather complex equation showed that the rate of adsorption at the surface of a growing drop with linear volume increase, as it is arranged in a usual drop volume experiment, is about 1/3 of that at a surface with constant area. This result is supported by experimental findings (Davies et al. 1957, Kloubek 1972, Miller & Schano 1986, 1990) and also by an approximate solution first discussed by Delahay & Trachtenberg (1957) and Delahay & Fike (1958):

$$\Gamma(t) = 2c_0 \sqrt{\frac{3Dt}{7\pi}}. \tag{4.46}$$

Different flow patterns, also a turbulent flow, were considered by Fainerman (1983) for the discussion of kinetic data obtained from dynamic drop volume experiments.

A more advanced theory of the adsorption process at growing drop surfaces was made by MacLeod & Radke (1994). In contrast to the theory discussed above they do not assume a point source at the beginning of the process but a finite drop size. On the basis of an arbitrary dependence R(t) a theory of diffusion- as well as kinetically-controlled adsorption was then derived. In addition to the diffusion equation (4.43) the following boundary condition is proposed:

$$\frac{\partial \Gamma}{\partial t} = \pm D \frac{\partial c}{\partial r} - \Gamma \frac{d \ln A}{dt}, r = R(t). \tag{4.47}$$

The sign of the first term on the right hand side of Eq. (4.45) depends on whether the diffusion takes place inside or outside the drop. The result is given by the equation

$$\frac{d\Gamma}{dt} = -(R(t))^2 \sqrt{\frac{D}{\pi}} \int_0^t \frac{\left(\frac{dc(0,t)}{dt}\right)_{t_0}}{\left(\int_{t_0}^t (R(\xi))^4 d\xi\right)^{1/2}} dt_0 - \Gamma \frac{d \ln A}{dt}. \tag{4.48}$$

Although this is a very complex equation, it allows to take into consideration any function of R(t), and consequently A(t), resulting from experiments with growing drops or bubbles. In combination with an adsorption isotherm (diffusion-controlled case) or a transfer mechanism (mixed diffusion-kinetic-controlled model) it describes the adsorption process at a growing or even receding drop. Eq. (4.48) can be applied in its present form only via numerical calculations and an algorithm is given by MacLeod & Radke (1994).

Assuming a spherical geometry the radius R(t) and the area A(t) of a drop can be calculated from the following relations:

$$R(t) = \frac{r_{cap}^2}{2h} + \frac{h}{2},$$

(4.49)

$$A(t) = 2\pi r_{cap} h.$$

(4.50)

The drop height h as the specific parameter in the two equations can be easily calculated from Eq. (4.51):

$$h = \sum_{i=1}^{2} (\frac{3}{\pi}(Qt + V_0) + (-1)^i \sqrt{r_{cap}^6 + \frac{9}{\pi^2}(Qt + V_0)^2})^{1/3},$$

(4.51)

where Q is the liquid volume flow rate, V_0 is the initial volume of the residual drop and r_{cap} is the capillary radius.

The pure kinetic-controlled adsorption process can of course be modelled by the simple combination of any transfer mechanism and the change of adsorption with time under the condition of surface area changes. Such models have been derived by Miller (1983):

$$\frac{d(\Gamma A)}{dt} = A(t)(j_{ad} - j_{des}).$$

(4.52)

Here, j_{ad} and j_{des} are the adsorption and desorption fluxes, respectively. Considering a Langmuir mechanism, Eq. (4.52) takes the final form:

$$\frac{d\Gamma}{dt} = k_{ad}(1 - \frac{\Gamma}{\Gamma_\infty}) - k_{des}\frac{\Gamma}{\Gamma_\infty} - \Gamma\frac{d\ln A}{dt}.$$

(4.53)

Dependent on the form of A(t), Eq. (4.48) can be solved analytically or has to be approximated by a numerical algorithm.

4.5.3. ADSORPTION KINETICS MODEL FOR THE MAXIMUM BUBBLE PRESSURE METHOD

The idea of a maximum bubble pressure instrument is that the pressure inside a growing bubble passes through a maximum. The pressure maximum is reached at a hemispherical bubble size. After the maximum has passed the bubble grows fast and finally detaches. For quantitative interpretation of bubble pressure experiments details are required on the time scale of the bubble growth (cf. Fig. 5.8). First of all, the dead time, needed by a bubble to detach after it has reached its hemispherical size, must be determined accurately. Under special experimental conditions (Fainerman 1990, Joos et al. 1992, Fainerman et al. 1993a, b, Makievski et al. 1994) a maximum bubble pressure experiment can be performed under definite conditions such that the dead time can be easily calculated from the parameters of the instrument (see for more details Chapter 5, paragraph 5.3),

$$\tau = \tau_b - \tau_d = \tau_b (1 - \frac{L p_c}{L_c p}). \tag{4.54}$$

The time for bubble growth to a hemisphere τ, is the difference between the total bubble time τ_b and the dead time τ_d, p and L are pressure and gas flow rate and p_c and L_c are the respective values at a critical point. From considerations discussed above about adsorption processes at the surface of growing drops it is concluded that the situation with the growing bubble is comparable, at least until the state of the hemisphere. Then, the process runs without specific control and leads to an almost bare residual bubble after detachment due to the very fast bubble growth.

Fainerman et al. (1993c, 1994b) presented an analysis of the adsorption process at the bubble surface in order to derive a relation between the bubble life time and the effective surface age τ_a. The basis of this analysis is the condition of constant pressure p at any moment of bubble life in the time interval $0 < t < \tau$,

$$p(t) = \frac{2\gamma(t)}{r_{cap}} \cos\varphi(t) = \frac{2\gamma(\tau)}{r_{cap}} \cos 0 = \frac{2\gamma_o}{r_{cap}} \cos\varphi_o = const. \tag{4.55}$$

Here, φ is the angle between the bubble surface and the front area of the capillary of tip diameter $2r_{cap}$. The bubble growth in the time after having reached the hemisphere is very fast, so one can assume an almost bare surface with a surface tension $\gamma(0) = \gamma_o$. If a spherical geometry is assumed the area of the bubble cap is given by

$$A(t) = \frac{2\pi r_{cap}}{1 + \sin\varphi(t)}. \tag{4.56}$$

The change of the area with time can be derived from Eqs (4.55) and (4.56), leading to

$$dA = \frac{2\pi r_{cap}^2}{(1 + \sin\varphi)^2} \cos\varphi \, d\varphi = \frac{2\pi r_{cap}^2}{(1 + \sin\varphi)^2} \cos\varphi \, (-\frac{\gamma(\tau)}{\gamma^2 \sin\varphi}) d\gamma. \tag{4.57}$$

γ and φ are the corresponding values during the growth of the bubble and $\gamma(\tau)$ is the surface tension at the state of the hemisphere. To estimate the effective age of the bubble surface the rate of deformation of the bubble surface area is necessary. From the relative surface area change,

$$v = \frac{d \ln A}{dt} \approx \xi \frac{1}{\tau},$$

(4.58)

with

$$\xi = -\gamma^2(\tau) \int_{\sigma_o}^{\sigma(\tau)} \frac{d\gamma}{\gamma^3(1+\sin\varphi)\sin\varphi} \approx \frac{\sin\varphi_o}{1+\sin\varphi_o},$$

(4.59)

with $\varphi_o = \arccos\left(\frac{\gamma(\tau)}{\gamma_o}\right)$. The transport in the bulk phase adjacent to the bubble surface is governed by a modified diffusion equation (Joos & Rillaerts, 1981),

$$\frac{\partial c}{\partial t} + vx\frac{\partial c}{\partial x} = D\frac{\partial^2 c}{\partial x^2}.$$

(4.60)

The following transformation

$$z = x\tau^{2\xi+1} \text{ and } t = \frac{\tau^{2\xi+1}}{2\xi+1}$$

(4.61)

leads to the same form of the diffusion equation, which is valid in absence of bulk and surface flow. Thus, the solution of the classic diffusion equation (4.9) can be transferred into the solution of the actual form Eq. (4.60) by a simple inverse coordinate transform, according to (4.61). This finally yields,

$$\tau_a = \frac{\tau}{2\xi+1}.$$

(4.62)

At surface tensions γ close to the one of the solvent γ_o, the effective surface age is coincides with the bubble life time. The function $1/(2\xi+1)$ changes from 1 down to 0.5 at low surface tensions, and therefore, the effective surface age τ_a at small surface tensions amounts to about 50% of the bubble life time τ.

4.6. ADSORPTION KINETICS OF SURFACTANTS AT LIQUID/LIQUID INTERFACES

The adsorption of surfactants at a liquid-liquid interface follows in general the same principles. There is only one remarkable peculiarity, the solubility of the surfactant molecules in both adjacent bulk phases. To model the adsorption kinetics at such an interface transport by diffusion has to be considered in both phases. Also a transfer of surfactants across the interface is possible and must be allowed for (cf. Fig. 4.10).

Liquid/Liquid Interface

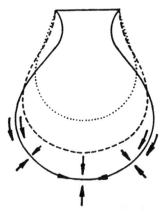

Fig. 4.10   Schematic of the diffusional transport inside and outside a drop

If the indices α and β stand for the two liquid phases having a common interface at x=0, the transport equations read,

$$D_\alpha \frac{\partial^2 c_\alpha}{\partial x^2} = \frac{\partial c_\alpha}{\partial t} \qquad \text{at } x > 0,\, t > 0, \qquad (4.63)$$

$$D_\beta \frac{\partial^2 c_\beta}{\partial x^2} = \frac{\partial c_\beta}{\partial t} \qquad \text{at } x < 0,\, t > 0. \qquad (4.64)$$

The boundary condition at the interface changes to

$$\frac{\partial \Gamma}{\partial t} = D_\alpha \frac{\partial c_\alpha}{\partial x} - D_\beta \frac{\partial c_\beta}{\partial x} - \Gamma \frac{d \ln A(t)}{dt}, \qquad x{=}0,\, t > 0. \qquad (4.65)$$

This is already the more general case with a time dependent interfacial area, again without taking into account any resulting flow. For A = const. the final term on the right hand side diminishes. MacLeod & Radke (1994) discussed the adsorption for growing drops in a second liquid. To do so, they used diffusion equations of the type (4.43) in both liquid phases as well as the boundary condition (4.65). The result has the structure of Eq. (4.46) but consists of two equivalent terms, one for each phase. This equation, of course, can be applied only via numerical calculations. A suitable algorithm is given by MacLeod & Radke (1994).

4.7. ADSORPTION KINETICS FROM MICELLAR SOLUTIONS

So far, all theoretical models are based on surfactant solutions with a distribution of surfactant molecules as monomers. One of the specific properties of surfactants is that they form aggregates once a certain concentration, the critical micelle concentration CMC, is reached. The shape and size of such aggregates are different and depend on the structure and chain length of the molecules. At higher concentrations, far beyond the CMC, the phase behaviour is often complex giving rise to novel physical properties (Hoffmann 1990).

The presence of micelles in the solution bulk can influence the adsorption kinetics remarkably this may be explained in the following way. After a fresh surface has been formed surfactant monomeric molecules are adsorbed resulting in a concentration gradient of monomers. This gradient will be equalised by diffusion to re-establish a homogeneous distribution. Simultaneously, the micelles are no longer in equilibrium with monomers within the range of the concentration gradient. This leads to a net process of micelle dissolution or rearrangement to re-establish the local equilibrium. As a consequence, a concentration gradient of micelles results, which is equalised by diffusion of micelles. On the basis of this idea, it is expected that the ratio of monomers c_{o1} to micelles c_{on}, the aggregation number n, rate of micelle formation k_f and dissolution k_d, influence the rate of the adsorption process. A physical picture of the rather complicated situation is shown in Fig. 4.11.

The schematic shows that the transport of monomers and micelles as well as the mechanism of micelle kinetics have to be taken into account in a reasonable physical model.

To account for the micelle effect, specific parameters of the respective surfactant micelles have to be known. Numerous papers on the determination of aggregation numbers and rate constants of micelle kinetics of many surfactants have been published (for example Aniansson et al. 1976, Hoffmann et al. 1976, Kahlweit & Teubner 1980). Different micelle kinetics mechanisms exist, for example that summarised by Zana (1974). Three of these mechanisms are demonstrated in Fig. 4.12.

The formation-dissolution mechanism assumes a total dissolution of a micelle in order to re-establish the local equilibrium monomer concentration. This model is based on an idealised distribution of only monomers and micelles with a definite aggregation number. Mechanism 2 is based on the existence of micelles of different size and therefore, a broad micelle size distribution has to be assumed. Model 3 is the most probable but the most complex one. It allows the aggregation or dissolution of individual monomers to or from the micelles. If a micellar size distribution is assumed, as expected from experiments, a large number of formation and dissolution rate constants, for each micellar size, are necessary.

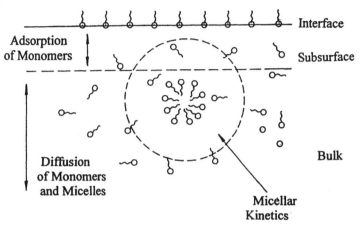

Fig. 4.11   Schematic of an adsorption process from micellar solutions

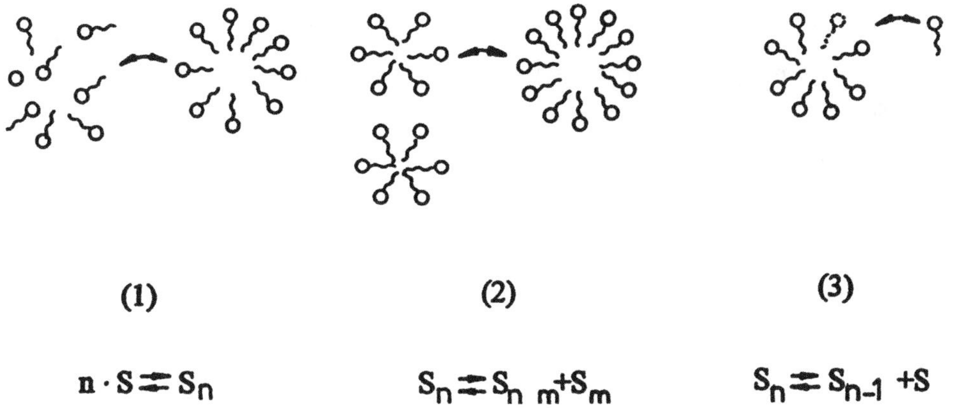

(1)   **(2)**   **(3)**

$$n \cdot S \rightleftharpoons S_n \qquad S_n \rightleftharpoons S_{n\,m} + S_m \qquad S_n \rightleftharpoons S_{n-1} + S$$

Fig. 4.12   Micelle kinetics mechanisms; 1- formation-dissolution, 2 - rearrangement, 3 - stepwise aggregation-disintegration

The physical model, based on the micelle kinetics mechanism 1, has the following form. The transport of monomers is given by,

$$D_1 \frac{\partial^2 c_1}{\partial x^2} = \frac{\partial c_1}{\partial t} + q_1 \text{ at } x > 0, t > 0, \tag{4.66}$$

the diffusion of the micelles, having the concentration c_n, is described by,

$$D_n \frac{\partial^2 c_n}{\partial x^2} = \frac{\partial c_n}{\partial t} + q_n \text{ at } x < 0, t > 0. \tag{4.67}$$

The terms q_1 and q_n are source terms, determined by the mechanism of micellar kinetics. If a mechanism of the formation-dissolution type is assumed (McQueen & Hermans 1972, Lucassen 1976), these terms read,

$$q_1 = -nk_f c_1^n + nk_d c_n, \tag{4.68}$$

$$q_n = k_f c_1^n - k_d c_n \tag{4.69}$$

For the monomers, the boundary condition at the interface is identical to Eq. (4.11), the micelles are assumed not to adsorb,

$$\frac{\partial c_n}{\partial x} = 0, t > 0, x = 0. \tag{4.70}$$

For small disturbances of the adsorption layer from equilibrium, Lucassen (1976) derived an analytical solution (cf. Section 6.1.1). An analysis of the effect of a micellar kinetics mechanism of stepwise aggregation-disintegration and the role of polydispersity of micelles was made by Dushkin & Ivanov (1991) and Dushkin et al. (1991). Although it results in analytical expressions, it is based on some restricting linearisations, for example with respect to adsorption isotherm, and therefore, it is valid only for states close to equilibrium.

A slow relaxation of micelles was described by Fainerman (1981), Fainerman & Makievski (1992a, b), and Fainerman et al. (1993d), using a linear source term and neglecting the transport of micelles,

$$q_1 = k(c_{cmc} - c); q_n = 0, \tag{4.71}$$

where c is the total surfactant concentration and c_{cmc} is the critical micelle concentration. k is the relaxation constant of the micellar kinetics, but it cannot be related easily to the mechanisms discussed above. As a result Fainerman et al. (1993d) obtained a relation for this rate constant, which can be applied to experimental data,

$$k = \frac{4}{\pi} \left(\frac{(d\gamma / dt^{-1/2})_{c=c_{cmc}}}{(d\gamma / dt^{-1})_{c>c_{cmc}}} \right)_{t \to \infty}. \tag{4.72}$$

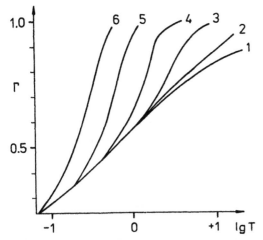

Fig. 4.13 Effect of the formation dissolution rate constant k_f on the adsorption kinetics of a micellar solution; $D_n / D_1 = 1$, $c_{on} / c_{ol} = 10$, n=20, $nk_f c_{ol}^{n-3} \Gamma_o^2 / D_1$ =0 (1), 0.2(2), 2.0 (3), 20 (4), 200 (5), 2000 (6)

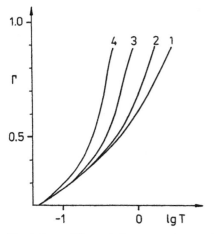

Fig. 4.14 Effect of the aggregation number n on the adsorption kinetics of a micellar solution; $D_n / D_1 = 1$, c_{on} / c_{ol} =10, $k_d \dfrac{\Gamma_{ol}^2}{c_{ol}^2 D}$ = 1; n = 2 (1), 5 (2), 20 (3), 100 (4)

A numerical solution, based on the model presented for a formation-dissolution mechanism, was derived by Miller (1981). The following two Figs 4.13 and 4.14 demonstrate the effect of micelles on adsorption kinetics. The effect of the rate of formation and dissolution of micelles, represented by the dimensionless coefficient $nk_f c_{ol}^{n-3} \Gamma_o^2 / D_1$, becomes remarkable for a value larger than 0.1. Under the given conditions (D_n / D_1 =1, c_{on} / c_{ol} =10, n=20) the fast micelle kinetics accelerates the adsorption kinetics by one order of magnitude.

The aggregation number also plays an important role in the total rate of the adsorption process from micellar solutions. The presence of dimers, which are assumed not to adsorb, increases the adsorption rate remarkably, although only 10% of the surfactant is aggregated in dimers (Fig. 4.14).

The ratio of diffusion coefficients D_n / D_1, and the part of surfactant molecules aggregated in micelles, represented by c_{on} / c_{o1}, can also change the rate of adsorption by orders of magnitude (Miller 1991).

4.8.   ADSORPTION PROCESS AT THE SURFACE OF LAMINAR FLOWING LIQUID FILMS

In coating processes the problem of controlling the flow of liquids down an inclined plate is a key question (Scriven 1960, Kretzschmar 1974). Therefore, the hydrodynamic flow of such films in combination with surface rheological and adsorption kinetics models were described. As the principle of a flowing film can be used also as a separate method to study adsorption processes in the range of milliseconds, the theory is presented here, while the experimental details are given in the next chapter.

The situation of a liquid film flowing down a plate is represented in Fig. 4.15.

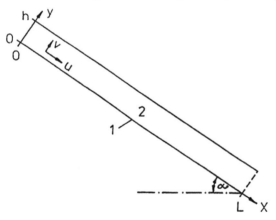

Fig. 4.15   Flow of a liquid film down an inclined plate - definition sketch; 1 plate, 2 film

The flow velocities u and v are directed along and normal to the flow direction, x, y and z are the corresponding spatial coordinates, α is the angle of inclination, and h is the thickness of the liquid film. The hydrodynamic problem may be simplified by neglecting the entrance region at x=0 (Van den Bogaert & Joos 1979),

$$\frac{dh}{dx} \ll 1, v \ll u, u \frac{\partial u}{\partial x} \ll g, \frac{\partial^2 u}{\partial x^2} \ll \frac{\partial^2 u}{\partial y^2}. \tag{4.73}$$

Thus, the Navier-Stokes equation reduces to

$$\eta \frac{d^2 u}{dy^2} + \rho g \sin \alpha = 0, \qquad\qquad 0 \le y \le h. \tag{4.74}$$

g, ρ, η and γ are the acceleration due to gravity, density, viscosity and surface tension of the liquid. Neglecting surface shear and dilational viscosity the surface momentum equation is given by (Scriven 1960),

$$\eta \frac{du}{dy} = \frac{d\gamma}{dx}, \qquad\qquad 0 \le x \le L, y = h. \tag{4.75}$$

The integration of the flow problem yields (van den Bogaert & Joos 1979)

$$u = -\frac{\rho g \sin \alpha}{2\eta} y^2 + \left(\frac{1}{\eta}\frac{d\gamma}{dx} + \frac{\rho g h}{\eta} \sin \alpha\right) y, \qquad 0 \le x \le L, 0 \le y \le h. \tag{4.76}$$

The description of the adsorption kinetics process at the surface of the liquid film, freshly formed at the inlet has to take into consideration this flow regime, given by Eq. (4.76). The transport equations in the bulk and at the surface y=h read,

$$\frac{\partial c}{\partial t} + u \frac{\partial c}{\partial x} = D \frac{\partial^2 c}{\partial x^2}, \qquad\qquad 0 < x < L, 0 < y < h, \tag{4.77}$$

$$\frac{\partial \Gamma}{\partial t} + \frac{\partial(u(x, y = 0) \Gamma)}{\partial x} = D_s \frac{\partial^2 \Gamma}{\partial x^2} - \frac{\partial c}{\partial y}, \qquad 0 < x < L, y = h. \tag{4.78}$$

A numerical solution, based on a linear adsorption isotherm, was performed by Ziller et al. (1985) using a finite difference technique. An example of the numeric results is given in Fig. 4.16. The dramatic drop of the surface concentration at x=L/K is explained by the boundary condition, which assumed a 10-fold expansion of the interface. Due to the complexity of the problem, it seems impossible to derive an analytical solution without further serious simplifications.

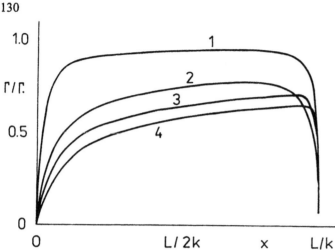

Fig. 4.16 Surface concentration along the surface of a flowing film; $k = \Gamma_o / c_o$, $u(x,0) = u_s = 5$ cm s[-1] (1,2), 15 cm s[-1] (3), 25 cm s[-1] (4); $D / k^2 = 1000$ s[-1] (1), 10 s[-1] (2,3,4); according to Ziller et al. (1985)

4.9. ADSORPTION KINETICS OF POLYMERS

There are many theories which describe the adsorption of polymers in general and proteins in particular. For model polymers, linear, flexible polymers with neutral chains in good solvents de Gennes (1987) gave an overview about the structure and dynamics of their adsorption. Using general scaling concepts the structure of polymer adsorption layers was discussed qualitatively (de Gennes 1980, 1981, 1982, 1985a, b). In contrast, the theory developed by Scheutjens & Fleer (1979, 1980, 1986, Fleer et al. 1982) is based on numerical procedures taking into account many details of polymer adsorption given in the literature. The disadvantage of the latter theory is that it exists only as a computer programme, and therefore, it is not possible to use the results in a case where a relation between bulk polymer concentration and amount adsorbed at the interface is unknown.

It seems impossible that general theories can be developed to describe polymer adsorption in an universal way. There are molecules which can unfold after adsorption at the interface while others remain in a compact structure (Fig. 4.17).

In the literature, attempts were made to find analytical expressions for equations of state to describe the adsorption of polymers at interfaces. At the same time, such relations can be used as the basis for the theoretical description of dynamic interfacial processes. Graham & Phillips (1979, 1980) used the well-known square-root-approximation for the kinetics of protein adsorption

$$\Gamma(t) = 2c_0 \sqrt{\frac{Dt}{\pi}}, \tag{4.79}$$

which is a first approximation of Eq. (4.1) at short adsorption times.

(a)   (b)

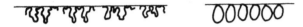

Fig. 4.17   Adsorption of macromolecules at liquid interfaces; a - unfolding molecules, b - molecules with compact structure

A very simple relation for the relaxation process caused by harmonic disturbances is also given by Graham & Phillips. This can be combined with any explicit adsorption equations. MacRitchie (1963, 1986, 1989, 1991) and Damodaran & Song (1988) also used this approximation and additionally some linear reaction equations taking into consideration the area ΔA needed for a molecule to adsorb were derived,

$$\frac{d\Gamma}{dt} = K_a c_0 \exp(-\frac{\Pi \Delta A}{RT}) - K_d \Gamma(t) \exp(-\frac{\Pi \Delta A}{RT}), \tag{4.80}$$

where Π is the surface pressure, defined by $\Pi(t) = \gamma_0 - \gamma(t)$. This equation is equivalent to the rate equations given by Eqs (4.32)-(4.34).

Another interesting but very complex model was discussed by Douillard & Lefebvre (1990). Their final linear differential equation system is able to describe many experimental data, but it contains eight independent rate constants and it is very time consuming obtaining such a large number of parameters from adsorption kinetics experiments. A more suitable kinetic relation was proposed by de Feijter et al. (1987) which can be easily applied to experimental adsorption data. Other adsorption kinetics and relaxation models for polymer solutions based on diffusion transport and kinetic equations are presented by Miller (1991).

4.10. ASYMPTOTIC SOLUTIONS

Due to complexity the application of the general Ward & Tordai equation (4.1) is usually connected with large numerical efforts. Therefore, many attempts have been made to find approximate solutions. Recently Fainerman et al. (1994a) and Miller et al. (1994b) published several approximate solutions useful for easy data interpretation. A first approximation was presented by Sutherland (1952) in the form of Eq. (4.22) using a linear adsorption isotherm. A further approximation for small surface coverage and small and large adsorption times was given by Hansen (1960), expressed as a function of the dimensionless time $\Theta = Dt / (\Gamma_0 / c_0)^2$,

$$\frac{\Gamma(t)}{\Gamma_0} \approx 2\sqrt{\frac{\Theta}{\pi}}\left(1 - \frac{\sqrt{\pi\Theta}}{2} \pm \ldots\right) \text{ for } \sqrt{\Theta} \leq 0.2 \tag{4.81}$$

$$\frac{\Gamma(t)}{\Gamma_0} \approx 1 - \frac{1}{\sqrt{\pi\Theta}}\left(1 - \frac{1}{2\Theta} \pm \ldots\right) \text{ for } \sqrt{\Theta} \geq 5.0 \tag{4.82}$$

Rillaerts & Joos (1982) derived an approximate solution at long time t by assuming that c(0,t) is almost constant and can be placed outside the integral of Eq. (4.1), which leads to

$$\Gamma(t) = 2\sqrt{\frac{D}{\pi}}(c_0\sqrt{t} - c(0,t)\sqrt{t}). \tag{4.83}$$

Replacing c(0,t) by $\frac{dc}{d\Gamma}\Gamma(t)$ and, after some rearrangements, a long time approximation is obtained

$$\Gamma(t) = \frac{2c_0\sqrt{\dfrac{Dt}{\pi}}}{1 + \dfrac{dc}{d\Gamma}\sqrt{\dfrac{4Dt}{\pi}}}. \tag{4.84}$$

The most simple and often used short time approximation is

$$\Gamma = 2c_0\sqrt{\frac{Dt}{\pi}}, \tag{4.85}$$

obtained from Eq. (4.1) by neglecting the integral. Using the linear relation between γ and Γ,

$$\gamma = \gamma_0 - nRT\Gamma, \tag{4.86}$$

the interfacial tension of a surfactant solution at $t \to 0$ results to

$$\gamma_{t\to0} = \gamma_o - 2nRTc_o\sqrt{\frac{Dt}{\pi}} \qquad (4.87)$$

where γ_o is the surface tension of the solvent and n is 1 for non-ionic and 2 for ionic surfactants, respectively. The derivative of Eq. (4.87) with respect to $t^{1/2}$ yields:

$$\left(\frac{d\gamma}{dt^{1/2}}\right)_{t\to0} = -2nRTc_o\sqrt{\frac{D}{\pi}}. \qquad (4.88)$$

Thus, for the diffusion-controlled adsorption on the non-deforming surface, experimental values of $\gamma(t^{1/2})$ must give a straight line having the slope equal to the expression on the right hand side of Eq. (4.88).

For a multi-component solution and the same diffusion-controlled mechanism as $t\to0$ we obtain the total adsorption,

$$\Gamma_T = \sum_{i=1}^{r}\Gamma_i = \sum_{i=1}^{r}2c_{oi}\sqrt{\frac{D_i t}{\pi}} \qquad (4.89)$$

and finally

$$\left(\frac{d\gamma}{dt^{1/2}}\right)_{t\to0} = -2RT\sum_{i=1}^{r}n_i c_{oi}\sqrt{\frac{D_i}{\pi}}. \qquad (4.90)$$

From Eqs (4.89) and (4.90) it follows that surface-active contaminants present in the solution of the main surfactant, do not influence the dependence of $\gamma(t)$ at $t\to0$.

If the adsorption mechanism is not diffusion-controlled but by the kinetic equation of the Langmuir type (4.31) at short time t one gets,

$$\Gamma = k_{ad}\, t\, c_o. \qquad (4.91)$$

Thus, together with Eq. (4.86) we obtain

$$\gamma = \gamma_o - nRTk_{ad}c_o t \qquad (4.92)$$

and $\left(\frac{d\gamma}{dt}\right)_{t\to0}$ should give a straight line if the mechanism holds. In a recent paper Fainerman et al. (1994a) summarised additional short and long time approximations using diffusion-controlled, barrier-controlled and mixed kinetic models.

For radially growing drops MacLeod & Radke (1994) derived a long time approximation for a drop growing into air,

$$\Gamma(t) = \Gamma(0) + 2\sqrt{\frac{3Dt}{7\pi}}(c_o - c(0,t)),$$

(4.93)

or into a second liquid,

$$\Gamma(t) = \Gamma(0) + 2\left(\sqrt{\frac{3D_\alpha t}{7\pi}}(c_{\alpha o} - c_\alpha(0,t)) + \sqrt{\frac{3D_\beta t}{7\pi}}(c_{\beta o} - c_\beta(0,t))\right).$$

(4.94)

Using an appropriate equation of state, for example a Langmuir-Szyszkowski equation,

$$\gamma = \gamma_o + RT\Gamma_\infty \log(1 - \frac{\Gamma(t)}{\Gamma_\infty}),$$

(4.95)

the change of surface tension with time at states close to equilibrium can be determined.

4.11. SUMMARY

Qualitative and quantitative models of adsorption kinetics of surfactants and polymers are described in this chapter. A comprehensive presentation of the most developed physical model, the diffusion-controlled adsorption and the desorption model, is given and different methods of solving the resulting differential equations are discussed (Miller & Kretzschmar 1991). A direct numerical integration enables us to consider any type of adsorption isotherm relating the surfactant bulk concentration with the adsorbed amount at the interface.

Besides the diffusion-controlled adsorption theory other mechanisms are presented, comprising transport processes in the bulk solution and transfer mechanisms from the subsurface to the interface (Chang & Franses 1992).

The consideration of time dependent interfacial area is of importance for many experimental investigations. Various models are described The linearised theories for small area changes, which belong to relaxation theories and are outlined in Chapter 6. The flow in the liquid layer adjacent to a time variant surface area is also incorporated into the theoretical models. Systems such as liquid films flowing down an inclined plane or growing bubbles and drops are described quantitatively, assuming laminar and radial flow patterns (MacLeod & Radke 1993).

In the discussion of the adsorption kinetics of micellar solutions, different micelle kinetics mechanisms are taken into account, such as formation/dissolution or stepwise aggregation/disaggregation (Dushkin & Ivanov 1991). It is clear that the presence of micelles in the solution influences the adsorption rate remarkably. Under certain conditions, the aggregation number, micelle concentration, and the rate constant of micelle kinetics become the rate controlling parameters of the whole adsorption process. Models, which consider solubilisation effects in surfactant systems, do not yet exist.

The kinetics of adsorption from solutions of surfactant mixtures are described on the basis of a generalised Langmuir isotherm. The simultaneous adsorption leads to the replacement of less surface active compounds by those of higher surface activity, which are usually present in the bulk at much lower concentration. More general descriptions of the process are possible on the basis of the Frumkin and Frumkin-Damaskin isotherms, which include specific interfacial properties of the individual surface active species. Quantitative studies of such very complex models can be performed only numerically.

The adsorption processes of polymers at liquid interfaces have been modelled, based on very simple assumptions. More sophisticated models contain a large number of independent system parameters, which are rarely available (Douillard & Lefebvre 1990). A quantitative description of polymer adsorption which takes into account peculiarities of the adsorbing molecules has still to be developed. It is not possible yet, to describe the polymer adsorption process on the basis of the theory of Scheutjens & Fleer (1979, 1980), which does not apply under dynamic conditions.

At the end of the chapter, a number of approximate and asymptotic formulae for different adsorption conditions are given (cf. Fainerman et al. 1994a). The asymptotic relations are especially valuable in the determination of adsorption mechanisms and characteristic adsorption parameters.

As discussed in the following sections, even soluble surfactants can form inhomogeneous adsorption layers. The origin of such inhomogeneities is not fully understood. It could be possible that it is a feature of mixed adsorption layers, formed by commercial surfactants of low purity or by mixtures of particular composition. Such domain-like structures could be formed due to remarkable differences in the interaction strength within the adsorption layer. The formation of 2- and 3-dimensional aggregates at interfaces is known from studies of insoluble monolayers (Möhwald et al. 1986 and Barraud et al. 1988, Vollhardt 1993). First dynamic models of aggregate formation in insoluble monolayers were theoretically described by Retter & Vollhardt (1993) and Vollhardt et al. (1993). These simple and new, more sophisticated theories combined with bulk transport processes give the first physical models to treat inhomogenous soluble adsorption layers, as observed by Berge et al. (1991), Flesselles et al. (1991) or Hénon & Meunier (1992). Many effects of even well known systems like fatty acid and alcohol monolayers are still to be studied, for example the effect of spreading solvent investigated by Gericke & Hühnerfuss (1993) and Gericke et al. (1993a, b) using the infrared reflection - absorption spectroscopy.

A new insight into the dynamic processes in the bulk and at the surface of surfactant solutions can be seen in molecular dynamics simulations. Only now are computers sufficiently powerful that such simulations can be performed without too many simplifications. The state of the art of molecular dynamics was recently summarised by van Os & Karaborni (1993), showing that complex processes such as micelle formation (Karaborni & O'Connell 1993), emulsion formation or solubilisation processes (Smit et al. 1993) can be simulated. Future improvements of computers and algorithms will provide a deep insight into even more complex processes connected with dynamics of interfacial phenomena, such as adsorption layer structure and formation, effects of molecular interfacial and bulk interactions in mixed systems of surfactants and polymers.

4.12. REFERENCES

Adamczyk, Z., J. Colloid Interface Sci., 120(1987)477

Aniansson, E.A.G., Wall, S.N., Almgren, M., Hoffmann, H., Kielmann, I., Ulbricht, W., Zana, R., Lang, J. and Tondre, C., J. Phys. Chem., 80(1976)905

Barraud, A., Flörsheimer, M., Möhwald, H., Richard, J., Ruaudel-Texier, A., and Vandervyver, M., J.Colloid Interface Sci., 121(1988)491

Baret, J.F., J. Phys. Chem., 72(1968a)2755

Baret, J.F., J. Chim. Phys. , 65(1968b)895

Baret, J.F., J Colloid Interface Sci., 30(1969)1

Berge, B., Faucheux, L., Schwab, K., and Libchaber, A., Nature (London), 350(1991)322

Blair, C.M., J. Chem. Phys., 16(1948)113

Borwankar, R.P. and Wasan, D.T., Chem. Eng. Sci., 41(1986)199

Bronstein, I.N. and Semendjajew, K.A., Taschenbuch der Mathematik, B.G. Teubner Verlagsgesellschaft Leipzig, 1968

Chang, C.H. and Franses, E.I., Colloids & Surfaces, 69(1992)189

Chang, C.H., Wang, N.-H.L. and Franses, E.I., Colloids & Surfaces, 62(1992)321

Cody, W.J., Math. Comp., 23(1968)631

Davies, J.T., Smith, J.A.C. and Humphreys, D.G., Proc. Int. Conf. Surf. Act. Subst., 2(1957)281

de Feijter, J. , Benjamins, J. and Tamboer M., Colloids & Surfaces, 27(1987)243 de Gennes, P.-G., Adv. Colloid Interface Sci., 27(1987)189

de Gennes, P.-G., Macromolecules, 13(1980)1069

de Gennes, P.-G., Macromolecules, 14(1981)1637

de Gennes, P.-G., Macromolecules, 15(1982)492

de Gennes, P.-G., C.R. Acad. Sc. Paris, 301-II(1985a)1399

de Gennes, P.-G., Scaling Concepts in Polymer Physics, Cornell University Press, 1985b

Damodaran,S. and Song, K.B., Biochim. Biophys. Acta, 954(1988)253-264

Delahay, P. and Trachtenberg, I., J. Amer. Chem. Soc., 79(1957)2355

Delahay, P. and Fike, C.T., J. Amer. Chem. Soc., 80(1958)2628

Doss, K.S.G., Koll. Z., 86(1939)205

Douillard, R. and J.Lefebvre, J.Colloid Interface Sci., 139(1990)488

Dushkin, C.D. and Ivanov, I.B., Colloids & Surfaces, 60(1991)213

Dushkin, C.D., Ivanov, I.B. and Kralchevsky, P.A., Colloids & Surfaces, 60(1991)235

Fainerman, V.B., Koll. Zh. 43(1981)94

Fainerman, V.B., Zh. Fiz. Khim., 42(1983)457

Fainerman, V.B., Rakita, Yu.M. and Zadara, V.M., Koll. Zh., 49(1987)80

Fainerman, V.B., Koll. Zh. 52(1990)921

Fainerman, V.B., Khodos, S.R., and Pomazova, L.N., Kolloidny Zh. (Russ), 53(1991)702

Fainerman, V.B., Colloids & Surfaces, 62 (1992) 333

Fainerman, V.B., Makievski, A.V. , Koll. Zh., 54(1992a)75

Fainerman, V.B., Makievski, A.V. , Koll. Zh., 54(1992b)84

Fainerman, V.B., Makievski, A.V. and Joos, P., Zh. Fiz. Khim., 67(1993a)452

Fainerman, V.B., Makievski, A.V. and Miller, R., Colloids & Surfaces A,   75(1993b)229

Fainerman, V.B., Makievski, A.V. and Joos, P., Zh.Fiz.Khim. 67(1993c)452

Fainerman, V.B., Makievski, A.V. and Joos, P., Zh.Fiz.Khim. 67(1993d)452

Fainerman, V.B., Makievski, A.V. and Miller, R., Colloids & Surfaces A, 87(1994a)61

Fainerman, V.B., Miller, R. and Joos, P., Colloids Polymer Sci., 272(1994b)731

Fleer; G.J. and Scheutjens; J.M.H.M., Adv. Colloid Interface Sci., 16(1982)341

Flesselles, J.M., Magnasco, M.O., and Libchaber, A., Phys.Rev.Lett., 67(1991)2489

Gericke, A. and Hühnerfuss, H., J. Phys. Chem., 97(1993)12899

Gericke, A., Simon-Kutscher, J. and Hühnerfuss, H., Langmuir, 9(1993a)2119

Gericke, A., Simon-Kutscher, J. and Hühnerfuss, H., Langmuir, 9(1993b)3115

Graham, D.E. and Phillips, M.C., J. Colloid Interface Sci., 70(1979)403

Graham, D.E. and Phillips, M.C., J. Colloid Interface Sci., 76(1980)227

Haase, R., in "Grundzüge der physikalischen Chemie, Band II - Transportvorgänge",
    Dr. Dietrich Steinkopff Verlag, Darmstadt 1973

Hansen, R.S. and Wallace, T., J. Phys. Chem., 63(1959)1085.

Hansen, R.S., J. Phys. Chem., 64(1960)637

Hansen, R.S., J. Colloid Sci., 16(1961)549

Hénon,S. and Meunier,J., Thin Solid Films, 210/211(1992)121

Hoffmann, H., Nagel, R., Platz, G. and Ulbricht, W., Colloid Polymer Sci., 254(1976)812

138

Hoffmann, H., Progr. Colloid Polymer Sci., 83(1990)16

Horozov, T., Danov, K., Kralschewsky, P., Ivanov, I. and Borwankar, R., 1st World
        Congress on Emulsion, Paris, Vol. 2, (1993)3-20-137

Ilkovic, D., Collec. Czechoslow. Commun., 6(1934)498

Ilkovic, D., J. Chim. Phys. Physicochem. Biol., 35(1938)129

Jiang, Q. Chiew, Y.C. and Valentini, J.E., Langmuir, 9(1993)273

Joos, P. and Rillaerts, E., J. Colloid Interface Sci. 79(1981)96

Joos, P., Fang, J.P. and Serrien, G., J. Colloid Interface Sci. 151, 144 (1992)

Joos, P. and Van Uffelen, M., J. Colloid Interface Sci., 155(1993a)271

Joos, P. and Van Uffelen, M., Colloids & Surfaces, 75(1993b)273

Kahlweit, M. and Teubner, M., Adv. Colloid Interface Sci., 13(1980)1

Karaborni, S. and O'Connell, J.P., Tenside Detergents, 30(1993)235

Kloubek, J., J. Colloid Interface Sci., 41(1972)1

Kiesewetter, H. and Maeß, G., Elementare Methoden der numerischen Mathematik, Akademie-
        Verlag, Berlin, 1974

Krawczyk, M.A., Wasan, D.T., and Shetty, C.S., I&EC Research 30(1991)367

Kretzschmar, G., Z. Chemie, 14(1974)261

Kretzschmar, G. and Miller, R., Adv. Colloid Interface Sci. 36(1991)65

Levich, V.G., Physicochemical Hydrodynamics, Prentice-Hall, Englewood Cliffs, New York,
        1962

Liggieri, L., Ravera, F., Ricci, E. and Passerone,A., Adv. Space Res. 11(1991)759

Lin, S.-Y., McKeigue, K. and Maldarelli, C., Langmuir, 7(1991)1055

Lucassen, J., J. Chem. Soc. Faraday I, 72(1976)76

Lucassen-Reynders, E.H. and Kuijpers, K.A., Colloids & Surfaces, 65(1992)175

MacLeod, C.A. and Radke, C.J., J. Colloid Interface Sci., 160(1993)435

MacLeod, C.A. and Radke, C.J., J. Colloid Interface Sci., 166(1994)73

MacRitchie, F. and Alexander, A.E., J. Colloid Sci., 18(1963)453

MacRitchie, F., Adv. Colloid Interface Sci., 25(1986)341

MacRitchie, F., Colloids & Surfaces, 41(1989)25

MacRitchie, F., Analytica Chimica Acta., 249(1991)241

Makievski, A.V., Fainerman, V.B. and Joos, P., J. Colloid Interface Sci., 166(1994)6

Malysa, K., Adv. Colloid Interface Sci., 40(1992)37

McCoy, B.J., Colloid Polymer Sci., 261(1983)535

McQueen, D.H. and Hermans, J.J., J. Colloid Interface Sci., 39(1972)289

Miller, R., Colloid Polymer Sci., 258(1980)179

Miller, R., Colloid Polymer Sci., 259(1981)375

Miller, R., Colloid Polymer Sci., 261(1983)441

Miller, R., Colloids & Surfaces, 46(1990)75

Miller, R. and Kretzschmar, G., Colloid Polymer Sci., 258(1980)85

Miller, R., in "Trends in Polymer Science", 2(1991)47

Miller, R. and Kretzschmar, G., Adv. Colloid Interface Sci. 37(1991)97

Miller, R. and Lunkenheimer, K., Z. phys. Chem., 259(1978)863

Miller, R. and Lunkenheimer, K., Colloid Polymer Sci., 264(1986)357

Miller,R. and Schano, K.-H., Colloid Polymer Sci. 264(1986)277

Miller, R. and Schano, K.-H., Tenside Detergents 27(1990)238

Miller, R., Loglio, G.; Tesei, U., and Schano, K.-H., Adv. Colloid Interface Sci., 37(1991)73

Miller, R., Lunkenheimer, K. and Kretzschmar, G., Colloid Polymer Sci., 257(1979)1118

Miller, R., Sedev., R., Schano, K.-H., Ng, C. and Neumann, A.W., Colloids & Surfaces,
        69 (1993) 209

Miller, R., Kretzschmar, G. and Dukhin, S.S., Colloid Polymer Sci., 272(1994a)548

Miller, R., Joos, P. and Fainerman, V.B., Adv. Colloid Interface Sci., 49(1994b)249

Miller, R., Fainerman, V.B., Schano, K.-H., Heyer, Wolf, Hofmann, A. and Hartmann, R.,
Labor Praxis, (1994c)56

Milner, S.R., Phil. Mag., 13(1907)96

Möhwald, H., Miller, A., Stich, W., Knoll, W., Ruaudel-Texier, A., Lehmann, T., and
        Fuhrhop, J.-H., Thin Solid Films, 141(1986)261

Nagarajan, R. and Wasan, D.T., J. Colloid Interface Sci., 159(1993)164

Oberhettinger, F. and Bardii, L., Tables of Laplace Transforms, Springer-Verlag, Berlin, 1973

Panaiotov, I. and Petrov, J.G., Ann. Univ. Sofia, Fac. Chem., 64(1968/69)385

Passerone,A., Liggieri, L. Rando, N., Ravera, F. and Ricci, E., J. Colloid Interface Sci.,
        146(1991)152

Petrov, J.G. and Miller, R., Colloid Polymer Sci., 255(1977)669

Pierson, F.W. and Whittaker, S., J. Colloid Interface Sci., 52(1976)203

Rakita, Yu.M. and Fainerman, V.B., Koll. Zh., 51(1989)714

Ravera, F., Liggieri, L. and Steinchen, A., J. Colloid Interface Sci., 156(1993)109

Ravera, F., Liggieri, L., Passerone, A. and Steinchen, A., J. Colloid Interface Sci., 163(1994)

Retter, U. and Vollhardt, D., Langmuir, 9(1993)2478

Rillaerts, E. and Joos, P., J. Phys. Chem., 86(1982)3471

Ross, S., Amer. Chem. Soc., 67(1945)990

Scheutjens, J.M.H.M. and Fleer, G.J., J. Phys. Chem., 83(1979)1619

Scheutjens, J.M.H.M. and Fleer, G.J., J. Phys. Chem.,84(1980)178

Scheutjens, J.M.H.M., Fleer, G.J. and Cohen Stuart, M.A., Colloids & Surfaces, 21(1986)285

Scriven, L.E., Chem. Eng. Sic., 12(1960)98

Serrien, G., Geeraerts, G., Ghosh, L., and Joos, P., Colloids & Surfaces, 68(1992)219

Smit, B., Hilbers, P.A.J., and Esselink, K., Tenside Detergents, 30(1993)287

Sutherland, K.L., Austr. J. Sci. Res., A5(1952)683

van den Bogaert, P. and Joos, P., J. Phys. Chem., 83(1979)2244

van Os, N.M. and Karaborni, S., Tenside Detergents, 30(1993)234

van Uffelen, M. and Joos, P., Colloids & Surfaces, 158(1993c)452

van Voorst Vader, F., Erkelens, Th.F. and Van den Tempel, M., Trans. Faraday Soc.,
        60(1964)1170

Vollhardt, D., Adv.Colloid Interface Sci., 47(1993)1

Vollhardt, D., Ziller, M., and Retter, U., Langmuir, 9(1993)3208

Ward, A.F.H. and L.Tordai, J. Phys. Chem. 14(1946)453

Zana, R., Chemical & Biol. Appl. Relaxation Spectroscopy, Proc. NATO Adv. Study Inst.,
        Ser. C, 18(1974)133

Ziller, M., Miller, R. and Kretzschmar, G., Z. phys. Chemie (Leipzig), 266(1985)721

Ziller, M. and Miller, R., Colloid Polymer Sci. 264(1986)611

CHAPTER 5

5. EXPERIMENTAL TECHNIQUE TO STUDY ADSORPTION KINETICS

The main objective of this chapter is to discuss different techniques for studying adsorption processes at liquid interfaces. These studies finally aim to determine the adsorption mechanism at liquid interfaces of surfactants and surfactants mixtures as well as of polymers and mixtures with other polymers or surfactants (cf. Kretzschmar & Miller 1991). Several experimental methods are described, either frequently used or recently developed, for studying adsorption kinetics: oscillating jet, drop volume, maximum bubble pressure, axisymmetric drop shape analysis, growing drop tensiometry, and some other methods. The theoretical background of these methods has been described in Chapter 4. The Table 5.1 give an overview of some characteristics of dynamic surface and interfacial tension methods described in this chapter. An overview of these and other dynamic methods to study adsorption kinetics was given by Miller et al. (1994b).

The drop methods, drop volume, drop shape and drop pressure, seem to be the most general ones, and they can usually be run, after small modification, with bubbles. These methods are applicable to both liquid-liquid and liquid-gas interfaces and need only small amounts of solvent and solute. In addition, the temperature control is easily arranged, even at higher temperatures.

For dynamic adsorption studies several methods are suitable. The selection of a certain method depends on the experimental conditions, temperature and time interval, of the investigations to be performed. These experimental conditions are governed by the surface activity of the surfactant or polymer under study and its concentration. Some characteristics of dynamic surface tension methods are summarised in Table 5.2.

The accuracy of all methods, over time intervals from several seconds to hours, is usually $\pm\ 0.1\ mN/m$. Special instruments enable measurements in the millisecond time scale. Such studies are performed with less accuracy, sometimes of the order of $\pm\ 1.0\ mN/m$.

Before describing the experimental set-ups and interpreting the data on the basis of actual theories, the main prerequisite of interfacial studies is discussed: the quality of the object under investigation. The most powerful technique is unable to yield reasonable results, if the necessary grade of purity of the surfactants and the solvents are not established. Both the effect of impurities and criteria to judge the purity of surfactant and solvent systems with respect to their purity will be described in this chapter.

Table 5.1   Characteristics of dynamic surface and interfacial tension methods

Method	Suitability for liqu./liqu.	Suitability for liqu./gas	Problems
Capillary rise	possible	good	no commercial set-up
Capillary waves	possible	possible	large efforts, no commercial set-up
Drop volume	good	good	hydrodynamic effects
Growing drops and bubbles	good	good	no commercial set-up
Inclined plate	bad	good	small time interval, no commercial set-up
Maximum Bubble pressure	possible	good	data interpretation
Oscillating jet	bad	good	small time interval, no commercial set-up
Pendent drop	good	good	no commercial set-up
Plate tensiometer	possible	good	contact angle
Ring tensiometer	bad	good	wetting
Sessile drop	possible	possible	no commercial set-up, low accuracy
Spinning drop	good	possible	small range of application
Static drop volume	good	good	time consuming

142

Table 5.2   Characteristics of dynamic surface and interfacial tension methods

Method	Time range	Temperature range	Suitability l/g	Suitability l/l
Drop volume	1 s - 20 min	10 - 90 °C	yes	yes
Maximum bubble pressure	1 ms - 100 s	10 - 90 °C	yes	no
Pendent drop	10 s - 24 h	20 -25 °C	yes	yes
Ring tensiometer	30 s - 24 h	20 - 25 °C	yes	no
Plate tensiometer	10 s - 24 h	20 - 25 °C	yes	no
Growing drops and bubbles	0.01s - 600 s	10 - 90 °C	yes	yes
Drop relaxations	1 s - 300 s	10 - 90 °C	yes	yes
Elastic ring	10 s - 24 h	20 - 25 °C	yes	no
Pulsating bubble	0.005 s - 0.2 s	20 - 25 °C	yes	no
Oscillating jet	0.001s - 0.01s	20 - 25 °C	yes	no

5.1.   CHARACTERISATION OF THE PURITY OF SURFACTANTS AND SOLVENTS

Overviews of attempts to quantify the effect of impurities and to derive criteria for the purity of a liquid system for interfacial studies was given by Lunkenheimer (1984) and Miller (1987). It was shown that the absence of a minimum in a surface tension isotherm, γ as a function of $\log(c_o)$ is the criterion most frequently used to judge the purity of a surfactant. On the other hand, it had been shown by Krotov & Rusanov (1971) that this is not sufficient. Under certain conditions such a minimum can disappear by some compensating effects as well as by addition of electrolyte (Weiner & Flynn 1974). In addition, there are surfactants which do not form micelles or are not sufficiently soluble so that higher concentrations cannot be reached.

Mysels & Florence (1970, 1973) were the first to study the effect of surface active impurities on interfacial properties systematically. They demonstrated the importance of the grade of purity of a surfactant solution, especially for dynamic interfacial investigations, and derived a quantitative relation to evaluate it. This derivation was based on a number of simplifications, so it is not sufficiently accurate to serve as a purity criterion. Later Lunkenheimer & Miller (1979, 1987) and Lunkenheimer et al. (1982b, 1984), Lunkenheimer & Czichocki (1993) discussed different situations and derived several criteria to judge the grade of purity of surfactant solutions.

5.1.1. PURITY OF SURFACTANT SOLUTIONS

The term "surface-chemically pure", defined by Lunkenheimer & Miller (1979), is based on the finding that small amounts of highly surface active contaminations can control the properties of an adsorption layer. The study of the ability of surfactants to change interfacial properties requires an adsorbed layer consisting mainly of the surfactant molecules under study. Therefore, any information about the percentage of a potential impurity, given for commercially available samples, is useless. The reason can be understood easily from the situations shown in Fig. 5.1.

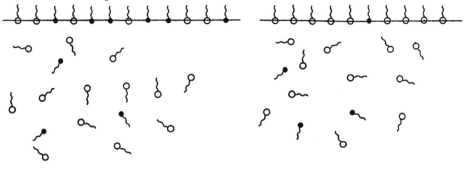

o~ substance A
•~ substance B

Fig. 5.1.   Schematic representation of purity of surfactants with respect to bulk and interfacial concentration of the main (substance A) and minor component (substance B), respectively

Although, there is a very small number of molecules of type B, compared to type A, in the solution, the equilibrium adsorption layer is formed by almost the same number of both molecule types (Fig. 5.1. a). In this case, the surface activity of the contaminant (substance B) is much higher than that of the main component (substance A). In contrast, Fig. 5.1. b demonstrates the situation where both components have a similar surface activity. Here, the adsorbed layer at the interface is mainly controlled by molecules of type A. This solution is in a state of sufficiently high grade of purity for interfacial studies.

Investigations of the solution, demonstrated in part (a) could lead to very confusing results, depending on the experimental conditions. For example, the change of temperature could change the surface activity of the two components in different way and the composition of the adsorbed layer can change with temperature. If dynamic interfacial studies are performed, the results will depend strongly on the time windows of the methods used. The three states of a dynamic adsorption layer (Fig. 5.2., a-c) demonstrate such a situation schematically.

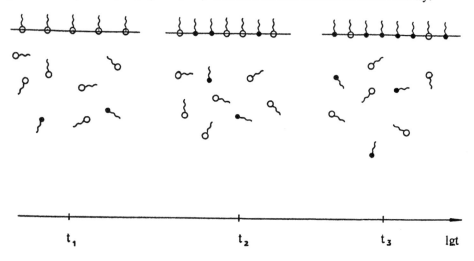

Fig. 5.2.   Change of the composition of a mixed adsorbed layer with time

In the present case, at time $t = t_1$, the adsorption layer has reached an intermediate coverage and consists of almost only molecules of type A. After a time $t = t_2$, equilibrium has not yet been reached, but component B occupies already a remarkable part of the interfacial area. The final equilibrium state of the adsorbed layer is established at time $t > t_3$, and the interfacial properties are mainly controlled by the contaminant B. The evolution of the adsorbed layer composition with time is on a logarithmic scale. The absolute time ranges are a function of the absolute concentration, the time differences and concentration ratios at the interface and the surface activities of the two components. In practical cases, surfactants are not only contaminated by one but often by several components of different surface activity. This complicates the analysis of purification procedures and the grade of purity of prepared surfactant solutions.

A thermodynamic criterion for judging the grade of purity of a surfactant solution was developed by Lunkenheimer & Miller (1987, 1988). It is based on any purification procedure to remove impurities from a surfactant solution. If ξ is the variable which denotes the extent of removal of surface active molecules from a solution, it was shown that the relation $\gamma_e(\xi)$ is useful for the definition of a purity criterion. Fig. 5.3. shows the change of the equilibrium surface tension as a function of purification cycles j (cf. paragraph 5.1.3. for details). First, the surface tension increases due to the removal of the contaminant and then $\gamma_e(j)$ levels off.

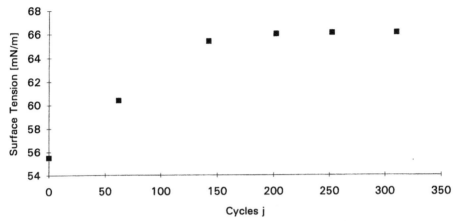

Fig. 5.3.   Change of surface tension of an aqueous 0.2 mol/l Triethylbenzyl ammonium chloride solution in dependence of the number of purification cycles j (according to Lunkenheimer & Miller 1988)

The necessary grade of purity has been reached if

$$\left(\frac{d\gamma_e}{d\xi}\right)_{p,T,c_1} = 0, \tag{5.1}$$

i.e., when the equilibrium surface tension is not changed by a proceeding purification. With the change of $\gamma_e(j)$, the time required to establish equilibrium t_e is also changed. This enables formulation of an equivalent criterion on the basis of the adsorption time,

$$\left(\frac{dt_e}{d\xi}\right)_{p,T,c_1} = 0. \tag{5.2}$$

Both relations are valid at constant pressure p, temperature T and concentration of the main surfactants c_1. For surfactant solutions with very low concentration of the main component, the function $\gamma_e(j)$ changes even after a high grade of purity has been reached. This change is more or less linear with j, and the criterion (5.1) changes to

$$\left(\frac{d\gamma_e}{d\xi}\right)_{p,T,c_1} = \alpha_\xi, \tag{5.3}$$

with

$$\alpha_\xi = \left(\frac{d\gamma_e}{dc_1}\right)\left(\frac{dc_1}{d\xi}\right)_{p,T},$$

(5.4)

If a generalised Langmuir isotherm,

$$\Gamma_i = \Gamma_\infty \frac{c_i / a_{Li}}{1 + \sum_{j=1}^{r} c_j / a_{Lj}} \, ; i = 1,2,...,r,$$

(5.5)

is used and the purification is performed by removing the adsorption layer of area A from a solution of volume V (cf. paragraph 5.1.3), the following relation is obtained,

$$\frac{d\gamma_e}{d\xi} = \frac{d\gamma_e}{dj} = -\frac{RT\Gamma_\infty \frac{A}{V} \frac{\Gamma_\infty}{a_{L1}}\left(1 + \frac{c_2}{c_1}(\frac{a_{L1}}{a_{L2}})^2\right)\frac{c_1}{a_{L1}}}{\left(1 + \frac{c_1}{a_{L1}}(1 + \frac{c_2}{c_1}\frac{a_{L1}}{a_{L2}})\right)\left(1 + (1 + \frac{c_2}{c_1}\frac{a_{L1}}{a_{L2}})\right)}.$$

(5.6)

From Eq. (5.6) it is possible to estimate the range of applicability of the purity criterion. If an experimental accuracy of ±0.1 mN/m is given, a surfactant solution is called "surface-chemically pure", if the condition

$$\frac{d\gamma_e}{d\xi} \le 0.1 \text{ mN / m}$$

(5.7)

is fulfilled. This finally leads to the sensitivity of the criterion, which depends on the surface activity of the main component as well as the contaminant. The higher the surface activity of the main component (characterised by the parameter $\frac{\Gamma_\infty}{a_{L1}}$) is, the easier is the detection of impurities. For example, at $\frac{\Gamma_\infty}{a_{L1}} \sim 10^{-3}$ cm, traces of remarkably higher surface active impurities can be detected, $\frac{a_{L1}}{a_{L2}} > 10$. The limits of impurities which can be detected by the criterion Eqs (5.6), (5.7) are summarised in Table 5.3. The larger the ratio of the surface activities of the main surfactant to that of the impurity, the more sensitive is the criterion. At low surface activity $\frac{\Gamma_\infty}{a_{L1}}$ and activity ratio $\frac{a_{L1}}{a_{L2}}$, the criterion fails.

The disadvantage of this thermodynamic criterion is that it can be used only in combination with a purification procedure. Thus, if such a procedure is not available, the purity of a surfactant solution with respect to its useful interfacial study cannot be checked. Therefore, a second criterion was developed, which is based on an adsorption kinetics model. If adsorption of both, the main component and the impurity is assumed to be diffusion-controlled, the difference between the surface tension values, measured after a definite time t_{ad} after adsorption and desorption, respectively, is a measure of the purity of the solution.

Table 5.3.   Limits of sensitivity of the purity criterion Eqs (5.6) and (5.7)

$\dfrac{\Gamma_\infty}{a_{L1}}$ [cm]	$\dfrac{a_{L1}}{a_{L2}}=10$	$\dfrac{a_{L1}}{a_{L2}}=100$	$\dfrac{a_{L1}}{a_{L2}}=1000$
$4\cdot10^{-5}$	-	-	10%
$4\cdot10^{-4}$	-	10%	1%
$4\cdot10^{-3}$	10%	1%	0.1%
$4\cdot10^{-2}$	1%	0.1%	0.01%

Table 5.4.   Surface tension difference $\Delta\gamma(t_{ad}) = \gamma_{ad}(t_{ad}) - \gamma_{des}(t_{ad})$ in [mN/m], calculated for a surfactant mixture; $t_{ad}= 60s$, $D_1 = D_2 = 5\cdot10^{-6}$ cm²/s, $\Gamma_\infty= 5\cdot10^{-10}$ mol/cm², $a_{L1} = 10^{-7}$ mol/cm³, $a_{L2} = 10^{-9}$ mol/cm³, compression ration $A_{ad} / A_{des} = 2$

c_2/a_{L2}	$c_1/a_{L1} = 0.1$	$c_1/a_{L1} = 0.2$	$c_1/a_{L1} = 0.4$	$c_1/a_{L1} = 0.8$
0.0	0.26	0.45	0.69	0.86
0.1	0.36	0.54	0.77	0.94
0.3	0.54	0.73	0.96	1.11
0.9	1.12	1.30	1.52	1.65

Some data are summarised in Table 5.4., calculated for a special surfactant mixture of two surfactants, having a ratio of the surface activities of $\dfrac{a_{L1}}{a_{L2}}=100$. The indices "ad" and "des" correspond to the surface tension measurements in the adsorption and desorption state (after a certain compression of the adsorption layer from A_{ad} to A_{des}), respectively.

Based on such tables, it is possible to check the purity of a surfactant solution. The use of such tables requires many calculations and, therefore, their use is not very convenient. Starting from such tables, an easy relation could be derived which is very simple to use. This relation is derived under the conditions $c_1 / a_{L1} = 1$ and $\gamma_e = 65 \, mN/m$ and calculates the time t_{ad} necessary to reach a surface tension difference

$$\Delta\gamma(t_{ad}) < 0.2 \, mN/m, \tag{5.8}$$

a value which corresponds to the usual accuracy of many experiments,

$$t_{ad} = t_o \exp\left(-18 - \frac{3}{2}\ln(\frac{a_{L1}}{a_o})\right). \tag{5.9}$$

with

$$t_o = 1s; \, a_o = 1 \, mol/cm^3. \tag{5.10}$$

To decide whether a surfactant is surface-chemically pure or not, the following steps have to be made. First the characteristic time t_{ad} of Eq. (5.9) is calculated and then, the surface tension difference $\Delta\gamma(t_{ad}) = \gamma_{ad}(t_{ad}) - \gamma_{des}(t_{ad})$ at $c_1 = a_{L1}$ is measured. If the condition of Eq. (5.8) is fulfilled the solution is surface-chemically pure.

The range of application of the criterion, given by Eqs (5.8) - (5.10), covers the following intervals,

$$3 \cdot 10^{-8} \, mol/cm^3 \le a_{L1} \le 10^{-6} \, mol/cm^3,$$

$$10 \le \frac{a_{L1}}{a_{L2}} \le 100, \, \frac{c_2}{a_{L2}} \le 0.1. \tag{5.11}$$

There is one important fact worth mentioning. The thermodynamic criterion as well as the kinetic criterion may be used to judge the purity of a surfactant solution only for time scales only up to the chosen adsorption/desorption time t_{ad}. If the measurements for checking the purity are performed in a time interval $0 < t < t_{ad}$, the result is valid only in a time interval $t < 10t_{ad}$. After longer times, the situation at an interface can change and none of the conditions hold.

To demonstrate the applicability of the criterion, given by Eqs (5.8) - (5.10), $\Delta\gamma(t_{ad})$-values of aqueous solutions containing decyl dimethyl phosphine oxide as the main component and dodecyl dimethyl phosphine oxide as the simulated impurity were measured. The results, shown in Fig. 5.4, confirm that the assumption works well.

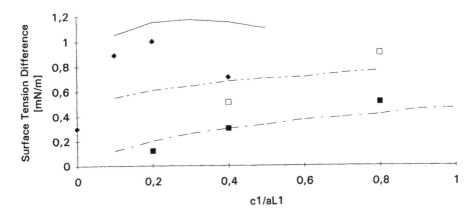

Fig. 5.4   Comparison of theoretical (lines) and experimental results (symbols) for a system containing decyl dimethyl phosphine oxide as main component and dodecyl dimethyl phosphine oxide as simulated impurity; $c_2/a_{L2} = 0.0$ (■, ‒‒‒), 0.1 (□, ·· ‒··), 0.3 (♦, ——), $D_1 = D_2 = 5 \cdot 10^{-6}$ cm²/s, $\Gamma_\infty = 5 \cdot 10^{-10}$ mol/cm², $a_{L1} = 5 \cdot 10^{-8}$ mol/cm³, $a_{L2} = 5 \cdot 10^{-9}$ mol/cm³, $A_{ad}/A_{des} = 2$; (cf. Miller & Lunkenheimer 1982)

The presence of impurities in surfactant solutions can give very misleading results. In a recent paper, experimental dynamic surface tensions of sodium dodecyl sulphate (SDS) solutions were interpreted by Fainerman (1977) on the basis of a mixed diffusion-kinetic-controlled adsorption model. As the result a rate constant of adsorption k_{ad} as a function of time was obtained (cf. Fig. 5.5, ■), although this parameter was assumed to be a constant.

In a "theoretical experiment", Miller & Lunkenheimer (1982) demonstrated that the observed change of k_{ad} with time can be simulated by the presence of an impurity, which is most probably the homologous alcohol, dodecanol (Vollhardt & Czichocki 1990). Using a diffusion controlled adsorption model of a mixed surfactant system consisting of a 10^{-6} mol/cm³ SDS with 1% dodecanol, a dependence $\gamma(t)$ was calculated. These data were then interpreted on the basis of a mixed diffusion-kinetics-controlled model of a single surfactant solution, as it is usually made, and almost identical results are obtained by this simulation (cf. Fig. 5.5, □). This example demonstrates, that the presence of a surface active impurity can give rise to data which may be misleading. In the present case it simulated an adsorption barrier, although the main surfactant as well as the contaminant adsorb via diffusion controlled mechanism.

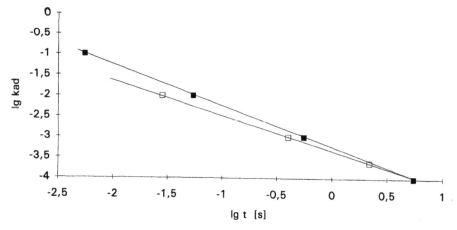

Fig. 5.5   Rate constant of adsorption k_{ad} in dependence on time t for a 10^{-6} mol/cm^3 SDS solution, (■) determined from experimental data by Fainerman (1977), (□) simulated data (cf. Miller & Lunkenheimer 1982)

5.1.2. PURITY OF SOLVENTS

The purity of solvents, especially in studies at liquid-liquid interfaces, is as important as the purity of the surfactant sample. The proven absence of any surface active contamination, both in the surfactant sample and in the solvents, is a necessary prerequisite for interfacial studies. The picture shown in Fig. 5.6 demonstrates the situation.

The analysis of the purity of liquid-liquid interfaces can also be performed by surface or interfacial tension measurements, as described above. If the two liquid phases are free of any interfacially active compounds, the interfacial tension should be time independent and identical with the literature value. Since many experimental instruments do not yield absolute values and therefore, this criterion is often not useful.

A better way of checking the purity of the bare interface is by a transient relaxation experiment. The change in the interfacial tension after surface area compressions or dilations is an excellent measure of the interfacial purity. No interfacial tension changes should be observed after an area change. Changes give information on the presence and dynamics of adsorbed impurities. The relaxation behaviour of two decane samples, studied with a modified pendent drop experiment (Miller et al. 1994a, d), is illustrated in Fig. 5.7. The commercial sample shows a remarkable interfacial tension change after area compressions and dilations, while the purified sample does not exhibit measurable changes.

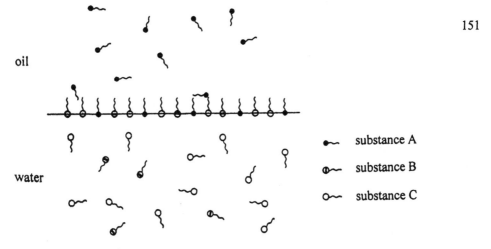

oil

water

substance A

substance B

substance C

Fig. 5.6   Adsorption at a liquid-liquid interface, taking into account impurities from the surfactant sample
and the organic solvent

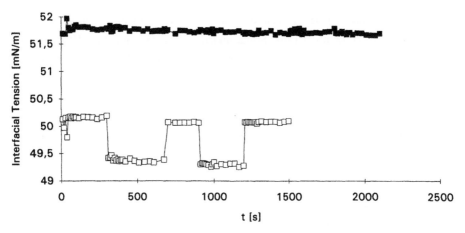

Fig. 5.7   Relaxation behaviour of the water-decane interface; commercial decane (□), specially purified
decane (■); according to Miller et al. (1994c)

There is one striking difference between the two sources of impurities. The contamination
present in the surfactant sample appears as a minor component in the solution bulk with a
concentration proportional to that of the main surfactant. If the origin of the impurity is the
organic phase, its concentration is constant and independent of the surfactant concentration.
Therefore, the effects of these two types of impurities on the adsorption kinetics and adsorption
equilibrium are different.

5.1.3. PROCEDURES TO PURIFY SURFACTANT SOLUTIONS

The purification of surfactants has to be performed in a way that addresses their specific effects on interfacial properties. Thus, chemical purification procedures, such as recrystallisation, distillation, flocculation, and washing with organic solvents, are important steps (Czichocki et al. 1981) but usually do not lead to a sufficient purity, i.e. surface-chemical purity. Therefore, efficient purification methods are based on interfacial principles.

One possibility consists in the use of dispersed solids, such as latex particles. Surface-chemically active impurities are preferentially adsorbed at the rather large surface area of these particles, and after removal of the disperse phase by filtration, a purified solution is obtained (Rosen 1981, Carroll 1982, Lunkenheimer et al. 1984). This method has two disadvantages: there is a large remarkable and unknown loss of the main surfactant, and the dispersion can introduce impurities into the surfactant solution.

A second purification method based on interfacial principles is foam flotation (Brady 1949, Elworthy & Mysels 1966, Lemlich 1968, Somasundaran 1975, Gilanyi et al. 1976). A column is filled with the surfactant solution to be purified. Bubbles of purified gas are formed at the bottom of the column and rise up in the solution, carrying an adsorption layer to the top. This adsorption layer is formed by the main surfactant and the contamination. Due to the higher surface activity of the impurity, it is enriched in the contaminant in the top layer and in any resulting foam. Liquid in the column is progressively purified by removing the foam or the top layer. The same procedure may be successfully used for the purification of water samples from any surface active material (Loglio et al. 1986, 1989).

A further and very efficient method of purifying aqueous surfactant solution consists in a cyclic removal of the adsorption layer. The general idea is to establish an almost equilibrium adsorption layer in which the highly surface active impurity is enriched with respect to the main surfactant. If the adsorption layer is removed, the ratio of surfactant/impurity is changed (cf. Eq. (5.6)). The idea has been realised in an apparatus which automatically performs the following steps (Lunkenheimer et al. 1987a):

1. formation of an adsorption layer at a large surface during time t_{ad};
2. compression of the adsorption layer;
3. removal of the compressed adsorption layer;
4. increase of the surface to its initial large area.

An overall view of the purification unit is shown in Fig. 5.8. In position 1 the surface of the solution inside the glass container is large and the surfactant and impurities adsorb. After the time t_{ad} the container is tilted by means of a motor so that the small outlet tube stands upright while the solution surface is compressed to the small cross section of the outlet tube. After the solution has become calm, the capillary tip is placed close to the surface (adjustment is achieved using the electro-conductivity and a feedback control) and the surface of the liquid removed by sucking. The whole instrument is computer controlled and allows a definite number of cycles with chosen adsorption times t_{ad}.

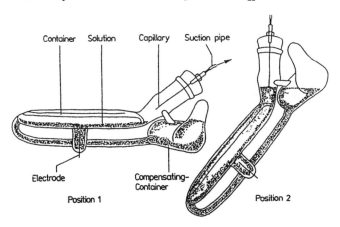

Fig. 5.8   Purification unit of an automatic apparatus, working with cyclic removal of the adsorption layer; position 1 - adsorption at a large area, position 2 - compressed adsorption layer for removal by aspiration; according to Lunkenheimer et al. (1987a)

The purity of the prepared solutions can be checked by the criteria discussed above. For special surfactants, a specific analysis of impurities by very sensitive methods is possible. Czichocki et al. (1981) and Vollhardt & Czichocki (1990) have shown the strong impurity effect of isomeric alcohols on the adsorption properties of sodium alkyl sulphates. The purity of the solutions could be analysed in this case by a highly sensitive HPLC method, which cannot simply be transferred to other surfactant systems.

5.2.   Drop Volume Technique

During this century the drop volume method has been developed as a standard method in surface and interfacial studies. Since the early theoretical work of Lohnstein (1906, 1907, 1908, 1913) and the extensive experimental investigations by Addison and coworkers (1946a, b, 1948, 1949a, b), many other authors used the general idea of the drop volume method (Tornberg 1977, 1978a, 1978b, Joos & Rillaerts 1981, Nunez-Tolin et al. 1982, Carroll & Doyle 1985, Miller & Schano 1986, Babu 1987, Doyle & Carroll 1989, Miller & Schano 1990, Wawrzynczak et al. 1991, Paulsson & Dejmek 1992, Fainerman et al. 1993b, Miller et al.

154

1994a) to measure dynamic surface and interfacial tensions. Over a long time only laboratory set-ups were used until a commercial set-up (TVT1 by the company LAUDA) was designed which operates in a wide range of temperature and adsorption time. Beside the usual version based on continuously growing drops with a constant liquid flow, also a so-called "quasi-static" measuring procedure is implemented. This version, which is comparable with other static methods, has been used by other authors for studies of comparatively slow adsorption processes (Addison 1946a, Tornberg 1978b, Van Hunsel et al. 1986, Van Hunsel & Joos 1989, Miller & Schano 1990, Fang & Joos 1994).

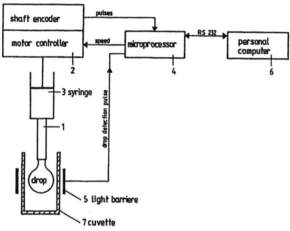

Fig. 5.9   Schematic representation of an automated drop volume experiment according to Miller et al. (1992)

The principle of the method is shown schematically in Fig. 5.9, referring specifically to the commercial set-up mentioned above. A metering system in form of a motor-driven syringe allows the formation of liquid drops at the tip of a capillary, which is positioned in a sealed cuvette. The cuvette is either filled with a small amount of the measuring liquid, to saturate the atmosphere, or with a second liquid in case of interfacial studies. A light barrier arranged below the forming drop enables the detection of drop detachment from the capillary. Both the syringe and the light barrier are computer-controlled and allow a fully automatic operation of the set-up. The syringe and the cuvette are temperature controlled by a water jacket which make interfacial tension studies possible in a temperature interval from 10°C up to 90°C.

The easy sample handling, the simple temperature control over a broad temperature interval, and the direct applicability without any modifications to liquid/air and liquid/liquid interfaces are reasons for the frequent use of the drop volume method. The accuracy and reproducibility are as high as those of other methods, ± 0.1 mN/m. In addition, it has the advantage that only small amounts of solute and solvent are needed for a series of measurements.

The principle of the drop volume method is of dynamic character and therefore, it can be used for studies of adsorption processes in the time interval of seconds up to some minutes. At small drop times a so-called hydrodynamic effect has to be considered, as discussed in many papers (Davies & Rideal 1969, Kloubek 1976, Jho & Burke 1983, Van Hunsel et al. 1986, Van Hunsel 1987, Miller et al. 1994a). This hydrodynamic effect appears at small drop times under the condition of constant liquid flow into the drop and gives rise to apparently higher surface tensions. Davies & Rideal (1969) discussed two factors influencing the drop formation at and its detachment from the tip of a capillary: the so-called "blow up" effect and a "circular current" effect inside the drop. The first effect increases the detaching drop volume and simulates a higher surface tension while the second process leads to an earlier break-off of the drop and results in an opposite effect. A schematic of these two effects on measured drop volumes is shown in Fig. 5.10.

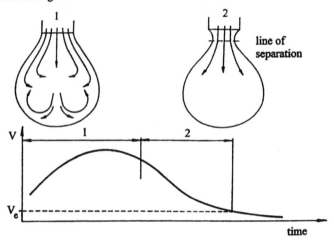

Fig. 5.10.   Schematisation of the "blow up" (1) and "circulation" (2) effects on experimentally determined drop volumes

In order to quantify the hydrodynamic effect Kloubek (1976) measured the drop volume of pure liquids at small drop times using capillaries with different design and diameter. As the result he presented an empirical relation for the measured drop volumes V(t).

It allows the calculation of the unaffected drop volume V_e from the measured drop volume dependence V(t) by the following equation,

$$V_e = V(t) - \frac{K_v}{t},$$

(5.12)

where K_v is a proportionality factor. Jho & Burke (1983) studied various liquids using capillaries with different diameters and interpreted their data with another empirical relation, which has previously been used by Davies & Rideal (1969)

$$V_e = V(t) - \frac{K_v}{t^{2/3}}.$$ (5.13)

They found that K_v depends on surface tension γ, density difference $\Delta\rho$ and tip radius r_{cap}. These findings were confirmed by Van Hunsel (1986, 1987) who performed similar measurements with a number of pure liquids. Van Hunsel et al. (1986) also studied the hydrodynamic effect on drop volume measurements at a liquid-liquid interface and described it on the basis of the same Eq. (5.13).

Miller et al. (1994a) described the error due to hydrodynamics in the drop volume determination only on the basis of the so-called "blow-up" effect. Neglecting the influence of turbulent flow is justified when capillary diameters and drop times are not too small and liquid viscosities are not too high. The apparently higher drop volumes of pure liquids at small drop times are explained by a finite drop detachment during t_o. During this time the detaching drop is still connected to the liquid pumped by the syringe and additional volume flows into the drop. The actual volume of a detached drop is then finally given by

$$V(t) = V_e + t_o F = V_e \left(1 + \frac{t_o}{t - t_o} \right),$$ (5.14)

F is the liquid flow per time defined by

$$F = \frac{V(t)}{t} = \frac{V_e}{t - t_o}.$$ (5.15)

From the data presented (Miller et al. 1994a) it can be concluded that the drop detachment time t_o is the characteristic parameter for the observed hydrodynamic effects at small drop times and is directly related to the tip radius r_{cap} and the capillary constant a $= \sqrt{\dfrac{2\sigma}{\Delta\rho\, g}}$ of the liquid. As a first approximation for low-viscosity liquids t_o can be calculated from a linear relation, which results from a regression of the data presented in (Miller at al. 1993a),

$$t_o = \alpha + \beta\, r_{cap}.$$ (5.16)

The coefficients α and β are functions of the capillary constant a. When averaging over the interval 0.19 cm < a < 0.39 cm the following mean values $\alpha = 0.08$ s and $\beta = 0.041$ s·cm^{-1} are obtained. The correction of measured drop volumes can be made using the following relation,

$$V_e = V(t)\left(1 - \frac{\alpha + \beta r_{cap}}{t}\right). \qquad (5.17)$$

From drop volume experiments dynamic interfacial tension data are obtained in time intervals from 10 s up to hours when using the so-called quasi-static measuring procedure. These data can be interpreted by the usual adsorption kinetics theories. When practising the dynamic drop volume method with continuously growing drops (metering systems with linear volume change) dynamic interfacial tensions result as a function of drop formation time. Such data have to be interpreted by theories where the dilation of the interface and the radial flow inside the drop are considered. The analysis of the theory has shown (c.f. Eq. (4.43) that a simple interpretation is possible by transforming the drop formation time t_{drop} into the effective surface age τ_a: $\tau_a = t_{drop}/3$ (Miller at al. 1992). The advantage of the dynamic drop volume technique is that it allows measurements from about 1 s up to 20-30 minutes and longer. Experimental data obtained by this technique are compared to results obtained with other measuring techniques in the next chapter. For the calculation of surface and interfacial tensions from drop volumes correction factors are needed which take into consideration the effective radius of detachment instead of the tip radius (cf. Wilkinson 1972). Tables of the correction factor are given in Appendix 5A.

Limits for the application of the drop volume technique are recently discussed by Fainerman & Miller (1994a). Under the conditions of fast drop formation and larger tip radii the drop formation shows irregular behaviour. Depending on the geometry bifurcations in the measured drop volume are observed, which are explained by wave effects along the liquid bridge between residing and detaching drop (Milliken et al. 1993).

5.3. MAXIMUM BUBBLE PRESSURE TECHNIQUE

The maximum bubble pressure technique is a classical method in interfacial science. Due to the fast development of new technique and the great interest in experiments at very small adsorption times in recent years, commercial set-ups were built to make the method available for a large number of researchers. Rehbinder (1924, 1927) was apparently the first who applied the maximum bubble pressure method for measurement of dynamic surface tension of surfactant solutions. Further developments of this method were described by several authors (Sugden 1924, Adam & Shute 1935, 1938, Kuffner 1961, Austin et al. 1967, Bendure 1971,

Finch & Smith 1973, Kragh 1964). It was established by Austin et al. (1967) that the time interval between two subsequent bubbles consists of the surface lifetime and the so-called "dead time", τ_d, which is the time required to detach after the bubble has reached a hemispherical shape. Kloubek (1968, 1972a, b) made a remarkable contribution for the further development of this method by deriving a simple experimental procedure for the determination of the dead time and giving an estimate of the effective bubble surface lifetime (1972b). The use of electric pressure sensors for the measurement of pressure and bubble formation frequency (Bendure 1971, Razour & Walmsley 1974, Miller & Meyer 1984, Woolfrey et al. 1986, Hua & Rosen 1988, Mysels 1989) simplified substantially the measurement procedure. Original designs for the apparatus for the measurement of dynamic surface tension by the maximum bubble pressure method were presented for example by Ramakrishnan et al. (1976), Tzykurina et al. (1977), Papeschi et al. (1981), Markina et al. (1987), Hua & Rosen (1988), Schramm & Green (1992), Ross et al. (1992), Holcomb & Zollweg (1992), Iliev & Dushkin (1992), for different regions of lifetimes and by Lunkenheimer et al. (1982a) using a modified method with two capillaries.

The design of a maximum bubble pressure method for high bubble formation frequencies must address three main problems: the measurement of bubble pressure, the measurement of bubble formation frequency, and the estimation of surface lifetime and effective surface age.

The first problem can be solved easily if the system volume, which is connected with the bubble, is big enough in comparison with the volume of the bubble separating from the capillary. In this case the system pressure is equal to the maximum bubble pressure. On the contrary, the use of an electric pressure transducer for measuring the bubble formation frequency presumes that pressure oscillations in the measuring system are distinct enough. This condition is fulfilled in systems with comparatively small volumes only. As shown by Mysels (1989), the determination of maximum bubble pressure values in systems with a small volume can be erroneous.

To separate the surface lifetime from the total time interval between subsequent bubbles an approximation of the dead time according to geometric parameters of capillary and bubble volume was derived Fainerman & Lylyk (1982) and Fainerman (1990). A substantial improvement for the exact determination of surface lifetime and its calculation was carried out by Fainerman (1992) who defined a critical point in the experimental curve in co-ordinates "pressure-gas flow rate". This point corresponds to a change in the flow regime from individual bubble formation to a gas jet regime. The calculation of the so-called effective age the surface (effective adsorption time) from the bubble surface lifetime was discussed by different authors:

Joos & Rillaerts (1981), Garrett & Ward (1989) Fainerman (1992), Joos et al. (1992). The definition of the different times characterising the life of a bubble is displayed in Fig. 5.11.

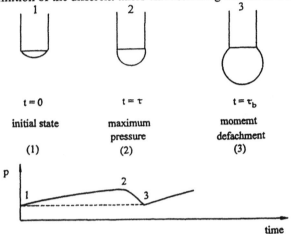

Fig. 5.11   Definition of characteristic bubble times and principal course of the pressure p in the bubble with time t

A schematic representation of the most advanced set-up, based on the latest scientific findings, the MPT1 from LAUDA, is shown in Fig. 5.12. The air coming from a micro-compressor flows first through the flow capillary. The air flow rate is determined by measuring the pressure difference at both ends of the flow capillary with the electric transducer PS1. Thereafter the air enters the measuring cell. The excess air pressure in the system is measured by a second electric sensor PS2. In the tube which leads the air to the measuring cell, a sensitive microphone is placed.

The measuring cell is equipped with a water jacket for temperature control, which simultaneously holds the measuring capillary and two platinum electrodes, one of which is immersed into the liquid under study. The second is situated exactly opposite to the capillary and controls the size of bubbles. The electric signals from the gas flow rate sensor PS1 and pressure transducer PS2, the microphone and the electrodes, as well as the compressor are connected to the PC which operates the apparatus and acquires the data.The value of τ_d, equivalent to the time interval necessary to form a bubble of radius R, can be calculated using Poiseuille's law, as long as the conditions $p = \text{const}$ in the system holds (Fainerman 1979, 1990, 1992),

$$\tau_d = \frac{\tau_b L}{Kp}\left(1 + \frac{3r_{cap}}{2R}\right),$$

(5.18)

where $K = \pi r^4/8l\eta$ is the Poiseuille law constant (for a non-immersed capillary Eq. (5.18) reads $L = Kp$), η is the gas viscosity, l is the length and r_{cap} the radius of the tip of the capillary. The calculation of the dead time τ_d can be simplified when taking into account the existence of two gas flow regimes for the gas flow leaving the capillary: bubble flow regime when $\tau > 0$ and jet regime, when $\tau = 0$ and hence $\tau_b = \tau_d$. A typical dependence of p on L is presented in Fig. 5.13 for two liquid systems.

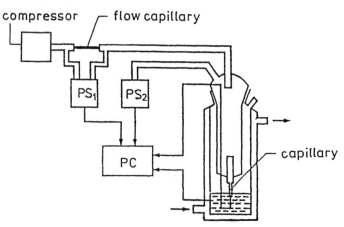

Fig. 5.12   Schematic representation of the automatic operating bubble pressure method according to Fainerman et al. (1993a, b); PS_1 and PS_2 are pressure sensors for determining the gas flow rate and the system pressure, M is the microphone

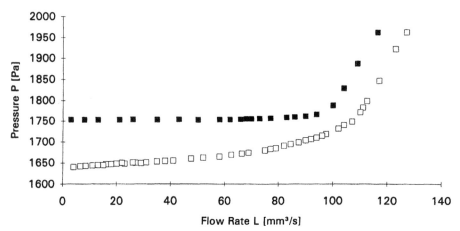

Fig. 5.13   Dependence of the pressure p in the system on the gas flow rate L according to Fainerman et al. (1993b); water (■), and a water / glycerine mixture (ratio 2:3) (□) at 30°C; r = 0.0824 mm

On the right hand side of the critical point the dependence of p on L is linear in accordance with the Poiseuille law. The existence of the electrode placed opposite the capillary controls the dimensions of bubbles. Under this condition the bubble radius, and consequently the bubble volume, is constant at any given L inside the bubble flow regime (Fainerman 1990). Therefore, instead of Eq. (5.13) the following simplified equation results to determine τ_d,

$$\tau_d = \tau_b \frac{L p_c}{L_c p},$$
(5.19)

where L_c and p_c are related to the critical point, and L and p are the actual values of the dependence left from the critical point.

The surface lifetime can be calculated using the formula

$$\tau = \tau_b - \tau_d = \tau_b (1 - \frac{L p_c}{L_c p}).$$
(5.20)

As one can see, Eq. (5.20) involves only experimentally available values. The critical point in the dependence p on L can easily be located. In the software the location is automatically calculated by an algorithm based on the Poiseuille law. The calculation of the effective surface age (effective adsorption time) can be made using Eq. 4.62, derived in Chapter 4. The derivation of the relative surface deformation rate is based on the condition p = const. For values of γ not too close to γ_o (for example, for aqueous solutions below 60 mN/m), ξ is approximately equal to 0.5, and consequently $\tau_a \approx \tau/2$.

Most of the instruments, based on the principle of maximum bubble pressure, do not allow the effective surface age to be calculated because the conditions during the bubble formation are unknown. These methods yield only a dependence of surface tension on bubble frequency or bubble time τ_b. The graph in Fig. 5.14 shows the remarkable differences of the three possible form of final data: $\gamma(\tau_b)$, $\gamma(\tau)$, and $\gamma(\tau_a)$.

The surface tension value in the maximum bubble pressure method is calculated via the Laplace equation. For the instrument under discussion, the capillary radius is small and the bubble shape is thus assumed to be spherical. Thus the deviation of the bubble shape from a spherical one can be neglected and needs no correction. Hence, the following equation results,

$$p = \frac{2\gamma}{r} + \rho gh + \Delta p$$
(5.21)

where ρ is the density of the liquid, g is the acceleration of gravity, h is the depth the capillary is immersed into the liquid, and Δp is a correction value to allow for hydrodynamic effects. $\Delta p < 0$ leads to a correction of $\Delta \gamma = \gamma_{app} - \gamma_{corr} > 0$ (indices "app" and "corr" stand for apparent and corrected surface tensions, respectively) which can be estimated according to the following relation,

$$\Delta\gamma = \frac{3}{2}\frac{\eta r}{\tau}.$$

(5.22)

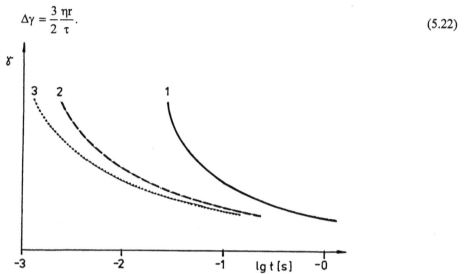

Fig. 5.14   Dynamic surface tension γ in dependence of bubble time τ_b (—), bubble life time τ (---), and effective surface age τ_a (···); (schematically)

Recent experimental studies (Fainerman et al. 1993b) corroborated qualitatively the validity of Eq. (5.22): the value increases with increasing liquid viscosity η, increasing capillary radius r and decreasing surface lifetime τ. Kao et al. (1993) and Edwards et al. (1993) made an analysis of surface and bulk rheological effects on the growth and rising of gas bubbles in a bubble pressure experiment and found a term to calculate the bulk viscosity influence comparable with Eq. (5.22). For aqueous surfactant solutions this effect should be negligible, which is in full agreement with the results of Fainerman et al. (1993b). At higher viscosities, this bulk viscosity effect has a remarkable effect on the bubble formation and has to be allowed for, also in the experiments to study dilational viscosity proposed by Kao et al. (1993). The tables in Appendices 5B and 5C provide density, viscosity and surface tension of some frequently used liquids.

The maximum bubble pressure method, realised as the set-up discussed above, allows measurements in a time interval from 1 ms up to several seconds and longer. At present, it is the only commercial apparatus which produces adsorption data in the millisecond and even sub-millisecond range (Fainerman & Miller 1994b, cf. Appendix G). Otherwise data in this time interval can be obtained only from laboratory set-ups of the oscillating jet, inclined plate or other, even more sophisticated, methods. The accuracy of surface tension measurements in the time interval of milliseconds is about ± 0.5 mN/m, at effective surface ages $\tau_a > 0.01$ s the accuracy is ± 0.1 mN/m. Experimental data obtained by this technique and comparisons with other measuring techniques are given below. Due to its time window, the maximum bubble pressure method even provides the possibility to study the effect of micelle kinetics on the adsorption process of surfactants and thus is an experimental source for information about rate constants of micelle formation or dissolution processes (Fainerman & Makievski 1992a, b).

5.4. PENDENT DROP TECHNIQUE

The pendent or sessile drop technique has been developed to determine interfacial tensions from the shape of drops without a direct contact of the interface. First experiments were performed by measuring characteristic drop diameters and interpreting them on the basis of different tables (Fordham 1948, Stauffer 1965). Later the direct fitting of drop shape co-ordinates to the Gauss-Laplace equation was used to determine interfacial tension and contact angle data (Maze & Burnet 1971, Rotenberg et al. 1983, Girault et al. 1984, Anastasiadis et al. 1987, Cheng 1990, Cheng et al. 1990). Reviews on the pendent and sessile drop techniques are given for example by Padday (1969), Neumann & Good (1979), Ambwami & Fort (1979) or Boucher (1987).

The Gauss-Laplace equation describing liquid menisci in general was discussed in detail by Padday & Russel (1960) and Padday et al. (1975). The profile of an axisymmetric drop can be calculated in dimensionless co-ordinates from the following equation (Rotenberg et al. 1983),

$$\frac{d\phi}{dS} = 2 - \beta Y - \frac{\sin\phi}{X}, \tag{5.23}$$

$$\frac{dX}{dS} = \cos\phi, \tag{5.24}$$

$$\frac{dY}{dS} = \sin\phi, \tag{5.25}$$

where X, Y and S are made dimensionless by dividing x, y, and s, respectively, by R_o. x and y are the horizontal and vertical coordinate, s is the arc length of the profile measured from the drop apex, R_o is the radius of curvature in the drop apex, and ϕ is the angle made by the radius of curvature and the y axis. $\beta = \dfrac{\Delta\rho g R_o^2}{\gamma}$ is a parameter which contains the density and surface tension of the liquid. The definition of all co-ordinates and characteristic parameters is given in Fig. 5.15.

164

Beside the fitting of drop profile co-ordinates to the Gauss-Laplace equation, based on least square algorithms, a relation exists which allows the surface tension to be calculated from the characteristic diameters D_E and D_S (Andreas et al. 1938, Girault et al. 1984, Hansen & Rødsrud 1991).

Girault et al. (1984) derived the following relation to calculate γ from the ratio D_S/D_E,

$$\frac{\Delta\rho g R_o^2}{\gamma} = 0.02664 + 0.62945(\frac{D_S}{D_E})^2.$$

(5.26)

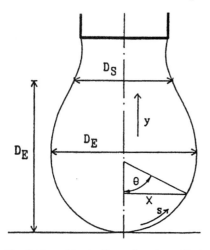

Fig. 5.15   Sketch of a pendent drop with the definition of drop parameters

An improved relation was derived by Hansen and Rødsrud (1991):

$$\frac{\Delta\rho g R_o^2}{\gamma} = 0.12836 - 0.7577(\frac{D_S}{D_E}) + 1.7713(\frac{D_S}{D_E})^2 - 0.5426(\frac{D_S}{D_E})^3.$$

(5.27)

The application of Eqs (5.26) or (5.27) needs the diameters D_E and D_S and the radius of curvature R_o. For values $0.1 < \beta < 0.5$ a polynomial was given by Hansen & Rødsrud (1991) to determine R_o,

$$\frac{D_E}{2R_o} = 0.9987 + 0.1971\,\beta - 0.0734\,\beta^2 + 0.34708\,\beta^3.$$

(5.28)

Thus, using Eqs (5.27) and (5.28) the surface tension γ can directly be calculated from the two drop diameters D_E and D_S.

Calculations with the characteristic diameters is fast but is less precise than with the fitting procedure. The latter is much more accurate and therefore the more commonly used. The graphs in Fig. 5.16 shows how the principle of the fitting procedure works.

Four parameters have to be adjusted in order the fit the experimental coordinates of the drop shape: the localisation of the drop apex X_o, Y_o, the radius of curvature R_o, and the parameter ß. A software package called ADSA does the detection of the drop edge coordinates and the fitting of the Gauss-Laplace equation to these data. A suitable algorithm to solve the Gauss Laplace equation is given in Appendix 5F.

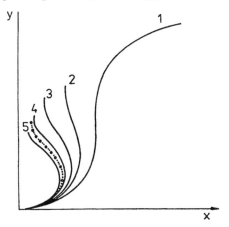

Fig. 5.16   Procedure of fitting the Gauss-Laplace equation to experimental drop shape coordinates (o); the family of curves represent different theoretical drop shapes calculated for different parameters ß

Different experimental set-ups have been developed to determine the surface and interfacial tension from the shape of pendent or sessile drops (Nahringbauer 1987, Bordi et al. 1989, Cheng et al. 1990, Hansen & Rødsrud 1991, Carla et al. 1991, MacMillan et al. 1992, 1994, Racca et al. 1992, Naidich & Grogerenko 1992). In Fig. 5.17 the experimental set-up is shown according to Cheng et al. (1990).

The video signal of the pendent or sessile drop is transmitted to a digital video processor which performs the frame grabbing and digitisation of the drop image. A computer is used to acquire the image. This computer also performs the image analysis and calculations. Depending on the efficiency of the hardware as well as the software, which has to analyse the images and to perform necessary calculations, the determination of interfacial tension needs 10 and more seconds when using complete drop-shape-fitting procedures on a PC. For faster measurements either faster computers, such as workstations, has to be used or the drop images have to be saved to a hard disk and are analysed later. Another fast possibility is the use of the set of

Eqs (5.19), (5.20). The accuracy of measurements with the pendent drop depends on the algorithm used for the drop shape analysis. While the accuracy is of the order of ± 1 mN/m when using characteristic drop diameters only, the analysis of the full drop profile by fitting the data to the Gauss-Laplace equation gives values with an accuracy of ± 0.1 mN/m.

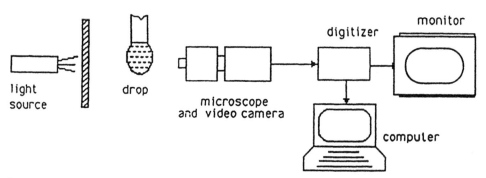

Fig. 5.17   Schematic of a video enhanced pendent drop experiment according to Cheng et al. (1990)

5.5.   GROWING DROP METHODS

An experiment, that acquires adsorption kinetics data much faster than the present pendent drop experiments, is based on the idea of measuring the pressure directly within growing drops and bubbles. Different instruments were developed very recently by Passerone et al. (1991), Liggieri et al. (1991, 1994), Horozov et al. (1993), MacLeod & Radke (1993), Nagarajan & Wasan (1993). All these methods have in common that the pressure inside a drop, formed at the tip of a small capillary, is measured directly by a very sensitive pressure transducer. The set-up of Passerone et al. (1991) uses a differential pressure transducer to measure the pressure difference between inside and outside the growing drop (Fig. 5.18). A motor driven syringe is used as a metering system to produce a highly precise constant liquid flow into the continuously growing drop.

The difference in the design of the other growing drop set-ups consists most of all in the use of a direct pressure transducer instead of a differential one. In all cases the data acquisition is made by an on-line coupled computer. In the instrument of Nagarajan & Wasan (1993) the syringe is also controlled by the computer allowing different types of volume, and consequently drop surface area changes to be measured. The instruments of MacLeod & Radke (1993) and Horozov et al. (1993) are additionally equipped with a video system to record the evolution of the drop and to determine accurate geometric drop data from the video image.

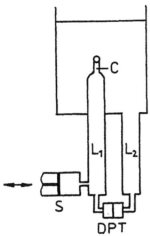

Fig. 5.18   Principle of a growing drop experiment according to Passerone et al. (1991); S - motor driven syringe, C - capillary, DPT - differential pressure transducer, L1 and L2 - the two liquids

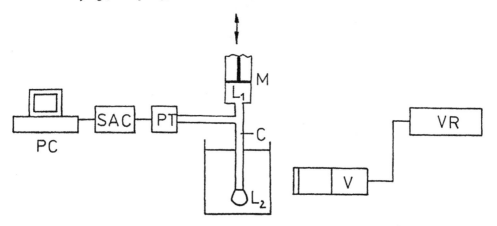

Fig. 5.19   Principle of a drop pressure experiment according to MacLeod & Radke (1993); C - capillary, M - motor driven syringe for liquid 1, L1 - liquid 1, L2 - reservoir of liquid 1, PC - computer, V - video camera with objective, VR - video recorder, PT - pressure transducer, SAC - signal amplifier and converter

Data acquisition and interpretation is comparatively fast, and processes from time scales of tens of milliseconds up to 100 seconds and more can be studied. The accuracy of the method depends directly on the pressure transducer. It amounts to ± 0.1 mN/m for example for the set-up described by Nagarajan & Wasan (1993). By applying definite volume changes, linear or step-type ones, this set-up can be used simultaneously for relaxation studies. Such experiments are described in the next chapter.

As discussed by MacLeod & Radke (1993) the growing drop instrument in the design shown in Fig. 5.19 provides three experimental techniques: a maximum drop pressure, a continuously growing drop and drop volume experiment. The recorded pressure as a function of time and the constant liquid flow, produced by the precise metering system, allows the determination of the maximum pressure in a drop and the volume of a detaching drop. The three sets of data, obtained from growing drop experiments at different liquid volume flow rates Q are shown in Fig. 5.20 for an aqueous 1-decanol solution.

Under certain conditions, the hydrodynamics of growing drops can cause flow pattern inside and outside the drop, dependent on the liquid flow rate. Such situations were described above in connection with the drop volume method.

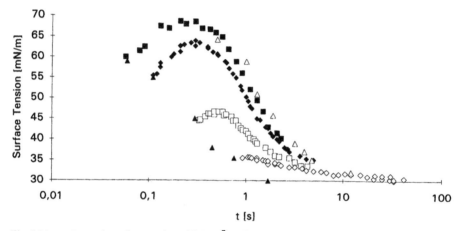

Fig. 5.20   Dynamic surface tension of $2.5 \cdot 10^{-7}$ mol/cm³ aqueous 1-decanol drops according to MacLeod & Radke (1993); maximum drop pressure data ($\blacktriangle$), drop volume data ($\triangle$), growing drop data; $Q = 100$ mm³/min ($\blacksquare$), 50 mm³/min ($\blacklozenge$), 20 mm³/min ($\square$), 5 mm³/min ($\lozenge$); $r_{cap} = 0.255$ mm

According to Liggieri et al. (1990), the process of bubble of drop growth is stable when a stability number BSN, defined by

$$BSN = \frac{27 \pi \, Pr_{cap}^2}{8 \gamma V},$$
(5.29)

is well above unity, BSN > 1. A detailed analysis of the drop growth and the process of adsorption at the drop surface was presented by MacLeod & Radke (1994) and was discussed above in Chapter 4.

5.6.   THE OSCILLATING JET METHOD

One of the oldest experimental methods for the measurement of dynamic surface tensions of surfactant solutions is the oscillating jet (OJ) method. The idea is based on the analysis of a stationary jet issuing from a capillary pipe into the atmosphere which oscillates about its equilibrium circular section. A detailed description of the method and the corresponding theory was given by Defay & Pétré (1971). The basic theory for an analysis of horizontal liquid jets was already given by Bohr (1909). Later, Hansen et al. (1958) extended the Bohr equation for vertical jets.

Fig. 5.21 shows a liquid jet ejected from a cylindrical capillary of diameter $2r_{cap}$. Prior to break-up into individual drops, the liquid jet surface oscillates with wavelength λ and amplitude $\Delta r = r_{max} - r_{min}$. The surface tension can be calculated from the wave parameters of two adjacent nodes using the Bohr equation,

$$\gamma = \frac{2 \cdot 10^{-3} \rho u^2 (1 + 1.542 \frac{b^2}{r^2}) \left(1 + 2(\frac{\eta\lambda}{\rho u})^{3/2} + 3(\frac{\eta\lambda}{\rho u})^2\right)}{3r\lambda^2 + 5\pi^2 r^3}.$$

(5.30)

ρ and η are the density and viscosity of the liquid and u is liquid flow rate in [ml/s]. The stream radius r and the amplitude of the fluctuation b are defined by

$$r = \frac{1}{2}(r_{max} + r_{min})(1 - \frac{1}{6}\frac{b^2}{r^2})$$

(5.31)

and

$$\frac{b}{r} = \frac{r_{max} - r_{min}}{r_{max} + r_{min}}.$$

(5.32)

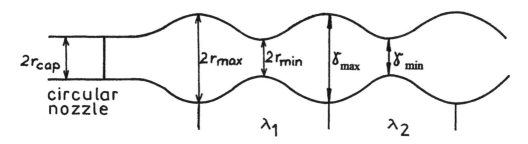

Fig. 5.21   Sketch of an oscillating jet

The analysis of the jet geometry yields surface tension values at different distances from the orifice. The corresponding time t to each of the surface tension values results from the distance d and the mean linear velocity of the jet,

170

$$t = \frac{d}{u / \pi r^2},$$

(5.33)

where d is the distance between the orifice and the midpoint of the wave.

The oscillating jet method provides dynamic surface tensions in a time interval from 3 ms - 50 ms and has been used by many authors (for example Bohr 1909, Addison 1943, 1944, 1945, Rideal & Sutherland 1952, Defay & Hommelen 1958, Thomas & Potter 1975a, b, Fainerman et al. 1993a, Miller et al. 1994d).

5.7. THE INCLINED PLATE TECHNIQUE

Another method applicable over longer times, from 50 ms up to seconds, is the so-called inclined plate (IP) method developed by Van den Bogaert & Joos (1979). A surfactant solution is allowed to flow over an inclined plate with an angle of inclination α. The solution is pumped over the plate and the flow rate is measured by the rotameter. The surface tension of the solution surface, freshly formed at the inlet, is measured by using a Wilhelmy plate oriented with its length parallel to the flow direction. The schematic of the set-up is shown in Fig. 5.22.

Wilhelmy Balance

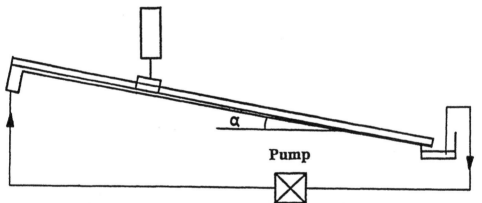

Fig. 5.22   Schematic of the inclined plate according to Van den Bogaert & Joos (1979); α - tilting angle of the plate

From the measured surface tension as a function of the distance from the inlet, adsorption kinetics data are obtained by recalculating the effective age of the surface at a distance x. Due to an analysis of the flow pattern and the effect of surface tension gradients on it given by Hansen (1964) the effective age is calculated from the surface velocity of the flowing liquid v_s by $\tau_a = x / v_s$.

An analysis of the complex problem of liquid flow and adsorption kinetics was given by Ziller et al. (1985) (cf. Section 4.8). The system of linearised hydrodynamic and transport equations was solved numerically under the assumption of different boundary conditions. The results show that a quantitative interpretation of inclined plate experiments requires a time consuming analysis. However, a rather good approximation is possible by neglecting the Marangoni flow induced by surface tension gradients against the flow direction.

The inclined plate technique is applicable to liquid/gas interfaces only. Temperature control seems to be very complicated. As the time is also limited to a small interval of 25 ms up to a few seconds, the application of this method is restricted only to some special cases. However, the comparison of experimental results of the inclined plate technique with those obtained by other methods exhibits good agreement (Van Hunsel & Joos 1987a, Fainerman et al. 1993a, Miller et al. 1994b, d).

5.8.   DESCRIPTION OF OTHER DYNAMIC METHODS FOR INTERFACIAL STUDIES

There are many other experimental method for studying the dynamics of adsorption at liquid interfaces. First of all, many other techniques exist to measure dynamic surface and interfacial tensions. Only a subjective selection of some more experimental developments are given in the following section. Moreover, other than surface and interfacial techniques are discussed in this chapter too, such as radiotracer, ellipsometer, electric potential, and spectroscopic methods.

5.8.1.   OTHER DYNAMIC SURFACE AND INTERFACIAL TENSION METHODS

The following so-called dynamic capillary method was developed by Van Hunsel & Joos (1987b) and complements the area of application with respect to other methods. This method allows measurements from 50 ms up about 1 s, similar to the inclined plate and growing drop techniques described above, and can be used at liquid/liquid and liquid/gas interfaces without modification. The principle of the experiment is schematically given in Fig. 5.23. Two fluids are contained in a tube of diameter R. The interface (or surface in case of studies at the water/air interface) is located in such a way that its interfacial tension can be measured by the capillary rise of the lower liquid in a narrow capillary c, which connects the both fluids. The height of the capillary rise h is determined via a cathetometer Cat.

When the liquid flows through the tube T with a constant flow rate the interface moves downwards with the velocity v. Under the assumption that the liquid film adheres to the wall of the capillary new surface is created by the receding liquid meniscus. Assuming a hemispherical meniscus, the dilation rate is given by

$$\Theta = \frac{1}{A}\frac{dA}{dt} = \frac{v}{R}. \tag{5.34}$$

Depending on the flow rate v a stationary state is reached (Van Voorst Vader et al. 1964) and the surface or interfacial tension in function of the dilation rate results

$$\gamma(\Theta) = \gamma_e + \frac{\Delta\rho g \Delta h R}{2}, \tag{5.35}$$

where Δh is the difference in capillary rise height with respect to the equilibrium of adsorption with the interfacial tension γ_e. The analysis of the flow system yields an effective interfacial age of $\tau_a = 1/2\Theta$.

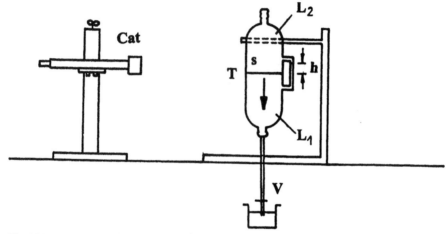

Fig. 5.23   Set-up of the dynamic capillary method according to Van Hunsel & Joos (1987b)]; Cat - cathetometer, c - capillary, L_1 and L_2 - two immiscible liquids, s - interface, h - capillary rise height, T - tube

Although many other experimental set-ups were developed to study the dynamics of adsorption, mainly via surface and interfacial tensions, of solutions of surface active compounds and polymers, they cannot all be described in detail here. More are given in textbooks of surface chemistry, e.g. by Adamson (1990) or Edwards et al. (1991). The last original technique, briefly discussed in this chapter, is the overflowing cylinder method used for example by Bergink-Martens et al. (1990).

It consists of a cylinder through which a liquid is pumped upwards (cf. Fig. 5.24). In order to ensure a laminar flow field, the liquid first passes through a conical tube having a small slope. After the liquid has reached the cylinder, it is allowed to flow over the top rim. This flow causes the circular liquid surface to be expanded continuously in a radial way. After a steady state has been established a definite surface dilation pattern exists. In the centre of the circular surface the dynamic surface tension is measured by the Wilhelmy plate technique.

These experiments also allow the investigation of adsorption dynamics. Berginck-Martens et al. (1990) call the coefficient they determine,

$$\frac{(\gamma(v_r) - \gamma_e)}{d \ln A / dt},$$ (5.36)

surface dilational viscosity, with v_r being the surface radial velocity. Although it has the dimension of a surface dilational viscosity [Ns/m], this coefficient is mainly controlled by the adsorption kinetics of the surfactant. At least, it cannot be distinguished between the intrinsic viscosity and the exchange of surfactant from the experiment.

(a)   **(b)**

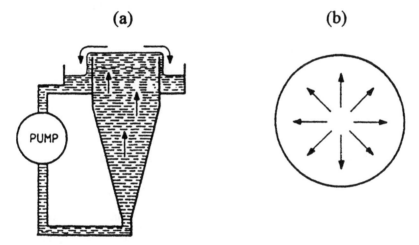

Fig. 5.24   Sketch of the overflowing cylinder set-up, according to Berginck-Martens et al. (1990)

5.8.2. STUDIES OF ADSORPTION KINETICS WITH ALTERNATIVE EXPERIMENTAL TECHNIQUE

Beside the very frequently used methods of dynamic surface and interfacial tension measurements, adsorption kinetics processes at liquid interfaces can also be studied by other methods, such as dynamic surface potentials, ellipsometry and other light scattering and reflection methods, X-ray techniques, neutron scattering, radiotracer techniques. These methods yield more or less relative information on the change of adsorption with time at different time resolutions.

The principles of the measurements of the surface potential in general, and the ΔV-potential in particular, based either on a vibrating capacitor plate or the use of radioactive probes are discussed in Chapter 2. It was shown that only changes in the potential, $\Delta(\Delta V)$, can be determined, as the bare surface already exhibits a finite ΔV-potential.

Dynamic surface potentials were measured by different authors studying surfactant adsorption layers (Kretzschmar & Vollhardt 1968, Kretzschmar 1972, Carroll & Haydon 1975, Sharma 1978, Dragcevic et al. 1986, Bois & Baret 1988, Vogel & Möbius 1988a, b) or proteins at liquid interfaces (Ghosh & Bull 1963, Graham & Phillips 1979b, Tagaya & Watanabe 1983). With a time resolution of several seconds to hours each experimental set-up can be used to measure $\Delta(\Delta V(t))$-dependencies. A special apparatus was developed by Kretzschmar (1965, 1976) using a rotating cylinder, producing dynamic data in a time scale of part of a second. Another experimental arrangement was proposed by Geeraerts et al. (1993) who constructed a special vibrating capacitor using a toothed wheel which rotates around an axis parallel to that of a liquid oscillating jet. The capacitor can be moved along the oscillating jet so that $\Delta(\Delta V(t))$-values in the ms-range are available.

The measurement of the variation of $\Delta(\Delta V(t))$ often does not give us a direct information on changes in the surfactant or polymer adsorption. So far the models are more or less qualitative. A quantitative model should be able to describe the complex structure of the electric double layer in the interfacial region and to explain peculiarities, such as the surface potential of a bare interface and changes in the potential due to adsorption of nonionic surfactants.

A more recently used method for studying surfactant and polymer adsorption at liquid interfaces is ellipsometry. The idea of ellipsometry is to measure the polarisation of light reflected from a surface. The polarisation of the light depends on the thickness of the adsorption layer, its refractive index and the anisotropy of the refractive index. Modern devices allow the measurement of the thickness and the refractive index of an adsorption layer approximately every second which makes this instrument valuable for adsorption kinetics investigations over an interesting time window of seconds to hours. Ellipsometry was applied to studies of surfactants (Privat et al. 1988, Hirtz et al. 1990, Meunier & Lee 1991, Jiang & Chiew 1993) as well as proteins (de Feijter et al. 1978, Kawaguchi et al. 1988a, b, Krisdhasima et al. 1992) at liquid interfaces. Although models for the computation of adsorbed amounts of surfactants or polymers per unit area have been derived (McCrackin 1969, de Feijter et al. 1978, Cuypers et al. 1983) the accuracy is still lower than for any tension measurements. It seems worth combining ellipsometry with apparatus constructed for dynamic surface or interfacial tension studies, such as the inclined plane or overflowing cylinder. Such types of apparatus would allow the surface tension and the adsorbed amount to be measured simultaneously.

Spectroscopic methods, such as fluorescence recovery and quenching, Fourier-transform infrared spectroscopy (FT-IR), and light reflection technique have been used for studies of adsorbed proteins (for example Burghardt & Axelrod 1981, Thompson et al. 1981, van Wagenen et al. 1982), and surfactant adsorption layers (for example Lösche et al. 1983, Lösche & Möhwald 1989, Daillant et al. 1991, Henon & Meunier 1992, Möhwald 1993). Considerable progress has been made in recent years with respect to the sensitivity of detectors and the efficiency of computers, so that the power of these methods has increased remarkably.

Other recently developed methods have also become available for adsorption studies. The availability of synchrotrons as excellent x-ray sources allows x-ray studies at liquid interfaces (Möhwald et al. 1990, Meunier & Lee 1991, Daillant et al. 1991). The same applies for small angle neutron scattering (SANS, based on the different scattering cross section of hydrogen and deuterium). The use of SANS for dynamic studies of structures in membranes and at interfaces has been shown by different authors (Grundy et al. 1988, Bayerl et al. 1990, Vaknin et al. 1991). The method is characterised by a fast data acquisition and should allow dynamic investigations at freshly formed surfaces, as discussed by several authors (Blake, Howe, Penfold, private communication).

One of the older techniques for measuring directly the adsorbed amount of surfactant molecules or polymers at liquid interfaces is the radiotracer technique. Its idea is the measurement of the radiation emitted by radio-labelled molecules, adsorbed at an interface (Sally et al. 1950, Flengas & Rideal 1959). Because of the background radiation the method yields relative data only. Using equilibrium adsorption isotherms, the dynamics of adsorption can also be followed by the radiotracer method. Experiments were performed with various surfactant systems (Matuura et al. 1958, 1959, 1961, 1962, Tajima 1970, Konya et al. 1973, Muramatsu et al. 1973, Okumura et al. 1974) and adsorbed polymers (Frommer & Miller 1966, Adams et al. 1971, Graham & Phillips 1979b). Due to the development of more efficient methods the use of this technique has been reduced.

In summary, in combination with facilities for producing fresh surfaces and interfaces, all these methods provide powerful tools for investigating the formation of the adsorption layer and the dynamics of its structuring. This combination will enable the interfacial dynamics of soluble adsorption layers to be investigated.

The formation of interfacial aggregates is known from studies of insoluble monolayers . For example, Möhwald et al. (1986) and Barraud et al. (1988) studied the formation of 3D-aggregates in monolayers. The dynamics of aggregate formation in insoluble monolayers was studied by Vollhardt (1993) and 2- and 3-dimensional nucleations are discussed and described theoretically using different models (Retter & Vollhardt, Vollhardt et al. 1993). Such theories, combined with bulk transport processes, would give the first physical models for treating inhomogenous soluble adsorption layers, as observed by Berge et al. (1991), Flesselles et al. (1991) or Hénon & Meunier (1992).

5.9.   EXPERIMENTAL RESULTS FROM STUDIES OF SURFACTANTS AT LIQUID INTERFACES

In this section a selection of experimental data is presented which demonstrate that different experimental technique partly overlap or complement each other in their range of application. Cases of partial overlap should not be interpreted as needless developments but more as additional resources to obtain information on dynamics of adsorption. On the one hand, experimental data of the same surfactant solution are performed to demonstrate the agreement between different experimental techniques. Many examples of experiments with different techniques are given in a recent paper by Miller et al. (1994b, d).

On the other hand, results are also presented to show the experimental limitations of methods and measuring procedures. As mentioned above, the knowledge of an adsorption isotherm of a surfactant is of fundamental importance for the study of adsorption dynamics. For the surfactants discussed in this chapter, the parameters of the Frumkin isotherm are summarised in the tables of Appendix 5D. In case the interfacial interaction parameter a' is zero, the Frumkin isotherm changes into a Langmuir isotherm.

As an experimental prerequisite for studies at liquid/liquid interfaces, the two liquids have to be mutually saturated. For example, even solvents like alkanes are remarkably soluble in water and a transfer of solvent molecules across the interface would influence surfactant adsorption kinetics. The table in Appendix 5E summarises the mutual solubility of some solvents with water.

5.9.1.   THE EFFECTIVE SURFACE AGE IN ADSORPTION KINETICS EXPERIMENTS

The determination of the effective surface age is the key for comparison of results obtained by different experimental techniques. If for example the drop volume technique is used in its "classical" version, which is based on continuously growing drops, dynamic surface tensions are obtained as a function of drop formation time. It was shown in the previous chapter, that the process of adsorption at the surface of a growing drop is overlapped by a radial flow inside the drop, which changes the diffusion profile. In addition, the drop area increases and simultaneously the adsorption decreases. This process is in opposition to the adsorption kinetics and slows down the adsorption rate.

On the contrary, the so-called "quasi-static" drop volume version is based on drops having almost constant volume and surface area. Thus, the result of these experiments is a dynamic surface tension as a function of adsorption time. In Fig. 5.25 a comparison of both types of experimental data is shown plotted as surface tension vs. $1/\sqrt{t}$.

If the classical and quasi-static results are compared, different times are necessary to reach one and the same surface tension value. The ratio of the two values amounts to about 3, in good agreement with the theory described above (cf. section 4.5.).

Similar situations exist for other methods. If quantitative theories are applied to data interpretation, all peculiarities of a method have to be considered. However, this does not allow a direct comparison of experimental results from different methods, because the time scales are different and depend on the specific experimental conditions. A solution to this problem is the determination of the effective surface age, which then allows a direct comparison between experimental data independent of their origin.

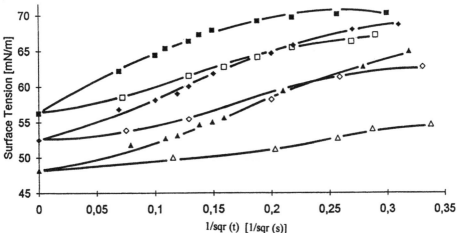

Fig. 5.25   Comparison of dynamic surface tension data obtained from "classical" and "static" drop volume experiments for three different dodecyl dimethyl phosphine oxide solutions; c_o=2·10^{-8} mol/cm^3 (□■), 3·10^{-8} mol/cm^3 (◊♦), 5·10^{-8} mol/cm^3 (△▲); dynamic drop - (■♦▲) and static drop - (□◊ △); data taken from Miller & Schano (1990)

The diagram in Fig. 5.26 shows schematically the ratio between the experimental time t, which is the life time of a drop or bubble in the respective experiment, and the effective adsorption time τ_a. It becomes clear that each experimental method works under specific conditions and, therefore, different relations between a specific experimental time and the effective adsorption time or surface age exist. While the effective age τ_a in a maximum bubble pressure experiment

178

is about 50% or more of the bubble life time at the moment of maximum pressure, the τ_a-values are smaller than 30% in a drop volume experiment. In growing drop experiments, due to the different changes in drop surface area the relation between time t and τ_a is rather complex and can be estimated only schematically so far.

A comparison of some experimental data, obtained from different techniques, is presented below. All data are shown as a function of the effective adsorption time and can therefore be interpreted by the theories elaborated for a standard situation, neglecting any liquid flow, surface area changes etc.

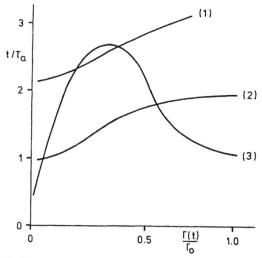

Fig. 5.26   Schematic relations between specific experimental time scales and the effective surface age (adsorption time) for several methods; drop volume (1), maximum bubble pressure (2), growing drop (3)

5.9.2. COMPARISON OF SEVERAL EXPERIMENTAL TECHNIQUE

The aim of the present section is the comparison of selected adsorption kinetics data obtained with different experimental methods. It will be demonstrated that results of different origin show good agreement if prerequisites discussed above are fulfilled.

In a recent paper Miller et al. (1994d) discussed parallel experiments with a maximum bubble pressure apparatus and a drop volume method (MPT1 and TVT1 from LAUDA, respectively), and oscillating jet and inclined plate instruments, performed with the same surfactant solutions. As shown in Fig. 5.27, these methods have different time windows. While the drop volume and bubble pressure methods show only a small overlap, the time windows of the inclined plate and oscillating jet methods are localised completely within that of the bubble pressure instrument.

At longer adsorption times a large variety of static and quasi-static methods exists, reaching into the time domain of hours and even days.

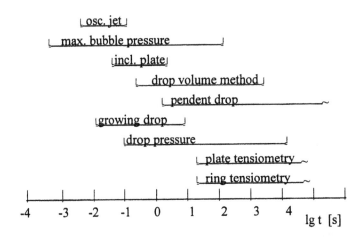

Fig. 5.27   Overlap of time windows of different dynamic experimental technique

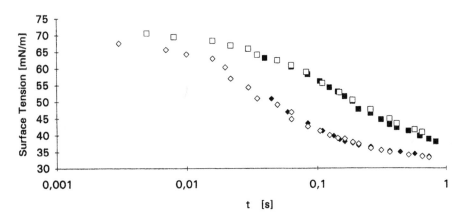

t [s]

Fig. 5.28   Dynamic surface tension of two TRITON X-100 solutions measured using the maximum bubble pressure (□◇) and inclined plate (■◆) methods; c_0= 0.2 (■□); 0.5 (◆◇) g/l; according to Fainerman et al. (1994a)

A comparison of the bubble pressure with the inclined plate method, performed with aqueous solutions of Triton X-100 (octylphenol polyglycol ether, $C_{14}H_{21}O(C_2H_4O)_{10}H$ from Serva), is displayed in Fig. 5.28. The time interval of the inclined plate from 50 ms to 1000 ms is completely contained in the one of the bubble pressure method, from 1 ms to 10 s. The $\gamma\,/\log \tau_a$ - plot, based on the calculated effective surface age τ_a for both methods, shows good agreement.

A comparison of the bubble pressure method with the oscillating jet method was also performed with aqueous Triton X-100 solutions. Some results are given in Fig. 5.29 as a $\gamma/\log \tau_a$ - plot. In contrast to the inclined plate, the oscillating jet only yields data in the time interval of few milliseconds. Also in this time interval the agreement with the maximum bubble pressure method is excellent and shows deviations only within the limits of the accuracy of the two methods.

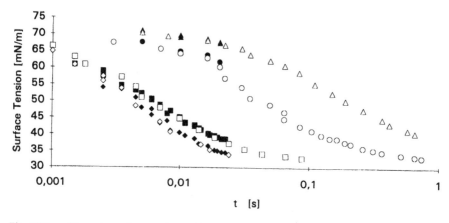

Fig. 5.29   Dynamic surface tension of four TRITON X-100 solutions measured using the maximum bubble pressure (□◇○△) and oscillating jet (■◆●▲) methods; c_o= 0.2 (△△), 0.5 (●○), 2.0 (■□), 5.0 (◆◇) g/l; according to Fainerman et al. (1994a)

Experiments with a maximum bubble pressure and a drop volume set-up were performed with aqueous solutions of an oxyethylated para-tertiary butyl phenol with 10 EO-groups, pt-BPh-EO10 (synthesised and purified by Dr. G .Czichocki, Max-Planck-Institut für Kolloid- und Grenzflächenforschung Berlin). The dynamic surface tension of a 0.025 mol/l solution of pt-BPh-EO10 is shown in Fig. 5.30.

The figure contains the original data as well as the recalculated results in form of surface tension as a function of the effective surface age τ_a. The original data $\gamma(t)$ of the bubble pressure method are transferred into $\gamma(\tau_a)$ by more or less a shift in the $\gamma/\log t$ - plot, according to Eq. (4.62). The drop volume data were corrected first with respect to the hydrodynamic effect at drop formation times $t < 30$ s using Eq. (5.17) and then the effective surface age τ_a was calculated from the drop formation time t_{drop} using the approximate relation $\tau_a = t_{drop}/3$ (cf. Section 5.9.1.).

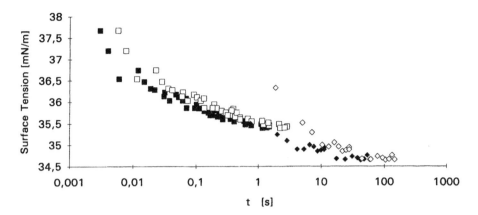

Fig. 5.30   Dynamic surface tension of a 0.025 mol/l pt-BPh-EO10 solution measured using the maximum
bubble pressure (■□) and drop volume (◆✧) methods; original data (□✧), corrected data
(■◆); according to Miller et al. (1994d)

It becomes clear that the apparent surface tension is significantly increased by up to 1 mN/m at
drop times up to about 10 s due to the hydrodynamic effect. Only the corrected dynamic
surface tensions γ as functions of the recalculated effective surface age τ_a are displayed in the
following Figs 5.31 and Fig. 5.32.

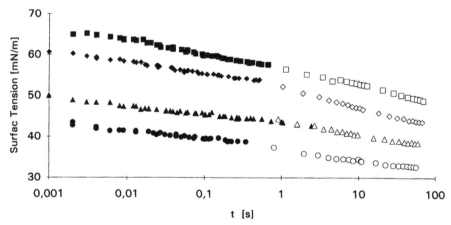

Fig. 5.31   Dynamic surface tension of five pt-BPh-EO10 solutions measured using the maximum bubble
pressure (■◆●▲) and drop volume (□○◊△) methods; c_o= 0.0001(■□); 0.0005 (◆✧); 0.001
(▲△); 0.0025 (●○) mol/l; according to Miller et al. (1994d)

The dynamic surface tensions of aqueous solutions of pt-BPh-EO10 at five concentrations are
shown in Fig. 5.31, obtained from complementary bubble pressure and drop volume
measurements. The curves show the typical course of $\gamma(\log \tau_a)$-behaviour for a diffusion

controlled adsorption process. At medium concentrations a slight shoulder is observed, which is not expected for a pure surfactant system. A diffusion controlled adsorption of a surfactant does not show this shape of curves in $\gamma(\log t)$-plot. Thus, the shoulder can be attributed either to a surface active contamination in the sample or to changes in the adsorption mechanism.

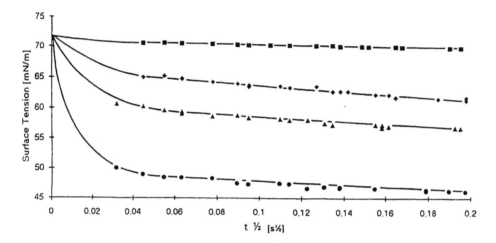

Fig. 5.32   Dynamic surface tension as a function of the square root of surface age for four pt-BPh-EO10 solutions measured using the maximum bubble pressure method; $c_o=$ 0.0001(■); 0.0005 (◆); 0.001 (△); 0.0025 (●) mol/l; according to Miller et al. (1994d)

The quantitative analysis of the adsorption mechanism (cf. Miller & Kretzschmar 1991) shows a diffusion controlled adsorption over the whole concentration range with a slight change of the diffusion coefficient D with adsorption time and surfactant concentration. A detailed data analysis with butyl phenols of different chemical structure is in progress.

When analysing the data in the range of short adsorption times, a $\gamma(\sqrt{t})$-plot is useful (Van Hunsel & Joos 1987a, Fainerman et al. 1994a). The results from Fig. 5.31 are shown in this form in Fig. 5.32. Only at low concentrations can a reasonable diffusion coefficient be calculated. For higher concentrations, the final slope of $\gamma(\sqrt{t})$ is located outside the experimental range, so that D cannot be determined from the plot.

The dynamic capillary method described above also yields experimental data in a time interval from 50 ms up to several seconds (cf. Fig. 5.27). A comparison with other methods is shown in Fig. 5.33 for an aqueous Triton X-100 solution. Again, the agreement is excellent and the deviations do not exceed the error in the individual methods.

Summarising, the agreement of experimental data, obtained with methods of different physical principles, was shown to be excellent when displayed as functions of the effective surface age.

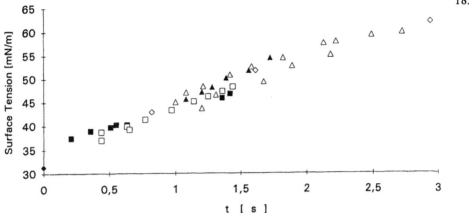

Fig. 5.33   Dynamic surface tensions of an aqueous $1.55 \cdot 10^{-7}$ mol/cm³ Triton X-100 solution measured with the dynamic capillary (◊), inclined plate (▲,△), drop volume (□), strip (■) and Wilhelmy plate (♦) methods; according to Rillaerts & Joos (1982)

5.9.3. SIMULTANEOUS PROCESS OF ADSORPTION KINETICS AND TRANSFER ACROSS THE INTERFACE

As pointed out above, peculiarities of the surfactant and solvent system have to be considered in experiments and their theoretical discussion. If a surfactant is lost from the interface by evaporation into the gas phase or by dissolution into a solvent, then further molecules have to adsorb from the bulk of the surfactant-rich phase.

In the first case, either the theoretical model has to allow for the evaporation process or evaporation has to be avoided by the establishment of special experimental conditions. MacLeod & Radke (1994) report on the adsorption kinetics of 1-decanol at the aqueous solution interface using the growing drop method. They distinguish between three cases: decanol in the aqueous phase only, decanol in the air phase only, decanol in both phases. The adsorption kinetics shows different behaviour and is fastest for the case of decanol in both phases (Fig. 5.34). The application of a proper theory (for example Miller 1980, MacLeod & Radke 1994) in all three cases is a diffusion-controlled mechanism of the decanol adsorption kinetics.

A different type of experiments was performed by Van Hunsel & Joos (1987b). They studied the steady state of adsorption of various alkanols at the alkane/water interface by means of the drop volume method (Fig. 5.35). The steady states differ remarkably from the equilibrium state. A description of the adsorption process has therefore to allow for a transfer of hexanol molecules across the hexane/water interface. The difference of the two studied steady states is

184

determined by the distribution of hexanol between water and hexane. As shown in section 4.6. and discussed in more detail by Miller (1980), the local distribution equilibrium of the surfactant at the interface, given by the bulk distribution coefficient, has a direct effect on the adsorption process.

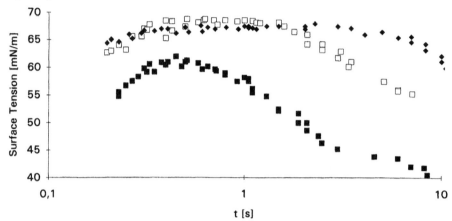

Fig. 5.34   1-decanol solution adsorption at the water/air interface, (■) - decanol in the aqueous phase, (□) - decanol in the air phase, (♦) - decanol in both phases; according to MacLeod & Radke (1994)

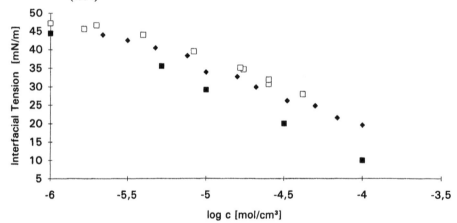

Fig. 5.35   Steady state experiments with hexanol at the hexane/water interface, (■) - equilibrium values, steady state when hexanol is dissolved in water (□) and hexane (♦), respectively; according to Van Hunsel & Joos (1987)

5.9.4. Determination of Equilibrium Adsorption Data for Slowly Adsorbing Surfactants

The adsorption isotherms play a central role in characterising surfactants. Based on the adsorption isotherm it is possible to classify surfactants with regard to their efficiency.

Moreover, adsorption isotherms, or equations of state, represent the basis for the evaluation of adsorption kinetics and rheological properties of adsorption layers. Exact equilibrium values of surface or interfacial tensions are necessary to determine adsorption isotherms. For surfactants of low surface activity (for example, sodium octyl or decyl sulphate, hexanol or hexanoic acid) the adsorption reaches its equilibrium state in a time of the order of seconds to minutes. Higher surface activity results in greater times for establishing the equilibrium state of adsorption which sometimes cannot be realised by available experimental methods. To avoid long-time experiments, extrapolations were often carried out in order to get equilibrium values. Different extrapolation procedures as well as criteria of an equilibrium state of adsorption are discussed in the literature (cf. Miller & Lunkenheimer 1983).

An obvious method is to define the equilibrium state as the condition of constant surface or interfacial tension over an appropriate time interval, taking into account the accuracy of the method used. Difficulties are inherent to such procedures as the time interval and the accuracy of the method chosen are subjective. For example, if a state is called as an equilibrium one after the surface tension has changed less than 0.1 mN/m within a time period of 10 minutes, this state is characterised by the relation,

$$\frac{d\gamma}{dt} \leq 1.7 \cdot 10^{-4} \frac{mN}{m\,s}. \tag{5.37}$$

Miller & Lunkenheimer (1983) showed by model calculations, that this condition is useful only for comparatively weak surfactants. If a surfactant is characterised, for example, by the parameters,

$$RT\Gamma_\infty = 10mN\,/\,m,\ D = 10^{-5}\,cm^2\,/\,s,\ \Gamma_\infty\,/\,a_L = 10^{-2}\,cm, \tag{5.38}$$

at a relative concentration $c_o\,/\,a_L = 0.5$ the condition (5.37) is already fulfilled at $\Gamma(t)\,/\,\Gamma_\infty = 0.95$ and the corresponding surface tension differs from the real equilibrium value by 0.25 mN/m. For surfactants with higher a activity of $\Gamma_\infty\,/\,a_L = 10^{-1}\,cm$, a difference to the real equilibrium surface tension of 0.95 mN/m would result if Eq. (5.37) is applied. From the calculations given, only for surfactants with a surface activity characterised by $a_L \geq 10^{-7}\,mol\,/\,cm^3$ equilibrium values are obtained by condition (5.37).

A completely erroneous procedure for obtaining data of an adsorption isotherm is the measurement of surface tensions as a function of concentration at a constant adsorption time. Such procedures can simulate a different adsorption behaviour, for example an additional interfacial interaction.

To get equilibrium surface tensions of surfactant solutions different procedures of extrapolation are used in the literature. For example, extrapolations of the following forms are used:

$$\lim_{t\to\infty} (\gamma_o - \gamma(t^{-1}))^{-1}. \text{ (Kloubek 1975),} \tag{5.39}$$

$$\lim_{t\to\infty} \gamma(1/\sqrt{t}) \text{ (Bendure 1971),} \tag{5.40}$$

$$\lim_{t\to\infty} (\gamma_o - \gamma(1/\sqrt{t}))^{-1} \text{ (Baret \& Roux 1968, Baret et al. 1968),} \tag{5.41}$$

$$\lim_{t\to\infty} \left(t\frac{d\gamma(t^{-1})}{dt} + 2\gamma(t^{-1})\right) \text{ (Nakamura \& Sasaki 1970).} \tag{5.42}$$

Based on diffusion controlled adsorption, it was shown that none of these extrapolation procedures have general validity and all yield results of sufficiently high accuracy only in special time and concentration intervals.

A successful way of determining equilibrium adsorption data consists in extrapolation of subsequent dilation/compression cycles (Miller & Lunkenheimer 1983). If γ_{ad} and γ_{des} denote the surface tension after adsorption or desorption (after a compression of the adsorption layer to half of its area) at time t_{ad} and t_{des}, respectively, a good approximate value of the equilibrium surface tension is given by

$$\gamma_e \approx \frac{\gamma_{ad} + \gamma_{des}}{2}. \tag{5.43}$$

Taking $t_{ad} = t_{des} = 900$ s, Eq. (5.43) yields accurate results for surfactants with $a_L \geq 10^{-8}$ mol / cm^3. The extrapolation,

$$\gamma_e = \lim_{1/\sqrt{t} \to 0} \frac{(\gamma_{ad} + \gamma_{des})}{2}, \tag{5.44}$$

gives excellent equilibrium data for highly surface active substances ($a_L \leq 10^{-8}$ mol / cm^3), for which equilibrium surface tension values are not available with conventional methods, as the adsorption would take days or weeks. Examples for the extrapolation (5.44) were given by Wüstneck et al. (1992). The results of three subsequent adsorption/desorption cycles for a sodium dodecyl sulphate solution are demonstrated in Fig. 5.36.

It is clear that a time interval of 90 minutes for each adsorption/compression cycle is not sufficient for the system to reach adsorption equilibrium, so that the difference amounts still to $\gamma_{ad} - \gamma_{des} \approx 0.9$ mN/m. The extrapolation based on Eq. (5.44) is shown in Fig. 5.37. The slow rate of adsorption of the alkyl sulphate ions cannot be explained by a diffusion mechanism. Also the electrostatic retardation, described in Chapter 7, does not justify the time necessary to establish the adsorption equilibrium. Again surface active contaminants are probably responsible for the long time dependence, although all possible care was taken to obtain surface-chemically pure surfactant solutions.

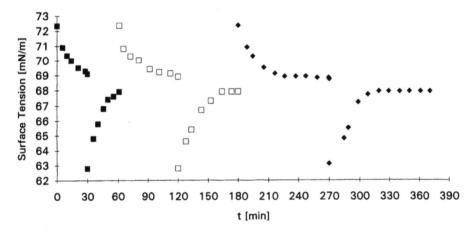

Fig. 5.36   Compression/dilation cycles for a solution of $5 \cdot 10^{-7}$ mol/cm³ sodium dodecyl sulphate; compressions at 30 (■), 120 (□), and 270 (♦) min, aspiration of the adsorption layer at 60 and 180 min; according to Wüstneck et al. (1992)

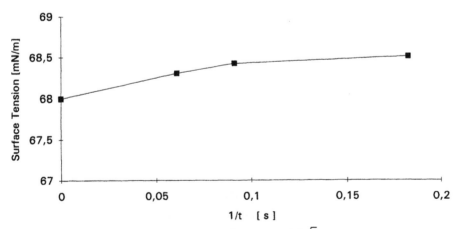

Fig. 5.37   Extrapolation $t \to \infty$ of $(\gamma_{ad} + \gamma_{des})/2$ versus $1/\sqrt{t}$; data from Fig. 5.36, according to Wüstneck et al. (1992)

The validity and high accuracy of the proposed extrapolation procedure calculated on the basis of a diffusion controlled adsorption, has also been proved for surfactant mixtures. For example Wüstneck et al. (1993, 1994) and Fiedler et al (1994) applied this procedure successfully to different mixed surfactant systems.

5.9.5. MEASUREMENTS OF THE FORMATION OF ADSORPTION LAYERS OF SOLUTIONS OF NONIONICS CONTAINING OXYETHYLENE GROUPS

In experiments on nonionic surfactants, namely Triton X-405 Geeraerts at al. (1993) performed simultaneously dynamic surface tension and potential measurements in order to discuss peculiarities of nonionic surfactants containing oxethylene chains of different lengths as hydrophilic part. Deviations from a diffusion controlled adsorption were explained by dipole relaxations. In recent papers by Fainerman et al. (1994b, c, d) and Fainerman & Miller (1994a, b) developed a new model to explain the adsorption kinetics of a series of Triton X molecules with 4 to 40 oxethylene groups. This model assumes two different orientations of the nonionic molecule and explains the observed deviations of the experimental data from a pure diffusion controlled adsorption very well. Measurements in a wide temperature interval and in presence of salts known as structure breaker were performed which supported the new idea of different molecular interfacial orientations. At small concentration and short adsorption times the kinetics can be described by a usual diffusion model. Experiments of Liggieri et al. (1994) on Triton X-100 at the hexane/water interface show the same results.

5.10. EXPERIMENTAL STUDIES OF THE ADSORPTION DYNAMICS OF BIOPOLYMERS AT LIQUID INTERFACES

There has been a long history of studies of protein adsorption at liquid/fluid interfaces. Most studies of the dynamics of adsorption of polymers in general, and proteins in particular, at liquid/fluid interfaces have been performed in measurements of interfacial tensions as a function of time or exchange of matter at periodically changed interfaces (cf. Chapter 6). Different experimental technique has been applied to determine the time dependence of interfacial tension $\gamma(t)$ or surface pressure $\pi(t)=\gamma_0 - \gamma(t)$. (for example Ghosh et al. 1964, Glass 1971, Tornberg 1978a, Sato & Ueberreiter 1979a, b, Ward & Regan 1980, Trapeznikov 1981, Schreiter et al. 1981, Kretzschmar 1994). Other techniques have been used, such as radiotracer methods to determine directly $\Gamma(t)$ (for example Frommer & Miller 1966, Khaiat & Miller 1969, Adams et al. 1971, Graham & Phillips 1979a). The most frequently studied proteins are bovine serum albumin (BSA), casein, lysozyme, ovalbumin, gelatine, human serum albumin

(HA) (Miller 1991). A very extensive study on the properties of several adsorbed proteins at water/air and water/oil interfaces, such as adsorption kinetics and isotherms and dilational and shear rheological behaviour, was given by Graham & Phillips (1979a, b, c, 1980 a, b). Even after such a systematic work, there is still a striking lack of fundamental understanding.

Early experiments of MacRitchie & Alexander (1963) showed that the adsorption behaviour of different proteins can be described by a simple reaction relation given by Eq. (4.80). Important investigation of the reversibility of protein adsorption

were performed by several authors. Very original attempts were made by MacRitchie (1991, 1993) using for example a Langmuir trough technique at constant surface pressure to record the loss of spread protein molecules. He also performed desorption experiments to determine the degree of denaturation of adsorbed proteins. As a result, MacRitchie (1986, 1989) showed the high degree of reversibility of adsorption of many proteins.

For many proteins a time lag of surface tension changes is observed after the formation of a fresh surface. De Feijter & Benjamins (1987) defined this time lag as induction period. Wei at al. (1990) discussed the induction period of several proteins, such as cytochrome-c (CYTC), myoglobin (MYG), superoxide dismutase (SOD), lysozyme and ribonuclease-A. Induction periods of up to 25 minutes are observed at low concentrations, as one can see in Fig. 5.38 for three of the five studied proteins.

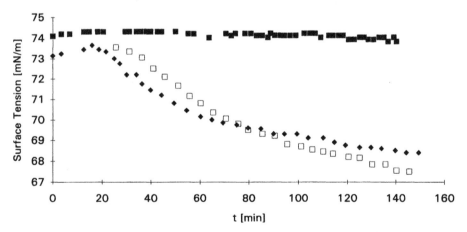

Fig. 5.38   Dynamic surface tensions of three proteins at the water/air interface at 0.01 mg/ml; SOD(■), MYG(□), CYTC(♦); according to Wei at al. (1990)

The long induction times are explained by Wei at al. (1990) as the consequence of denaturation processes of adsorbed protein molecules. At higher concentration, the same proteins adsorb without a visible induction period. The experimental data were described by empirical rate equations based on two independent processes (Wei et al. 1990).

Similar induction times were observed for low lysozyme, β-casein and HA concentrations at the aqueous solutions/air interface (for example Douillard et al. 1994, Xu & Damodaran 1994). While β-casein shows an induction time of less than 10 minutes, the time lag for lysozyme amounts to about 1 hour (Graham & Phillips 1979a; cf. Fig. 5.39).

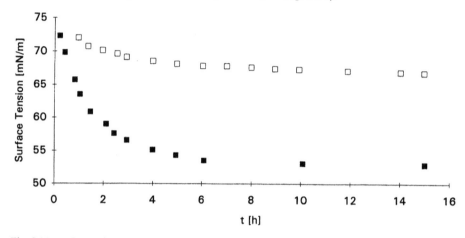

Fig. 5.39   Dynamic surface tension of two protein solutions at the water/air interface; $7.3 \cdot 10^{-5}$wt% ß-casein (■), $7.6 \cdot 10^{-5}$wt% lysozyme (□); according to Graham & Phillips (1979a)

To date there is no experimental evidence of an induction period for protein solutions at higher concentrations. The application of experimental technique to investigate lower surface ages, such as the maximum bubble pressure method, might give more information and help to clarify the process of protein adsorption.

Using the ADSA technique (cf. Rotenberg et al. 1983, Cheng et al. 1990), the adsorption of HA at the water/decane interface was studied (Miller et al. 1993a). At low concentrations an induction time is detectable (Fig. 5.40). At higher concentrations, of 0.015 and 0.02 mg/ml, the interfacial tensions tend to the same equilibrium value. A similar phenomenon was also observed at the water air interface. Even at extremely high HA concentrations of about 50 mg/ml the equilibrium surface tension values are not less than about 51 mN/m, a value which has already been reached at a concentration as low as 1 mg/ml.

One peculiarity should be mentioned in the dependencies given in Fig. 5.38, which can be also seen in Fig. 5.40 for the two lower HA concentrations at the water/decane interface. Just after the formation of a fresh surface, the surface tension is higher than that of pure water/air or water/decane interface. This effect, which is much higher than the accuracy of the methods

(Wilhelmy plate used by Wei et al. 1990, pendent drop technique used by Miller et al. 1993a), is not understood so far.

Dynamic surface tension behaviour of BSA intermediates was measured by Damodaran & Song (1988) with a Langmuir trough technique. The different intermediates were thought to trap the structure of the albumin molecules at different degrees of refolding. The data are interpreted on the basis of a simple diffusion model (cf. Eqs (4.79), (4.80)) and it comes out that the resulting diffusion coefficient is a function of the degree of refolding of the BSA molecule. This is a very important piece of evidence that the structure of adsorbing molecules has a significant influence on their adsorption and relaxation mechanism.

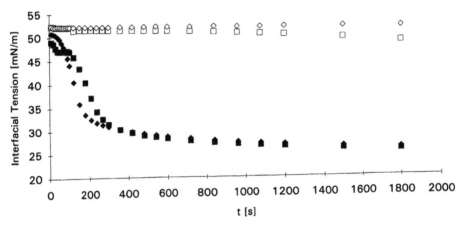

Fig. 5.40   Dynamic surface tension of HA at the water/decane interface at 24°C, $c_o = 0.0075$ mg/ml ($\lozenge$), 0.01 mg/ml ($\square$), 0.015 mg/ml ($\blacksquare$), 0.02 mg/ml ($\blacklozenge$); according to Miller et al. (1993a)

The kinetics of adsorption of BSA water/air interface was also investigated by Serrien et al. 1992), using a quasi-static drop volume experiment (cf. Fig. 5.41). The results at four different concentrations are described by an approximate equation which takes into account a diffusion and reorientation process (Serrien & Joos 1990).

There is one more peculiarity in protein adsorption kinetics which is worthy of note. As shown in Fig. 5.42 (Miller et al. 1993a) slight compressions of the adsorption layers of a 0.1 mg/ml HA solution surprisingly lead to surface tensions as low as 42 mN/m, although the lowest equilibrium surface tension at 50 mg/ml amounts to about 52 mN/m.

This effect is also not fully understood. Possibly, the temporary compression leads not necessarily to a higher surface coverage but may lead to an increase in the adsorption layer thickness or to macromolecular rearrangements, which then result in such a surface tension decrease. More extensively, the relaxation behaviour of proteins at liquid interfaces is discussed in Chapter 6.

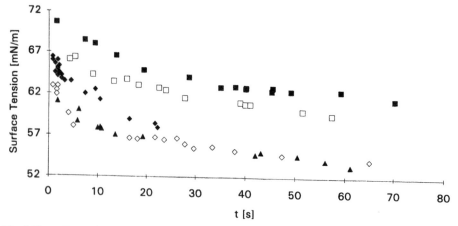

Fig. 5.41   Dynamic surface tension of BSA at the water/air interface at 22°C, c_o = 0.0335 mg/ml (■), 0.1 mg/ml (□), 1.0 mg/ml (♦), 2.3 mg/ml (◊), 4.8 mg/ml (▲); according to Serrien et al. (1992)

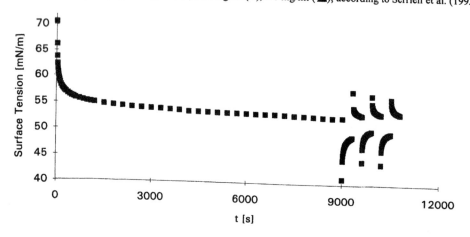

Fig. 5.42   Dynamic surface tension of HA at the water/air interface at 24°C, c_o = 0.05 mg/ml (■), 3 square pulse are changes of about 20% were performed, according to Miller et al. (1992)

5.11. SUMMARY

This chapter presents the state of the art of experimental equipment available for dynamic surface and interfacial tension measurements. The introductory section demonstrates the crucial importance of a special purity of surfactants and solvents, called surface-chemical purity, for interfacial studies. It turns out that studies of surfactant adsorption layers could be tremendously affected by surface active impurities. Results obtained on adsorption layer properties could, therefore, be dominated by the adsorbed contaminants instead of the main surfactant under investigation.

The development of on-line computer-coupled measuring technique has ushered in a new area of experiments. Most of the newly developed apparatus for measuring interfacial properties are equipped with interfaces allowing direct control of the measurement and an efficient data acquisition and interpretation. For example, classical methods such as the drop volume and maximum bubble pressure methods are now running on-line, with a considerably improved accuracy and efficiency. Methods like the pendent and sessile drop technique became useful only after the drop shape detection and the fitting to the Gauss-Laplace equation of capillary has been computerised. The recently developed software package ADSA (Rotenberg et al. 1993 and Chen et al. 1990) allows the interfacial tension, the drop volume and area to be determined simultaneously with high precision. The data when used in combination with an accurate metering system would lead to the first device for measuring interfacial tensions, keeping the interfacial area constant. At the same time it would allow for controlled definite interfacial area changes.

A new group of experiments became possible after the development of highly precise pressure transducers which were directly applicable to liquids. These experiments are based on the measurement of the capillary pressure inside a drop. The technique exists so far only in form of laboratory set-ups but provides many advantages compared with others (MacLeod & Radke 1993, Nagarajan & Wasan 1993). Again combined with a highly precise metering system this technique could be developed as a very efficient tool for liquid surface characterisations.

As pointed out in the preceding chapter phase transitions in adsorption layers as observed in insoluble monolayers (Möhwald 1993), can also be obtained for soluble surfactants too. Structures in mixed monolayers in particular were investigated by Hénon & Meunier (1993). These overlapping processes are now starting to be considered in new studies of equilibrium properties (cf. Lin et al. 1991, Lunkenheimer & Hirte 1992) or in dynamics of insoluble monolayers in terms of nucleation processes (Vollhardt et al. 1993). So far no attempts have been made to consider such transition or structure formation processes are considered in the dynamics of soluble adsorption layers.

One of the future challenges will of course be the further development of faster methods shifting the available time windows into the µs range. Recent improvements of the bubble pressure technique already allow definite information in the ms range to be obtained (Fainerman et al. 1994a). As research in technology calls for experimental tools to look at even smaller times a further refinement of the bubble pressure technique can provide dynamic adsorption data in the range of some hundreds or even tens of μs in the near future.

Future developments will also focus on the combination of different techniques, such as drop pressure and drop shape methods. A more efficient approach would be to combine macroscopic with microscopic or molecular methods, for example drop shape or pressure experiments with ellipsometric or spectroscopic techniques. Another useful possibility involves linking, for example, the inclinded plate or overflowing cylinder technique with scattering experiments, which would allow studies of structure formation under dynamic conditions and at freshly formed surfaces (Howe et al. 1993).

More or less systematic studies have been carried out on nonionic surfactants below the CMC but there is lack of systematic studies on micellar and mixed surfactant solutions. Moreover, there is an almost complete lack of studies on ionic surfactants, as discussed in Chapter 7. It seems that comprehensive experiments on adsorption dynamics can be performed on the basis of the recent theories and considerably improved experimental technique in order to understand the formation and action of dynamic adsorption layers better. This of course applies unrestrictedly to proteins, and mixed surfactant/protein systems where the level of understanding is even lower than for surfactants solutions (de Feijter & Benjamins 1987, Serrien et al. 1992).

5.12. REFERENCES

Adam, N.K. and Shute, H.L., Trans. Faraday Soc., 31(1935)204
Adam, N.K. and Shute, H.L., Trans. Faraday Soc., 34(1938)758
Adams, D.T., Evans, M.T.A., Mitchell, J.R., Phillips, M.C. and Rees, P.M., J. Polymer Sci., Part C, 34(1971)167
Adamson, A.W., Physical Chemistry of Surfaces, 5th Edition, John Wiley & Sons, Inc., New York, 1990
Addison, C.C., J. Chem. Soc., (1943)535
Addison, C.C., J. Chem. Soc., (1944)252
Addison, C.C., J. Chem. Soc., (1945)98
Addison, C.C., J. Chem. Soc., (1946a)570
Addison, C.C., J. Chem. Soc., (1946b)579
Addison, C.C., Bagot, J., McCauley, H.S., J. Chem. Soc., (1948)936
Addison, C.C. and Hutchinson, S.K., J. Chem. Soc., (1948)943
Addison, C.C. and Hutchinson, S.K., J. Chem. Soc., (1949a)3387
Addison, C.C. and Hutchinson, S.K., J. Chem. Soc., (1949b)3406
Ambwami, D.S. and Fort, Jr., T., in "Surface and Colloid Science" (R.J. Good and R.R. Stromberg, Eds.), Vol. 11, pp. 93-119, Plenum, New York 1979

Anastasiadis, S.H., Chen. J.-K., Koberstein, J.T., Siegel, A.F., Sohn, J.E. and Emerson, J.A., J. Colloid Interface Sci., 119(1987)55

Andreas, J.M., Hauser, E.A. and Tucker, W.B., J. Phys. Chem., 42(1938)1001

Austin, M., Bright, B.B. and Simpson, E.A., J. Colloid Interface Sci., 23(1967)108

Babu, S.R., J. Colloid Interface Sci., 115(1987)551

Baret, J.F. and Roux, R.A., Koll. Z. & Z. Polymere, 225(1968)139

Baret, J.F., Armand, L., Bernard, M. and Danoy, G., Trans. Faraday Soc., 64(1968)2539

Barraud, A., Flörsheimer, M., Möhwald, H., Richard, J., Ruaudel-Texier, A., and Vandervyver, M., J.Colloid Interface Sci., 121(1988)491

Bayerl, T.M., Thomas, R.K., Penfold, J., Rennie, A., and Sackmann, E., Biophys.J., 57(1990)1095

Bendure, R.L., J. Colloid Interface Sci., 35(1971)238

Berge, B., Faucheux, L., Schwab, K., and Libchaber, A., Nature (London), 350(1991)322

Bergink-Martens, D.J.M., Bos, H.J., Prins, A. and Schulte, B.C., J. Colloid Interface Sci., 138(1990)1

Bohr, N., Phil. Trans. Roy. Soc., London Ser. A, 209(1909)281

Bois, A.G. and Baret, J.F., Langmuir, 4(1988)1358

Bordi, S., Carla, M. and Cecchini, R., Electrochimica Acta, 34(1989)1673

Boucher, E.A., Evans, M.J.B. and Jones, T.G.J., Adv. Colloid Interface Sci., 27(1987)43

Brady, A.P., J. Phys. Chem., 53(1949)56

Burghardt, T.P. and Axelrod, D., Biophys. J., 33(1981)455

Carla, M., Cecchini, R. and Bordi, S., Rev. Sci. Instrum., 62(1991)1088

Carroll, B.J. and Haydon, D.A., J. Chem.Soc.Faraday Trans.1, 71(1975)361

Carroll, B.J., J. Colloid Interface Sci., 86(1982)586

Carroll, B.J. and Doyle, P.J., J. Chem.Soc.Faraday Trans.1, 81(1985)2975

Cheng, P., PhD Thesis, University of Toronto, 1990

Cheng, P., Li, D., Boruvka, L., Rotenberg Y. and Neumann, A.W., Colloids &Surfaces, 43(1990)151

Cuypers, P.A., Corsel, J.W., Janssen, M.P., Kop, J.M.M., Hermens, W.T., and Hemker, H.C., J. Biol. Chem., 258(1983)2426

Czichocki, G., Vollhardt, D. and Seibt, H., Tenside Detergents, 18(1981)320

Daillant,J., Bosio,L., Benattar,J.J., and Blot,C., Langmuir, 7(1991),611

Damodaran,S. and Song, K.B., Biochim. Biophys. Acta, 954(1988)253

Davies, J.T. and Rideal, E.K., "Interfacial Phenomena", Academic Press, New York, 1969

Defay, R. and Hommelen, J.R., J. Colloid Sci., 13(1958)553

Defay, R. and Pétré, G., in "Surface and Colloid Science", Vol. 3, E.Matijevic (Ed.), Wiley-Interscience, New York, 1971

de Feijter, J.A., Benjamins, J. and Veer, F.A., Biopolymers, 17(1978)1759

de Feijter, J. and Benjamins, J., in "Food Emulsions and Foams", E. Dickinson (Ed.), Special Publ. no. 58, Royal Soc. Chem., (1987)72

de Feijter, J. , Benjamins, J. and Tamboer M., Colloids & Surfaces, 27(1987)243

Douillard, R., Daoud, M., Lefebvre, J., Minier, C., Lecannus, G. and Coutret, J., J.Colloid Interface Sci., 163(1994)277

Doyle, P.J. and Carroll, B.J., J. Phys. E: Sci. Instrum., 22(1989)431

Dragcevic, D., Milunovic, M. and Pravdic, V., Croat.Chem.Acta, 59(1986)397

Edwards, D.A., Brenner, H. and Wasan, D.T., Interfacial Transport Processes and
    Rheology, Butterworth-Heineman Publishers, Stoneham, 1991

Edwards, D.A., Kao, R.L. and Wasan, D.T., J. Colloid Interface Sci., 155(1993)518

Elworthy, P.H. and Mysels, K.J., J. Colloid Interface Sci., 21(1966)331

Fainerman, V.B., Koll. Zh., 39(1977)106

Fainerman, V.B., Koll. Zh., 41(1979) 111

Fainerman, V.B. and Lylyk,S.V., Koll. Zh., 44(1982)598

Fainerman, V.B., Koll. Zh., 52(1990)921

Fainerman, V.B., Colloids & Surfaces, 62(1992) 333

Fainerman, V.B., Makievski, A.V., Koll. Zh., 54(1992a)75

Fainerman, V.B., Makievski, A.V. , Koll. Zh., 54(1992b)84

Fainerman, V.B., Makievski, A.V. and Joos, P., Zh. Fiz. Khim., 67(1993a)452

Fainerman, V.B., Makievski, A.V. and Miller, R., Colloids & Surfaces A, 75(1993b)229

Fainerman, V.B., Miller, R. and Joos, P., Colloid Polymer Sci., 272(1994a)731

Fainerman, V.B., Makievski, A.V. and Joos, P., Colloids & Surfaces A, 90(1994b)213

Fainerman, V.B., Miller, R. and Makievski, A.V., Langmuir, (1994c), submitted

Fainerman, V.B., Makievski, A.V. and Miller, R., Colloids & Surfaces A, 87(1994d)61

Fainerman, V.B. and Miller, R., Colloids & Surfaces A, (1994a), submitted

Fainerman, V.B. and Miller, R., J. Colloid Interface Sci., (1994b), submitted

Fang, J.P. and Joos, P., Colloids & Surfaces A, 83(1994)63

Fiedler, H., Wüstneck, R., Weiland, B., Miller, R. and Haage, K., Colloids & Surfaces, (1994),
    in press

Finch, J.A. and Smith, G.W., J. Colloid Interface Sci., 45(1973)81

Flengas, S.N. and Rideal, E., Trans. Faraday Soc., 55(1959)339

Flesselles, J.M., Magnasco, M.O., and Libchaber, A., Phys.Rev.Lett., 67(1991)2489

Fordham, S., Proc. Roy. Soc. London, A194(1948)1

Frommer, M.A. and Miller, I.R., J. Colloid Interface Sci., 21(1966)245

Garrett, P.R. and Ward, D.R., J. Colloid Interface Sci., 132(1989)475

Geeraerts, G., Joos, P. and Villé, F., Colloids & Surfaces A, 75(1993)243

Ghosh, S. and Bull, H.B., Biochemistry, 2(1963)411

Ghosh, S., Breese, K. and Bull, H.B., J. Colloid Sci., 19(1964)457

Gilanyi, T., Stergiopulos, C. and Wolfram, E., Colloids Polymer Sci., 254(1976)1018

Girault, H.H.J., Schiffrin, D.J. and Smith, B.D.V., J. Colloid Interface Sci., 101(1984)257

Glass, J.E., J. Polymer Sci., Part C., 44(1971)141

Graham, D.E. and Phillips, M.C., J. Colloid Interface Sci., 70(1979a)403

Graham, D.E. and Phillips, M.C., J. Colloid Interface Sci., 70(1979b)415

Graham, D.E. and Phillips, M.C., J. Colloid Interface Sci., 70(1979c)427

Graham, D.E. and Phillips, M.C., J. Colloid Interface Sci., 76(1980a)227

Graham, D.E. and Phillips, M.C., J. Colloid Interface Sci., 76(1980b)240

Grundy, M., Richardson, R.M., Roser, S.J., Penfold, J., and Ward, R.C., Thin Solid Films,
    159(1988)43

Handbook of Chemistry and Physics, 60th Edition, 1983

Hansen, F.K. and Rødsrud, G., J. Colloid Interface Sci., 141(1991)1

Hansen, R.S., Rurchase, M.E., Wallace, T.C. and Woody, R.W., J. Phys. Chem., 62(1958)210

Hansen, R.S., J. Phys. Chem., 68(1964)2012

Hénon, S. and Meunier, J., Thin Solid Films, 210/211(1992)121

Hénon, S. and Meunier, J., J. Chem. Phys., 98(1993)9148

Hirtz,A., Lawnik,W. and Findenegg,G.H., Colloids & Surfaces, 51(1990)405

Holcomb, C.D. and Zollweg, J.A., J. Colloid Interface Sci., 154(1992)51

Horozov, T., Danov, K., Kralschewsky, P., Ivanov, I. and Borwankar, R., 1st World Congress on Emulsion, Paris, Vol. 2, (1993)3-20-137

Howe, A., 1993, private communication

Hua, X.Y. and Rosen, M.J., J. Colloid Interface Sci., 124(1988)652

Iliev, T.H. and Dushkin, C.D., Colloid Polymer Sci., 270(1992)370

Jho; C. and Burke, R., J. Colloid Interface Sci., 95(1983)61

Jiang, Q. and Chiew, Y.C., Langmuir, 9(1993)273

Joos, P. and Rillaerts, E., J. Colloid Interface Sci., 79(1981)96

Joos, P., Fang, J.P. and Serrien, G., J. Colloid Interface Sci., 151(1992)144

Joos, P., Vollhardt, D. and Vermeulen, M., Langmuir, 6(1990)524

Kao, R.L., Edwards, D.A., Wasan, D.T. and Chen, E., J. Colloid Interface Sci., 148(1993)247

Kawaguchi, M., Tohyama, M., Mutoh, Y., Takahashi, A., Langmuir, 4(1988a)407

Kawaguchi,M., Tohyama,M., andTakahashi,A., Langmuir, 4(1988b)411

Khaiat, A. and Miller, I.R., Biochim. Biophys. Acta, 183(1969)309

Kloubek, J., Tenside, 5(1968)317

Kloubek, J., J. Colloid Interface Sci., 41(1972a)1

Kloubek, J., J. Colloid Interface Sci., 41(1972b)7

Kloubek, J., Colloid Polymer Sci., 253(1975)754

Kloubek ; J.; K. Friml and F. Krejci, Czech. Chem. Commun., 41(1976)1845

Konya, J., Kovacs, Z., Joo, P. and Madi, I., Acta Phys.Chim. Debrecina, 18(1973)203

Kragh, A.M., Trans. Faraday Soc., 60(1964)225

Kretzschmar, G., Kolloid-Z. Z. Polymere, 206(1965)60

Kretzschmar, G. and Vollhardt, D., Monatsber. DAW, 10(1968)203

Kretzschmar, G., Tenside Detergents, 9(1972)267

Kretzschmar, G., Int. Konferenz über Grenzflächenaktive Stoffe, Vol. 1, (1976)567

Kretzschmar, G. and Miller, R., Adv. Colloid Interface Sci., 36(1991)65

Kretzschmar, G., J. Inf. Rec. Mats., 21(1994)335

Krisdhasima,V., McGuire,J., Sproull,R., Surface and Interface Analysis, 18(1992)453

Krotov, V.V. and Rusanov, A.I., Koll. Zh., 39(1971)58

Kuffner, R.J., J. Colloid. Sci., 16(1961)797

Lemlich, R. Ind. Eng. Chem., 60(1968)16

Liggieri, L., Ravera, F. and Passerone,A., J. Colloid Interface Sci., 140(1990)436

Liggieri, L., Ravera, F., Ricci, E. and Passerone, A., Adv. Space Res., 11(1991)759

Liggieri, L., Ravera, F. and Passerone, A., J. Colloid Interface Sci., 168(1994), in press

Lin, S.-Y., McKeigue, K. and Maldarelli, C., Langmuir, 7(1991)1055

Loglio, G., Legittimo, P.C., Mori, G., Tesei, U. and Cini, R., Chemistry in Ecology, 2(1986)89

Loglio, G., Degli-Innocenti, N., Tesei, U., Cini, R. and Wang, Q.-S., Il Nuovo Cimento, 12(1989)289

Lohnstein, T., Ann. Physik, 20(1906a)237

Lohnstein, T., Ann. Physik, 20(1906b)606

Lohnstein, T., Ann. Physik, 21(1907)1030

Lohnstein, T., Z. phys. Chem., 64(1908)686

Lohnstein, T., Z. phys. Chem., 84(1913)410

Lösche, M. and Möhwald, H., J.Colloid Interface Sci., 131(1989)56

Lösche, M., Sackmann,E. and Möhwald,H., Ber.Bunsengesellschaft Phys.Chem., 87(1983)848

Lunkenheimer, K. and Miller, R., Tenside Detergents, 16(1979)312

Lunkenheimer, K., Miller, R. and Becht, J., Colloid Polymer Sci., 260(1982a)1145

Lunkenheimer, K., Miller, R. and Fruhner, H., Colloid Polymer Sci., 260(1982b)599

Lunkenheimer, K., Habilitation, Academy of Sciences of the GDR, Berlin 1984

Lunkenheimer, K., Miller, R., Kretzschmar, G., Lerche, K.-H. and Becht, J., Colloid Polymer Sci., 262(1984)662

Lunkenheimer, K. and Miller, R., J. Colloid Interface Sci., 120(1987)176

Lunkenheimer, K., Pergande, H.-J. and Krüger, H., Rev. Sci. Instrum., 58(1987a)2313

Lunkenheimer, K., Haage, K. and Miller, R., Colloids Surfaces, 22(1987b)215

Lunkenheimer, K. and Miller, R., Material Science Forum, 25-26(1988)351

Lunkenheimer, K. and Hirte, R., J. Phys. Che., 96(1992)8683

Lunkenheimer, K. and Czichocki, G., J. Colloid Interface Sci., 160(1993)509

MacLeod, C.A. and Radke, C.J., J. Colloid Interface Sci., 160(1993)435

MacLeod, C.A. and Radke, C.J., J. Colloid Interface Sci., 166(1994)73

MacRitchie, F. and Alexander, A.E., J. Colloid Sci., 18(1963)453

MacRitchie, F., Adv. Colloid Interface Sci., 25(1986)341

MacRitchie, F., Colloids & Surfaces, 41(1989)25

MacRitchie, F., Analytica Chimica Acta., 249(1991)241

MacRitchie, F., Colloids & Surfaces, 76(1993)159

Makievski, A.V., Fainerman, V.B. and Joos, P., J. Colloid Interface Sci., in press ???

Markina, Z.N., Zadymova, N.M. and Bovkun, O.P., Colloids & Surfaces, 22(1987)9

Matuura, R., Kimizuka, H., Miyamoto, S. and Shimozawa, R., Bull. Chem. Soc. Jpn., 31(1958)532

Matuura, R., Kimizuka, H., Miyamoto, S., Shimozawa, R. and Yatsunami , K., Bull. Chem. Soc. Jpn., 32(1959)404

Matuura, R., Kimizuka, H. and Matsubara, A., Bull. Chem. Soc. Jpn., 34(1961)1512

Matuura, R., Kimizuka, H., Matsubara, A., Matsunobu, K. and Matsuda , T., Bull. Chem. Soc. Jpn., 35(1962)552

Maze, C. and Burnet, G., Surface Sci., 24(1971)335

McCrackin, F.L., Natl. Bur. Stand., (US), Tech. Note, 479(1969)

McMillan, N.D., Fortune, F.J.M., Finlayson, O.E., McMillan, D.D.G., Townsend, D.E., Daly, D.M., Fingleton, M.J., Dalton, M.G. and Cryan, C.V., Rev. Sci. Instrum., 63(1992)3431

McMillan, N.D., O'Mongain, E., Walsh, J.E., Orr, D., Ge, Z.C. and Lawlor, V., SPIE, 2005(1994)216

Meunier, J. and Lee, L.T., Langmuir, 7(1991)1855

Miller, R., Koll. Zh., 42(1980)1107

Miller,R. and Lunkenheimer, K., Colloid Polymer Sci., 260(1982)1148

Miller,R. and Lunkenheimer, K., Colloid Polymer Sci., 261(1983)585

Miller,R. and Lunkenheimer, K., Colloid Polymer Sci., 264(1986)273

Miller,R. and Schano, K.-H., Colloid Polymer Sci., 264(1986)277

Miller, R., Habilitation, Academy of Sciences of the GDR, Berlin 1987

Miller, R. and Schano, K.-H., Tenside Detergents, 27(1990)238

Miller, R. and Kretzschmar, G., Adv. Colloid Interface Sci., 37(1991)97

Miller, R., in "Trends in Polymer Science", 2(1991)47

Miller, R., Hoffmann, A., Hartmann, R., Schano and K.-H., Halbig, A., Advanced Materials, 4(1992)370

Miller, R., Policova, Z., Sedev., R. and Neumann, A.W., Colloids & Surfaces, 76(1993b)179

Miller, R., Sedev., R., Schano, K.-H., Ng, C. and Neumann, A.W., Colloids & Surfaces, 69(1993c)209

Miller; R., Schano, K.-H. and Hofmann, A., Colloids & Surfaces A, (1994a) in press

Miller, R., Joos, P. and Fainerman, V.B., Adv. Colloid Interface Sci., 49(1994b)249

Miller, R., Policova, Z. and Neumann, A.W., Colloid Polymer Sci., (1994c), submitted

Miller, R., Joos, P. and Fainerman, V.B., Progr. Colloid Polymer Sci., (1994d) in press

Miller,T.E. and Meyer,W.C., American Laboratory, February, 91 (1984)

Milliken, W.J., Stone, H.A. and Leal, L.G., Phys. Fluids, A5(1993)69

Möhwald, H., Miller, A., Stich, W., Knoll, W., Ruaudel-Texier, A., Lehmann, T., and Fuhrhop, J.-H., Thin Solid Films, 141(1986)261

Möhwald, H., Kenn, R.M., Degenhardt, D., Kjaer, K., and Als-Nielsen, J., Physica, 168(1990)127

Möhwald, H., Rep. Prog. Phys., 56(1993)653

Muramatsu, M., Tajima, K., Iwahashi, M. and Nukina , K., J. Colloid Interface Sci., 43(1973)499

Mysels, K.J. and Florence, A.T., in "Clean Surfaces: Their Preparation and Characterization for Interfacial Studies" (G.Glodfinger, Ed.), Marcel Dekker, New York, 1970

Mysels, K.J. and Florence, A.T., J. Colloid Interface Sci., 43(1973)577

Mysels, K.J., Langmuir, 5(1989)442

Nagarajan, R. and Wasan, D.T., J. Colloid Interface Sci., 159(1993)164

Nahringbauer, I., Acta Pharm. Suec., 24(1987)247

Naidich, Y.V. and Grigorenko, N.F., J. Mater. Sci., 27(1992)3092

Nakamura, M. and Sasaki, T., Bull. Chem. Soc. Japan, 43(1970)3667

Neumann, A.W. and Good, R.J., in "Surface and Colloid Science" (R.J. Good and R.R. Stromberg, Eds.), Vol. 11, pp. 31-91, Plenum, New York 1979

200

Nunez-Tolin, V., Hoebregs, H., Leonis, J. and Paredes, S., J. Colloid Interface Sci., 85(1982)597

Okumura , T., Nakumura , A., Tajima, K. and Sasaki, T., Bull. Chem. Soc. Jpn., 47(1974)2986

Padday; J.F. and Russell; D.R., J. Colloid Sci., 15(1960)503

Padday, J.F., in "Surface and Colloid Science" (E.Matijević and F.R. Eirich, Eds.), Vol. 1, Wiley-Interscience, New York 1969

Padday ; J.F.; Pitt ; A.R.; Pashley; R.M., J. Chem.Soc.Faraday Trans.1, 61(1975)1919

Passerone, A., Liggieri, L., Rando, N., Ravera, F. and Ricci, E., J. Colloid Interface Sci., 146(1991)152

Papeschi, G., Bordi, S. and Costa, M., Naa. Chim., (1981)407

Paulsson, M. and Dejmek, P., Journal Colloid Interface Sci., 150(1992) 394

Privat, M., Bennes, R., Tronel-Peyroz, E. and Douillard, J.-M., J.Colloid Interface Sci., 12(1988)198

Racca, R.G., Stephenson, O. and Clements, R.M., Opt.l Eng., 31(1992)1369

Ramakrishnan, S., Mailliet, K. and Hartland, S., Proc. Indian Acad. Sci., 83A(1976)107

Razouk, R. and Walmsley, D., J. Colloid Interface Sci., 47(1974)515

Rehbinder, P.A., Z. Phys. Chem., 111(1924)447

Rehbinder, P.A., Biochem. Z., 187(1927)19

Retter, U. and Vollhardt, D., Langmuir, 9(1993)2478

Rideal, E.K. and Sutherland, K.L., Trans. Faraday Soc., 48(1952)1109

Rillaerts, E. and Joos, P., J. Colloid Interface Sci., 88(1982)1

Rosen, M.J., J. Colloid Interface Sci., 79(1981)587

Ross, J.L., Bruce, W.D. and Janna, W.S., Langmuir, 8(1992)2644

Rotenberg , Y., Boruvka, L. and Neumann, A.W., J. Colloid Interface Sci., 93(1983)169

Sally, D.J., Weith, A.J. jr. , Argyle, A.A. and Dixon, J.K., Proc. Roy. Soc., Ser. A, 203(1950)42

Sato, H. and Ueberreiter, K., Macromol. Chem., 180(1979)829

Sato, H. and Ueberreiter, K., Macromol. Chem., 180(1979)1107

Schramm, L.L. and Green, W.H.F., Colloid Polymer Sci., 270(1992)694

Schreiter, W., Wolf, F. and Walther, W., J. Signal AM, 9(1981)63

Serrien, G. and Joos, P., J.Colloid Interface Sci., 139(1990)149

Serrien, G., Geeraerts, G., Ghosh, L. and Joos, P., Colloids & Surfaces, 68(1992)219

Sharma, M.K., Indian J. Chem., 16(1978)803

Somasundaran, P., AIChE, Symposium Ser., No. 150, 71(1975)1

Sörensen, J.M. and Arlt, W., in "Liquid-liquid equilibrium data collection", Vol. V, Part I: Binary Systems, Frankfurt/M., 1979

Stauffer, C.A., J. Phys. Chem., 69(1965)1933

Sugden, S., J. Chem. Soc., 125(1924)27

Tagaya, H. and Watanabe, A., Membrane, 8(1983)31

Tajima, K., Bull. Chem. Soc. Jpn., 43(1970)3063

Thomas, W.D.E. and Potter, L. J. Colloid Interface Sci., 50(1975a)397

Thomas, W.D.E. and Potter, L. J. Colloid Interface Sci., 51(1975b)328

Thompson , N.L., Burghardt, T. P. and Axelrod, D., Biophys. J., 33(1981)435

Timmerman, J., Physico-Chemical Constants of Pure Organic Compounds, Elsevier, New York 1950

Tornberg, E., J. Colloid Interface Sci., 60(1977)50

Tornberg, E., J. Colloid Interface Sci., 64(1978a)391

Tornberg, E., J. Sci. Fd Agric., 29(1978b)762

Trapeznikov, A.A., Koll. Zh., 43(1981)322

Tzykurina, N.N., Zadymova, N.M., Pugachevich, P.P., Rabinovich, I.I. and Markina, Z.N., Koll. Zh., 39(1977)513

Vaknin, D., Kjaer, K., Als-Nielsen, J., and Lösche, M., Biophys.J., 59(1991)1325

Van den Bogaert, P. and Joos, P., J. Phys. Chem., 83(1979)2244

Van Hunsel, J., Bleys, G. and Joos, P., J. Colloid Interface Sci., 114(1986)432

Van Hunsel, J., "Dynamic Interfacial Tension at Oil-Water Interfaces", Thesis 1987, University of Antwerp

Van Hunsel, J. and Joos, P., Colloids & Surfaces, 24(1987a)139

Van Hunsel, J. and Joos, P., Langmuir, 3(1987b)1069

Van Hunsel, J. and Joos, P., Colloid Polymer Sci., 267(1989)1026

van Os, N.M., Haak, J.R. and Rupert, L.A.M., Physico-chemical Properties of Selected Anionic, Cationic and Nonionic Surfactants, Elsevier, Amsterdam, London, New York, Tokyo, (1993)

Van Voorst Vader, F., Erkelens, Th.F. and Van den Tempel, M., Trans. Faraday Soc., 60(1964)1170

van Wagenen, R.A., Rockhold, S. and Andrade , J.D., Adv. Chem. Ser., 199(1982)351

Vogel, V. and Möbius, D., Thin Solid Films, 159(1988a)73

Vogel, V. and Möbius, D., J.Colloid Interface Sci., 126(1988b)408

Vollhardt, D. and Czichocki, G., Langmuir, 6(1990)317

Vollhardt, D., Ziller, M., and Retter, U., Langmuir, 9(1993)3208

Vollhardt, D., Adv.Colloid Interface Sci., 47(1993)1

Ward, A.J.F. and Regan, L.H., J.Colloid Interface Sci., 78(1980)389

Wawrzynczak, W.S., Paleska, I. and Figaszewski, Z., Journal of Electroanalytical Chemistry and Interfacial Electrochemistry, 319(1991)291

Wei, A.P., Herron, J.N. and Andrade, J.D., in "From Clone to Clinics", D.J.A. Crommelin and H.Schellekens (Eds), Kluwer Academic Publishers, The Netherlands, (1990)305

Weiner, N.D. and Flynn, G.L., Chem. Pharm. Bull., 22(1974)2480

Wilkinson, M.C., J. Colloid Interface Sci., 40(1972)14

Woolfrey, S.G., Banzon, G.M. and Groves, M.J., J. Colloid Interface Sci., 112(1986)583

Wüstneck, R., Miller, R. and Czichocki, G., Tenside Detergents 29(1992)265

Wüstneck, R., Miller, R. and Kriwanek, J., Colloids & Surfaces A, 81(1993)1

Wüstneck, R., Miller, R., Kriwanek, J. and Holzbauer, R., Colloids & Surface A, (1994) in press

Xu, S. and Damodaran, S., Langmuir, 10(1994)427

Ziller, M., Miller, R. and Kretzschmar, G., Z. Phys. Chem. (Leipzig), 266(1985)721

CHAPTER 6

6. RELAXATION STUDIES AT LIQUID INTERFACES

In this chapter specific theories and experimental set-ups for interfacial relaxation studies of soluble adsorption layers are presented. A general discussion of relaxation processes, in bulk and interfacial phases, was given in Chapter 3. After a short introduction, in which the important role of mechanical properties of adsorption layers and the exchange of matter for practical applications are discussed, the main differences between adsorption kinetics studies and relaxation investigations are explained. Then, general theories of exchange of matter and specific theories for different experimental techniques are presented. Finally, experimental set-ups, based on harmonic and transient interfacial area deformations, are described and results for surfactant and polymer adsorption layers discussed.

6.1. INTRODUCTION TO INTERFACIAL RELAXATION STUDIES

Knowledge of the mechanism of adsorption and desorption kinetics and exchange of matter has great practical significance in many technologies and a range of natural phenomena are also effected by non-equilibrium properties of interfaces. For example, heterogeneous catalysis and electrochemical reactions (Adamczyk et al. 1987), or liquid extractions (Szymanowski & Prohaska 1988) are controlled by the dynamics of adsorption. In coating processes of photographic emulsions containing gelatine and surfactants dilational and shear mechanical properties as well as exchange of matter at the interface are of importance. The same situation exists for the process of enhanced oil recovery as described by Dorshow & Swafford (1989) or Kovscek et al. (1993). The relevance of these properties to phenomena occurring at the atmosphere-ocean interface, such as aerosol formation, evaporation, wave propagation and damping, is discussed for example by Loglio et al. (1985, 1987, 1989).

6.1.1. PRACTICAL IMPORTANCE OF MECHANICAL PROPERTIES AND EXCHANGE OF MATTER OF SOLUBLE ADSORPTION LAYERS

Basic knowledge has been increased to a level that allows more and more qualitative and even quantitative attempts to describe the effect of surface active compounds in complex liquid disperse systems like foam and emulsion formation and their stabilisation. A general discussion of the role of dynamic surface properties in different areas of application was given by van den Tempel & Lucassen-Reynders (1983). Foam and emulsion formation and stability depend on various specific parameters and the discussions are based on elementary processes, such as stability of thin liquid films, coalescence of two drops or bubbles, break-up process of drops in flow fields, motion of bubbles and drops in a solution (Stebe & Maldarelli 1994). The relevance of dilational properties and exchange of matter mechanisms, along with shear rheological parameters, to phenomena such as foaming and emulsification has already been emphasised and is explained by the rapid expansion of surfaces during the generation of foams or emulsions (Garrett & Joos 1976, Lucassen-Reynders & Kuijpers 1992, Garrett & Moore 1993).

The motion of bubbles in a surfactant solution depends on the bubble size and the properties of the solution/bubble interface (Małysa 1992). Adsorbed surfactants control the bubble motion in a way not fully understood. With increasing surfactant concentration the floating speed increases, passes a maximum and then decreases again (Loglio et al. 1989). The maximum of the speed of floating is passed at very low surfactant concentration so that the phenomenon can even be used as a measure for the contamination of water by surface active compounds.

In a recent review Malysa (1992) discussed the role of effective dilational elasticity and exchange of matter as a contribution to the resistance of wet foam films against disturbances. The stability of wet foams results as a complex process, in which surface activity, adsorption kinetics and effective elasticity of present surfactants have a complementary effect. A qualitative correlation between wet foam stability and effective dilational elasticity was found for some simple surfactants.

The rheology of wet foams was analysed by Wasan et al. (1992) and a relation was derived (Edwards et al. 1991) connecting the foam dilational viscosity K_F to the surface dilational elasticity E_o

$$K_F = \frac{4E_o}{3h} \frac{\dfrac{\gamma_F}{\gamma} - \cos \vartheta}{1 + \dfrac{\cos \vartheta \tan^2 \vartheta}{\dfrac{3}{2\pi} - \tan^2 \vartheta}} \tag{6.1}$$

with

$$\cos \vartheta = \frac{\gamma_F}{\gamma} - \frac{h\Pi(h)}{2\gamma}. \tag{6.2}$$

$\Pi(h)$ is the disjoining pressure, γ_F is the film tension, and h is the film thickness. Wasan et al. (1992) also discussed the stability of foams of surfactants solutions containing oil droplets in the Plateau borders and discussed it in terms of dynamic interfacial parameters. In a recent paper Lucassen-Reynders & Kuijpers (1992) distinguished three stages in the emulsification process: extension of initial drops of the disperse phase in an applied flow field, the break-up of extended drops, and the re-coalescence of drops in the emulsion. The first two stages are qualitatively and to some extend quantitatively discusses by Lucassen-Reynders & Kuijpers (1992), while the third is a matter linked with emulsion film stratification and stability, analysed by Wasan et al. (1992). The drop size of an emulsion directly depends on the effective interfacial tension. If one assumes a diffusional transport of stabilising surfactant and defines the characteristic time of this process

$$\tau_{\text{diff}} = \frac{2}{D}(\frac{d\Gamma}{dc})^2, \tag{6.3}$$

two regions of interfacial tensions in an emulsification process can be distinguished. Under the condition

$$\frac{\partial \ln A}{\partial t} \gg (\tau_{\text{diff}})^{-1}, \tag{6.4}$$

the interfacial tension is close to the one of the pure solvents system, if

$$\frac{\partial \ln A}{\partial t} \ll (\tau_{\text{diff}})^{-1}, \tag{6.5}$$

the interfacial tension is close to the equilibrium value. From the analysis of these conditions the drop size of an emulsion and hence its stability can be estimated. A direct effect of the interfacial viscoelastic behaviour on emulsification process was also discussed by these authors. Although many individual facts are known, there is no comprehensive theory for the foam or emulsion formation process.

These subjects will be described explicitly in the subsequent chapters. It is evident that both the mechanical properties and the exchange of matter at interfaces are of general interest for an improved understanding of the foaming and emulsification processes. Experimental access to these properties of adsorption layers is given by relaxations experiments which provide simultaneously information about the exchange of matter and the dilational interfacial elasticity defined by

$$E_o = -\frac{d\gamma}{d\ln\Gamma}. \tag{6.6}$$

Before the theories of different relaxation methods and the experimental details of several set-ups are discussed, one important fact should be clarified beforehand. There is a significant difference in the study of adsorption kinetics and the relaxation behaviour. Experiments of adsorption kinetics are usually started from an almost bare surface and follow the change of adsorption with time. In contrast, relaxation experiments are based on small deviations from equilibrium. This equilibrium has to be established before. The composition, the structure and the nature of interactions can therefore differ tremendously and diverging results can be obtained.

In the present chapter current relaxation theories will be described first: both damping of harmonically generated disturbances and relaxations to transient perturbations. Thereafter, experiments are described, based on the damping of capillary and longitudinal waves, oscillation behaviour of bubbles. Also transient relaxations with pendent drop and drop and bubble pressure measurements are shown. Finally, applications to different interfaces, using surfactants, surfactant mixtures, polymers and polymer/surfactant mixtures are discussed.

6.1.2. THE DIFFERENT SUBJECTS OF ADSORPTION KINETICS AND RELAXATIONS AT INTERFACES

Before starting with the description of present theories of interfacial relaxations, the difference from adsorption kinetics studies has to be pointed out. The general difference lies in the composition of the adsorption layer. Adsorption kinetics processes, described in previous chapters usually start from an uncovered interface. The species with the highest concentration and surface activity adsorb first. The best measure to estimate the rate of adsorption at the beginning of the process is the ratio of surface concentration Γ_o over bulk concentration c_o. To compare the adsorption rate of two surface active compounds a simplification of Eq. (4.85) can be used, from which the time t_{ad} needed by a surfactant to reach 95% of the equilibrium adsorption results,

$$t_{ad} = \frac{\pi}{4D}\left(\frac{\Gamma_o}{c_o}\right)^2. \tag{6.7}$$

This adsorption time of course, does not represent the time necessary to establish adsorption equilibrium but is only a characteristic adsorption time for comparisons. The ratio $\frac{\Gamma_o}{c_o}$ can be calculated from the Langmuir isotherm, given by Eq. (2.37),

$$\frac{\Gamma_o}{c_o} = \frac{\Gamma_\infty / a_L}{1 + c_o / a_L} \le \frac{\Gamma_\infty}{a_L}. \tag{6.8}$$

Thus, the first estimation of the characteristic adsorption time is obtained from

$$t_{ad} \approx \frac{\pi}{4D} \left(\frac{\Gamma_\infty}{a_L} \right)^2. \tag{6.9}$$

Typical values of the diffusion coefficient and the maximum surface concentration are $D = 5 \cdot 10^{-6} \, cm^2 / s$ and $\Gamma_\infty = 5 \cdot 10^{-10} \, mol/cm^2$. The characteristic adsorption time of a sodium octyl sulphate with $a_L = 3.5 \cdot 10^{-6} \, mol/cm^3$ results to $t_{ad} \approx 0.00004$ s, for sodium dodecylsulfate $t_{ad} \approx 0.009$ s, more than two orders of magnitude higher (cf. Appendix 5D).

From these approximations one can conclude that in time intervals up to the order $O(100 \cdot t_{ad})$ the adsorption kinetics are controlled by the corresponding surfactant, for sodium dodecylsulfate up to some seconds. At longer times, the effect of surface active contaminants have to be taken into consideration. At times $t \gg 100 \cdot t_{ad}$, the properties of the adsorption layer can already be dominated by impurities.

In contrast to adsorption, the process of desorption starts from a more or less pre-established equilibrium adsorption layer. As the compressions performed to initialise desorption processes are usually significant, the situation is comparable to the process of adsorption, although it is much more affected by surface active impurities.

If kinetic processes of adsorption or desorption were observed in time scales $t \gg 100 \cdot t_{ad}$, it is difficult to distinguish between impurity effects and specific effects of the surfactant system, such as electrostatic retardation, phase transitions in adsorption layers, conformational changes, structure formations etc.

In relaxation experiments, the situation is completely different. These experiments are performed after the equilibrium state of adsorption has been established. Relaxations are defined as processes occurring after the equilibrium state has been disturbed by a small amount. The disturbance of an adsorption layer can be produced by changing the surface concentration, the pressure or temperature or other parameters. Relaxations then start to establish an equilibrium state of the system. More general discussions of relaxation processes are given in Chapter 3.

If the situation of the presence of surface active contaminants is reviewed again, the relaxation behaviour after an interfacial area disturbance is controlled simultaneously by the main surfactant, the concentration of contaminant and the ratio between them. The latter depends on the time allowed for the establishment of equilibrium. In most cases only a quasi-equilibrium is reached in real systems.

In the ideal case, where no contamination is present in the system, relaxations enable the study of dynamic behaviour of the adsorption layer. Such investigations yield information about adsorption mechanisms as well as interfacial interactions and transitions of co-existing phases. Due to the small deviation from equilibrium, theories of relaxation experiments are usually easier to derive, because linearisations are justified. Thus, complex processes are better studied by relaxations than by adsorption kinetics.

6.2. INTERFACIAL RELAXATION TECHNIQUES

The two types of relaxations methods, based on harmonic and transient area changes respectively, will be described in this paragraph first from the theoretical and then from the experimental point of view. It will be shown that the exchange of matter functions are generally applicable to both types of relaxations, as pointed out by Loglio et al. (1991b). Finally, experimental details of several methods and examples of results from literature, for surfactants as well as biopolymers, will be discussed.

6.2.1. TECHNIQUES BASED ON HARMONIC INTERFACIAL DISTURBANCES

The method of studying the damping of capillary waves at interfaces is the classical version of all relaxation methods at interfaces. Harmonic waves are generated by mechanical, thermal or electrical means. The response of the system is usually measured mechanically, in terms of a relative damping of the propagated wave. Detailed descriptions of capillary wave techniques including specific theories are given in numerous reviews (for example by Lucassen-Reynders 1973, Lucassen-Reynders et al. 1975, van den Tempel & Lucassen-Reynders 1983, Kretzschmar & Miller 1991). Recent work has focused on theoretical modifications for this technique (de Voeght & Joos 1984, Hård & Neuman 1987, Stenvot & Langevin 1988, Earnshaw et al. 1990, Earnshaw & Hughes, 1991, Henneberg et al. 1992, Jiang et al. 1992, Romero-Rochin et al. 1992, Sun & Shen 1993) as well as on experimental improvements and new equipment (Hühnerfuss et al. 1985, Thominet et al. 1988, McGivern & Earnshaw 1989, Earnshaw & Robinson 1990, Earnshaw & Winch 1990, Sakai et al. 1991, 1992, Jiang et al . 1992).

One of the more recently developed methods to investigate the surface relaxation of soluble adsorption layers due to harmonic disturbances is the oscillating bubble method. The technique involves the generation of radial oscillations of a gas bubble at the top of a capillary immersed into the solution under study. The first set-up was described by Lunkenheimer & Kretzschmar 1975 and Wantke et al. 1980 followed by a new design of apparatus using a novel pressure transducer technique to monitor the pressure changes inside bubbles or drops (MacLeod & Radke 1992, 1993, Stebe & Johnson 1994, Wasan et al. 1993, Chang & Franses 1993, Nagarajan & Wasan 1993, Horozov et al. 1993.

If area changes are performed in an anisotropic way, for example in trough experiments, the theoretical model has to take into account the lateral transport of adsorbed molecules (Lucassen & Giles 1975, Dimitrov et al. 1978, Kretzschmar & König 1981). Assuming isotropic area deformations, the diffusional flux at the interface is given by

$$\frac{1}{A}\frac{d(\Gamma A)}{dt} = D\frac{\partial c}{\partial x} \text{ at } x = 0, \tag{6.7}$$

where $A(t)$ is the time function of the interfacial area. The transport of molecules in the bulk by diffusion is given by Fick's diffusion law (4.9). The solution to the problem has the general form (Lucassen & van den Tempel 1972a)

$$c(x,t) = c_0 + \alpha \exp(\beta x)\exp(i\omega t). \tag{6.8}$$

The boundary condition (6.7) can be rearranged to

$$\frac{d\ln\Gamma}{d\ln A} = -\left(1 + D\frac{\partial c/\partial x}{(d\Gamma/dc)(\partial c/\partial t)}\right)^{-1}. \tag{6.9}$$

Using the definition of the dilational elasticity (cf. Section 3.6.),

$$E_c = \frac{d\gamma}{d\ln\Gamma}\frac{d\ln\Gamma}{d\ln A}, \tag{6.10}$$

the following relation is obtained

$$E_c = -\frac{d\gamma}{d\ln\Gamma}\left(1 + D\frac{dc}{d\Gamma}\frac{\partial c/\partial x}{(\partial c/\partial t)}\right)^{-1}. \tag{6.11}$$

The introduction of Eq. (6.8) leads to the expression for $E(i\omega)$ as the final result,

$$E_c(i\omega) = E_0\frac{\sqrt{i\omega}}{\sqrt{i\omega} + \sqrt{2\omega_0}} \tag{6.12}$$

with

$$E_o = -(\frac{d\gamma}{d\ln\Gamma})_A \text{ and } \omega_o = (\frac{dc}{d\Gamma})^2 \frac{D}{2}.$$ (6.13)

If we assume a Langmuir type equation of state (2.37) to describe the equilibrium adsorption behaviour of the system then the following exchange of matter function is obtained,

$$E_c(i\omega) = RT\Gamma_\infty \frac{c_o}{a_L} \frac{\sqrt{i\omega}}{\sqrt{i\omega} + \frac{(1+c_o/a_L)^2}{\Gamma_\infty/a_L}\sqrt{D}}.$$ (6.14)

To model the exchange of matter at the interface of a mixed surfactant solution the same principle can be used as for the system containing only one surfactant. However, one term for each of the r surface active compounds in the system is needed. Garrett & Joos (1976) derived the complex elasticity modulus which is given by

$$E_c = \frac{d\gamma}{d\ln\Gamma_T} \sum_{i=1}^{r} \frac{\Gamma_i}{\Gamma_T} \frac{d\ln\Gamma_i}{d\ln A}$$ (6.15)

with the total interfacial concentration defined by

$$\Gamma_T = \sum_{i=1}^{r} \Gamma_i.$$ (6.16)

The solution to Eq. (6.15) is found in the same way as described for a single surfactant system and was given in its general form also by Garrett & Joos (1976),

$$E_c(i\omega) = E_o \sum_{i=1}^{r} \frac{\Gamma_i}{\Gamma_T} \frac{\sqrt{i\omega}}{\sqrt{i\omega} + \sqrt{2\omega_{io}}}$$ (6.17)

with

$$\omega_{io} = (\frac{dc_i}{d\Gamma_i})^2 \frac{D_i}{2}.$$ (6.18)

Only for a generalised linear adsorption isotherm the adsorption of the components i are independent and consequently the functions ω_{io}. Assuming a generalised Langmuir type adsorption isotherm Eq. (2.47) the following complex elasticity modulus results,

$$E_c(i\omega) = E_o \sum_{i=1}^{r} \frac{c_i / a_{Li}}{1 + \sum_{j=1}^{r} c_j / a_{Lj}} \frac{\sqrt{i\omega}}{\sqrt{i\omega} + \sqrt{2\omega_{io}}} \qquad (6.19)$$

with

$$\sqrt{2\omega_{io}} = \frac{1 + \sum_{j=1}^{r} c_j / a_{Lj}}{\frac{\Gamma_\infty}{a_{Li}} (1 + \sum_{j \neq i}^{r} c_j / a_{Lj})} \sqrt{D_i} . \qquad (6.20)$$

As mentioned above in Section 4.6., on adsorption kinetics at liquid/liquid interfaces, one must consider that the surfactant is usually soluble in both adjacent phases. Therefore, the exchange of matter takes place in both bulk phases and the diffusion laws must also be applied in both bulk phases.

The result is obtained in an analogous way to the method described above, but the exchange function contains both the diffusion coefficients of the surfactant in the respective phase and the distribution coefficient of the surfactant between the two liquids

$$c_\alpha(0,t) = K c_\beta(0,t). \qquad (6.21)$$

The result is of the same form as before, Eq. (6.12), but the characteristic frequency ω_o is defined by

$$\omega_o = (\frac{dc}{d\Gamma})^2 \frac{(\sqrt{D_\alpha} + \sqrt{D_\beta}/K)^2}{2} . \qquad (6.22)$$

If we assume again a Langmuir type adsorption isotherm (2.37), and note that it must be valid with respect to both liquid phases, we obtain

$$\Gamma_\infty \frac{c_\alpha / a_{L\alpha}}{1 + c_\alpha / a_{L\alpha}} = \Gamma_\infty \frac{c_\beta / a_{L\beta}}{1 + c_\beta / a_{L\beta}} . \qquad (6.23)$$

The distribution coefficient, defined by Eq. (6.21), now reads

$$K = a_{L\alpha} / a_{L\beta} . \qquad (6.24)$$

Usually, the distribution coefficient of surfactants cannot be easily be determined. However, it seems to be reasonable to derive K from relaxation measurements, though there is no experimental evidence to support this.

The models of the complex elasticity, given by Eqs (6.12) and (6.17) are derived for surfactant solutions well below the critical micelle concentration, CMC. Thus it is assumed that no aggregates exist in the solution bulk. Lucassen (1976) derived the respective function for the case of micellar solutions and obtained

$$E_c = E_o \left(1 + (1-i)\zeta \frac{\sqrt{(1-ix)(1+\delta\sqrt{1-ix})^2 + ik(\delta^2 - 1)}}{1 - ik + \delta\sqrt{1-ix}} \right). \tag{6.25}$$

with

$$x = k(1 + n^2 \frac{c_n}{c_1}), \ k = \frac{k_d}{\omega}, \ \delta = \sqrt{\frac{D_n}{D_1}}, \ \zeta = \left(\frac{dc}{d\Gamma}\right)\sqrt{\frac{D_o}{2\omega}}. \tag{6.26}$$

This function has to be used to describe the relaxation behaviour of surfactant solutions above the CMC, or in solutions containing premicelles or other aggregates.

The method of capillary waves was developed more than five decades ago. Many authors contributed to the theory, such as Levich (1941a, b, 1962), Hansen (1964), Hansen & Mann (1964), van den Tempel & van de Riet (1965), Lucassen & Hansen (1966), Stenvot & Langevin (1988), Noskov (1993). New generalisations were made with respect to experimental peculiarities, for example to cylindrical waves (Jiang et al. 1992) or with respect to the analysis of the wave propagation (Lemaire & Langevin 1992). As the damping of capillary waves is a useful method only for small values of the dilational elasticity, a second wave method was developed, the longitudinal wave method. The theory of the damping effects of longitudinal waves was developed by van den Tempel & van de Riet (1965), Lucassen (1968a, b), van den Tempel 1971, Lucassen & van den Tempel 1972a, b). Both theories allow a quantitative interpretation of experimental data and provide information about the dilational elasticity modulus and the amount of the exchange of matter. The exchange of matter was erroneously named dilational viscosity because it has the same physical dimension. When introducing a complete rheological model, an intrinsic dilational viscosity term results, which has to be distinguished from the exchange of matter term. More details on wave damping methods and a discussion of this fact was given above in Chapter 3.

The theory of pulsating bubbles is also a well developed one (Wantke et al. 1980, 1993). Recently, it was generalised by Johnson & Stebe (1994). According to their results, the determination of the frequency dependence of the phase lag between oscillation generation and pressure response should allow the exchange of surfactant and the surface dilational viscosity to be differentiated. There are no experiments available so far to check this hypothesis.

6.2.2. TECHNIQUES BASED ON TRANSIENT INTERFACIAL DISTURBANCES

Both the capillary wave and the oscillating bubble methods use harmonic disturbances of the equilibrium adsorption layer to generate relaxation processes. Their frequency intervals are very different and sometime reach only to few oscillations per minute (Chang & Franses 1993, Wasan et al. 1993). Methods applicable with arbitrary area changes to induce relaxation processes are the Langmuir trough technique (Dimitrov et al. 1978), the elastic ring (Loglio et al. 1986, 1988), different surface dilational methods (Joos & van Uffelen 1992, 1993a, b, Van Uffelen & Joos 1993), or the modified pendent drop experiment (Miller et al. 1993a, b). By moving a barrier at the trough, changing the shape of the elastic ring, lifting the funnel or the strip, or increasing/decreasing the volume of a pendent drop, a variety of area changes can be performed, such as jumps, square waves, ramp type, trapezoidal and harmonic area changes or continuous linear and non-linear expansions. Some of the possible transient area changes are graphed in Fig. 6.1.

The whole theoretical treatment of the derivation of interfacial response functions was discussed recently by Miller et al. (1991). It was shown by Loglio et al. (1991a, b, 1994) that exchange of matter functions derived for harmonic disturbances can be applied to transient ones.

As the result for a diffusion-controlled exchange of matter, using the theory of Lucassen & van den Tempel (1972a), the following functions result when assuming a trapezoidal area change (Loglio et al. 1991a):

$$\Delta\gamma_1(t) = \frac{\Omega E_o}{2\omega_o}\exp(2\omega_o t)\,\mathrm{erfc}(\sqrt{2\omega_o t}) + \frac{2\Omega E_o\sqrt{t}}{\sqrt{2\pi\omega_o}} - \frac{\Omega E_o}{2\omega_o}, \quad 0<t<t_1, \qquad (6.27)$$

$$\Delta\gamma_2(t) = \Delta\gamma_1(t) - \Delta\gamma_1(t-t_1), \qquad\qquad\qquad t_1<t<t_1+t_2, \qquad (6.28)$$

$$\Delta\gamma_3(t) = \Delta\gamma_2(t) - \Delta\gamma_1(t-t_1-t_2), \qquad\qquad t_1+t_2<t<2t_1+t_2, \qquad (6.29)$$

$$\Delta\gamma_4(t) = \Delta\gamma_3(t) - \Delta\gamma_1(t - 2t_1 - t_2), \qquad\qquad t > 2t_1 + t_2, \qquad (6.30)$$

Here, the relative area change is denoted by $\Omega = \dfrac{d\ln A}{dt} = \dfrac{1}{t_1}\ln(1 - \dfrac{\Delta A}{A_o})$. t_1 and t_2 are the characteristic times of the trapezoidal perturbation (c.f. Fig. 6.1). The characteristic frequency is denoted ω_o. For a diffusion controlled exchange of matter ω_o is defined by Eq. (6.13), as it was defined for a harmonic area change. The system theory used to derive the surface tension response functions is explained in Appendix 6A. Response functions to other types of area deformation functions are given in Appendix 6B.

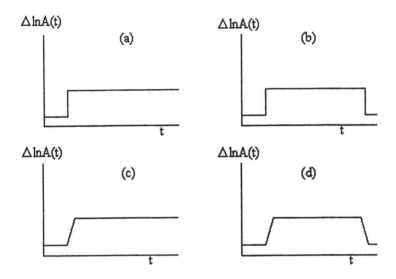

Fig. 6.1.   Schematic of different transient area changes (solid lines) and the corresponding interfacial tension responses (dotted lines): (a) - step type, (b) - ramp type, (c) - square pulse, (d) - trapezoidal change

The derivation of $\Delta\gamma(t)$ for surfactant mixtures and the consideration of specific peculiarities of l/l interfaces was described by Miller et al. (1991). For a surfactant mixture, consisting of r compounds, the functions $\Delta\gamma(t)$ results from the exchange of matter relations Eqs (6.19) and (6.20). For a square pulse the following relations result,

$$\Delta\gamma_1(t) = E_o\frac{\Delta A}{A_o}\sum_{i=1}^{r}\frac{\Gamma_i}{\Gamma_T}\exp(2\omega_{io}t)\,\text{erfc}\left(\sqrt{2\omega_{io}t}\right) \qquad\qquad \text{at } 0 < t < t_2, \qquad (6.31)$$

$$\Delta\gamma_2(t) = \Delta\gamma_1(t) - \Delta\gamma_1(t - t_2) \qquad\qquad \text{at } t > t_2, \qquad (6.32)$$

with the total adsorption Γ_T defined by

$$\Gamma_T = \sum_{i=1}^{r} \Gamma_i, \tag{6.33}$$

and the characteristic frequencies, defined for each of the components by

$$\omega_{io} = \left(\frac{dc_i}{d\Gamma_i}\right)^2 \frac{D_i}{2}. \tag{6.34}$$

The surface tension response function $\Delta\gamma(t)$ of relaxations at the liquid/liquid interface has exactly the same form as the one at the liquid/air interface, except the characteristic frequency is defined in a different way, taking into account the peculiarity of solubility of the surfactant in both adjacent phases (cf. Eq. (6.22)).

As mentioned above, beside the diffusion-controlled models, others exist to describe the adsorption kinetics and exchange of matter. De Feijter et al. (1987) have developed a relation taking into consideration simultaneous adsorption of proteins and surfactants at an interface. As a special case a relation results which describes the equilibrium state of adsorption of polymer molecules at a liquid interface,

$$\frac{\dfrac{\Gamma}{\Gamma_\infty}}{\left(1 - \dfrac{\Gamma}{\Gamma_\infty}\right)^f} = c / a_{dF}, \tag{6.35}$$

with a_{dF} the concentration characterising the surface activity of the polymer, and f the number of adsorption sites per adsorbing molecule.

The isotherm Eq. (6.32) results as equilibrium state of the following kinetic relation

$$\frac{d\Gamma}{dt} = k_{ad} c_o \left(1 - \frac{\Gamma}{\Gamma_\infty}\right)^f - k_{des} \frac{\Gamma}{\Gamma_\infty}, \tag{6.36}$$

where k_{ad} and k_{des} are the rate constants of adsorption and desorption, respectively. This equation can be used to describe the relaxation process of adsorbed polymer molecules. The special case $f = 1$, known as the Langmuir reaction mechanism, was given in Eq. (4.31). Although k_{ad} and k_{des} have the same physical meaning in Eqs (4.31) and (6.36) their definition is different. According to the used isotherm (6.36) its definition results from $\dfrac{d\Gamma}{dt} = 0$ at the equilibrium state of adsorption. Based on the rate equations a model for an adsorption layer relaxation can be derived by considering the time-dependent area A(t). For the case of Eq. (6.36) this modification leads to:

$$\frac{d(\Gamma A)}{A dt} = k_{ad} c_o \left(1 - \frac{\Gamma}{\Gamma_\infty}\right)^f - k_{des} \frac{\Gamma}{\Gamma_\infty}. \tag{6.37}$$

After some rearrangements and using the solution to the problem in the general form (Lucassen & van den Tempel, 1972a)

$$\Gamma = \Gamma_o + \Delta\Gamma \exp(i\omega t), \tag{6.38}$$

the exchange of matter equation reads (Miller et al. 1994b),

$$E_c = E_o \frac{i\omega}{K + i\omega} \tag{6.39}$$

with

$$K = f k_{ad} \frac{c_o}{\Gamma_\infty} (1 - \frac{\Gamma_o}{\Gamma_\infty})^{(f-1)} + \frac{k_{des}}{\Gamma_\infty}. \tag{6.40}$$

By applying the inverse Fourier-Transform, according to the calculus discussed in (Miller et al., 1991), the interfacial tension response can be calculated. Under the assumption of a step-type disturbance the response function reads:

$$\Delta\gamma(t) = E_o \frac{\Delta A}{A_o} \exp(-Kt). \tag{6.41}$$

The surface dilational modulus E_o defined by Eq. (6.10), is related to the surface coverage by the equation

$$E_o = RT\Gamma_\infty \left((1 - f)\frac{\Gamma_o}{\Gamma_\infty} + f \frac{\dfrac{\Gamma_o}{\Gamma_\infty}}{1 - \dfrac{\Gamma_o}{\Gamma_\infty}} \right), \tag{6.42}$$

If we use Eq. (4.31) instead of Eq. (6.36) the same response function results with only a different definition of the parameters K and ε_o:

$$K = k_{ad} \frac{c_o}{\Gamma_\infty} + \frac{k_{des}}{\Gamma_\infty}; \quad E_o = RT\Gamma_\infty \frac{\dfrac{\Gamma_o}{\Gamma_\infty}}{1 - \dfrac{\Gamma_o}{\Gamma_\infty}}. \tag{6.43}$$

It was shown by Loglio et al. (1991a) that the most useful disturbance for interfacial relaxation experiments is the trapezoidal area change. For time regimes realised in most of the transient relaxation experiments the trapezoidal area change can be approximated adequately by a square pulse. For the square pulse area change we obtain:

$$\Delta\gamma_1(t) = E_o \frac{\Delta A}{A_o} \exp(-Kt) \qquad\qquad \text{at } 0 < t \leq t_2, \qquad (6.44)$$

$$\Delta\gamma_2(t) = \Delta\gamma_1(t) - \Delta\gamma_1(t-t_2) \qquad\qquad \text{at } t > t_2, \qquad (6.45)$$

where t_2 is the duration of the pulse and $\dfrac{\Delta A}{A_o}$ is the relative area change (cf. Fig. 6.1).

Eqs (6.41) and (6.44), (6.45), respectively, can be used to interpret experimental data by use of a fitting procedure. The resulting values for ε_o and K are obtained independently of any knowledge of the parameters of the equilibrium adsorption isotherm. On the other hand both parameters contain specific constants of the adsorption isotherm and are therefore of interest also from this point of view.

A very simple relation for the description of relaxation processes was given by Graham & Phillips (1979, 1980),

$$\frac{d\Gamma}{dt} = k_{ad} c_o \exp\left(-\frac{\Pi \Delta A}{RT}\right) - k_{des} \Gamma(t) \exp\left(-\frac{\Pi \Delta A}{RT}\right). \qquad (6.46)$$

This relation can be used instead of Eqs (6.36) or (4.31) to derive a response function equivalent to Eqs (6.44), (6.45).

6.3. INTERFACIAL RELAXATION METHODS

There are many experimental techniques for studying interfacial relaxations of soluble adsorption layers. Except for the wave damping techniques, these methods are developed and used only by individual research groups. Up to now, no commercial set-up exists and therefore, relaxation experiments are not so wide spread. New developments in this field will probably increase the number of investigators studying the dynamic and mechanical properties of adsorption layers, since instruments are easy to construct and data handling is relatively simple. In this section, wave damping and other harmonic methods as well as transient relaxation techniques will be described.

6.3.1. Damping of Capillary and Longitudinal Waves

The oldest relaxation techniques, developed to measure dilational elasticities and exchange of matter, are the methods of wave damping. If a wave is generated by mechanical or other means, it propagates along the surface and is damped by hydrodynamic and surface mechanical properties (Goodrich 1962).

The construction of experimental set-ups differ in the way of wave generation, wave propagation and damping detection. One of the possible designs is shown in Fig. 6.2. The capillary waves are produced by a vibrator attached to a drive unit of a loudspeaker. The wave damping is determined via a microscope and stroboscope. The set-up works in a frequency range from 25 Hz to 4 kHz.

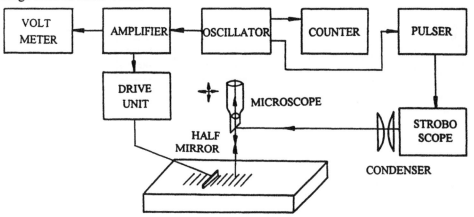

Fig. 6.2.   Schematic diagram of the apparatus for measuring the propagation characteristics of capillary waves; according to Yasunaga & Sasaki (1979)

A new type of capillary waves, cylindrical waves, is used by Jiang et al. (1992). A sharp needle is used as a point source to excite cylindrical waves through electrocapillarity. The wavelength and the damping coefficient are detected with specular reflection of a laser beam. Unlike plane waves, cylindrical waves propagate isotropically in the radial direction and no shear deformation caused by edge effects are expected. The schematic diagram of an experimental set-up designed by Jiang et al. (1992) is shown in Fig. 6.3. The motor allows wave profiler to move the surface along the surface enabling the distance to the wave generation origin to be scanned. Both wave generation and detection avoid direct contact with the interface. The technique is applicable in a frequency range from 50 Hz to 1000 Hz.

To perform measurements of longitudinal wave propagation and damping characteristics, a device designed by Lemaire & Langevin (1992) can be used. While the longitudinal wave is

218

generated by a piezo element, its propagation is measured by analysing the propagation of high-frequency capillary waves, generated by electrocapillarity and analysed via the specular reflection of a laser beam from the interface (Wielebinski & Findenegg 1988, Lemaire & Langevin 1992),. The apparatus, using two types of waves, capillary wave damping to analyse the propagation of longitudinal waves, is schematically shown in Fig. 6.4.

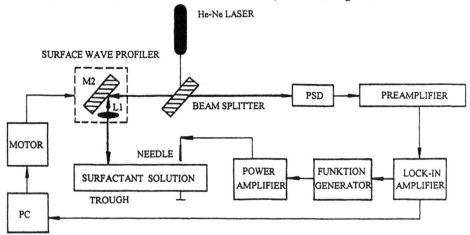

Fig. 6.3   Schematic diagram of the apparatus for the generation and detection of propagating cylindrical waves; according to Jiang et al. (1992)

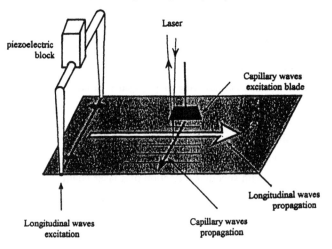

Fig. 6.4   Schematic diagram of the apparatus for the generation and detection of longitudinal waves; according to Lemaire & Langevin (1992)

The set-ups designed by Wielebinski & Findenegg (1988) or Lemaire & Langevin (1992) are also applicable to liquid/liquid interfaces without significant modification.

Wielebinski & Findenegg (1988) used a light scattering method to measure the characteristic of very low amplitude, thermally-induced capillary waves. These ever-present waves, are of very high frequency and usually the exchange of matter can be ignored. As the accuracy of these measurements is very high, light scattering technique is an attractive method to study dynamic interfacial properties.

6.3.2. THE OSCILLATING BUBBLE METHOD

Beside the capillary wave techniques, the oscillating bubble method belongs to the first experiments for measuring the surface dilational elasticity (Lunkenheimer & Kretzschmar 1975, Wantke et al. 1980, 1993). For soluble adsorption layers it allows of the exchange of matter at a harmonically deformed bubble surface to be determined.

The principle of the experiment is shown in Fig. 6.5. A small air bubble is formed at the tip of a capillary which is immersed in the solution. Via an electrodynamic excitation system and a membrane, a gas volume directly connected with the capillary is excited to harmonic oscillations. From the excitation voltage of the system in dependence on frequency, while keeping the bubble oscillation amplitude constant, the dilational elasticity and the exchange of matter can be calculated. The comparatively complex theory for data interpretation was described recently by Wantke et al. (1980, 1993). The method can be applied in a frequency interval from 5 Hz up to about 150 Hz.

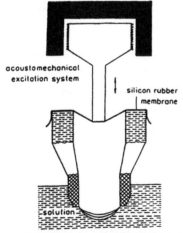

Fig. 6.5   Schematic of a pulsating bubble method according to Lunkenheimer & Kretzschmar (1975) and Wantke et al. (1980)

6.3.3. THE ELASTIC RING METHOD

An experimental set-up for the study of surface relaxation processes in soluble adsorption layers after transient surface area disturbances was developed by Loglio et al. (1986, 1988, 1991a, b). The main feature of the apparatus is the elastic circular ring, which confines the surface area in the sample vessel and substitute for the traditional barrier and trough (Fig. 6.6).

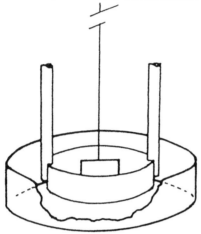

Fig. 6.6   Perspective view of sample vessel, area-confining ring and measuring plate according to Loglio et al. 1988

By changing the shape of the ring, which is immersed into the solution, small area changes of the solution surface can be applied. The surface tension response after such deformations are registered via force measurements using a Wilhelmy plate.

The software driven apparatus allows different types of area changes: step and ramp type, square pulse and trapezoidal as well as sinusoidal area deformations. The construction ensures that area changes are almost isotropic. Area changes used in transient and harmonic relaxation experiments are of the order of 1 to 5%. The surface tension response measured via the Wilhelmy balance has an accuracy of better than ± 0.1 mN/m.

6.3.4. THE MODIFIED PENDENT DROP TECHNIQUE

A recently developed modification of the pendent drop method gives definite area changes of the drop surface, which can be used to initiate transient relaxation processes (Miller et al. 1993a, b). A metering system consisting of two syringes (cf. Fig. 6.7) is used to form a drop with a definite volume (syringe 1) and then to change this volume by small increments by using the highly precise syringe 2.

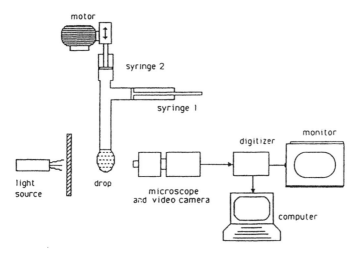

Fig. 6.7   Video enhanced pendent drop method modified for relaxation experiments according to Miller et al. (1993a)

In this way transient relaxation experiments with any type of area disturbances are possible. The high resolution and excellent accuracy of better than ± 0.1 mN/m provides a useful tool for studies of dynamics of soluble adsorption layers. The method can be applied to liquid/gas as well as liquid/liquid interfaces and is easily temperature controlled. Experimental data obtained by this technique on aqueous solutions of surfactants and proteins are given below (Miller et al. 1993a, b, c).

Very recently this technique was also applied for studies of monolayer isotherms of insoluble surfactants (Kwok et al. 1994) and the dynamic behaviour of insoluble monolayers. In Appendix 6C these investigations are discussed in more detail.

6.3.5.   DROP PRESSURE RELAXATION EXPERIMENTS

The drop pressure experiment developed by Nagarajan & Wasan (1993), and Horozov et al. (1993) is a modification of the pendent drop technique, which also allows relaxation studies to different area deformations. The directly coupled pressure transducer can register the change of the pressure inside the drop (or bubble) and therefore, the interfacial tension with time. The time resolution is very high and fast computers are able the acquire data in time scales less than

222

10 µs. Wasan et al. (1993) demonstrated different types of surface area changes of surfactant solution drops and the respective surface tension responses.

The high accuracy of modern pressure transducer and metering systems, the high speed of simple desk computers will make this type of experiments very successful. Its design is easy and the application of the method is very general.

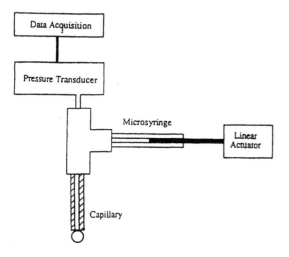

Fig. 6.8   Principle of a drop pressure experiment according to Nagarajan & Wasan (1993)

6.3.6. OTHER RELAXATION EXPERIMENTS

A very recently developed method for transient and low-frequency harmonic relaxation experiments was published by Kokelaar et al. (1991), which appears as a modification of the elastic ring principle of Loglio et al. (1986, 1988, 1991a, b). A cylindrical ring is placed vertically in the liquid surface (cf. Fig. 6.9). The area inside the ring can be changed by moving it up and down. The resulting changes in surface tension can be measured by using a Wilhelmy plate, being in permanent contact with the surface in the centre of the ring. The amount of relative surface area increase or decrease is in the range of up to 15%, which is sufficiently large. Usual relaxation experiments need only very small disturbances to ensure the validity of the previously elaborated linearised theories.

Another experimental set-up also applicable for relaxation studies is the overflowing cylinder designed for example by Bergink-Martens et al. (1990) and Prins (1992). Similar set-ups were used for the measurement of dilational properties by de Rijcker & Defay (1956) or Joos & de Kayser (1980), or for desorption kinetics studies by Kretzschmar & Vollhardt (1967).

Various stress relaxation experiments with different, slightly soluble adsorption layers were performed by Joos et al. (1992), Joos & Van Uffelen (1993a, b), and Van Uffelen & Joos (1993). Such investigations provide also the dilational elasticity along with the exchange of matter mechanism, although it was not discussed specifically in these papers.

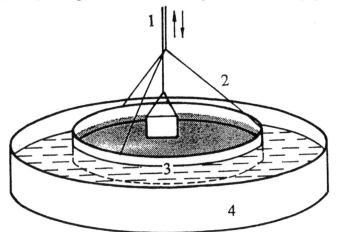

Fig. 6.9   Perspective view of the cylindrical ring method, 1 - Wilhelmy plate, 2 - appliance to move the ring up and down, 3 - cylindrical ring, 4 - vessel; according to Kokelaar et al. 1991

6.4.   EXPERIMENTAL RESULTS OF RELAXATION STUDIES

The present section is dedicated to a selection of experimental results obtained by different techniques. So far there are only few experiments performed to compare results of different methods for the same surfactant solution, as in Chapter 5 with respect to adsorption kinetics investigations. Thus, experimental data have to be interpreted with regard to the specific experimental conditions. Several experimental and theoretical problems of surface relaxations where discussed recently (Miller et al. 1994a) and experimental examples were given in order to demonstrate the importance of such studies.

6.4.1. SURFACTANT ADSORPTION LAYERS

This section aims to demonstrate the variety of relaxation studies performed with surfactant solutions. Results for single surfactant and surfactant mixtures as well as micellar solutions will be shown. A newly available technique for liquid/liquid interfaces is also discussed here.

As mentioned above, only few comparisons of different techniques have been reported. For example, Jiang et al. (1992) studied the dilational elasticity of Triton X-100 adsorption layers with plane and cylindrical capillary wave. The results of the two measurements are given in Fig. 6.10.

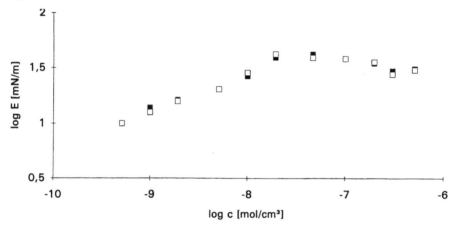

Fig. 6.10   Comparison of surface dilation elasticities determined from two different wave damping experiments with Triton X-100 solutions at 150 Hz; plane waves (□), cylindrical waves (■); according to Jiang et al. (1992)

Although the experimental conditions and the theories differ from each other, the agreement between the results is excellent.

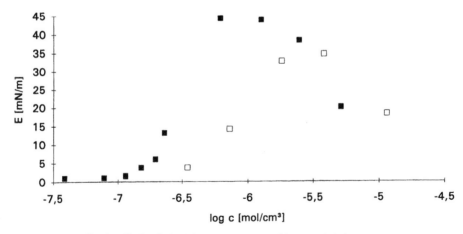

Fig. 6.11   Effective dilational elasticity of n-octanoic acid (■) and dodecyl amine hydrochloride (□) at 200 Hz, measured with a capillary wave technique; according to Lucassen & Hansen (1966)

Lucassen & Hansen (1966) were one of the first to investigate the damping of capillary waves of surfactant solutions. In Fig. 6.11 their results for n-octanoic acid in 0.005 N HCl and dodecyl amine hydrochloride, both nonionic surfactants, are shown in form of effective dilational elasticity as a function of concentration.

The curves have a typical shape with a maximum at a certain concentration which corresponds to a surface concentration of about 50% of the maximum surface coverage. The maximum is caused by the competitive effect of two phenomena: increase of the dilational elasticity modulus E_0 with concentration and increase of the exchange of matter with increasing concentration, diminishing the effective elasticity.

The first studies of the effect of micelles on the exchange of matter to interfaces subject to harmonical disturbances of the surface area were performed by Lucassen (1976). He used an aqueous solution of hexadecyl dimethyl ammonium propanesulfonate (HDPS) below and above the CMC. The exchange of matter, shown by the effective dilational elasticity E, is affected considerably by the presence of micelles, as discussed above. The lower the frequency of disturbance, the more pronounced is the influence of micelle kinetics on the exchange of matter (cf. Fig. 6.12), the line marks the CMC of HDPS. Lucassen was able to describe the behaviour by using the theory given by Eqs (6.25), (6.26)).

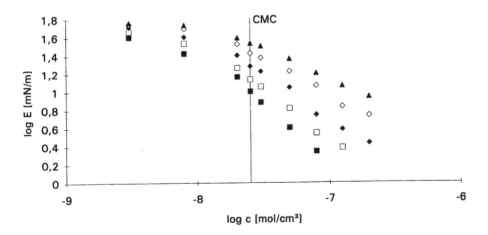

Fig. 6.12   Concentration dependence of the effective dilational elasticity of hexadecyl dimethyl ammoniopropane sulfonate solutions at different frequencies f = 10 (▲), 5(◊), 2(♦), 1(□), 0.5(■) c.p.m., vertical line - CMC; according to Lucassen (1976)

Wantke et al. (1993) discussed the exchange of matter at the surface of surfactant solutions subject to harmonical disturbances using the oscillating bubble technique. As an example,

Fig. 6.13 shows the agreement between the elasticity modulus, derived form the adsorption isotherm and from relaxation experiments with n-dodecyl dimethyl phosphine oxide solutions.

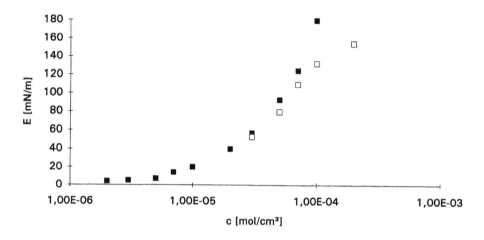

Fig. 6.13   Dilational elasticity modulus of n-dodecyl dimethyl phosphine oxide determined for oscillating bubble experiments ($\square$), and calculated from the adsorption isotherm ($\blacksquare$); according to Wantke et al. (1993)

The surface relaxation behaviour of mixed surfactant solutions was studied by Garrett (1976) using the longitudinal wave technique. By keeping the total surfactant concentration constant at 10^{-8} mol/cm^3, different compositions of a mixture of nonionic dodecyl tri(ethylenglycol) ($C_{12}E_3$) and dodecyl hexa(ethylenglycol) ($C_{12}E_6$) were investigated. Although the surface activity of the two components is similar, the frequency dependence of the effective dilational elasticity changes remarkably with the composition. The theory for the exchange of matter for mixed surfactant solutions (cf. Eqs (6.17), (6.18)) well agrees with the experimental data shown in Fig. 6.14.

So far, there are only a few experimental results available on surfactant mixtures, although in practice mixtures of surfactants are usually used in order to reach optimal effects. The lack of experimental data seems to be evidence that the composition of such mixtures is arrived more or less empirically.

Recently, reliable relaxation experiments at liquid/liquid have been performed by different techniques. Bonfillon & Langevin (1993) measured the dilational elasticity of different surfactants (Triton X-100, SDS in water and in 0.1 M NaCl) at water/air as well as alkane/water interfaces by a modified longitudinal wave damping method (cf. Section 6.3.1). While at the water/air interface, the behaviour of Triton X-100 could be described by a diffusion controlled exchange of matter (cf. Fig. 6.15) but SDS showed a strong deviation from a diffusional exchange mechanism. On the contrary a good agreement between the diffusion theory and experiments was reached at the dodecane/0.1 M NaCl interface.

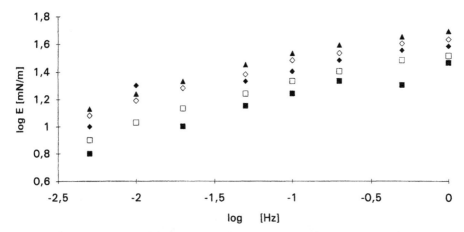

Fig. 6.14   Frequency dependence of the efficient dilational elasticity of dodecyl tri(ethylenglycol) ($C_{12}E_3$) and dodecyl hexa(ethylenglycol) ($C_{12}E_6$) mixtures at a total concentration of 10^{-8} mol/cm³; ratio $[C_{12}E_3]/[C_{12}E_6] = \infty$ (▲) 3 (◊), 1 (♦), 0.333 (□), 0 (■) according to Garrett (1976)

This phenomenon can be explained allowing for the usual presence of dodecanol in SDS solutions. At first the salt leads to an increase in surface activity (shift of about one order of magnitude of the adsorption isotherm to lower concentration), and secondly the potential impurity dodecanol, which strongly adsorbs at the interface water/air, will more or less transfer to the dodecane phase after it has been adsorbed at the water/dodecane interface. Thus, no different mechanism is needed to describe the relaxation behaviour, as done by Bonfillon & Langevin (1993).

It should be possible to describe the exchange of matter by a diffusion model for surfactant mixtures, as shown by Miller et al (1993b). They also developed another new relaxation technique, based on the pendent drop method (cf. Section 6.3.4), and studied the relaxation behaviour of SDS at the water/air interface. It could be shown that surface active impurities can alter the relaxation behaviour of the adsorption layer tremendously. The same method was also applied to detect impurities in organic solvents (cf. Section 5.1.2., Fig. 5.7.).

Another very recently developed method, the drop pressure method by Nagarajan & Wasan (1993), is distinguished by its applicability to both liquid/gas and liquid/liquid interfaces. The results of Fig. 6.16 for the adsorption behaviour of BRIJ58 at the surface of a aqueous solution drop continuously growing into dodecane demonstrate that the technique yields very reliable

228

data. Thus, the technique is useful for obtaining information about dynamics of adsorption as well as relaxation to any form of interfacial area disturbances.

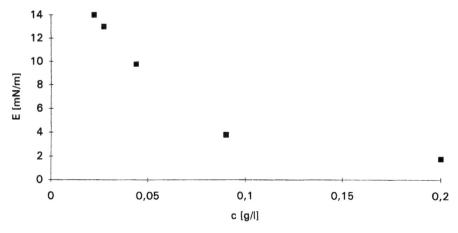

Fig. 6.15   Dilational elasticity modulus of Triton X-100 solutions at the water/dodecane interface determined from longitudinal wave damping; according to Bonfillon & Langevin (1993)

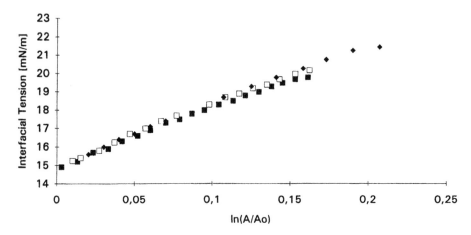

Fig. 6.16   Dynamic Interfacial tension at the surface of a growing drop at different volume flows; $v = 1.69 \ 10^{-5}$ cm³/s (■), $1.95 \ 10^{-5}$ cm³/s (□), $3.3 \ 10^{-5}$ cm³/s (◆); according to Nagarajan & Wasan (1993)

6.4.2. POLYMER ADSORPTION LAYERS

Graham & Phillips (1980a) measured the dilational elasticity of BSA and lysozyme by longitudinal wave damping (2 to 10 cpm). The elasticity values calculated from π-A isotherms significantly differ from those obtained by relaxation experiments. BSA shows an expected behaviour, lysozyme has an irregular peak at 10^{-3} wt% (cf. Fig. 6.17).

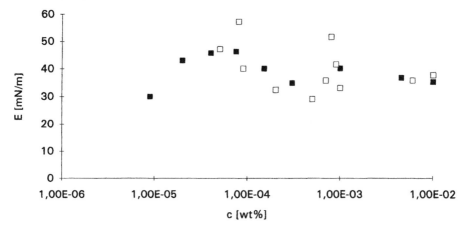

Fig. 6.17   Dilational elasticity modulus determined from longitudinal wave damping experiments,BSA(■), lysozyme(□); according to Graham & Phillips (1980a)

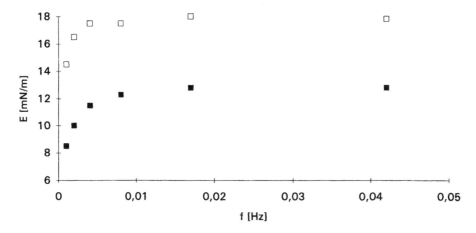

Fig. 6.18   Effective dilational elasticity of protein adsorption layers as a function of deformation frequency in longitudinal wave experiments, 0.1 mg/ml BSA(■), 0.1 mg/ml BSA + 6 mol/l urea (□); according to Serrien et al. (1992)

Serrien et al. (1992) measured the damping of planar longitudinal waves, generated with a barrier on a Langmuir trough and detected with a Wilhelmy balance. The longitudinal waves correspond to slow periodic compressions/dilations of the protein adsorption layer (BSA, casein). The results for BSA are shown in Fig. 6.18. Further experiments using a stress relaxation technique were performed with BSA and buttermilk. The results show that the characteristic parameters of the used models depend on the age of the adsorption layer. The discussion of the results comprises explanations with respect of the established structure, to the distance from the equilibrium state, and the amount of denaturated molecules.

The dynamics of mixtures of surfactants with proteins is of great importance for many practical processes, such as coating of photographic films, where gelatine in mixtures with surfactants and surface active dyes adsorb at the interface. Hempt et al. (1985) studied the relaxation behaviour of gelatine solutions in presence and absence of surfactants (SDS, tetradecyl dimethyl phosphine oxide, cetyltrimethyl ammonium bromide, n-decanoic acid, perfluoro octanoic acid tetraethyl ammonium salt).

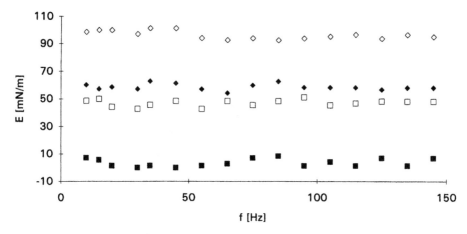

Fig. 6.19   Effective dilational elasticity determined from oscillating bubble experiments of gelatine solutions at different concentration $c_o = 0.01$ wt%($\blacksquare$), 0.1 wt%($\square$), 0.5 wt%($\blacklozenge$), 1 wt%($\lozenge$), according to Hempt et al. (1985)

Adsorbed gelatine molecules alone do not show a frequency dependence of surface elasticity (Fig. 6.19), which corresponds to a behaviour of an insoluble monolayers. The presence of surfactants changes the elastic and relaxation behaviour dramatically. With increasing SDS concentration the elasticity modulus (frequency independent plateau value of the elasticity) first increases and then decreases. The dynamic behaviour of the mixed adsorption layer changes from one completely formed by gelatine molecules to an adsorption layer completely controlled by surfactant molecules (Fig. 6.20). A similar behaviour can be observed for CTAB and a perfluorinated surfactant (Hempt et al. 1985).

Transient relaxation experiments of protein adsorption layers were published by Miller et al. (1993a, c, d). The experiments were performed using a modified pendent drop technique described in Section 6.3.4. The surface tension response to three subsequent square pulse perturbations of 0.1 mg/ml HA adsorbed at the aqueous solution/air interfaces (Miller et al. 1993a) are shown in Fig. 6.21.

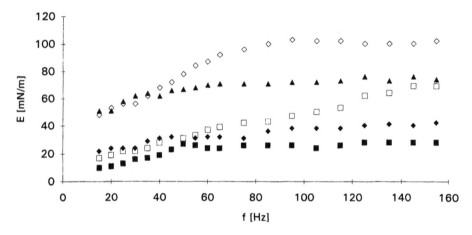

Fig. 6.20   Effective dilational elasticity determined from oscillating bubble experiments of gelatine/SDS mixtures at 0.5 wt% gelatine, SDS concentration $c_0 = 2 \cdot 10^{-7}$ ($\blacktriangle$), $5 \cdot 10^{-7}$ ($\lozenge$), $1 \cdot 10^{-6}$ ($\square$), $1.5 \cdot 10^{-6}$ ($\blacklozenge$), $3 \cdot 10^{-6}$ ($\blacksquare$) mol/cm^3, according to Hempt et al. (1985)

If the results are discussed on the basis of a model of diffusion-controlled exchange of matter, the resulting diffusion coefficients were found to be two to three orders of magnitude higher than those expected from the Stokes-Einstein relation. The same experimental data were interpreted on the basis of Eqs (6.42), (6.43) (Miller et al. (1993c)). Using a least square fitting algorithm values of E_0 and K are obtained. The quality of agreement between experimental data and the fitting curve is also demonstrated in Fig. 6.21.

The calculated values of E_0 and K for the individual runs differ from each other more than expected from the experimental accuracy. This inconsistency is probably caused by different experimental conditions. Although the relative area change $\dfrac{\Delta A}{A_0}$ should not effect the relaxation behaviour of an adsorption layer, an adsorbed HA layer seems to be very sensitive to it. This phenomenon can be explained by the very high slope of the adsorption isotherm leading to non-linearity effects at very small deviations from equilibrium. These effects are not considered in the present theory.

Interfacial relaxations of HA at the water/decane interface are performed with the same technique (Miller et al. 1993c, d). The results for three subsequent square pulse disturbances of the adsorption layer of 0.02 mg/ml HA are shown in Fig. 6.22.

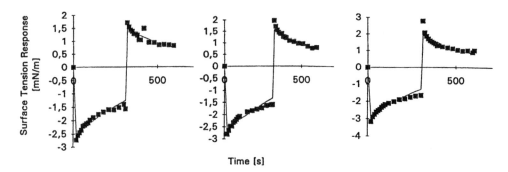

Fig. 6.21   Surface tension relaxation to a surface area square pulse of an aqueous 0.1 mg/ml HA solutions with the modified pendent drop method at the water/air interface; symbols - experimental data, solid lines - theory Eqs (6.42), (6.43); $\dfrac{\Delta A}{A_o}$ = 0.041(a); 0.084 (b); 0.091(c); according to Miller et al. (1993a)

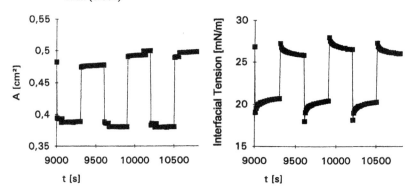

Fig. 6.22   Interfacial tension relaxation of 0.02 mg/ml HA to three square pulses at 24°C at the water/decane interface; a) drop area change, b) surface tension response; according to Miller et al. (1993c)

The interpretation of the experimental data is done on the basis of the two models given above (Eqs (6.25) - (6.28), and (6.42), (6.43)). The results of fitting the models to the data are summarised in Table 6.1. It becomes evident that the kinetic model yields values of the parameters which are rather consistent while those obtained from the diffusion model are more scattered.

Table 6.1   Results from fitting the data of Fig. 6.22, 0.02 mg/ml HA at the water/decane interface to two relaxations models

square pulse number	$\dfrac{\Delta A}{A_o}$ [/]	Diffusion model Eqs(6.25)-(6.28)			Kinetic model Eqs (6.42), (6.43)		
		E_o [mN/m]	ω_o [1/s]	standard deviation [mN/m]	E_o [mN/m]	K [s^{-1}]	standard deviation [mN/m]
1	0.22	28.7	$4.7 \cdot 10^{-5}$	1.25	23.8	$1.2 \cdot 10^{-3}$	0.216
2	0.28	25.8	$3.7 \cdot 10^{-5}$	0.266	21.7	$1.1 \cdot 10^{-3}$	0.239
3	0.29	23.6	$1.6 \cdot 10^{-5}$	0.265	20.2	$1.1 \cdot 10^{-3}$	0.213

6.5   SUMMARY

This chapter describes theoretical and experimental interfacial relaxation studies. It gives the basis of the theoretical description of exchange of matter models and presents the most frequently used experimental techniques. While quantitative exchange of matter theories exist for surfactant and mixed surfactant adsorption layers, similar models for polymers are still missing. Experimentally, the damping of capillary and longitudinal waves is the most frequently used technique, and has even been applied to liquid/liquid interfaces (Lemaire & Langevin 1992). This technique allows studies up to several hundred thousand Hz (Earnshaw & Hughes, 1990, Earnshaw et al. 1991), and can be applied to surfactant mixtures (Garrett 1976) and micellar solutions (Lucassen 1976), the most complex surfactant systems

Transient relaxation technique are used to investigate predominantly slower processes. New developments have been made, such as the elastic ring, the modified pendent drop or different drop/bubble pressure experiments. Also newly designed instruments based on oscillating bubbles and drops (Wantke et al. 1993, Nagarajan & Wasan 1993) have been reported. The stress relaxation experiments as performed by Serrien et al. (1992) provide simultaneous information about exchange of matter and dilational rheological parameters.

The disadvantage of relaxation experiments is that they are usually applied to equilibrium adsorption layers. Thus, the presence of surface active contaminants, present in almost all commercial samples and difficult to remove even from carefully synthesised products, can falsify results substantially. Relaxation experiments of adsorption layers at freshly formed interfaces are, therefore, of great interest. Such studies are possible by producing ripples on flowing films using, for example, the inclined plane or overflowing cylinder technique. Also the analysis of oscillations on freshly formed bubbles can be a source of additional information. The important prerequisite of such investigations is that the characteristic time of the disturbance is much smaller than the characteristic adsorption time. As stressed in the preceding chapter, the combination of different techniques will provide a new and deeper insight into the structure and structural forces in adsorption layers and their evolution.

There is a lack of systematic experimental studies. Only on the basis of a large data base can further improvements in the theories be made. Thus, the more effective instruments and newly developed techniques should enable surface scientists to produce more systematic data. There is also a deficiency in respect to relaxation theories for polymers and mixed polymer/surfactant systems. First ideas about dilational properties of composite adsorption layers were published by Lucassen (1992). A theory has also been developed by Johnson & Stebe (1994) to differentiate exchange of matter from dilational viscosity. This calls for new experimental developments.

6.6.   REFERENCES

Adamczyk, Z., J. Colloid Interface Sci., 120(1987)477
Adamczyk, Z. and Petlicki, J., J. Colloid Interface Sci., 118(1987)20
Adamczyk, Z., Wandzilak, P. and Pomianowski, A., Bull. Polish Acad. Sci., Chem.,
        35(1987)479
Balbaert, I. and Joos, P., Colloids & Surfaces, 23(1987)259
Bergink-Martens, D.J.M., Bos, H.J., Prins, A. and Schulte, B.C., J. Colloid Interface Sci.,
        138(1990)1
Bonfillon, A. and Langevin, D., Langmuir, 9(1993)2172
Chang, C.H. and Franses, E.I., 67th Annual Colloid and Surface Science Symposium, Toronto ,
        1993, 84
Chang, C.H. and Franses, E.I., Chem. Engineering Sci., 49(1994a)313
Chang, C.H. and Franses, E.I., J. Colloid Interface Sci., 164(1994b)107
Cheng, P., PhD Thesis, University of Toronto, 1990
Cheng, P., Li, D., Boruvka, L., Rotenberg Y. and Neumann, A.W., Colloids & Surfaces,
        43(1990)151
Defay, R., Prigogine, I. and Sanfeld, A., J. Colloid Interface Sci., 58(1977)498
de Feijter, J. , Benjamins, J. and Tamboer M., Colloids Surfaces, 27(1987)243
de Rijcker, M. and Defay, R., Boll. Soc. Chim. Belg., 65(1956)794
de Voeght, F. and Joos, P., J. Colloid Interface Sci., 98(1984)20

Dimitrov, D.S., Panaiotov, I., Richmond, P. and Ter-Minassian-Saraga, L., J. Colloid Interface Sci., 65(1978)483

Dorshow, R.B. and Swofford, R.L., J. Appl. Phys. 65(1989)3756

Dorshow, R.B. and Swofford, R.L., Colloids & Surfaces, 43(1990)133

Earnshaw, J.C. and Robinson, D.J., J. Phys.: Condens. Matter 2(1990)9199

Earnshaw, J.C. and Winch, P.J., J. Phys.: Condens. Matter 2(1990)8499

Earnshaw, J.C. and Hughes, C.J., Langmuir, 7(1991)2419

Earnshaw, J.C., McGivern, R.C., McLaughlin, A.C. and Winch, P.J., Langmuir, 6(1990)649

Edwards, D.A., Brenner, H. and Wasan, D.T., Interfacial Transport Processes and Rheology, Butterworth-Heineman Publishers, 1991

Garrett, P.R. and Joos, P., J. Chem. Soc. Faraday Trans. 1, 69(1976)2161

Garrett, P.R., J. Chem. Soc. Faraday Trans. 1, 69(1976)2174

Garrett, P. and Moore, P.R., J. Colloid Interface Sci., 159(1993)214

Goodrich, F.C., J. Phys. Chem., 66(1962)1858

Graham, D.E. and Phillips, M.C., J. Colloid Interface Sci., 70(1979a)403

Graham, D.E. and Phillips, M.C., J. Colloid Interface Sci., 70(1979b)415

Graham, D.E. and Phillips, M.C., J. Colloid Interface Sci., 70(1979c)427

Graham, D.E. and Phillips, M.C., J. Colloid Interface Sci., 76(1980a)227

Graham, D.E. and Phillips, M.C., J. Colloid Interface Sci., 76(1980b)240

Hansen, R.S., J. Colloid Sci., 16(1961)549

Hansen, R.S., J. Appl. Phys., 35(1964)1983

Hansen, R.S. and Mann, J.A., J. Appl. Phys., 35(1964)152

Hård, S. and Neuman, R.D., J. Colloid Interface Sci., 115(1987)73

Hempt, C., Lunkenheimer, K. and Miller, R., Z. Phys. Chem. (Leipzig), 266(1985)713

Hennenberg, M., Chu, X.-L., Sanfeld, A. and Velarde, M.G., J. Colloid Interface Sci., 150(1992)7

Horozov, T., Danov, K., Kralschewsky, P., Ivanov, I. and Borwankar, R., 1st World Congress on Emulsion, Paris, Vol. 2, (1993)3-20-137

Hühnerfuss, H., Lange, P.A. and Walter, W., J. Colloid Interface Sci., 108(1985)442

Jiang, Q., Chiew, Y.C. and Valentini, J.E., Langmuir, 8(1992)2747

Johnson, D.O. and Stebe, K.J., J. Colloid Interface Sci., (1994) in press

Joos, P. and de Kayser, P., "The overflowing funnel as a method for measuring surface dilational properties", Levich Birthday Conference, Madrid, 1980

Joos, P., Van Uffelen, M. and Serrien, G., J. Colloid Interface Sci., 152(1992)521

Joos, P. and Van Uffelen, M., J. Colloid Interface Sci., 155(1993a)271

Joos, P. and Van Uffelen, M., Colloids Surfaces, 75(1993b)273

Kokelaar, J.J., Prins, A. and de Gee, M., J. Colloid Interface Sci., 146(1991)507

Kovscek, A.R., Wong, H. and Radke, C.J., AIChE Journal, 39(1993) in press ???

Kretzschmar, G. and König, J. SAM, 9(1981)203

Kretzschmar, G. and Vollhardt, D., Ber. Bunsenges. phys. Chem., 71(1967)410

Kretzschmar, G. and Miller, R., Adv. Colloid Interface Sci. 36(1991)65

Kwok, D.Y., Vollhardt, D., Miller, R., Li, D. and Neumann, A.W., Colloids and Surfaces A, 88(1994a)51

236

Kwok, D.Y., Cabreizo-Vilchez, M.A., Gomez, Y., Susnar, S.S., del Rio, O., Vollhardt, D., Miller, R. and Neumann, A.W., J. Amer. Oil Chem. Soc., (1994b), submitted

Li, J.B., Kwok, D.Y., Miller, R., Möhwald, H. and Neumann, A.W., Colloids and Surfaces A, (1994a), submitted

Li, J.B., Miller, R., Wüstneck, R., Möhwald, H. and Neumann, A.W., Langmuir, (1994b), submitted

Lemaire, C. and Langevin, D., Colloids & Surfaces, 65(1992)101

Levich, V.G., Acta Physicochimica URSS, 14(1941a)307

Levich, V.G., Acta Physicochimica URSS, 14(1941b)322

Levich, V.G., Physicochemical Hydrodynamics, Prentice Hall, Englewood Cliffs, (1962)

Loglio, G., Tesei, U. and Cini, R., J. Colloid Interface Sci., 71(1979)316

Loglio, G., Tesei, U., Mori, G., Cini, R. and Pantani, F., Nuovo Cimento 8(1985)704

Loglio, G., Tesei, U. and Cini, R., Colloid Polymer Sci., 264(1986)712

Loglio, G., Tesei, U. and Cini, R., Bollettino di Oceanologia Teorica ed Applicata 3(1987)195

Loglio, G., Tesei, U., and Cini, R., Rev. Sci. Instrum., 59(1988)2045

Loglio, G., Degli-Innocenti, N., Tesei, U., Stortini, A.M. and Cini, R., Ann. Chim. (Rome), 79(1989)571

Loglio, G., Degli Innocenti, N., Tesei, U., Cini, R. and Wang Qi-Shan, Il Nuovo Cimento, 12(1989)289

Loglio, G., Tesei, U., Degli-Innocenti, N., Miller, R. and Cini, R., Colloids & Surfaces, 57(1991a)335

Loglio, G., Tesei, U., Miller, R. and Cini, R., Colloids & Surfaces, 61(1991b)219

Loglio, G., Miller, R., Stortini, A., Degli-Innocenti, N., Tesei, U. and Cini, R., Colloids & Surfaces, (1994) submitted

Lucassen, J. and Hansen, R.S., J. Colloids Interface Sci., 22(1966)32

Lucassen, J., Trans. Faraday Soc., 64(1968a)2221

Lucassen, J., Trans. Faraday Soc., 64(1968b)2230

Lucassen J. and van den Tempel, M., Chem. Eng. Sci., 27(1972a)1283

Lucassen J. and van den Tempel, M., J. Colloid Interface Sci., 41(1972b)491

Lucassen, J. and Giles, D., J. Chem. Soc. Faraday Trans. 1, 71(1975)217

Lucassen, J., Faraday Discussion Chem. Soc., 59(1976)76

Lucassen-Reynders, E.H., J. Colloids Interface Sci., 42(1973)573

Lucassen-Reynders, E.H., J. Colloid Interface Sci., 42(1973)573

Lucassen-Reynders, E.H., Lucassen, J., Garrett, P.R., Giles, D. and Hollway, F., Adv. in Chemical Series, Ed. E.D. Goddard, 144(1975)272

Lucassen-Reynders, E.H. and Kuijpers, K.A., Colloids Surfaces, 65(1992)175

Lunkenheimer, K. and Kretzschmar, G., Z. Phys. Chem. (Leipzig), 256(1975)593

MacLeod, C.A. and Radke, C.J., 9th Intern. Symposium "Surfactants in Solution", Varna, 1992, T2.A3.2

MacLeod, C.A. and Radke, C.J., J. Colloid Interface Sci., 160(1993)435

Malysa, K., Adv. Colloid Interface Sci., 40(1992)37

McGivern, R.C. and Earnshaw, J.C., Langmuir, 5(1989)545

Miller, R., Loglio, G.; Tesei, U., and Schano, K.-H., Adv. Colloid Interface Sci., 37(1991)73

Miller, R., Policova, Z., Sedev., R. and Neumann, A.W., Colloids & Surfaces, 76(1993a)179

Miller, R., Sedev., R., Schano, K.-H., Ng, C. and Neumann, A.W., Colloids &Surfaces, 69(1993b)209

Miller, R., Krägel, J., Loglio, G. and Neumann, A.W., Proceedings of the 1st World Congress on Emulsion, Paris, Vol. 2, (1993c) 3-20-143

Miller, R., Joos, P. and Fainerman, V.B., Adv. Colloid Interface Sci., 49(1994a)249

Miller, R., Loglio, G. and Neumann, A.W., Colloids & Surfaces A, (1994b), submitted

Nagarajan, R. and Wasan, D.T., J. Colloid Interface Sci., 159(1993)164

Noskov, B.A., Koll. Zh., 45(1983)689

Noskov, B.A. and Vasilev, A.A., Koll. Zh., 50(1988)909

Noskov, B.A. and Schinova, M.A., Koll. Zh., 51(1989)69

Noskov, B.A., Anikieva, O.A. and Makarova, N.V., Koll. Zh., 52(1990)1091

Noskov, B.A., Mech. Zhidk. Gasa, 1(1991)129

Noskov, B.A., Colloids & Surfaces A, 71(1993)99

Oberhettinger, F. and Berdii, L., Tables of Laplace Transforms, Springer-Verlag, Berlin, 1973

Park, S.Y., Chang, C.H., Ahn, D.J. and Franses, E.I., Langmuir, 9(1993)3640

Prins, A., Chem.-Ing.-Tech., 64(1992)73

Romero-Rochin, V., Varea, C. and Robledo, A., Physica A, 184(1992)367

Rotenberg , Y., Boruvka, L. and Neumann, A.W., J. Colloid Interface Sci. 93(1983)169

Sakai, K., Choi, P.-K., Tanaka, H. and Takagi, K., Rev. Sci. Instrum., 62(1991)1192

Sakai, K., Kikuchi, H. and Takagi, K., Rev. Sci. Instrum., 63(1992)5377

Serrien, G., Geeraerts, G., Ghosh, L. and Joos, P., Colloids & Surfaces, 68(1992)219

Stebe, K.J. and Johnson, D.O., 67th Colloid Surface Science Symposium, Toronto, 1993, 404

Stebe, K.J. and Maldarelli, C., J. Colloid Interface Sci., 163(1994)177

Stenvot, C. and Langevin, D., Langmuir, 4(1988)1179

Sun, S.M. and Shen, M.C., J. Math. Analysis Appl., 172(1993)533

Szymanowski, J. and Prochaska, K., Progr. Colloid Polymer Sci., 76(1988)260

Thoma, M. and Möhwald, H., Colloids and Surfaces A, (1994), in press

Thominet, V., Stenvot, C. and Langevin, D., J. Colloid Interface Sci., 126(1988)54

van den Tempel, M. and van de Riet, R.P., J. Che. Phys., 42(1965)2769

van den Tempel, M, Chemie-Ing.-Techn., 43(1971)1260

van den Tempel, M. and Lucassen-Reynders, E.H., Adv. Colloid Interface Sci., 18(1983)281

Van Uffelen, M. and Joos, P., J. Colloid Interface Sci., 158(1993)452

Wantke, K.D., Miller, R. and Lunkenheimer, K., Z. Phys. Chem. (Leipzig), 261(1980)1177

Wantke, K.D., Lunkenheimer, K. and Hempt, C., J. Colloid Interface Sci., 159(1993)28

Wasan, D.T., Nikolov, A.D., Lobo, L.A., Koczo, K. and Edwards, D.A., Progr. Surface Sci., 39(1992)119

Wasan, D.T., Koszo, K. and Nagarajan, R., 67th Annual Colloid and Surface Science Symposium, Toronto , 1993, 285

Wielebinski, D. and Findenegg, G.H., Progr. Colloid Polymer Sci., 77(1988)100

Yasunaga, T. and Sasaki, M., in "Techniques and Application of Fast Reactions in Solutions", Gettins, W.J. and Wyn-Jones, E. (Eds.), D.Reidel Publ. Comp., (1979)579

CHAPTER 7

7. EFFECT OF SURFACTANT CHARGE ON THE DYNAMICS OF ADSORPTION

The surface concentration changes with the ionisation of specific groups in the adsorbing molecule. In Chapter 2 is was shown that ionisation of surfactant molecules leads to a decrease of surface concentration due to increasing electrostatic repulsion in the adsorption layer. Similar dramatic changes occur in respect of adsorption kinetics of ionised surfactants compared to the non-ionised molecules. The electrostatic retardation is the respective analogue in adsorption kinetics of ionics (Dukhin et al. 1973). First qualitative models and approximate analytical solutions were derived by Kretzschmar et al. (1980), Dukhin et al. (1983), Fainerman & Jamilova (1986), Joos et al. (1986), Borwankar & Wasan (1988a, b). A first quantitative attempt is proposed by MacLeod & Radke (1994) giving numeric calculations of the transport problem. This chapter deals with the peculiarities adsorption dynamics of ionic surfactants and presents the actual state of the art.

7.1. INTRODUCTION

To draw a clear picture of the physical mechanism of the electrostatic retardation effect any existence of additional non-electrostatic barriers, such as discussed in Section 4.4, or other mechanisms influencing the transfer of surfactant molecules between the subsurface and the interface, are avoided. Thus, the electrostatic retardation effect is considered under the assumption of local equilibrium between the subsurface, defined as the bulk layer adjacent to the interface (cf. Chapter 2), and the interface.

The electrostatic retardation effect is the consequence of the existence of the electric double layer, or more exactly the diffuse part of the DL, near the interface. Naturally, this effect only arises if the sign of the surface charge and the adsorbing ions is the same. This is not a restriction of the generality of the approach to strongly adsorbing ions. Even if the signs differ in the initial state of the adsorption process, the interface will be recharged during the progressing adsorption. After the interface has become recharged, the electrostatic repulsion sets in and so retards the further adsorption. Only this part of the whole process will be considered in the following sections of this chapter. The adsorbing ions are considered to be inside the DL as coions.

Due to the electrostatic repulsion the coion concentration inside the diffuse part of the DL is decreased, which is the reason for the relatively low concentration of coions in the sublayer compared with the concentration at the boundary between the diffuse and diffusion layer, i.e. at $x = \kappa^{-1}$,

$$c(0,t) \ll c(\kappa^{-1},t). \tag{7.1}$$

An important step in the development of a theoretical model is the expression of $c(0,t)$ in terms of $c(\kappa^{-1},t)$ using the Boltzmann law, as discussed in Section 2.8,

$$c(0,t) = c(\kappa^{-1},t)\exp(-z\overline{\psi}_{St}), \tag{7.2}$$

where $\overline{\psi}_{St} = e\psi_{St}/kT$, e is the electron charge, k is the Boltzmann constant, T is the absolute temperature, and ψ_{St} is the Stern potential. The relative drop of the sublayer concentration can be rather large at high Stern potentials, e.g. ψ_{St} = 200 mV. Thus, the slow-down of the adsorption rate can be remarkable. However, it is not easy to make an estimate of the characteristic adsorption time τ_{ad}, however, does not follow from these qualitative findings.

To estimate τ_{ad}, the decrease in equilibrium adsorption and the actual adsorption rate according to the electrostatic phenomena, have to be considered. The application of Boltzmann's law assumes equilibrium condition of the DL and neglects any transport within the diffuse layer. Thus, the classic Boltzmann law cannot be used to describe the distribution of adsorbing ions within the double layer in non-equilibrium systems. The presence of any ionic flux is connected with a non-equilibrium state of the DL and the approach given by Overbeek (1943) in his theory of electrophoresis has to be considered. In that theory, the non-equilibrium of the DL causes non-linear dependencies of electrophoresis on the electrokinetic potential, in contrast to the theory of Smoluchowski where this effect is not allocated for. The importance of the non-equilibrium state of the DL for many other surface phenomena was emphasised by Dukhin & Derjaguin (1974), Dukhin & Shilov (1974), and Dukhin (1993).

The electrostatic retardation of the adsorption kinetics of ionic surfactants is one of these non-equilibrium surface phenomena to be described on the basis of this physical model, consisting of the electrochemical macro-kinetic equations used in theoretical and colloid electrochemistry. This approach describes the flux of ions in terms of their spatial distribution. The equations were first developed by Overbeek (1943) and later proved to be valid for the theory of different non-equilibrium surface phenomena (Dukhin 1993). At small deviations of the DL from equilibrium transport processes can be described using the equilibrium state as zero approximation.

From the theoretical point of view a similarity exists between electrostatic retardation of ion transport and coagulation retardation, known as slow coagulation (Fuchs, 1934). Both phenomena arise from electrostatic repulsion caused by the existence of the diffuse part of the DL. In the slow coagulation theory, the electric field if the DL is derived from the Gouy-Chapman model (cf. Chapter 2). This model does not consider a deviation of the diffuse layer from equilibrium. Initially, the same simplification was used by Dukhin et al. (1973) in describing the DL effect on the electrostatic retardation of adsorption.

During adsorption processes ions have to move under the action of a given electric field of the double layer. In the slow coagulation theory the electric potential distribution in the gap between two particles is determined by the overlap of the diffuse part of their double layers. The flux of particles to calculate the rate of slow coagulation, and the flux controlling the retardation of adsorption rate are given by the same physical equations (Dukhin et al. 1983). The principle equation contains two terms, the first of them describing the diffusional transport (which is the Brownian motion in slow coagulation), the second the particle flux caused by the given electric field. The total flux is proportional to the local ion (particle) concentration and local ion (particle) velocity, caused by forces of electric origin. The velocity is given by the ratio of force over the drag coefficient. It means, the inertial effects in slow coagulation is neglected. This holds for the movements of ions too. The energy of particle interaction and the electrostatic energy of ions in the diffuse layer are functions of the distance and their derivatives give the local force applied to the particles (ions).

For the description of the flux, all functions and parameters are known, except the ion (particle) distribution and therefore, it is a linear first order differential equation of $c(x)$,

$$j^- = -D^- \frac{\partial c}{\partial x} + z^- D^- c^- \frac{\partial \psi}{\partial x}, \quad 0 < x < \kappa^{-1}, \tag{7.3}$$

where x is the distance to the surface, D^- is the ion diffusivity, and $\psi(x)$ is the electric potential distribution in the DL.

The integration of Eq. (7.3) by common methods needs boundary conditions of the concentration distribution, drag coefficient and the potential distribution, which finally leads to the expression of the ion (particle) flux similar to Fick's first law,

$$j^- = \frac{D^- c_0}{K(\psi_{St})}, \quad K(\psi_{St}) = \int_0^{\kappa^{-1}} \exp(z\overline{\psi}(x))dx. \tag{7.4}$$

The denominator $K(\psi_{St})$ is the equivalent to the characteristic dimension in Fick's law, proportional to the diffuse layer thickness κ^{-1} and depends exponentially on the dimensionless interaction energy $\overline{\psi}(x)$. The physical idea is that slow coagulation, as well as ion adsorption, are controlled by diffusion through an electrostatic barrier, which slows down with increasing barrier height.

In the following, we comment on some peculiarities and restrictions of the slow coagulation theory since it can also be of importance for the adsorption kinetics. The elementary act of slow coagulation consists of the pairwise interaction of particles and their mutual diffusion as a non-steady process. The concentration distribution of pairwise interacting particles is a function of distance and time and is described by the equation of non-steady electro-diffusion

$$\frac{\partial n(t)}{\partial t} = \text{div } j(t) \tag{7.5}$$

Fuchs' theory is restricted to a quasi-steady process, where the time dependence is neglected. The quasi-steady state approach of electrostatic retardation in adsorption will be described in Section 7.3.

In the theory of adsorption retardation the energy of an ion approaching the interface is described by the very simple law $ze\psi(x)$, the maximum value of which is proportional to the Stern potential (cf. Chapter 2). Each ion is screened by counterions (Kortüm 1966) and ions pass the diffuse part of the DL together with their screening counterions. In Section 7.5. it will be shown, that under certain conditions this effect can be neglected.

Large differences between slow coagulation and ion adsorption consist in the chosen geometry and in the boundary conditions of ion (particle) distribution. Slow coagulation is assumed to be an irreversible process and particles which overcame the electrostatic barrier and occupy places in the potential well, are not considered any longer in the particle distribution. The particle distribution only contains mobile particles which can move in any direction, while particles in the potential well are considered as bounded. Thus, Fuchs' theory does not describe particles in the coagulated state and the boundary condition chosen sets the particle concentration in the well equal to zero. Consequently, the Eq. (7.4) contains only the bulk concentration c_0. For the process of ion adsorption a completely different situation exists. The concentration of ion is not assumed to be zero in the subsurface and the ions can move in any direction. The state of ions in the subsurface does not differ from the one in the bulk and the equation of electrostatic retardation holds also for the subsurface. The boundary condition is given by

$$c\big|_{x=0} = c(0,t). \tag{7.6}$$

The unknown function $c(0,t)$ is expressed by the theory of electrostatic retardation, containing the bulk concentration c_0.

First models of electrostatic retardation of ion adsorption used the boundary condition of slow coagulation (Dukhin et al. 1973, Glasman et al. 1974, Michailovskij et al. 1974, Dukhin et al. 1976). These models are discussed in Section 7.5. In later models the derivation of the theory was performed by expressing $c(0,t)$ through c_0, which is a more general case (Dukhin et al. 1983, 1990). This approach is described in detail in the following Section 7.2. The more complicated non-steady ion adsorption is considered in Section 7.3.

Favourable conditions for the manifestation of electrostatic retardation effects are met in the exchange of ionic surfactants at surfaces under harmonic disturbances. This process is outlined in Section 7.4.

In the models described in Sections 7.2. through 7.5., equilibrium between the adsorption layer and the adjacent subsurface is assumed. A generalisation, taking into account a Henry transfer mechanism as the relation between surface and subsurface concentration (cf. Section 4.4), is given in Section 7.6. The special problems connected with the adsorption model of ionic surfactants as well as macro-ions is discussed in Sections 7.7. and 7.8. and an attempt to solve the boundary value problem numerically is demonstrated in Section 7.9. The few experimental results on ionic adsorption kinetics are reported in Section 7.10.

7.2. THE RETARDATION OF THE STEADY TRANSPORT OF ADSORBING IONS THROUGH THE DIFFUSE PART OF DOUBLE LAYER

The effect of retardation of the transport of adsorbing ions passing the diffuse part of the double layer becomes clearer when considering a steady transport in one dimension. It is difficult to realise a strictly one-dimensional steady transport in an experiment. But the conditions of a one dimensional steady transport are approximately realised under the conditions of convective diffusion when the thickness of the diffusion layer δ_D is much less than the characteristic geometric dimension of the system L

$$\delta_D \ll L \tag{7.7}$$

In the case of buoyant bubbles (cf. Chapter 9), this dimension will be the bubble radius a_b. Thus, Eq. (7.7) reads

$$\delta_D \ll a_b \tag{7.8}$$

This condition is fulfilled for bubbles of sufficiently large diameter (cf. Chapter 8).

The electrolyte concentration can vary along the boundary between the diffuse and diffusion layers. However, despite the availability of this angular dependence, the ion transport is associated with surface element $\delta L \ll L$ which leads to an one-dimensional steady process as long as Eqs (7.7) or (7.8) hold. The present chapter deals with liquid interfaces in general. Some results are derived especially for bubble surfaces, as these will be needed in Chapter 9.

Three types of ions are assumed to be present in the solution: the surface active ions, which will be considered as anions, and inorganic cations and anions. This corresponds to the situation when an ionic surfactant and inorganic electrolyte with a common cation are present in the solution. In addition to the assumptions formulated in the preceding section, we will assume that the adsorption layer at the surface of a bubble consists of an infinitely thin monolayer of surface active ions and of a diffuse layer (DL) with a thickness κ^{-1} much less than the thickness of the diffusion layer δ_D,

$$\kappa\delta_D \gg 1 \tag{7.9}$$

In addition, it is supposed that the approximate equation equivalent to Eq. (2.61) applies even under non-equilibrium conditions (Dukhin & Lyklema 1987)

$$\Gamma = \Delta \exp(-W^-)c|_{x=0}, \tag{7.10}$$

where Γ is the surface concentration of surfactant anions, W^- is the adsorption energy, and $c|_{x=0}$ is the subsurface concentration of the surface active ions (concentration at the inner boundary of the diffuse layer). Finally, the electric potential at the bubble surface ψ_{Sto} is assumed to be sufficiently high, so that the inequalities are fulfilled

$$\frac{z^-e}{kT}|\psi_{Sto}| \gg 1, \qquad \frac{ze}{kT}|\psi_{Sto}| \gg 1 \tag{7.11}$$

Thus, the surface excess of coions in the DL can be neglected in the following calculations.

7.2.1. EQUATIONS AND BOUNDARY CONDITIONS DESCRIBING ONE-DIMENSIONAL STEADY TRANSPORT OF ADSORBING IONS

At first, the equations of convective diffusion for a system of three ions has to be considered. Under the condition (7.9), the diffuse layer is approximately electrically neutral:

$$zc + z^- c_- - z^+ c_+ \approx 0 \tag{7.12}$$

Besides, under condition (7.11), the charge of the monolayer is approximately equal to the uncompensated charge of the diffuse layer, i.e. the adsorption layer as a whole is electrically neutral. This means the density of the electric current I is approximately equal to zero:

$$I = F\left(-zj - z^- j^- + z^+ j^+\right) \approx 0 \tag{7.13}$$

where

$$j = -D\left(\nabla c + z \frac{e}{kT} Ec\right), \tag{7.14}$$

$$j^- = -D^-\left(\nabla c^- + z^- \frac{e}{kT} Ec^-\right), \tag{7.15}$$

$$j^+ = -D^+\left(\nabla c^+ - z^+ \frac{e}{kT} Ec^+\right) \tag{7.16}$$

are flux densities; D, D^- and D^+ are diffusion coefficients of the surface active ions, and anions, and cations of inorganic electrolyte, respectively; E is electric field strength. If the electric current I would deviate from zero, the DL would deviate from electroneutrality.

Let us consider first a system consisting only of surface active anions and inorganic cations. Using Eqs (7.12) and (7.13), and the formulas for the ion fluxes, the electric field strength E can be expressed in terms of the concentration gradient c. In this case j is given by

$$j_s = - D_{eff} \nabla c_s \tag{7.17}$$

where

$$D_{eff} = \frac{\left(z^+ + z\right) DD_+}{z^+ D^+ + zD}. \tag{7.18}$$

is the effective coefficient of ionic surfactant diffusion. At small D, this quantity can be much greater than the individual diffusion coefficient of surfactant anions.

This result can be qualitatively explained in the following way. Since the density of the electric current in the diffusion layer is approximately equal to zero, densities of the fluxes of surfactant anions and inorganic cations must be equal to each another. Since $D \ll D^+$, an electric field should arise at this stage that accelerates organic anions and retard cations. Therefore, the total flux of surfactant anions is greater than their pure diffusion flux, and $D_{eff} > D$.

Now, the system containing three ions is considered again. The rate of diffusion of surfactant anions must decrease with increasing electrolyte concentration and at

$$c_0^+ \gg c_0 \tag{7.19}$$

the effective diffusion coefficient of these ions becomes equal to D (here c_0 is the bulk concentration of surfactant anions, c_0^+ is the bulk concentration of cations in the solution). This is caused by the fact that additions of electrolyte suppress the electric field which arises due to the difference in ion mobilities (Levich 1962, Listovnichij & Dukhin 1984). In order to obtain analytical results, condition (7.19) is assumed to be fulfilled and the diffusion coefficient D is used.

At equilibrium, electrochemical potentials of ions are constant over the diffuse layer and these relations hold within the DL:

$$c = \exp\left(z \frac{e\psi(x)}{kT} \right) c(\kappa^{-1}),$$

$$c^- = \exp\left(z^- \frac{e\psi(x)}{kT} \right) c^-(\kappa^{-1}), \tag{7.20}$$

$$c^+ = \exp\left(-z^+ \frac{e\psi(x)}{kT} \right) c^+(\kappa^{-1}),$$

where $c(\kappa^{-1}), c^+(\kappa^{-1}), c^-(\kappa^{-1})$ are concentrations of the respective ions at the external boundary of the DL. It should be mentioned here that it is difficult to introduce the concept of the external boundary of the diffuse layer when Eq. (7.9) is not valid.

The same relations are valid also at small deviations of the adsorption layer from equilibrium. In this case the DL is in a quasi-equilibrium state (Dukhin & Derjaguin 1974, Dukhin & Shilov

1974), i.e. jumps of electro-chemical potentials of ions along the DL are much less than kT. In the following the conditions are analysed under which the DL of ionic surfactants is in a quasi-equilibrium state. Under the condition (7.19) and $|c^+ - c^-| \ll c^+$ it is possible to introduce a so-called electrolyte concentration outside the DL, which holds for symmetric electrolytes (Levich 1962) $c_{el} = c^+ = c^-$.

7.2.2. NON-EQUILIBRIUM DISTRIBUTION OF ADSORBING IONS ALONG THE DIFFUSE LAYER AND QUASI-EQUILIBRIUM DISTRIBUTION OF COUNTERIONS

We will consider the stationary distribution of concentration of surfactant anions within DL under conditions when the flux of these ions to the surface differs from zero. Under these conditions $c(x)$ is the solution of the equation

$$D\left(\frac{dc}{dx} - zc\frac{e}{kT}\frac{d\psi}{dx}\right) = j. \tag{7.21}$$

From Eq. (7.21) it follows

$$c(0) = \exp\left(z\frac{e\psi_{St}}{kT}\right)\left[c(\kappa^{-1}) + \frac{j}{D}\int_{\kappa^{-1}}^{0} \exp\left(-z\frac{e\psi(x)}{kT}\right)dx\right]. \tag{7.22}$$

The concentration distribution of surfactant anions is described by Eq. (7.20) in the case when the second term in brackets on the right hand side of (7.22) can be neglected. Clearly, at large j and high bubble surface potentials this is not allowed. This is connected with the fact that the anion concentration is very small deep within the DL and large gradients of the electrochemical potential are required to maintain the flow of surfactant anions through this region. For inorganic cations a similar expression can also be obtained,

$$c^+(0) = \exp\left(-z^+\frac{e\psi_{St}}{kT}\right)\left[c^+(\kappa^{-1}) + \frac{j^+}{D^+}\int_{\kappa^{-1}}^{0} \exp\left(z^+\frac{e\psi_{St}(x)}{kT}\right)dx\right]. \tag{7.23}$$

Taking into account the obvious estimates

$$j < D\frac{c_0}{\delta_D}; \quad j^+ \sim j, \tag{7.24}$$

and the different signs in the exponential functions of the integrals in Eqs (7.22) and (7.23), it can be concluded that the second term in the right hand side of formula (7.23) can always be neglected. Therefore, a condition should exist at which the differential of the electrochemical potential of inorganic cations in the DL is negligible, but the differential of electrochemical potential of potential determining surfactant ions in the DL is significant. Under these conditions, c_+ can be described within the DL by Eq. (7.20), and the more general Eq. (7.22) should be used for the calculation of $c(0)$. In other words, the DL is in quasi-equilibrium with respect to inorganic cations but not surfactant anions.

7.2.3. COEFFICIENT OF ELECTROSTATIC RETARDATION OF ADSORPTION

The aim now is to derive an equation for the flow of surfactant anions to the surface of a rising bubble. To do so, we have to calculate the integral on the right hand side of Eq. (7.22)

$$\int_{\kappa^{-1}}^{0} \exp\left(-z\frac{e\psi(x)}{kT}\right) dx = -K(\psi_{St}).$$ (7.25)

If the concentration of coions deep inside DL is neglected, the distribution of the electric potential this region is described by the simplified Poisson-Boltzmann equation (cf. Section 2.8.5.).

Substituting Eq. (2.57) into (7.25) yields

$$K(\psi_{St}) \approx \frac{z^+}{\kappa} \frac{\exp\left(\left(-z+\frac{z^+}{2}\right)\overline{\psi}_{St}\right)}{z-z^+/2}.$$ (7.26)

This result is valid under condition (7.11) and at $(z - z^+/2) > 0$. The flux of surfactant anions under discussion can be expressed in terms of $K(\psi_{St})$,

$$j = D\frac{c(\kappa^{-1}) - c(0)\exp(-z\psi_{St})}{K(\psi_{St})}.$$ (7.27)

On the other hand,

$$j = D\frac{c_o - c(\kappa^{-1})}{\delta_D}.$$ (7.28)

Electrostatic retardation of adsorption kinetics of surfactant anions is expected in the case

$$\frac{K(\psi_{St})}{\delta_D} \gg 1.$$ (7.29)

Under this condition (7.29) it follows from Eqs (7.27) and (7.28) that

$$c_o - c(\kappa^{-1}) \ll c(\kappa^{-1}) - c(0) \exp(-z\overline{\psi}_{Sto}), \tag{7.30}$$

i.e., the limiting step of the process of adsorption of surfactant anions is the passing through the DL.

It follows from (7.26) that at small Stern potentials at the bubble surface, the ratio $K(\overline{\psi}_{St})/\delta_D$ is a quantity of the order of $(1/(\kappa\delta_D))$, i.e., it is much less than unity. The new effect, the electrostatic retardation of kinetics of surfactant anions adsorption becomes visible when the dimensionless parameter $\exp(-\psi_{St})$ equals or exceeds $(\kappa\delta_D)$.

7.2.4. THE RATE EQUATION OF ION ADSORPTION

The calculation of the flow of surfactant anions to the surface of a bubble can be performed under the condition of electrostatic retardation of adsorption kinetics. It follows from Eq. (7.30),

$$c(\kappa^{-1}) \approx c_o. \tag{7.31}$$

The values ψ_{Sto} and Γ_o refer to the equilibrium values of the Stern potential and the surface concentration, respectively. To replace the concentration $c(0)$ from Eq. (7.27), the adsorption isotherm (7.10) can be used,

$$\Gamma_o = c_o \Delta \exp(-W^-) \exp(z\psi_{Sto}), \tag{7.32}$$

which yields

$$j = Dc_o \frac{1 - \dfrac{\Gamma}{\Gamma_o} \exp\left(z\left(\overline{\psi}_{Sto} - \overline{\psi}_{St}\right)\right)}{K(\psi_{St})}. \tag{7.33}$$

ψ_{St} and ψ_{Sto} can be expressed by the approximate relation

$$\overline{\psi}_{St} \approx -\frac{2}{z^+} \ln\left(\frac{z \kappa \Gamma}{z^+ 2 c_o^+}\right). \tag{7.34}$$

Substituting (7.34) into (7.26) and (7.33), the following equations result,

$$K(\psi_{St}) = K(\psi_{Sto}) \left(\frac{\Gamma}{\Gamma_o} \right)^{2z/z^+ - 1}, \tag{7.35}$$

$$j = \frac{Dc_o}{K(\psi_{Sto})} \frac{1 - \left[\dfrac{\Gamma}{\Gamma_o} \right]^{2z/z^+ + 1}}{\left[\dfrac{\Gamma}{\Gamma_o} \right]^{2z/z^+ - 1}}. \tag{7.36}$$

7.3. THE MANIFESTATION OF ELECTROSTATIC RETARDATION IN TRANSIENT ADSORPTION PROCESSES

7.3.1. DIFFERENTIAL EQUATION OF ADSORPTION KINETICS. GENERAL CONSIDERATION

The non-steady diffusion of surfactant ions is a problem similar to the non-steady diffusion of non-ionic surfactant, which was described in Chapter 4. There is a specific distinction caused by the electrostatic retardation effect. The non-steady transport of ionic surfactants to the adsorption layer is a two-step process, consisting of the diffusion outside and inside the DL.

The estimation, given by Eq. (7.30) cannot be used to discriminate between the two limiting cases of electrostatic retardation and diffusion outside the DL as the controlling steps of the adsorption kinetics process. The whole adsorption process can be seen as an overlap of these two steps, as it was also assumed for a mixed diffusion-kinetic-controlled adsorption (cf. Section 4.4). Initially, the thickness of the non-steady diffusion layer is very small and the diffusion outside the DL is a fast process. Thus, in the initial stage the electrostatic retardation dominates and the comparative fast diffusion outside the DL can be neglected as a time limiting step. This process is described by Eq. (7.36). After the thickness of the non-steady diffusion layer increases, the ion transport outside the DL comes slower and finally the step governed by the electrostatic retardation is negligible.

For simplification, in the following an equilibrium between the adsorption layer and the subsurface is assumed. The retardation by the electric double layer can be considered as a process analogous to the kinetic retardation of molecular adsorption discussed in Section 4.4. With respect to the latter mechanism, the physical picture of the electrostatic retardation is clear, while the nature of the adsorption barrier, leading to a deviation from equilibrium between the surface and the subsurface, has multiple origins.

In the following an estimate will be given to distinguish between electrostatic retardation and bulk diffusion as time-controlling steps of the adsorption process.

After substituting the concentration $c(0,t)$ from Eq. (7.22) into the Stern-Martynov adsorption isotherm Eq. (7.10) the equation of adsorption kinetics follows,

$$\Gamma(t) = \Delta \exp[-W^{\cdot} + z\,\overline{\psi}_{St}(t)]\left(c(\kappa^{-1},t) - \frac{d\Gamma}{dt}\frac{K(\psi_{St})}{D}\right). \qquad (7.37)$$

The integral in Eq. (7.22) is expressed by the coefficient of electrostatic retardation, given in Eq. (7.26) and the flux j is replaced by

$$j = \frac{d\Gamma}{dt}. \qquad (7.38)$$

By introducing the following dimensionless variables,

$$\overline{\Gamma} = \frac{\Gamma}{\Gamma_o} \text{ and } \theta = \frac{Dtc_o}{\Gamma_o K(\overline{\psi}_{St})}, \qquad (7.39)$$

Eq. (7.37) transforms to (Miller et al. 1994),

$$\frac{d\overline{\Gamma}}{d\theta} + \frac{\exp(-z(\overline{\psi}_{St} - \overline{\psi}_{Sto}))\overline{\Gamma}(\theta)K(\psi_{Sto})}{K(\psi_{St})} = \frac{c(\kappa^{-1},t)}{c_o}. \qquad (7.40)$$

Details about the derivation are given in Appendix 7A.

7.3.2. THE ESTIMATION OF THE INFLUENCE OF A NON-EQUILIBRIUM ELECTRIC DOUBLE LAYER ON ADSORPTION KINETICS

The function $\overline{\Gamma}(\theta)$ is monotonically increasing and asymptotically approaches to 1 with $\theta \to \infty$. The final stage of the adsorption kinetics process can be approximated by the term $1-\overline{\Gamma}(\theta)$. To consider this final stage we use a dimensionless time θ of the order of 1. From Eq. (7.39) the following characteristic adsorption time results,

$$T_2 = \frac{\Delta \exp(-W^{-} + z\overline{\psi}_{Sto})K(\psi_{Sto})}{D}. \qquad (7.41)$$

In a similar way a characteristic time can be derived for a theoretical model neglecting the importance of a non-equilibrium DL during the whole adsorption process. Starting from Eq. (7.37) and omitting the second term on the right hand side, representing the retardation coefficient, the following relation is obtained,

$$\Gamma(t) = \Delta \exp[-W^{\cdot} + z\,\overline{\psi}_{St}(t)]c(\kappa^{-1},t).$$ (7.42)

In this case the characteristic time is connected with the non-steady bulk diffusion leading to the following dimensionless time (cf. Appendix 7A),

$$\Theta = \frac{4 D t c_o^2}{\pi \Gamma_o^2}.$$ (7.43)

In an approximate relation of the equilibrium diffuse layer, the characteristic time T_1 of adsorption kinetics of ionic surfactants is estimated from

$$\Theta \approx 1,$$ (7.44)

leading to the relation

$$T_1 \approx \frac{\Gamma_o^2}{c_o^2 D}.$$ (7.45)

From Eqs (7.41) and (7.45) we get for the time ratio of both models

$$\frac{T_2}{T_1} = \frac{z}{2z - z^+}\frac{c_o}{c_{el}}\exp\left(-(z - z^+)\overline{\psi}_{Sto}\right).$$ (7.46)

The ratio T_2/T_1 characterises the role of a non-equilibrium DL on the adsorption kinetics of ionic surfactants. If for example $\overline{\psi}_{Sto} \approx 8-10$, $z=2$ and $c_o \sim c_{el}$ the deviation of $c(x, t)$ from equilibrium can retard the adsorption kinetics by two orders of magnitude. However, an addition of electrolyte can suppress this effect.

At first glance, when taking into account Eq. (7.19), it seems to be ill-defined to set $c_o \sim c_{el}$. However, this condition does not effect the given estimate. Although the relation for κ becomes more complex, its value does not change so much. A more exact analysis of the non-equilibrium DL and its influence on adsorption kinetics is given in Appendix 7A.

7.4. DYNAMICS OF ADSORPTION AT HARMONICALLY DISTURBED SURFACES

Due to the great interest in harmonically disturbed surfaces, such as capillary or longitudinal waves or oscillating bubbles and drops, models of periodic adsorption-desorption processes are of importance (cf. Chapter 6). This section deals with peculiarities of periodic adsorption-desorption processes in respect of ionic adsorption layers. Because of the small amplitudes of applied oscillations a linear analysis of the problem is possible and simplifies the theory enormously. Thus, the main equation, which describes the adsorption-desorption process of ionic surfactants, Eq. (7.35) can be linearised for small disturbances $\delta\Gamma \ll \Gamma_o$ by,

$$j = \frac{Dc(\kappa^{-1})}{K(\psi_o)}(2z/z^+ +1)\delta\bar{\Gamma} \qquad (7.47)$$

with

$$\delta\bar{\Gamma} = \frac{\delta\Gamma}{\Gamma_o} \ll 1. \qquad (7.48)$$

As $\delta\bar{\Gamma}$ changes its sign during the harmonic surface area changes, the direction of the mass transfer, given by Eq. (7.47), changes too.

The concentration at the boundary between the diffuse layer and diffusion layer, $c(\kappa^{-1}, t)$, is unknown. To express this concentration in terms of the surfactant bulk concentration c_o the Eq. (7.28) describing the steady transport through the diffusion layer can be used. This leads to

$$c(\kappa^{-1}) = \frac{c_o}{1 + (2z/z^+ +1)\delta\bar{\Gamma}\delta_D / K(\psi_{Sto})}. \qquad (7.49)$$

The substitution of $c(\kappa^{-1}, t)$ into Eq. (7.47) results in a relation for the surfactant flux accounting for the influence of both the diffuse and diffusion layer,

$$j_n = Dc_o \frac{(2z/z^+ +1)\delta\bar{\Gamma} / K(\psi_{Sto})}{1 + (2z/z^+ +1)\delta\bar{\Gamma}\delta_D / K(\psi_{Sto})}. \qquad (7.50)$$

Two extreme cases can be classified:

$$\delta_D \gg \frac{K(\psi_{Sto})}{\delta\bar{\Gamma}(2z/z^+ +1)} \qquad (7.51)$$

and

$$\delta_D \ll \frac{K(\psi_{Sto})}{\delta\bar{\Gamma}(2z/z^+ +1)} \qquad (7.52)$$

If condition (7.51) holds the influence of the diffusion layer dominates and the retardation of the electric double layer can be neglected and Eq. (7.50) transforms into Eq. (7.28). In the opposite case, the existence of the diffusion layer is irrelevant and Eq. (7.50) transforms into (7.47) with $c(\kappa^{-1}, t) = c_o$.

Analysis of the two relations (7.51) and (7.52) allow favourable conditions for a manifestation of the electrostatic retardation effects at oscillating surfaces to be defined. At first, the diffusion layer thickness has to be small and decreases with increasing frequency. Secondly, the deviation from equilibrium during the oscillation process must be small, $\delta\overline{\Gamma} \ll 1$.

To derive the sufficient condition for discarding the diffusion layer influence, the diffusion layer thickness must be estimated. The maximum thickness δ_D of the oscillating layer thickness results from the Einstein relation,

$$\delta_D \approx \sqrt{\frac{D}{\omega}}.$$
(7.53)

The substitution of δ_D into Eq. (7.52) allows to derive a dimensionless criterion,

$$\alpha = \delta\overline{\Gamma}\frac{\delta_D}{K(\psi_{Sto})}(2z/z^+ + 1) = \frac{\delta\overline{\Gamma}\sqrt{D/\omega}}{K(\psi_{Sto})}(2z/z^+ + 1).$$
(7.54)

Under the condition $\alpha \ll 1$ the electrostatic retardation controls the periodic transport of the surfactant ions. During the period of the oscillation the diffusion layer thickness is even smaller than δ_D and, therefore, Eq. (7.54) holds for the whole oscillation process.

At sufficiently high frequencies ω Eq. (7.54) is always fulfilled as δ_D and $\delta\overline{\Gamma}$ decrease with increasing ω. However, the most interesting range of frequency is around 100 Hz, available by many wave damping experiments (for example Lin et al. 1991). For oscillating bubbles, Johnson and Stebe (1993) pointed out, that frequencies also of the order of 100 Hz should be efficient. In that case, $\delta_D \approx 1\mu m$ and $\kappa^{-1} \approx 1\mu m$ in absence of an background electrolyte and the condition (7.54) is fulfilled for $(z - z^+/2)\overline{\psi}_{Sto} \approx 1$ and one-valent surfactant ions.

In presence of an background electrolyte, the adsorption kinetics can also be controlled by electric retardation effects. This conclusion holds for strongly charged surfaces and high electrolyte concentration. High Stern potentials, of about 200 mV, can be achieved at common surfactant and electrolyte concentrations. This becomes clear from the data in Fig. 7.1 representing results from electrophoresis measurements of n-heptane droplets in water in the presence of sodium dodecylsulphate and 10^{-2} M NaCl.

To incorporate the electrostatic retardation effect into the adsorption-desorption exchange of matter at harmonically disturbed surfaces, the amount of ions is described by

$$N(t) = \Gamma(t)A(t).$$
(7.55)

The ionic flux j(t)A(t) changes the amount of adsorbed ions,

$$j(t)Adt = dN.$$
(7.56)

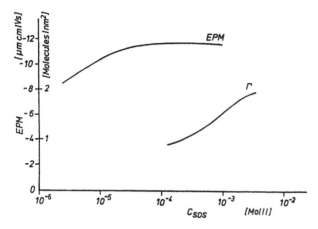

Fig. 7.1   Electrophoretic mobility (EPM) of n-heptane droplets in aqueous 10^{-2} M NaCl solution in dependence of SDS concentration; pH=5.5, t=25°C

Combination of these two equations results in

$$\frac{d\delta\overline{\Gamma}}{dt} + \frac{d\ln A}{dt} = j = \frac{Dc}{K(\psi_{Sto})\Gamma_o}\,\delta\overline{\Gamma}.$$

(7.57)

From the analysis of this expression, comparison of the terms containing $\delta\overline{\Gamma}$, the condition can be derived at which the exchange of matter is prevented by electrostatic retardation. If the term on the right hand side of Eq. (7.57) is negligible, the exchange of matter does not influence the adsorbed amount of ions at the interface. For a harmonic deformation of the surface, the relation

$$\frac{d\delta\overline{\Gamma}}{dt} = \omega\delta\overline{\Gamma}$$

(7.58)

holds. Thus, using the definition

$$k_{ad}^{el} = \frac{Dc(\kappa^{-1})}{K(\psi_{Sto})}\left(\frac{2z}{z^+}+1\right)$$

(7.59)

any exchange of matter at a harmonically disturbed surface is absent if the condition

$$\frac{k_{ad}^{el}}{\omega} \ll 1$$

(7.60)

is fulfilled. In this case, the adsorption layer, formed by soluble surfactant ions, behaves like an insoluble monolayer.

7.5. STAGES OF ADSORPTION KINETICS OF IONICS UNDER THE CONDITION OF CONVECTIVE DIFFUSION

If adsorption processes proceed under conditions of convective diffusion, an important characteristic parameter, the diffusion layer thickness, δ_D enters the discussion (cf. Section 7.2). This simplifies the description of the whole process. The flux of adsorbing ions can be determined from the two Eqs (7.27) and 7.28), leading to an expression for $c(\kappa^{-1}, t)$,

$$c(\kappa^{-1}, t) = \frac{c_o}{\left(1 + \dfrac{\delta_D}{K(\psi_{Sto})\overline{\Gamma}^{2z/z^+ - 1}}\right)}, \tag{7.61}$$

which is an approximate at $\overline{\Gamma}^{2z/z^+ - 1} \ll 1$. Inserting Eq. (7.61) into (7.36) yields

$$j(\Gamma(t)) = \frac{Dc_o}{K(\psi_{Sto})\overline{\Gamma}^{2z/z^+ - 1}}\left(1 + \frac{\delta_D}{K(\psi_{Sto})\overline{\Gamma}^{2z/z^+ - 1}}\right)^{-1}\left(1 - \overline{\Gamma}^{2z/z^+ + 1}\right). \tag{7.62}$$

The initial stage of the adsorption process is complicated and difficult to describe, while Eq. (7.62) was derived under steady state conditions.

For the case

$$\frac{K(\psi_{Sto})}{\delta_D} \gg 1 \tag{7.63}$$

the relaxation time of adsorption becomes very large in comparison with the time necessary to establish a steady diffusion layer outside the DL.

Initially, the surface potential can be small and the simplification of the retardation coefficient does not hold (cf. Eq. 7.11). Simultaneously, the initial value of adsorption is small (cf. Eq. 7A.9). If

$$\overline{\Gamma}\Big|_{t=0} \leq \frac{\delta_D}{K(\psi_{Sto})} \tag{7.64}$$

the concentration decrease at the boundary between the diffuse and diffusion layer becomes possible, which can be seen form Eq. (7.61). Thus, the diffusion process outside the DL influences the early stage of adsorption where surface concentration, surface charge and surface potential are small and consequently the retardation is weak. With increasing time the surface concentration increases and $c(\kappa^{-1}, t)$ approaches 1. Consequently, there is a stronger retardation and a smaller flux, and so the concentration gradient in the diffusion layer decreases. In this second stage the strengthening of the retardation means that diffusion outside the DL becomes less important. During both stages the condition (7A.9) holds and Eq. (7.62) can be simplified, for the first stage to

$$j(\Gamma(t)) = \frac{Dc_o}{K(\psi_{Sto})}\left(\left(1 + \frac{\delta_D}{K(\psi_{Sto})\overline{\Gamma}^{2z/z^*-1}}\right)\overline{\Gamma}^{(2z/z^*-1)}\right)^{-1},$$

(7.65)

and for the second stage to

$$j(\Gamma(t)) = \frac{Dc_o}{K(\psi_{Sto})\overline{\Gamma}^{2z/z^*-1}}.$$

(7.66)

The third corresponds to the region where $\overline{\Gamma}$ approaches 1. The last multiplier in Eq. (7.62) becomes negligible and the relation can be linearised to Eq. (7.47).

If the first fast stage is neglected, Eq. (7.62) is identical with Eq. (7A.8) and the second and third stage of the adsorption process can be described by the Eqs (7A.11) and (7A.14). The given approximations allow to simplify the description of the more complicated problem discussed in the following section.

7.6. ADSORPTION KINETICS MODEL, TAKING INTO ACCOUNT THE ELECTROSTATIC RETARDATION AND A SPECIFIC ADSORPTION BARRIER

On the basis of the Henry mechanism, given in Section 4.4, and the classification of stages of the adsorption process of the previous section, the simultaneous influence of electrostatic retardation and a specific barrier can be regarded for. To do so, the expression for $c(0,t)$ given in Eq. (7.22) is inserted into the Henry rate equation (4.32), which leads to,

$$\frac{d\Gamma}{dt} = k_{ad}\exp(z\overline{\psi}_{St}(t))\left(c(\kappa^{-1},t) - \frac{d\Gamma}{dt}\frac{K(\psi_{St}(t))}{D}\right) - k_{des}\Gamma.$$

(7.67)

Rearrangements yield

$$\frac{d\Gamma}{dt} = \frac{k_{ad}\exp(z\overline{\psi}_{St})c(\kappa^{-1},t) - k_{des}\Gamma}{1 + k_{ad}\exp(z\overline{\psi}_{St})K(\psi_{St})/D}.$$

(7.68)

In analogy to Eq. (7.61), an expression for $c(\kappa^{-1},t)$ can be obtained from Eqs (7.68) and (7.28). This relation allows to describe the first stage of a quasi-steady state adsorption process, in which the influence of the two retardation mechanisms are negligible. The electrostatic retardation is weak due to low surface concentration, surface charge, and surface potential. Far from equilibrium the specific retardation can be weak. As the result the initial adsorption rate can be sufficiently fast and the concentration gradient in the diffusion layer large. With increasing adsorption, its rate slows down and $c(\kappa^{-1},t) \to c_0$. Then, Eq. (7.68) can be simplified to

$$\frac{d\Gamma}{dt} = \frac{k_{ad} \exp(z\overline{\psi}_{St})c_0 - k_{des}\Gamma}{1 + k_{ad} \exp(z\overline{\psi}_{St})K(\psi_{St})/D}. \tag{7.69}$$

According to the electroneutrality condition, the dimensionless potential $\overline{\psi}_{Sto}$ can be expressed in terms of adsorption and electrolyte concentration,

$$\exp(z\overline{\psi}_{St}) = \left(\frac{2z^+}{z\kappa} c^+ / \Gamma\right)^{2z/z^+}. \tag{7.70}$$

At small values of Γ the second term in Eq. (7.69) disappears, i.e. desorption can be neglected and the following relation is obtained,

$$\frac{d\Gamma}{dt} = \frac{k_{ad} \exp(z\overline{\psi}_{St})c_0}{1 + k_{ad} \exp(z\overline{\psi}_{St})K(\psi_{St})/D}. \tag{7.71}$$

Due to Eq. (7.26), the product $\exp(z\overline{\psi}_{St})K(\psi_{St})$ is proportional to $\exp(-z^+\overline{\psi}_{St})$ and the second term in the denominator grows with succeeding adsorption. At small enough values for k_{ad}, the second and the third stage, discussed in the preceding section, can be introduced into Eq. (7.71). Using the second stage, Eq. (7.71) reads

$$\frac{d\Gamma}{dt} = k_{ad} \exp(z\overline{\psi}_{St})c_0, \tag{7.72}$$

which now contains two retardation processes: the first (k_{ad}) represents the specific retardation, the second ($\exp(z\overline{\psi}_{St})$) is caused by electrostatic retardation. The two processes are multiplied with each other and can lead to very strong changes of the adsorption rate. However, only at sufficiently low values of k_{ad} is the donation large. Thus, strong specific retardation, owing to many reasons (cf. Section 4.4), can be enhanced by electrostatic retardation of ionic surfactant adsorption kinetics.

If the specific retardation is weak, the change of surface concentration leads to a predomination of the second term in the denominator of Eq. (7.71) and the adsorption kinetics is controlled by the electrostatic effect only,

$$\frac{d\Gamma}{dt} = \frac{Dc_0}{K(\psi_{St})}.$$

(7.73)

Depending on the strength of specific retardation, three type of kinetics curves exist. At very strong specific retardation, the effects of both mechanisms multiply each other. At smaller adsorption barriers a two-step process results. First, both retardations act, and later the electrostatic effect outweighs the specific barrier. Finally, at very low adsorption barriers, only the electrostatic effect controls the retardation.

After adsorption succeeds, the desorption flux becomes significant and Eq. (7.47) has to be generalised.

7.7. THE PROBLEM OF ION ADSORPTION MODELS

In the presented theories of electrostatic retardation, very simple models are used to enable an analytical solution of the different problems and to clarify the physics of the mechanisms. The objective of further work is of course the generalisation of models with respect to the adsorption isotherm, content of background electrolyte, and ion transport properties.

As adsorption isotherm the equation of Stern-Martynov (1979) was used for simplification (cf. Chapter 2). This isotherm describes the surfactant adsorption only in a restricted concentration interval. A much more advantageous model of adsorption site dissociation-binding model was developed by Koopal (1993) and could be used preferentially.

The importance of electrostatic retardation increases with the surface potential, i.e. with the adsorption surfactant molecules. Especially in some practical systems of high background electrolyte, only at densely packed adsorption layers the electrostatic retardation will set in. This state of adsorption has not been taken into consideration so far. With increasing background electrolyte concentration counterions build the Stern layer. The charge of the adsorption layer is compensated partially by the diffuse layer and the Stern layer (Eq. 2.5) which decrease with the increased amount of counterions in the Stern layer. Simultaneously, the Stern potential is lowered and the electrostatic retardation becomes less effective. This aspect was discussed already by Kretzschmar et al. (1980). Consequently, the electrostatic retardation can exist in NaCl solution while it can disappear under certain conditions in $CaCl_2$ solutions.

The Stern-Martynov isotherm does not take into account the intermolecular interaction of adsorbed molecules, as it is considered in the Frumkin isotherm (cf. Eq. (2.43)). The classical version of the Frumkin isotherm was derived for nonionic surfactant adsorption layers. The incorporation of electrostatic interaction was proposed by Borwankar & Wasan (1986, 1988).

On the basis of this generalised adsorption isotherm Borwankar & Wasan developed a macro-kinetic approach of ionics adsorption kinetics taking into account the interaction between adsorbed ions within the adsorption layer. On the other hand, the deviation of the diffuse part of the DL from the equilibrium state was ignored in their models.

Both, the theories of Dukhin et al. (1983) and Miller et al. (1994), and Borwankar & Wasan (1986, 1988) assume that the continuum approach for the ion transport is valid and the transport equations can be used with constant transport coefficients up to the interface. Another very significant restriction is the macro-kinetic approach itself, ignoring the discrete nature and finite dimensions of adsorbing ions. Adsorption is impossible not only on that part of the interface occupied by ions adsorbed before, but also in some neighbourhood of this area. An ion approaching pre-adsorbed ions will be affected by a strong repulsion at a distance of several diameters of an ion. As the degree of coverage of the surface with adsorbed ions increases transport of ions from the sublayer to the surface will be slowed down. This is not considered within the framework of the macro-kinetic approach and the theory should be developed in this direction.

A first attempt to consider the role of the Debye counterion atmosphere on the transport of a surfactant ion through the DL was made by Mikhailovskij (1976, 1980) (cf. Kortüm 1966, Lyklema 1991). In contrast to a macro-kinetic model, Mikhailovskij derived kinetic equations for a multi-component system under the influence of an external electric field. The basis of this derivation was the set of Bogolubow equations for the partial distribution functions. As the result of the model derivation the following set of electro-diffusion equations is obtained,

$$\frac{\partial n_i(\vec{\Gamma})}{\partial t} - \vec{\nabla}\left[D_i\vec{\nabla}n_i(\vec{\Gamma}) + \frac{D_i n_i(\vec{\Gamma})}{kT}\vec{\nabla}(z_i e \psi_i(\vec{\Gamma}))\right] = 0, \qquad (7.74)$$

where the index i corresponds to different types of ions and the notation of the mean field theory is used. Due to the correlations in the ion distributions the mean electric field in each point is different for different ion types, given by the potential ψ_i of the mean field of the ion type i. The decisive step is the description of the transport of one ion type through the diffuse part of the DL. It could be proved that under condition (7.19) the transport of adsorbing ions can be described by the diffusion in the mean field formed by the collective action of all other ions,

$$\frac{\partial c}{\partial t} - \vec{\nabla}\left[D_{eff}\,\nabla c + \frac{e D_{eff}}{kT}\,zc\,\nabla\psi\right] = 0. \qquad (7.75)$$

The transfer to a quasi-steady adsorption allows to transfer Eq. (7.75) into Eq. (7.5). Thus, the attempt of Mikhailovskij is the non-equilibrium statistical foundation of the macro-kinetic approach used throughout this chapter.

The investigations of Mikhailovskij are significant for the discussion of the limits of the macro-kinetic approach and their extension, which can be established by the analysis of the transition from Eqs (7.74) to Eq. (7.75). To do so, Mikhailovskij concluded, that the influence of the Debye atmosphere of counterions on the transport of a higher valency ion through the diffuse layer can be neglected at sufficiently low background electrolyte concentrations. This limit, of course, has to be defined for different valences of surfactant ions, different surface activity etc.

7.8.   ELECTROSTATIC RETARDATION IN MACRO-ION ADSORPTION

From the physical point of view it is obvious that the electric double layer will affect the adsorption kinetics of polyelectrolytes. This aspect is of special interest for example in studies of the DL influence on protein adsorption. In general, the Debye atmosphere of a macro-ion cannot be ignored when considering its transport through the diffuse layer. However, under extreme conditions the macro-kinetic approach can be applied: very low background electrolyte concentration, small Stern potential, rigid macromolecule structure of small size. If the diameter of the macro-ion is small compared with the Debye length the Debye atmosphere can be neglected in the ion transport (Hückel 1924). In this case, the force originally applied to the counterion atmosphere is transferred to a small extend to the macro-ion. This phenomenon is well-known from electrophoresis theories (Debye & Hückel 1924), but it is valid only up to small potential drops within the counterion atmosphere of ~25 mV. The higher the macro-ion potential is, the more pronounced is the part of counterion distribution near the macro-ion. Simultaneously, the influence of an external field on counterion transfer becomes more efficiently with macro-ion.

For small adsorption and macro-ion potentials the superposition approximate is valid and the potential distribution around the macro-ion, transported through the diffuse layer, is given by the superposition of the spherical symmetrical potential distribution of the macro-ion and the one-dimensional potential distribution of the diffuse layer near the flat charged surface. Using this superposition the electrostatic energy of a rigid spherical macro-ion in the comparatively thick diffuse layer can be calculated in function of its distance from the surface. The thicker the diffuse layer is, the more correct is the calculated energy. Thus, the force acting on the macro-ion can be determined from the derivative of the electrostatic energy and inserted into the adsorption flux instead of the term $ze(\partial\overline{\psi}/\partial x)$ in Eq. (7.21). Finally, the coefficient of electrostatic retardation of the adsorption kinetics of weakly charged, small-size spherical macro-ions results from Eq. (7.25) by simply replacing $z\overline{\psi}$ in the integrand by the electrostatic energy. A strong electrostatic retardation of macro-ion adsorption can result even under the used restriction of $\overline{\psi} < 1$ because $z\overline{\psi}$ can be very large. Summarising, the electrostatic retardation of macro-ionic adsorption is remarkable even under very extreme conditions and an extension of the consideration with respect to higher background electrolyte concentration and higher macro-ion charge is important although connected with many difficulties.

In some cases the protein adsorption was found to be slower than expected from diffusion-controlled adsorption kinetics, which was attributed by van Dulm & Norde (1983) to some barrier the adsorbing molecules have to break through before they adsorb at the interface. This barrier has been considered to be caused by electrostatic repulsion between protein molecules and the surface. This phenomenon can be associated with the objective of the present section and the observed barrier could be caused by long range forces.

As discussed already in Section 7.6. the transfer of molecules or ions from the subsurface to the interface can be controlled by another, specific barrier which does not coincide with the electrostatic retardation arising from the transport of ions through the electric double layer. The maximum of electrostatic repulsion is located at a distance to the interface approximately equal to the radius of the adsorbing macro-ions.

In several experimental investigations no electrostatic retardation of adsorbing proteins was observed. For example, Hasegawa & Kitano (1992) did not obtain any change in protein adsorption kinetics within the pH range of 3 to 7 and electrolyte concentration interval of 0.01 to 0.1 M. However, this result cannot be generalised and from the presented results in this chapter could be expected that at lower electrolyte concentration electrostatic retardation arises. In contrast to this, Elgersmo et al. (1992) studied adsorption in low concentrations of a NaH_2PO_4 buffer (0.005 M) and observed the manifestation of electrostatic interactions.

7.9. A NUMERIC SOLUTION TO THE PROBLEM

A numeric solution of a model for diffusive transport of ionic surfactants to an adsorbing interface was recently proposed by MacLeod & Radke (1994). The model considers both the diffusion and migration of surfactant, counterions, and background electrolyte in the electric field that develops as the charged surfactant adsorbs at the interface.

The transport of the three charged species (1 - surfactant, 2 - counterion, 3 - coion) with valence z_i under the effect of an electric potential is given by Eq. (7.75), which reads for the present situation,

$$\frac{\partial c_i}{\partial t} = D_i \frac{\partial}{\partial x}\left[\frac{\partial c_i}{\partial x} + z_i c_i \frac{F}{RT}\frac{\partial \psi}{\partial x} \right], \; i=1, 2, 3. \tag{7.76}$$

The electrical potential is related to the ion distribution through the Poisson equation

$$\frac{\partial^2 \psi}{\partial x^2} = -\frac{F}{\varepsilon}\sum_{i=1}^{3} z_i c_i . \tag{7.77}$$

The initial condition and the boundary conditions far from the interface are,

$$c_i(x,0) = c_{io}, \tag{7.78}$$

$$c_i(\infty,t) = c_{io} \tag{7.79}$$

and

$$\psi(\infty,t) = 0. \tag{7.80}$$

The mass balance at the interface $x = 0$ for the adsorbing ion is given by

$$\frac{d\Gamma_1}{dt} = D_1\left[\frac{\partial c_1}{\partial x}\Big|_{x=0} + z_1 c_1 \frac{F}{RT}\frac{\partial \psi}{\partial x}\Big|_{x=0} \right], \; \text{at } x = 0, \tag{7.81}$$

and the balance for the species with no specific adsorption, counterions and coions, reads

$$\frac{\partial c_i}{\partial x}\Big|_{x=0} + z_i c_i \frac{F}{RT}\frac{\partial \psi}{\partial x}\Big|_{x=0} = 0, \text{ at } x = 0 \text{ and } i=2, 3. \tag{7.82}$$

As interrelation between potential and surface concentration it is assumed that both the gradient in potential and the surfactant surface concentration are proportional the total surface charge, which gives the boundary condition,

$$\frac{\partial \psi}{\partial x}\Big|_{x=0} = -z_1 \frac{F}{\varepsilon}\Gamma_1. \tag{7.83}$$

Based on a Frumkin type adsorption isotherm (cf. Chapter 2), which after rearrangement into a $c(\Gamma)$-form reads

$$\Gamma_1 = \Gamma_\infty \frac{c_1/a}{c_1/a + \exp\left[-\frac{2\alpha'}{RT\Gamma_\infty}\frac{\Gamma_1}{\Gamma_\infty} \right]}, \tag{7.84}$$

numerical calculations of the complete partial differential equation system were performed using a finite difference procedure. As the first result MacLeod & Radke (1994) showed that the electrical potential decays to zero within the DL thickness, as demonstrated in Fig. 7.2. The development of the diffusion layer thickness is shown in Fig. 7.3, calculated for the same set of parameters as in Fig. 7.2.

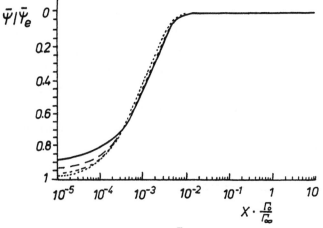

Fig. 7.2   Potential profile $\overline{\Psi}(X\frac{\Gamma_o}{\Gamma_\infty})$ calculated from Eqs (7.76) - (7.84) with $c_{1o}/a = 10$, $\alpha'=0$, $\sum_{i=1}^{3}z_i^2c_{io}/c_{1o} = 2$, $\kappa\Gamma_\infty/c_{1o} = 350$ and $D_i/D_1 = 1$, $\Theta\left[\frac{\Gamma_o}{\Gamma_\infty}\right]^2 = 10^{-4}(\text{——})$, $10^{-3}(\text{--})$, $10^{-2}(\cdot \cdot \cdot)$, $10^{-1}(\cdots)$, according to MacLeod & Radke (1994)

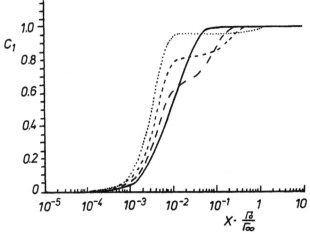

Fig. 7.3   Concentration profile $c(X,\Theta)/c_{1o}$ calculated from Eqs (7.76) - (7.84) with $c_{1o}/a = 10$, $\alpha'=0$, $\sum_{i=1}^{3}z_i^2c_{io}/c_{1o} = 2$, $\kappa\Gamma_\infty/c_{1o} = 350$ and $D_i/D_1 = 1$, $\Theta\left[\frac{\Gamma_o}{\Gamma_\infty}\right]^2 = 10^{-4}(\text{——})$, $10^{-3}(\text{--})$, $10^{-2}(\cdot \cdot \cdot)$, $10^{-1}(\cdots)$, according to MacLeod & Radke (1994)

With increasing time, the concentration at the outer double layer boundary increases while the diffusion layer thickness increases. The Fig. 7.4 plots the non-dimensional surface concentration versus dimensionless time.

As expected, the transport of an ionic surfactant ($\sum_{i=1}^{3} z_i^2 c_{io} / c_{lo} = 2$ means no additional background electrolyte) is slower than the transport on a nonionic surfactant. As background electrolyte is added, the thickness of the electrical double layer becomes smaller and consequently the overall effect of charge on surfactant transport is reduced. Further calculations by MacLeod & Radke (1994) showed also the influence of relative concentration, ion valence and diffusivity. This way of analysing the transport effect of ions on their adsorption kinetics is very straight forward. Further improvements must be made especially with respect to the equilibrium state of the ionic adsorption layer and the adjacent bulk phase, as all these model calculations are based on the Gouy-Chapman model of the electric double layer.

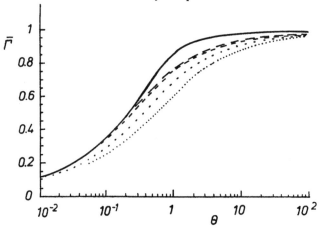

Fig. 7.4   Surface concentration as a function of time calculated from Eqs (7.76) - (7.84) with $c_{lo} / a = 10$, $\alpha'=0$, $\kappa\Gamma_\infty / c_{lo} = 350$ and $D_i / D_1 = 1$, $\sum_{i=1}^{3} z_i^2 c_{io} / c_{lo} = 202$ (– –), 22(--), 4($\cdot\cdot\cdot$), 2($\cdots$), the solid line corresponds to a nonionic adsorption calculated from the Ward & Tordai equation (4.1), according to MacLeod & Radke (1994)

7.10.   EXPERIMENTAL INVESTIGATIONS OF ADSORPTION KINETICS OF IONIC SURFACTANTS

Systematic investigations of the adsorption kinetics of ionic surfactants does not exist. In comparison to nonionics very small attention was paid to the peculiarities of ionic surfactants adsorption. Therefore, a quantitative comparison of the developed theory with experimental data is as yet impossible. However, some data exist which are in a qualitative agreement with the theory.

A strong influence of the ion valence was established by Kretzschmar et al. (1980). According to their findings, four-valent ions (Na_4-nonyl diphosphonate) adsorb much slower than two-valent ions (Na_2-octyl phosphate). The time dependence of adsorption can be obtained from Eqs (7.4), (7.22) and (7.26) which yield

$$\frac{d\Gamma}{dt} = Dc(\kappa^{-1})\exp\left[(-z+\frac{z^+}{2})|\psi_{st}|\right]\left(\frac{z}{z^+}-\frac{1}{2}\right).$$ (7.85)

Under the conditions (7.11) the relation between surface charge and potential can be simplfied to (cf. chapter 2)

$$\Gamma = \frac{\sigma}{z^+ F} = 2c_{el}\kappa^{-1}sh\left(\frac{z^+\overline{\psi}_{st}}{2}\right) \approx c_{el}\kappa^{-1}\exp\left(\frac{z^+\overline{\psi}_{st}}{2}\right).$$ (7.86)

Combining Eqs (7.85) and (7.86) one obtains

$$D\kappa c(\kappa^{-1})\left(\frac{z}{z^+}-\frac{1}{2}\right)dt = \exp\left(\left[z-\frac{z^+}{2}\right]|\psi_{st}|\right)\frac{d\Gamma}{d\overline{\psi}_{st}}d\overline{\psi}_{st} = \exp\left[z|\psi_{st}|\right]c_{el}\kappa^{-1},$$ (7.87)

which after integration yields

$$t = \exp\left[z|\psi_{st}|\right]\frac{zc_{el}}{Dc(\kappa^{-1})}\frac{\kappa^{-2}}{\left(\frac{z}{z^+}-\frac{1}{2}\right)}.$$ (7.88)

Combining Eqs (7.86) and (7.88) we can conclude that

$$\Gamma(t) \sim t^{z^+/2z},$$ (7.89)

which of course holds only under the assumptions made.

Changing from z=2 to z=4 the power of the function decreases according to Eq. (7.85). A more quantitative interpretation of experimental data by this theory would be sensible after the location of counterions in the Stern layer could be taken into account. Interpretation of experiments using the numerical solution of the transport problem is the other alternative provided again a considerable improvement of the involved electric double layer model.

The influence of background electrolyte concentration on the adsorption kinetics of SDS was studied by Fainerman (1978, 1986, 1991). With increasing electrolyte concentration the surface activity of the SDS is enhanced. However, a quantitative interpretation of the kinetic data is not possible as the electrolyte concentration was too high ($c_{el} > 0.01$ M) and the dimension of the surfactant ions were of the same order of magnitude as the diffuse layer thickness (3 nm). As mentioned earlier, in this case the continuum approach is not exact.

7.11. Summary

Similar to the equilibrium adsorption behaviour, the adsorption kinetics of ionic surfactants has some peculiarities compared with nonionics. It is well known that under equilibrium conditions the electrostatic repulsion between adsorbed ions decreases the surface concentration. Furthermore, under non-equilibrium conditions the electrostatic repulsion between ions inside the adsorption layer and ions in the adsorption flux through the diffuse layer retards the adsorption kinetics in comparison with nonionic surfactants. Any condition which then increase the electrostatic repulsion in the adsorption and double layers simultaneously enhance the retardation. The retardation efficiency increases during the adsorption process due to increasing surface charge, and also with surfactant surface activity, valency and concentration. In contrast, increasing concentration of background electrolyte and counterion valence leads to weaker retardation, as the adsorption of higher valence ions can neutralise the surface charge.

A first attempt was made by deriving a macro-kinetic theory of electrostatic retardation. At first the rate of adsorption kinetics is decreased due to a decrease in concentration of adsorbing surfactant coions in the subsurface. This influence was described under the assumption of a quasi-equilibrium inside the electric double layer in any time during the transport process. This assumption includes a double layer which establishes quasi-equilibrium conditions instantaneously after any change of the surface charge during the adsorption process. Additionally, the deviation of the distribution of adsorbing coion from equilibrium within the diffuse part of the DL contributes to the retardation effect. The deviation from equilibrium is caused by the adsorption flux, which is derived as a quasi-steady-state approach. This approach shows many similarities with the theory of slow coagulation of disperse particles and the flux is described in both cases by one and the same equation. The specific differences are in the spatial functions, which reflect the interaction energy between particles or between ions and the interface, respectively. The similar integration of the adsorption flux equation in the same way as proposed by Fuchs in his theory of slow coagulation. The integration results in the same form, while the stability ratio in the coagulation theory is replaced by coefficient of electrostatic retardation in the adsorption theory. In spite of the strong similarity between these two different processes, the adsorption model is only an approximate and valid under certain conditions. The main distinction between the two theories consists in the irreversibility of the coagulation and the reversibility of the adsorption process caused by the difference in the used boundary conditions.

The equation derived for the transport of surfactant ions through the DL describes the adsorption kinetics as a reversible process. The qualitatively new result in the theory of ionic adsorption kinetics is the incorporation of electrostatic retardation for both the adsorption and desorption process, which is of essential importance for processes close to equilibrium. Such a situation exists at harmonically disturbed surfaces, used in investigations of adsorption dynamics like the damping of capillary waves or oscillating bubbles. At sufficiently high frequencies the diffusion layer becomes very thin and the adsorption-desorption exchange is controlled only by the ion transport through the DL, i.e. by the electrostatic retardation. At strong electrostatic retardation the exchange of matter during the small period of surface area oscillations is negligible and the soluble adsorption layer behaves like an insoluble monolayer. With respect to the dynamics of adsorption, the ionic surfactant adsorption layer comprises the properties of soluble and insoluble surfactants.

Any type of specific barriers hinder the adsorption/desorption exchange of ionic and nonionic surfactants between the subsurface and the adsorption layer. In contrast to these specific barriers the electrostatic retardation influences the exchange at the boundary between diffuse and diffusion layer. Thus, the effects of specific and electrostatic retardation do not overlap but multiply each other. This means, that the collective effect of both can be measurable even when the separate effects are insignificant. The amplification of one retardation by the other is quantitatively expressed in the present theory. Varying the electrostatic conditions by electrolyte concentration and pH changes systematic studies of specific barriers and electrostatic retardations can be performed.

Finally it should be mentioned that a direct numerical solution of the transport equations will allow us to obtain more exact quantitative results. This way of solving the complex partial differential equation system is not trivial. First results are obtained by MacLeod & Radke (1994) using a comparatively simple model of the electric double layer.

7.12. REFERENCES

Borwankar, R.P. and Wasan, D.T., Chem. Eng. Sci., 41(1986)199

Borwankar, R.P. and Wasan, D.T., Chem. Eng. Sci., 43(1988)1323

Debye, P. and Hückel, E., Phys. Z., 25(1924)49

Dukhin, S.S., Glasman, J.M. and Michailovskij, V.N., Koll. Zh., 35(1973)1013

Dukhin, S.S. and Derjaguin, B.V., "Electrokinetic Phenomena", in Surface and Colloid
  Science, E.Matijevic (Ed.), Vol. 7, Wiley, New York, 1974

Dukhin, S.S. and Shilov, V.N., "Dielectric Phenomena and Double Layer in Disperse Systems and
  Polyelectrolytes", in Surface and Colloid Science, E.Matijevic (Ed.), Vol. 7, Wiley, NY, 1974

Dukhin, S.S., Malkin, E.S. and Mikhailovskij, V.N., Koll. Zh., 38(1976)37

Dukhin, S.S., Miller, R. and Kretzschmar, G., Colloid Polymer Sci, 261(1983)335

Dukhin, S.S. and Lyklema, J., Langmuir, 3(1987)95

Dukhin, S.S. and Miller, R., Colloid Polymer Sci, 269(1991)923

Dukhin, S.S., Adv. Colloid Interface Sci., 44(1993)1

Elgersmo, A.V., Zsom, R.L.I., Lyklema, J. and Norde, W., Colloids Surfaces, 65(1992)17

Fainerman, V.B., Koll. Zh., 40(1978)769

Fainerman, V.B. and Jamilova, V.D. Zh. Fiz. Khim., 60(1986)1184

Fainerman, V.B., Colloids Surfaces, 57(1991)249

Fuchs, N.A., Z. Phys., 89(1934)736

Glasman, J.M., Michailovskij, V.N. and Dukhin, S.S., Koll. Zh., 36(1974)226

Hasegawa, M. and Kitano, H., Langmuir, 8(1992)1582

Hückel, E., Phys. Z., 25(1924)204

Johnson, D.O. and Stebe, K.J., J. Colloid Interface Sci., (1993) in press

Joos, P., van Hunsel, J. and Bleys, G., J. Phys. Chem., 90(1986)3386

Koopal, L.K., in "Coagulation and Flocculation", B. Dobias (Ed.), Marcel Dekker, New York
    1993

Kortüm, G., Lehrbuch der Elektrochemie, Verlag Chemie, Weinhein/Bergstr., 1966

Kretzschmar, G., Dukhin, S.S., Genais, C., Mikhailovskij, V.N., Koll. Zh., 42(1980)644

Levich, V.G., Physico-chemical Hydrodynamics, Prentice Hall, Englewood Hill, NewYork, 1962,
    Chapter VI

Lin, S.-Y., McKeigue, K. and Maldarelli, C., Langmuir, 7(1991)1055

Listovnichij, A.V. and Dukhin, S.S., Koll. Zh., 46(1984)1094

Lyklema, J., Fundamentals of Interface and Colloid Science, Academic Press, 1991, Vol. 1

MacLeod, C. and Radke, C., Langmuir (1994), in press

Martynov, G.A., Elektrochimiya, 15(1979)494

Michailovskij, V.N., Dukhin, S.S. and Glasman, J.M., Koll. Zh., 36(1974)579

Michailovskij, V.N., Thesis "Kinetics of Adsorption of High Valency Ions", Moscow University,
    1980

Michailovskij, V.N. and Malkin, E.S., ITF-76-4ER, Institute of Theoretical Physics, Kiev,
    1976

Miller, R., Dukhin, S.S. and Kretzschmar, G., Colloid Polymer Sci., 272(1994)548

Overbeek, T., Kolloid Beihefte, 59(1943)287

van Dulm, P. and Norde, W., J. Colloid Interface Sci., 91(1983)248

CHAPTER 8

8. DYNAMIC ADSORPTION LAYER OF BUOYANT BUBBLES. DIFFUSION-CONTROLLED TRANSPORT OF NONIONIC SURFACTANTS

8.1. BASIC PROBLEMS

The movement of bubbles and drops in a liquid is influenced by the kinetics of adsorption and desorption of surfactant molecules at the liquid surface. After a certain time the motion levels off and a steady state is reached, in which the hydrodynamic field around the bubble, and in particular the motion of its surface, appears as the driving force of the process. The transient process is more difficult to consider since it entails the necessity to solve non-stationary convective diffusion equations. Investigators have paid most attention to the steady-state process of convective diffusion and adsorption/desorption. In contrast to the adsorption/desorption caused by the deviation from equilibrium between the adsorption layer and the subphase in the initial moment, the total amount of adsorbed substance does not change after the stationary state of buoyant bubbles has been reached. Thus, rather than a time dependence of adsorption, a qualitatively different characteristic parameter determines the specific nature of adsorption kinetics caused by the hydrodynamic field of a buoyant bubble. Steady-state motion of bubbles induces adsorption-desorption exchange with the subsurface, with the amount of substance adsorbed on one part of the bubble surface being equal to the amount desorbed from another part. Thus, surface concentration varies along the surface of a buoyant bubble taking a maximum value at the rear stagnation point and a minimum one at the leading pole (Frumkin & Levich, 1947).

The surface concentration difference between the poles of a bubble is a consequence of its movement. The difference increases with faster movements and disappears in the case of a resting bubble. Therefore, the state of the adsorption layer on a moving bubble surface is qualitatively different from that on a resting one. Such adsorption layers are called dynamic adsorption layers (Dukhin, 1965). Thus, a dynamic adsorption layer (DAL) is an analogy of the time dependence (kinetics) of adsorption initiated by a deviation from equilibrium at a freshly formed surface.

The problem of adsorption kinetics in the case of a buoyant bubble can be transformed into the problem of a dynamic adsorption layer. As a result of such transformation, the adsorption kinetics is coupled with other processes caused by the angular dependence of adsorption. In addition to the diffusion transport normal to the bubble surface, lateral transport by surface diffusion along the adsorption layer arises leading to local adsorption decrease. Surface tension gradients generating the Marangoni effect are connected with a non-uniform adsorption and a feedback arises in this case. The hydrodynamic field around a bubble initiates the adsorption-desorption exchange and leads to a dynamic adsorption layer which retards the motion of the surface and so acts on the hydrodynamic field and the adsorption-desorption exchange.

Dynamic adsorption layers differ from equilibrium layers not only by the existence of an angular dependence but also by the difference in the adsorbed amount averaged over the bubble surface (Sadhal & Johnson, 1983). Usually, in foam flotation, the surfactant yield is calculated under the assumption of equilibrium adsorption at the surface of buoyant bubbles. The theory of dynamic adsorption layers lead to substantial changes in the notion of surfactant flotation. Thus, the mechanism of transport at the bubble-solution interface has a substantial effect on the transport process at the surfactant solution-foam boundary.

Surfactant flotation is only one among the variety of technological processes for whose optimisation the mechanisms of adsorption-desorption exchange between the bubble and the surfactant solution are important. Therefore, there is increasing interest in this and in related problems. The appearance of special monographs is significant in this respect. As well as the well-known work of Clift et al. (1978), the recently published collective monograph by Chabra & De Kee (1990) and the review by Quintana (1990) should be mentioned.

8.1.1.   QUALITATIVE DESCRIPTION OF DYNAMIC ADSORPTION LAYERS AND SURFACE RETARDATION

The state of the surface of a floating bubble depends on its size. Surfaces of reasonably large bubbles are mobile. As a result, adsorbed surfactants are pulled down to the rear of the bubble, i.e. even under steady-state conditions the value of adsorption on a mobile bubble surface is different from that on an immobile one, Γ_o (at the same surfactant concentration in the bulk).

The leading part of the mobile surface of a floating bubble is stretched, the lowest part is compressed (Levich, 1962). The newly created sections of the surface are being filled with adsorbed substance, in the compressed part of surface the substance desorbs. The surface concentration on the leading surface of the buoyant bubble is lower than Γ_o which provides a continuous supply of surfactant (or adsorbing inorganic ions) from the bulk to the stretched surface. The surface concentration on the rear part of the buoyant bubble is higher than Γ_o

which initiates desorption of surfactant. Thus, $\Gamma(\theta)$ increases in direction opposite to the bubble motion, i.e. from the leading pole ($\theta = 0$) to the rear one ($\theta = \pi$), where the angle θ is counted from the leading pole.

Supply of surfactant to the leading surface of the bubble and withdrawal of desorbing surfactant into the bulk from the lower half is governed by diffusion and leads to the formation of a so-called diffusion boundary layer adjacent to the surface. Its thickness δ_D is much smaller than the bubble radius a_b,

$$\delta_D = a_b / \sqrt{Pe}, \tag{8.1}$$

where Pe is the Peclet number defined by

$$Pe = a_b v / D = Re\, v / D. \tag{8.2}$$

Re is the Reynolds number,

$$Re = 2a_b v / v, \tag{8.3}$$

where v is the bubble velocity of buoyancy; $v = \eta / \rho$; ρ and η are density and viscosity of the liquid, respectively, D is the diffusion coefficient of surfactant molecules.

The concentration in the diffusion layer on the leading surface of the bubble is less than in the bulk and it increases from the leading to the rear pole like in a dynamic adsorption layer. At a given θ value, local equilibrium is preserved between the surface concentration $\Gamma(\theta)$ and the concentration adjacent to the surface $c(a_b, \theta)$,

$$\Gamma(\theta) / c(a_b, \theta) = \Gamma_o / c_o = K_H, \tag{8.4}$$

where Γ_o is the equilibrium concentration on an immobile bubble surface at a surfactant bulk concentration of c_o, K_H is the so-called Henry constant and a measure of the surface activity of the surfactant.

The situation is rather fine balanced since the motion of the surface has an effect on the formation of the dynamic adsorption layer, and vice versa.

Adsorption increases in direction of the liquid motion while the surface tension decreases. This results in the appearance of forces directed against the flow and retards the surface motion. Thus, the dynamic layer theory should be based on the common solution of the diffusion equation, which takes into account the effect of surface motion on adsorption-desorption

272

processes and of hydrodynamics equations combined with the effect of adsorption layers on the liquid interfacial motion (Levich 1962).

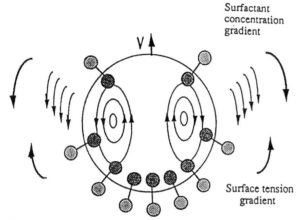

Fig. 8.1.a.   Surface concentration gradient created by interfacial convection

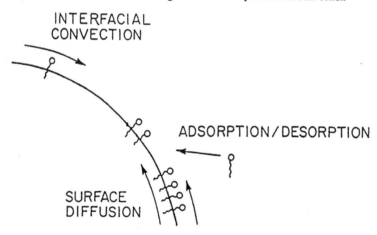

Fig. 8.1.b.   Transport process at the bubble interface

8.1.2.   BUBBLE HYDRODYNAMICS AND INTERFACIAL RHEOLOGY

The theoretical description of a diffusion process of a surfactant to, or from, the surface of a floating bubble is impossible without information on the floating velocity and the hydrodynamic field around the bubble. The first of these quantities can be found comparatively easily experimentally, whereas the Navier-Stokes equation is used to define the hydrodynamic field around the floating bubble. A solution of the equation must satisfy all boundary conditions at the bubble surface. It should be stated that a general analytical solution of this problem is impossible. This difficulty can be overcome in two extreme cases: when the bubbles rise at small Reynolds numbers (Re < 1, Stokes conditions) or at very large Reynolds numbers (Re » 1, potential conditions). In the first case, the inertial term can be eliminated from the Navier-Stokes equation, in the second the viscous term.

If the surface of a drop or bubble is immobile for any reason and the coordinate system is moving together with the bubble, the floating velocity is the same as that of a solid sphere. In particular, at small Reynolds numbers, the drop movement can be described by Stokes' equation,

$$v_{St} = \frac{2a_b^2}{9\eta} g,$$
(8.5)

where g is the acceleration due to gravity. If the drop surface is mobile, a velocity distribution also arises inside the drop. The velocity distribution over the drop surface can be found by a collective solution of the respective Navier-Stokes equation both inside the drop and in the surrounding liquid. The continuity conditions must be fulfilled both for the velocity and the tensor of viscous stresses when passing through the phase boundary. Under these boundary conditions, the Stokes equation (linearised Navier-Stokes equation at very small Reynolds numbers Re«1) for drops floating up in a liquid was solved by Rybczinski (1911) and independently by Hadamard (1911). Of course, this solution applies also to the case of a floating bubble when its viscosity is set equal to zero (neglecting the effects of the order of the ratio of the gas and liquid viscosities). According to the Hadamard-Rybczinski approximation the velocity of floating bubbles is expressed by the equation

$$v = \frac{1}{3} \frac{a_b^2 \rho}{\eta} g.$$
(8.6)

Any mobility of the surface decreases the velocity difference and the viscous stresses. The result is that the hydrodynamic resistance becomes smaller and the floating velocity of a bubble according to (8.6) increases by a factor of 3/2 as compared to Stokes' Eq. (8.5). In early experiments, under the condition of Re < 1, it was found (Lebedev 1916) that small bubbles of a diameters less than 0.01 cm behave like rigid spheres since their velocity is described by Stokes' formula (8.5). At the same time, Bond (1927) has found that drops of a sufficiently large size fall at velocities described by Eq. (8.6). To overcome contradictions with the Hadamard-Rybczynski theory, Boussinesq (1913) considered the hypothetical influence of the surface viscosity and derived the following relation,

$$v = \frac{2}{3} \frac{\rho - \rho'}{\eta} ga_b^2 \frac{\eta + \eta' + 2\eta_s / a_b}{2\eta + 3\eta' + 2\eta_s / a_b},$$
(8.7)

where η and η', ρ and ρ' are viscosities and densities of the liquid inside and outside the droplet, and η_s is the surface viscosity.

Reopening the discussion around the Hadamard-Rybczynski theory, Levich (1962) claimed that the introduction of a surface viscosity is not important, as suggested by Boussinesq, rather it is the Marangoni effect. Frumkin & Levich (1947) postulated that the anomalous experimental results of the movement of fluid droplets as solid spheres can be explained by the presence of surface-active agents which were swept to the rear of the droplet. The surface concentration gradients, and concomitant interfacial-tension gradients, created thereby were assumed to be responsible for the retardation of the drop velocity. Experimental findings reported later (Gorodetskaya 1949) qualitatively justified Levich's theory.

As both effects arise from the presence of surfactant, the controversy of "surface viscosity effect" against "surface tension gradient effect" was regarded as being nontrivial. This controversy was further developed in solving the classical problem of wave damping by surface-active agents. Data on the velocity of rising bubbles in aqueous surfactant solutions (sodium dodecylsulphate) published by Okazaki (1964) are presented in Fig. 8.2.

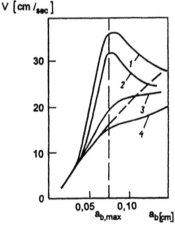

Fig. 8.2. Velocity of bubbles floating up in aqueous (distilled water) solutions of sodium dodecylsulphate of different concentrations (mole/l): 1-0; 2-10^{-6}; 3-$1.2 \cdot 10^{-5}$; dotted curve - solid bubble in distilled water

The dotted curve corresponds to solid bubbles. It was obtained by recalculations from experiments on spherical glass balls with a density of 2.37 g/m³. Significant surfactant effects appear for large bubbles. Small bubbles rise as solid spheres, even in thoroughly cleaned liquids: from $a_b = 0.03$ cm (Re = 36) and at c > 10^{-3} M up to $a_b = 0.065$ cm (Re = 182). It is typical that even at such low concentrations as 10^{-6} M the velocity of bubbles is essentially decelerated.

8.1.3. FOUNDATION OF THE THEORY OF DIFFUSION BOUNDARY LAYER AND DYNAMIC ADSORPTION LAYER OF MOVING BUBBLES

In connection with the development of the theory of convective diffusion in liquids the foundation of the theory of diffusion boundary layers and dynamic adsorption layers are given by Levich (1962) in his works on physico-chemical hydrodynamics. A variety of problems of convective diffusion in liquids was solved which are of essential interest for the description of different heterogeneous processes in liquids the rate of which is limited by diffusion kinetics. In connection with the objectives of the present chapter, only a general approach to problems of diffusion boundary layers and their concrete results (Levich 1962) are reported. These are of direct interest for the theory of dynamic adsorption layers of bubble.

The mathematical description of the convective diffusion process is based on the solution of the convective diffusion equation

$$\frac{\partial c(\theta,z,t)}{\partial t} + v(\theta,z)\,\text{grad}\,c(\theta,z,t) = \Delta\,Dc(\theta,z,t), \tag{8.8}$$

where $c(\theta,z,t)$ and $v(\theta,z)$ are the distributions of concentration and velocity of liquid, respectively. If we reduce this equation to a dimensionless form and assume stationary conditions the process can be characterised by a dimensionless Peclet number $Pe = \dfrac{v_0 L}{D}$; v_0 and L are the characteristic velocity of the flow and characteristic length, respectively. It is convenient to express Pe (cf. Eq. 8.2) in terms of the Reynolds number Re and the Prandtl number Pr which depend only on the properties of the medium,

$$Pe = Re\,Pr, \tag{8.9}$$

where $Pr = v/D \sim 10^3$. Like in the case of a viscous liquid flow at large Reynolds numbers, the role of molecular viscosity arises in a thin boundary layer, where the liquid is retarded, and the velocity difference is localised. At large Peclet numbers molecular diffusion manifests itself in a thin boundary layer which is called the diffusion layer.

While the thickness of the hydrodynamic boundary layer δ_G can be estimated by L/Re^{n_1}, a similar estimate L/Pe^{n_2} is obtained for the thickness of the diffusion boundary layer δ_D, with n_1 and n_2 having values less than unity. Therefore, a rough estimate of the ratio δ_D/δ_G is Pr^{-n_1} at $n_1 = n_2$. The kinematic viscosity of mobile water-like liquids is of the order of $v \approx 10^{-2}$ cm²/s, the diffusion coefficients of molecules and ions in aqueous solutions are of the order of $D \approx 10^{-5}$ cm²/s, those of macromolecules $D \approx 10^{-6}$ cm²/s. Thus, in water and in similar liquids $Pr \approx 10^3$ and δ_D is much smaller than δ_G. This means the diffusion boundary layer can arise also under the conditions of an absent hydrodynamic boundary layer and conditions for the existence of a diffusion boundary layer $Pe \gg 1$ can be fulfilled not only at $Re \gg 1$ but also at $Re \ll 1$.

It should be pointed out that in the formulation of the problem of convective diffusion one should know the velocity distribution $v(z)$ since it appears on the left-hand side of Eq. (8.8). A solution of the hydrodynamic problem is possible in the limiting cases of small and large Reynolds numbers. Therefore, the theory of the boundary diffusion layer for the two limiting cases $Re \ll 1$, $Pe \ll 1$ and $Re \gg 1$, $Pe \gg 1$ has to be developed.

In the following interest is focused on the dynamic adsorption layer and the boundary diffusion layer, respectively, which show a strictly stationary character due to the interplay of surfactant adsorption at one part of the mobile surface of the bubble (drop) and its desorption from the other part. Boundary conditions must take into account the convective transfer of surfactant along the surface and exchange between the surface and the bulk. The corresponding boundary condition reads

$$\text{div}_s\left(\Gamma(\theta)v_\theta(\theta) - D_s\text{grad}\Gamma(\theta)\right) = -j_n(\theta),$$
(8.10)

where D_s is the surface diffusion coefficient, j_n is the normal component of the flow density between the bulk to the surface; v_o is velocity distribution over the surface of the bubble. Thus, this boundary condition takes into account the process of mass transfer along the surface both due to convection and surface diffusion. At small adsorption times (high rate of adsorption) it can be expected that local equilibrium between adsorption $\Gamma(\theta)$ and bulk concentration at the bubble surface $c(a_b,\theta)$ exists. This means that the same functional relation between $\Gamma(\theta)$ and $c(a_b,\theta)$ as between Γ_o and c_o exists. Far from saturation, this functional relation can be considered as linear, Eq. (8.4). The adsorption rate j_n is determined by the diffusion rate,

$$j_n = -D\frac{\partial c(z,\theta)}{\partial z}\bigg|_{z=a_b}.$$
(8.11)

Thus, at high adsorption rates the quantities appearing on the right-hand and left-hand sides of Eq. (8.10) are expressed in terms of the concentration distribution so that (8.10) becomes a boundary condition for Eq. (8.8).

In Eq. (8.8) the components of the velocity field are generally very complex functions. Partial differential equations with variable coefficients are difficult to solve. The problem is simplified by approximating the velocity components within the diffusion layer by taking into account that the thickness of the diffusion layer δ_0 is much less than the typical distance over which the velocity components change significantly, i.e., $\delta_D \ll \delta_G$.

The problem can be solved effectively by converting the convection-diffusion equation into the well studied heat conduction equation by introducing the stream function Ψ as a new variable. In terms of the stream function the velocity components in spherical coordinates z and θ are,

$$v_\theta = -\frac{1}{z\sin\theta}\frac{\partial\Psi}{\partial z}; \ v_z = -\frac{1}{z^2\sin\theta}\frac{\partial\Psi}{\partial\theta}. \tag{8.12}$$

If these expressions are substituted into the left hand side of Eq. (8.8),

$$v_z\frac{\partial c}{\partial z}+\frac{v_\theta}{z}\frac{\partial c}{\partial\theta}=D\left(\frac{\partial^2 c}{\partial z^2}+\frac{2}{z}\frac{\partial c}{\partial z}\right), \tag{8.13}$$

this part of the equation is transformed into $v_\theta(\partial c/\partial z)$ for a steady process.

On the right hand side of equation (8.8) the second derivative of Ψ appears and one obtains an equation similar to that describing steady heat conduction in the variables θ, Ψ. There is also an analogy to the coefficient of thermal conductivity whose magnitude is determined by the velocity field.

Let us analyse the diffusion stream on a bubble o drop surface moving in another liquid with Re$\ll$1. According to Hadamard-Rybczinski, the drop surface is moving and the surface velocity field can be expressed by

$$v_\theta(a_b,\theta)=v_o\sin\theta, \tag{8.14}$$

The formula of Hadamard and Rybczynski are also valid for the "moving bubble" problem with $\eta'\ll\eta$. Using the Hadamard-Rybczynski velocity field, it is easy to show that the difference between the tangential component of the velocity in the diffusion boundary layer and the surface velocity field is negligible. This is the reason why the reduction of equation (8.8) to variables θ, Ψ leads to a coefficient on the right hand side which independent of $\Psi = x\sin^2\theta$,

$$\frac{\partial c}{\partial\theta}=\frac{na_b^2\sin^3\theta}{Pe}\frac{\partial^2 c}{\partial\Psi^2}, \tag{8.15}$$

with $n = v / v_o$ and $x = z - a_b$. With the introduction of a new variable the heat conduction equation with constant coefficient na_b^2 / Pe is obtained,

$$\frac{\partial c}{\partial t}=Kna_b^2\frac{\partial^2 c}{\partial\Psi^2}. \tag{8.16}$$

with $K = Pe^{-1}$.

In order to complete the analogy with the heat flow equation, it is necessary to choose the integration constant a_1 in such a way that t is everywhere positive and $\theta = 0$ coincides with t = 0. This happens at $a_1 = 2/3$, so that explicitly

$$t = \int \sin^3 \theta d\theta = \frac{2 + \cos\theta}{(1 + \cos\theta)^2} \frac{\sin^4 \theta}{3}. \tag{8.17}$$

8.1.4. MAIN STAGES IN THE DEVELOPMENT OF PHYSICO-CHEMICAL HYDRODYNAMICS OF BUBBLE

Physico-chemical hydrodynamics of bubble and drops attract the attention of many investigators in different countries over the last fifty years. Despite the obvious difficulties in contact between groups in East and West, the main results of investigations published in Russian and English agree well and complement each other. A prerequisite for this agreement is the fact that all the theories were developed on the same basis given by the works of Frumkin and Levich (Sections 8.1.2 and 8.1.3).

The founders of the physico-chemical hydrodynamics of bubbles (Frumkin & Levich, 1947) have derived only the simplest solution of the problem which was formulated at a time where the conditions of its applicability were not specified. The theory was simplified by three assumptions: i) the relative adsorption difference along the mobile bubble surface is small, ii) the dynamic adsorption layer retards the surface uniformly (i.e., the velocity of any section of the surface is decreased under its effect by the same factor which makes it possible to introduce the retardation coefficient χ_b), and iii) the variation of adsorption is strictly antisymmetric with respect to the equatorial plane which is expressed by a very simple angular dependence of adsorption. A somewhat paradoxical picture emerges when comparing the above-mentioned simplifications and the strict solution of the equation of convective diffusion given by Levich (1962). The angular dependence characterising the diffusion layer is much more complex than the simple dependence postulated for the adsorption layer. Clearly the dynamic adsorption layer in Frumkin-Levich's description is only a limiting case which has impelled Derjaguin and Dukhin to carry out systematic studies of that problem (Derjaguin et al., 1959, 1960, 1961; Dukhin & Derjaguin, 1961 a, b; Derjaguin & Dukhin, 1960; Dukhin & Buikov, 1965; Dukhin 1965, 1981, 1982, 1983; Dukhin et al., 1986).

Special attention to limiting cases is generally characteristic in the theory of transport phenomena, but here it is caused by extreme difficulties in constructing a general theory without any simplifications. A substantial difference to the work of Frumkin & Levich (1947) and Levich (1962) is the elimination of the three simplifying conditions mentioned above. Thus the problem was formulated by revealing the diversity of the states of dynamic adsorption layer differing in the complex angular dependence of adsorption and, respectively, in non-uniform surface retardation and the possibility of a substantial relative variation of adsorption along the bubble surface.

One of the unexpected results of these investigations was that the totality of simplifying assumptions made by Levich is fulfilled fairly well not at small rather than large Peclet numbers (cf. Section 8.2). Considering the conditions Re « 1, Pe » 1, the main difficulty in coupling the transport of momentum and the transport of matter was avoided in the limiting cases of weak (cf. Section 8.4) and very strong retardation (cf. Section 8.3) of the bubble surface. In the former case the hydrodynamic field of velocities is given and the convective diffusion equation is defined specifically. The solution is rather difficult, owing to the complexity of the boundary conditions (cf. Section 8.1.3). A further simplification is attained by restricting to low or high surface activity of a surfactant. In the first case the relative adsorption difference is small but its angular dependence is very complex. The second case turns out to be especially interesting since almost the whole surface is free from surfactant which is pulled down to the rear and forms a stagnant cap within which adsorption far exceeds the equilibrium value. In parallel with these limiting cases Dukhin (1965, 1981) derived an integro-differential equation which describes the structure of dynamic adsorption layers for any surface activity.

In the other limiting case of strong retardation, an efficient approximate analytical method was proposed to enable the coupling of transport of momentum and matter (see Section 8.3). Here it is also possible to discriminate between the conditions of a slight deviation of adsorption from equilibrium, and the appearance of two zones on the bubble surface showing strong and weak retardation. At strong retardation of the entire bubble surface, the relative adsorption differential is small and the angular dependence is close to that postulated by Levich (1962). Thus the simple solution by Levich can be realised also at Pe » 1, but it is valid only at strong retardation and not too high surface activity.

Summarising the work done in Russia, we can conclude that physico-chemical hydrodynamics of bubbles at Re « 1 was developed for low and high surfactant concentration where the entire bubble surface in the former case is slightly and in the latter case strongly retarded.

The intermediate concentration region is characterised by heterogeneities of the dynamic adsorption layer structure (Derjaguin and Dukhin, 1959 - 1961). When increasing the surfactant concentration a stagnant cap is formed and expands, while at decreasing concentration the weakly retarded zone expands. Thus, a double-zone dynamic adsorption layer arises in the intermediate concentration region. This problem was first formulated in a work by Savich (1953) and later much effort has been made for its solution.

In the dynamic adsorption layer theory of Derjaguin-Dukhin, the hydrodynamic field of a bubble can be assumed to be known as first approximate, while the more difficult stagnant cap problem has still to be solved. For the solution of this hydrodynamic problem unusual and very difficult boundary conditions exist which are very inconvenient even after essential simplifications. The hydrodynamic field of a bubble is studied under the assumption that the stagnant cap is completely immobilised and any motion of the surface beyond the stagnant cap is ignored. Since the description of the stagnant cap is to a large extent a hydrodynamic problem, it has received less attention (cf. Section 8.7).

Three stages or types of a dynamic adsorption layer can be distinguished: i) the dynamic adsorption layer slightly retards bubble motion; ii) dynamic adsorption layer strongly retards bubble motion; and iii) a stagnant cap exists. This classification discussed in investigations under the restriction Re « 1 is probably correct also at Re » 1. At the transition to Re » 1 in the case of a stagnant cap it is necessary to solve the hydrodynamic problem. The non-linear Navier-Stokes equation which is connected with difficult boundary conditions at the stagnant cap has not been solved yet. Therefore, the consideration of a stagnant cup was restricted up to now by the condition Re « 1. The dynamic adsorption layer theory in limiting cases of strong and weak retardation of almost the whole surface is developed on the basis of the well-known hydrodynamic fields which allows a generalisation for the important case Re»1 (Section 8.6).

8.1.5.   ROLE OF DYNAMIC ADSORPTION LAYER IN FOAMS, EMULSIONS, TECHNOLOGIES

A non-equilibrium state of the adsorption layer of bubbles or drops initiates adsorption processes in foams and emulsions. Transport of surface active substances in foams and emulsions can additionally be complicated by this dynamic adsorption layer. Specifically, water purification from surfactant or their fractionation are strongly influenced by a dynamic state of the adsorption layer (Section 8.8.3).

The problem of dynamic adsorption layers does not only arise in connection with adsorption-desorption processes, it can have a substantial effect on processes of interaction between bubbles or drops and thus on coagulation and coalescence processes in foams and emulsions (Chapter 12).

Consider flotation (Chapter 10), dynamic adsorption layers effect all stages of the elementary flotation process, since they affect the hydrodynamic field of a bubble and thus the trajectory of particles in the neighbourhood as well as the process of its deposition. The next stage is the thinning of the liquid interlayer between the surfaces of the bubble and the particle approaching it. Mechanisms of the surfactant influence on film thinning will be considered in Chapter 11. Finally, in case of overlapping fields of equilibrium surface forces of a bubble and a particle it is again important to take into account the dynamic state of adsorption layers of bubble since it controls the local values of equilibrium surface forces.

The description of dynamic adsorption layers under the condition of bubble/bubble or bubble/particle interaction is much more complex than the consideration of dynamic adsorption layers of individual bubbles discussed in the present chapter. It is still more difficult to control dynamic adsorption layers experimentally under conditions of the above-mentioned interactions. Because of these experimental difficulties, the role of mathematical modelling is extremely important in studying coagulation and heterocoagulation processes in foams and emulsions.

The problem of dynamic adsorption layers becomes very complicated in concentrated foams and emulsions and is probably of interest for rheology of foams and emulsions. Thus, a systematic study of the dynamic adsorption layer of an individual bubble is important not only by itself, but to a larger degree as investigation of fundamental processes in a variety of more complex situations sparring a range of fluid technologies.

8.2. Dynamic Adsorption Layer under Condition of Uniform Surface Retardation

8.2.1. The Case Pe«1

If the Peclet number is small, Pe«1, two simplifying approximations are possible (Dukhin, 1965, 1981). As it will be shown in the following, if Pe«1 the surface concentrations differ negligibly from their equilibrium values,

$$(\Gamma(\Theta)-\Gamma_o)/\Gamma_o =[c(a_b,\Theta)-c_o]/c_o \ll 1, \tag{8.18}$$

where a_b is radius of droplet or bubble, and the convection-diffusion equation (8.16) reduces to the Laplace equation. Thus, the hydrodynamic and diffusion fields may by truncated to the first two spherical harmonics and their derivatives (terms in $\sin\theta$ and $\cos\theta$ only),

$$c(z,\Theta) = c_o + \Delta c \frac{a_b^2}{z^2}\cos\Theta, \Gamma(\Theta) = \Gamma_o + \Delta\Gamma\cos\Theta \tag{8.19}$$

with

$$\Delta\Gamma = K_H \Delta c. \tag{8.20}$$

With these simplifications, it has been shown that the effect of the adsorption layer on the droplet hydrodynamic is quantitatively described by the introduction of a retardation coefficient. The velocity fields inside and outside the drop are respectively

$$v_\theta'(z,\Theta) = v_o(2z^2/a_b^2 - 1)\sin\Theta, \; v_z'(\Theta) = v_o(1 - z^2/a_b^2)\cos\Theta \tag{8.21}$$

$$v_\theta(z,\Theta) = v_0\left(1 - \frac{2\eta + 3\eta' + 3\chi_b}{4(\eta + \eta' + \chi_b)}\frac{a_b}{z} - \frac{\eta' + \chi_b}{4(\eta + \eta' + \chi_b)}\frac{a_b^3}{z^3}\right)\sin\Theta \tag{8.22}$$

$$v_z = v\left(1 - \frac{2\eta + 3\eta' + 3\chi_b}{2(\eta + \eta' + \chi_b)}\frac{a_b}{z} + \frac{\eta' + \gamma}{2(\eta + \eta' + \chi_b)}\frac{a_b^3}{z^3}\right)\cos\Theta \tag{8.23}$$

in which v is the translational velocity of the droplet

$$v = \frac{2}{3}\frac{\Delta\rho g a_b^2}{\eta}\frac{\eta + \eta' + \chi_b}{2\eta + 3\eta' + 3\chi_b} \tag{8.24}$$

and

$$v_o = \frac{1}{3}\frac{\Delta\rho g a_b^2}{2\eta + 3\eta' + 3\chi_b}. \tag{8.25}$$

χ_b is the retardation coefficient defined by,

$$\chi_b = \frac{\partial\gamma}{\partial c}\frac{\Delta c}{3v_o}. \tag{8.26}$$

Explicit expressions for Δc, $\Delta\Gamma$, χ_b and v_o can be derived from the boundary condition (8.10), which under the present simplifications reads

$$D\frac{\partial c}{\partial z}(a_b,\Theta) = \frac{1}{a_b\sin\Theta}\frac{\partial}{\partial\Theta}\left[\sin\Theta\left(\Gamma(\Theta)v_o\sin\Theta - D_s\frac{\partial\Gamma}{\partial\Theta}\right)\right]. \tag{8.27}$$

In view of Eqs (8.18) and (8.19) it is trivial to show that

$$\Delta c = \frac{2\Gamma_o v_o}{D + 2D_s K_H / a_b} \tag{8.28}$$

and

$$\chi_b = \frac{2}{3} \frac{\Gamma_o RTK_H}{D + 2D_s K_H / a_b}, \tag{8.29}$$

where R is the gas constant and T is the absolute temperature. If $\chi_b \to \infty$ Eq. (8.24) reduces to the terminal velocity of a falling spherical particle.

At Pe «1 condition (8.18) is satisfied for any surface activity

$$\frac{\Delta c}{c_o} = \frac{2\Gamma_o}{c_o a_b} \frac{v_o a_b}{D} \frac{1}{1 + 2D_s / D \cdot \Gamma_o / c_o a_b} < \frac{v_o}{v} \frac{D}{D_s} Pe«1. \tag{8.30}$$

The final result for the surface concentration distribution at Pe«1 and the establishment of practically instantaneous adsorption equilibrium is derived from Eqs (8.20), (8.27), (8.28), and (8.29),

$$\Gamma(\Theta) = \Gamma_o - \Gamma_o \frac{2\Delta\rho g a_b^2 K_H}{3\left[D + 2D_s \frac{K_H}{a_b}\right]\left[2\eta + 3\eta' + \frac{2RT\Gamma_o K_H}{D + 2D_s K_H / a_b}\right]} \cos\Theta \tag{8.31}$$

For low surface activity

$$K_H « a_b \tag{8.32}$$

the surface diffusion is negligible with respect to convective surface transport, provided that D and D_s have the same order of magnitude. For high surface activity,

$$K_H » a_b, \tag{8.33}$$

surface diffusion is the dominant factor and determines adsorption distribution and surface retardation while the convective surface transport is negligible.

8.2.2. PE » 1, RE « 1

Condition (8.18) can or cannot be fulfilled at Pe»1 and the angular distribution dependence of adsorption and surfactant concentration is more complex than the functions given in Eq. (8.19). The concentration distribution must obey the convective diffusion equation. In spite of this, Frumkin & Levich (1947) have proposed an approximate theory of the diffusion boundary layer and dynamic adsorption layer at Pe»1, and Re«1 based on condition (8.18) as well as the simple relation (8.20) characterising the dynamic adsorption layer, and a uniform surface retardation. This means that it is permitted to introduce the retardation coefficient.

At large Peclet numbers a thin diffusion boundary layer is formed, so that

$$j_n = D \frac{\Delta c \cos\theta}{\delta_D(\theta)}, \tag{8.34}$$

where the postulated angular dependence of the concentration distribution Eq. (8.19) is taken into account.

Expressing the right-hand side of the boundary condition (8.10) by means of (8.34), it is easy to conclude that the assumption of simple angular dependencies of concentration and adsorption (8.19) are not compatible, if we also take into account the angular dependence of the diffusion layer thickness $\delta_D(\Theta)$. In other words, the Frumkin-Levich approximation corresponds to simplifications of the boundary condition (8.10) by substituting the averaged value of the diffusion layer thickness into Eq. (8.34). With the mentioned assumptions and simplifications it is possible to derive the retardation coefficient for Pe»1

$$\chi_b = 2RT\Gamma_o^2\delta_D / 3Da_b c_o. \tag{8.35}$$

If in addition to Pe»1 the condition $\chi_b \ll 3/2\eta$ is valid, the condition for low surface retardation can be stated in the form

$$(\Gamma_o / c_o)^2 c_o \ll (27/4)\eta Da_b / RT\delta_D. \tag{8.36}$$

The Frumkin-Levich theory was of great methodological importance and was the basis for more rigorous considerations at Pe»1 by Derjaguin & Dukhin (1959 - 1961). These results are discussed in the next sections. It turns out that uniform retardation and relatively small variation of adsorption exist along with other conditions which radically differ from predictions of the Frumkin-Levich theory.

8.3. THEORY OF DYNAMIC ADSORPTION LAYER OF BUBBLE (DROP) AT RE«1 AND STRONG SURFACE RETARDATION

The condition $v_\theta(a_b,\theta) \ll v$ enables the application of a successive approximation method to calculate the velocity, concentration and adsorption fields (Dukhin & Derjaguin, 1961). Since the velocity of the flow relative to the bubble is much higher than the velocity of its surface, the following boundary conditions can be taken as the first approximate,

$$v_\theta(a_b, \theta) = 0. \tag{8.37}$$

This relation coincides with the boundary condition for a viscous flow around solid spheres. In this approximation the velocity distribution at $Re \ll 1$ is expressed by Stokes' formula. From Stokes' velocity distribution $v(z, \theta)$ it is easy to calculate the viscous stresses acting on the surface of the sphere and the equilibrating surface tension gradient

$$\frac{1}{a_b} \frac{d\gamma}{d\theta} = \eta \left(\frac{1}{a} \frac{\partial v_z}{\partial \theta} + \frac{\partial v_\theta}{\partial z} - \frac{v_\theta}{z} \right)_{z=a_b} = \frac{3}{2a_b} \eta v_o \sin\theta. \tag{8.38}$$

Under the assumption of adsorption equilibrium, the establishment of the viscous stress is promoted by surface retardation, and the equation follows

$$\frac{\partial c}{\partial \theta}(a_b, \theta) = \frac{3}{2} \eta \left(\frac{\partial \gamma}{\partial c} \right)^{-1} v_o \sin\theta. \tag{8.39}$$

Here $c(a_b, \theta)$ characterises the concentration distribution along the surface. Integrating both sides of the equation and assuming a linear relation between local adsorption $\Gamma(\theta)$ and $c(a_b, \theta)$ we obtain the distribution $c(a_b, \theta)$ and adsorption $\Gamma(\theta)$ on a strongly retarded bubble surface,

$$c(a_b, \theta) = c_o + \delta c - \frac{3}{2} \eta v_o \left(\frac{\partial \gamma}{\partial c} \right)^{-1} \cos\theta, \tag{8.40}$$

$$\Gamma(a_b, \theta) = \Gamma_o + \delta\Gamma - \frac{3}{2} \eta v_o \left(\frac{\partial \gamma}{\partial \Gamma} \right) \cos\theta. \tag{8.41}$$

δc and $\delta\Gamma$ are unknown quantities. Keeping in mind that a diffusion boundary layer is formed at $Pe \gg 1$ (Section 8.1.3) and that Stokes' velocity field can be used at high retardation, the convective diffusion equation at $Re \ll 1$ and $Pe \gg 1$ can be transformed into the following form (Levich 1962)

$$\left(\frac{\partial c}{\partial t} \right)_\psi = \frac{\partial}{\partial \psi} \left(\sqrt{\psi} \frac{\partial c}{\partial \psi} \right)_t. \tag{8.42}$$

where

$$t = \frac{Da_b^2 \sqrt{3v_o}}{2} f(\theta),$$

(8.43)

$$f(\theta) = \theta - \frac{\sin 2\theta}{2},$$

(8.44)

and the stream function Ψ within the diffusion boundary layer can be given as $\Psi = 3/4v_o x^2 \sin^2\theta$.

Solving equation (8.43) under the condition of constant concentration beyond the diffusion boundary layer we can determine $c(\psi,\theta)$ and an expression for the complete diffusion flow on the surface of the sphere can be derived. This flow equals to zero under stationary conditions and leads to a condition for determining $\delta\Gamma$ from Eq. (8.41)

$$\int_0^\pi \left(\frac{\partial c(\psi,\theta)}{\partial x} \right)_{x\to 0} \sin\theta d\theta = 0.$$

(8.45)

After $c(x,\theta)$ and $\Gamma(x,\theta)$ have been determined we can calculate the velocity distribution on a strongly retarded surface using the boundary condition

$$\mathrm{div}_s \left[\Gamma(\theta)v_\theta(a_b,\theta) \right] = D\left(\frac{\partial c(\psi,\theta)}{\partial x} \right)_{x\to 0}.$$

(8.46)

According to Eq. (8.46) and the expression for $c(x,\theta)$, we obtain the following result for the velocity distribution on a retarded surface,

$$v_\theta(a_b,\theta) = \frac{NF(\theta)}{1+\delta\Gamma/\Gamma_o+(A/c_o)\cos\theta},$$

(8.47)

with

$$N = \frac{mADPe^{1/3}}{3^{1/3}\Gamma_o}, \quad A = \frac{3}{2}\eta v_o \left(\frac{\partial\gamma}{\partial c} \right)^{-1}$$

(8.48)

To determine δc from Eq. (8.45) the following equation is obtained,

$$(1-\delta c/A) = \int_0^\pi \sin\theta'(1-f(\theta')/\pi)^{2/3}d\theta'.$$

(8.49)

The numerical value of this integral is 1.17 so that $\delta c / A = -0,17$. The function $F(\theta)$ and the ratio $F(\theta) / \sin \theta$ calculated by the numeric method are presented in Fig. 8.3.

At strong retardation the velocity distribution is approximately proportional to $\sin\theta$ since the ratio $F(\theta) / \sin\theta$ is nearly constant. This assumption was used by Frumkin & Levich (1947) as the basis of the theory for a retardation coefficient concept and leads to the velocity distribution Eq. (8.14) and the retardation coefficient Eq. (8.35).

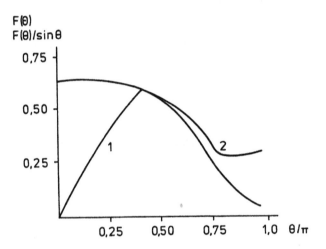

Fig. 8.3.   Functions characterising the velocity distribution on the strongly retarded bubble (drop) surface: 1 - function $F(\theta)$; 2 - function $F(\theta)/\sin \theta$.

If the denominator in Eq. (8.47) is set equal to 1, $F(\theta)$ is given by $\alpha \cdot \sin\theta$, with $\alpha = F(\theta) / \sin(\theta)$, and an equation for χ_b is obtained which only differs from Eq. (8.35) by a numerical factor. Hence, if we introduce an additional condition for a slight variation of the adsorption along the surface, both the sinusoidal velocity distribution and the relation for the retardation coefficient proposed by Levich (1962) can be verified. To compare this additional condition with the condition of strong retardation of the surface $\eta \ll \chi_b$, the following conditions result (cf. Eqs (8.40), (8.47) and (8.48)):

$$v\eta \ll \frac{\partial \gamma}{\partial c} c_o, \tag{8.50}$$

$$v\eta \left[\frac{a_b}{K_H Pe^{2/3}} \right] \ll \frac{\partial \gamma}{\partial c} c_o. \tag{8.51}$$

Comparing these two conditions and considering a not too high surface activity,

$$\Gamma_o / c_o \ll a_b / Pe^{2/3},$$
(8.52)

condition (8.51) is more severe than (8.50). This means that under condition (8.52) a weak adsorption variation always takes place if the surface is strongly retarded and the criterion for the applicability of the theory is given by (8.51).

If condition (8.52) is not fulfilled, an essential variation of adsorption is possible also at a strong retardation so that the applicability of the theory can be defined more specifically,

$$\eta \frac{\Gamma_o}{\Gamma_o - \frac{3}{2} v\eta RT} \ll \chi_b.$$
(8.53)

In the formulation of the convective diffusion equation a Stokes' velocity field can be taken into account and the weak motion of the bubble surface is neglected. This can be justified only at a sufficiently high degree of retardation. This can be estimated from the results as follows,

$$\eta \ll \chi_b Pe^{-1/3}.$$
(8.54)

The results obtained can be generalised for drops instead of bubbles by complementing the boundary condition (8.38) by the component of viscous tension tensor inside the drop. This is a minor correction when the following condition is fulfilled,

$$2\eta'/\eta \ll \chi_b / \eta \gg 1,$$
(8.55)

which means that the results are applicable not only to the case of a rising bubble but also to a liquid drop of not too high viscosity compared with the medium viscosity. It should be pointed out that condition (8.55) does not restrict the generality of the consideration significantly. Under a condition directly opposite to Eq. (8.55) the adsorption layer causes a weak retardation and the adsorption distribution can be derived by other methods (Dukhin 1965).

The described theory of strong retardation of a bubble surface by DAL at small Reynolds numbers was developed by Dukhin & Derjaguin (1961) and Dukhin & Buikov (1965) and confirmed by Saville (1973). The balance of Marangoni and viscous stresses, given by Eq. (8.38) as the basis for the determination of the surface concentration distribution was used later in the theory of a stagnant cap (Section 8.7). This stress balance is often characterised qualitatively by a dimensionless number, the Marangoni number (cf. He et al. 1991),

$$Ma = RT\Gamma_o / \eta v,$$
(8.56)

which characterises the ratio of the surface pressure that surfactant molecules exert under compression to the viscous forces tending to compress the surfactant layer.

A formula for the concentration c_{cr}^h above which the bubble surface as whole is strongly retarded

$$c_{cr}^h = \frac{3}{2} \frac{\eta v}{RTK_h}$$
(8.57)

follows from Eq. (8.40).

8.4. THEORY OF DYNAMIC ADSORPTION AND DIFFUSION BOUNDARY LAYERS OF A BUBBLE WITH PE»1, RE«1 AND WEAK SURFACE RETARDATION

When the surface retardation is weak and Eq. (8.36) is fulfilled, the convective diffusion equation (8.16) can be transformed via $t = 2/3 + 1/3\cos^3\theta - \cos\theta$; $\psi = x\sin^2\theta$; $(x = z - a_b)$ into the heat transport equation (8.15) with constant coefficient. The field equation is thus a simple one, but the complexity of Eq. (8.10) requires the investigation of extreme cases which simplifies the boundary condition. The ratio of the second to the first terms of the right-hand side of Eq. (8.27) may be written as $(D_s v / Dv_o \sin\theta\, Pe)\,(\alpha\,\ln\Gamma / \partial\theta)$, so that at Pe»1 (except near the poles $\theta=0$, $\theta = \pi$) Eq. (8.27) can be simplified by dropping the surface diffusion term

$$D\frac{\partial c}{\partial z}(a_b,\theta) = \frac{v_o}{a_b \sin\theta} \frac{\partial}{\partial\theta}\left[\sin^2\theta \cdot \Gamma(\theta)\right].$$
(8.58)

For lower values of K_H, surface concentration changes become relatively small, and the equilibrium concentration along the droplet surface $c(a_b,\theta)$ is derived from Eq. (8.58) (Derjaguin et al., 1959, 1960),

$$\frac{\partial c}{\partial z}(a_b,\theta) = \Lambda\cos\theta,$$
(8.59)

where

$$\Lambda = \frac{2v_o\Gamma_o}{Da_b}.$$
(8.60)

Transforming to the variables ψ, t, the boundary condition (8.59) reads

$$\frac{\partial c}{\partial \psi}\bigg|_{\psi=0} = \Lambda \frac{\cos\theta}{\sin^2\theta} = \Lambda \frac{f(t)}{1-f^2(t)}, \tag{8.61}$$

with

$$f(t) = \cos\theta. \tag{8.62}$$

To avoid the determination of the roots of the cubic equation (8.17), a root in terms of $f(\theta)$ is represented, the exact form of which is not required. In addition, we assume that in regions far from the bubble

$$c\big|_{\psi\to 0} = c_o. \tag{8.63}$$

Using a source function

$$\frac{1}{2} \frac{\exp\left[-\psi^2 / 4\pi na_b^2 (t-t')\right]}{\sqrt{\pi na_b^2 (t-t') / Pe}}, \tag{8.64}$$

and an initially homogeneous concentration field, and taking into account the boundary condition (8.61) equivalent to a non-steady source with a capacity $\dfrac{na_b^2}{Pe}\dfrac{\partial c(t)}{\partial \psi}$, the solution of equation (8.16) is

$$c(t,\psi) = \left(Kna_b^2 / 4\pi\right)^{1/2} \int_0^t \frac{\partial c(t)}{\partial \psi}\bigg|_{\psi=0} \frac{\exp\left[-\Psi^2 / 4Kna_b^2 (t-t')\right]}{(t-t')^{1/2}} dt' + c_o$$

$$= \Lambda\left(Kna_b^2 / 4\pi\right)^{1/2} \int_0^t \frac{f(t')}{1-f^2(t')} \frac{\exp\left[-\Psi^2 / 4Kna_b^2 (t-t')\right]}{(t-t')^{1/2}} dt' + c_o. \tag{8.65}$$

Because the functional form of $f(t)$ is unknown, we change the variable of integration from t' to θ' making use of equation (8.17),

$$t' = \frac{2}{3} + \frac{\cos^3\theta'}{3} - \cos\theta'. \tag{8.66}$$

By definition in Eqs (8.17) and (8.62) $f(t') = \cos\theta'$ and

$$\delta c(\theta,x) = c_o - c(\theta,x) = \frac{2\Gamma_o v_o n^{1/2}}{D\sqrt{\pi Pe}} \int_0^\theta \frac{\exp\left[-x^2 \sin^4 \theta Pe / 4a_b^2 n\chi_1(\theta,\theta')\right]d\theta'}{\chi_1(\theta,\theta')^{1/2}}, \qquad (8.67)$$

where

$$\chi(\theta,\theta') = \cos\theta' - \cos\theta - (\cos^3\theta' - \cos^3\theta)/3. \qquad (8.68)$$

From Eq. (8.67) it is evident that the concentration has been reduced exponentially at a distance from the surface $\delta_D = a_b / Pe^{1/2}$, and the local adsorption equilibrium results to

$$\frac{\Gamma_o - \Gamma(\theta)}{\Gamma_o} = \frac{c_o - c(a_b,\theta)}{c_o} = \frac{2}{\sqrt{\pi}} \frac{\Gamma_o}{c_o} \frac{Pe^{1/2}}{a_b n^{1/2}} \cdot I(\theta), \qquad (8.69)$$

where the function $I(\theta)$ is given by

$$I(\theta) = \int_0^\theta \frac{\cos\theta'\sin\theta'd\theta'}{\chi_1^{1/2}(\theta,\theta')}. \qquad (8.70)$$

For low surface activity

$$K_H \ll \delta_D, \qquad (8.71)$$

Eqs (8.69) and (8.70) lead to the relation (8.18), which specifies a relatively insignificant surface concentration change, is satisfied.

In the opposite case,

$$K_H \gg \delta_D, \qquad (8.72)$$

the surface concentration over the major part of the bubble surface is considerably smaller than Γ_o. In other words, if (8.72) is valid,

$$c(a_b,\theta) \ll c_o, \qquad (8.73)$$

and to a first approximation the simplified boundary condition reads

$$c(a_b,\theta) = 0. \qquad (8.74)$$

A solution to Eq. (8.16) with the boundary conditions (8.74) and (8.63) is

$$c(x,\theta) = \frac{2c_o}{\sqrt{\pi}} \int_0^N c^{-z^2} dz,$$ (8.75)

with

$$N = \sqrt{\frac{v_o}{a_b D}} \frac{\sqrt{3}}{2} \frac{1+\cos\theta}{\sqrt{2+\cos\theta}} x.$$ (8.76)

The last formula yields an equation for the angular dependence of the thickness of the diffusion boundary layer,

$$\delta_D(\theta) = \sqrt{\frac{\pi n}{3}} \delta_D \frac{\sqrt{2+\cos\theta}}{1+\cos\theta}.$$ (8.77)

Using (8.75) to derive $\frac{\partial c}{\partial r}(a_b,\theta)$ and introducing it into Eq. (8.58), a surface concentration distribution results,

$$\frac{1}{a_b \sin\theta} \frac{\partial}{\partial\theta} \left[\Gamma(\theta) v_o \sin^2\theta \right] = \frac{Dc_o}{\delta_D} \sqrt{\frac{3}{\pi n}} \frac{1+\cos\theta}{\sqrt{2+\cos\theta}},$$ (8.78)

which is easily integrated to

$$\Gamma(\theta) = \frac{2\delta_D n^{1/2}}{\sqrt{3\pi}} c_o \frac{1-\cos\theta}{\sin^2\theta} \sqrt{2+\cos\theta}.$$ (8.79)

The integration constant has been adjusted to make $\Gamma(\theta)$ finite at $\theta\to0$. The singularity at $\theta\to\pi$ is eliminated when surface diffusion is taken into account. From Eq. (8.79) we can see that condition (8.73) is satisfied everywhere except near $\theta=\pi$ when Eq. (8.72) is valid.

The results for these extreme cases have a simple physical interpretation. The tangential flux of adsorbed material along a drop or bubble surface is proportional to Γ_o, and the rate of bulk to surface exchange is proportional to c_o. It is furthermore clear that at sufficiently small ratios Γ_o/c_o, the tangential transport is so small and the exchange so large that the adsorption is practically always in equilibrium. At sufficiently large K_H, the exchange of matter is small and the steady state surface concentration distribution differs significantly from the equilibrium.

Now the dynamic adsorption layer is examined without restricting the value of the surface activity of the surfactant (Dukhin, 1965, 1981). By operational methods the solution of Eq. (8.16) together with the boundary condition of infinity (8.63) is,

$$c(t,\Psi) = \frac{2}{\sqrt{\pi}} \int\limits_{\Psi/2\sqrt{na_b^2 t/Pe}}^{\infty} c\left(t - \frac{Pe\Psi^2}{4na_b^2\zeta^2}, 0\right) e^{-\zeta^2} d\zeta - c_o \operatorname{erfc}\left(\frac{\Psi}{2\sqrt{na_b^2 t/Pe}}\right). \tag{8.80}$$

From Eq. (8.80) the derivative $\left.\dfrac{\partial c}{\partial \Psi}\right|_{\psi=0}$ is determined and an ambiguity in the limit $\psi \to \infty$ can be avoided by transforming to the variable $t' = t - Pe\Psi^2 / 4na_b^2\zeta^2$,

$$\frac{\sqrt{\pi}}{2} \left.\frac{\partial c}{\partial \Psi}\right|_{\psi=0} = \frac{c_o - c(0,0)}{2\sqrt{na_b^2 t/Pe}} + \frac{1}{2\sqrt{na_b^2/Pe}} \int\limits_0^t \frac{\partial c(t',0)}{\partial t'} \frac{dt'}{\sqrt{t-t'}}. \tag{8.81}$$

Substituting $\dfrac{\partial c}{\partial x} = \dfrac{\partial c}{\partial \psi} \dfrac{\partial \psi}{\partial x}$ into Eq. (8.58), recalling that $\dfrac{\partial \psi}{\partial x} = av_o \sin^2\theta$, $\dfrac{\partial t}{\partial \theta} = \sin^3\theta$, and after some rearrangements Eq. (8.58) becomes

$$\frac{\sqrt{\pi}}{2} \frac{d}{dt}\left[\Gamma(\theta)\sin^2\theta\right] = \frac{\delta_D \sqrt{n}}{2}\left[\frac{c_o - c(0,0)}{\sqrt{t}} + \int\limits_0^t \frac{dc(t',0)}{dt'} \frac{dt'}{\sqrt{t-t'}}\right]. \tag{8.82}$$

Eq. (8.82) is integrated from 0 to t,

$$\frac{\sqrt{\pi}}{2}\Gamma(\theta)\sin^2\theta = \frac{\delta_D \sqrt{n}}{2}\left[(c_o - c(0,0))\,2\sqrt{t} + \int\limits_0^t dt' \int\limits_0^{t'} \frac{dc}{dt''} \frac{dt''}{\sqrt{t'-t''}}\right], \tag{8.83}$$

the order of integration in the double integral is inverted so that the subsequent integration by parts yields

$$\int\limits_0^t dt' \int\limits_0^{t'} \frac{dc(t'',0)}{dt''} \frac{dt''}{\sqrt{t'-t''}} = -2c(0,0)\sqrt{t} + \int\limits_0^t \frac{c(t',0)}{\sqrt{t-t'}}dt'. \tag{8.84}$$

Upon substitution of Eq. (8.84) into Eq. (8.83) the following relation finally results,

$$K_H^{-1}\Gamma(\theta)\sin^2\theta = c(a_b\theta)\sin^2\theta = m\left[c_o\sqrt{t} - \frac{1}{2}\int\limits_0^t \frac{c(t',0)}{\sqrt{t-t'}}dt'\right], \tag{8.85}$$

with $m = 2\delta_D\sqrt{\pi n} / K_H$.

Before reducing the integral Eq. (8.85) to its canonical form, the special cases $m \ll 1$ and $m \gg 1$ are considered. With $m \ll 1$ the integral in Eq. (8.85) may be neglected, and the result agrees with Eq. (8.79), because the condition m « 1 is identical with Eq. (8.72). If c(t,0) in the integral term of Eq. (8.85) is substituted by $\Gamma(\theta) = K_H \, c(a_b,\theta)$ one can see that in the right hand side of this equation the ratio of the second term to the first one is of the order of m everywhere except near $\theta = \pi$. The logarithmic divergence at $\theta = \pi$ can be eliminated by taking into account the surface diffusion term. Therefore the integral of Eq. (8.85) may be neglected.

If m»1 the solution must be of the form $c(a_b,\theta) = c_o + \delta c(a_b,\theta)$ with $\delta c(a_b,\theta)$«c_o except at $\theta = \pi$ (justification will be given later). Inserting $c(a_b,\theta)$ into Eq. (8.85) $\delta c(a_b,\theta)$ can be obtained from

$$c_o \sin^2 \theta = -\frac{m}{2} \int_0^t \frac{\delta c(a_b,\theta)}{\sqrt{t-t'}} dt'. \qquad (8.86)$$

Writing

$$\phi(t) = \sin^2 \theta, \qquad (8.87)$$

in the left hand side of Eq. (8.86) an integral equation results,

$$\delta c(a_b,t) = \frac{2c_o \sin \pi/2}{m\pi} \left[\frac{\phi(0)}{t^{-1/2}} + \int_0^t \frac{\phi(\tau)d\tau}{\sqrt{t-\tau}} \right]. \qquad (8.88)$$

Eq. (8.87) yields $\phi(0)=0$. Because $d\phi/d\tau = 2\sin\theta'\cos\theta'(d\theta'/d\tau)$ and after insertion of this quantity into Eq. (8.88) a relation is recovered which is identical with Eq. (8.70).

It can be concluded that the dynamic adsorption equation (8.85) in the limiting cases leads to the same results as obtained by other methods. Expressed in terms of the variables $x = \cos\theta$; $y(x) = \sin^2\theta \cdot c(a_b,\theta)$, Eq. (8.85) is transformed into the Volterra integral equation,

$$y(x) = mc_o x \sqrt{\frac{2}{3} + \frac{x^3}{3} - x} + \frac{m}{2} \int_1^x \frac{y(x')dx'}{\left[x - x' - (x^3 - x'^3)/3\right]^{1/2}}. \qquad (8.89)$$

The theory of dynamic adsorption layer on slightly retarded surfaces presented by Derjaguin et al. (1959; 1960) and Dukhin (1961; 1966) was basically confirmed by Harper (1973), Saville

(1973) has found an additional derivation. The cases of high and low surface activity have been individually considered in these works. In the formulation of conditions (8.71) and (8.72) a dimensionless surface activity parameter K_H was introduced as the ratio of the amount of surfactant adsorbed on the surface to amount dissolved in the diffusion boundary layer. Harper (1973) and Saville (1973) confirm agreement of their results with those obtained by Derjaguin and Dukhin (1959, 1960, 1961, 1966) for the limiting cases $K_H \to 0$ and $K_H \to \infty$.

The results in the present paragraph have limits of applicability, in particular those connected with the application of the boundary layer method. It is well known that results obtained by this method are unsuitable in the neighbourhood of the rear stagnant point.

Saville (1973) solved the convective diffusion equation numerically and gave the same value of retardation coefficient as obtained by Dukhin (1965, 1981). Listovnichii (1985) has succeeded in obtaining simple approximation formulas for the concentration distribution not only along a bubble surface but also across the diffusion layer, based on numerical solution of Eq. (8.85). He has also shown that the analytical solutions Eqs (8.69) and (8.79) deviate from the exact solution less than 1%, at m > 10 and m < 0.1.

8.5. Hypothesis of Incomplete Retardation of a Bubble Surface at Re < 1 and Presence of a Dynamic Adsorption Layer

8.5.1. Hypothesis of Incomplete Retardation

Up to a_b = 0.03 cm bubbles float up like solid spheres even in distilled water. This is possible both at a complete and at a strong (incomplete) retardation of the bubble surface. At a complete retardation the velocity at any point of the surface (in a system associated with the bubble movement) is equal to zero. A strong retardation means a small surface velocity as compared with the bubble velocity,

$$v_\theta (a_b, \theta) \ll v. \tag{8.90}$$

Under condition (8.90), the viscous stress tensor, and therefore, the bubble rising velocity depends very weakly on the motion of the surface. Thus, the merging of curves 1 to 4 in Fig. 8.3 at a_b < 0.03 cm does not exclude the possibility of slow bubble surface motion. At the same time it points to the presence of some uncontrollable factors causing a strong or complete retardation of the surface at a_b < 0.03 cm. The fact that at a_b > 0.03 cm this factor is weak is difficult to understand along with its unexpectedly strong effect, which supposedly results in a complete retardation. It is more natural to assume that with decreasing a_b this factor provides a gradual growth of retardation of the surface, so that at $a_b < 0.03$ cm the motion of the surface lasts but becomes hardly observable (Dukhin, 1982).

Frumkin and Levich have shown that bubble surface retardation at $a_b > 0.03$ cm is controlled by the effect of surfactants which has reduced the interest in studying rheological properties of clean interfaces. However, it is important to recall the situations shown in Fig. 8.3. It is proved that i) rheological constants of a clean water-air interface are so small that the surface of a bubble with $a_b > 0.03$ cm is mobile to a large extent, and ii) the retarding effect of surfactant, on the contrary, can be very important at sufficiently high adsorption. However, retarding effects of a surfactant decrease indefinitely with decreasing adsorption and can become weaker than any unknown factors that would retard the surface of a bubble with a size < 0.03 cm. Thus, works by Frumkin and Levich do not prove strictly the prevailing role of surfactant in surface retardation of bubbles > 0.03 cm. The assumption that the same mechanism that causes surface retardation at $a_b > 0.03$ cm can also act at $a_b < 0.03$ cm (but now due to uncontrolled impurities which are always present in very low concentrations) seems to be a good one.

Papers by Frumkin & Levich (1947) and by Derjaguin & Dukhin (1961) are apparently only of methodical nature, since they do not take into account the bubble surface effect which causes retardation at $a_b < 0.03$ cm. Indeed if at $a_b < 0.03$ cm a strong but not complete retardation of the bubble surface takes place, the discussed theories cannot be applied directly, but can be transformed to the hypothesis of incomplete retardation.

To approach real conditions in the dynamic adsorption layer theory at small Reynolds numbers, one has to take into account that the motion of a surface is strongly retarded already without the presence of a surfactant, i.e. one should introduce the phenomenological retardation coefficient χ_o. The velocity distribution along a retarded surface is given by,

$$v_o(\theta) = \frac{\eta v}{\chi_o} \sin\theta. \tag{8.91}$$

The concept of a retardation coefficient is associated with the idea of a uniform surface retardation, since in the absence of retardation the velocity distribution along the surface is also expressed by a sinusoidal relation. The considerations of surface viscosity in Boussinesq's theory and of the retarding effects of surfactants in Frumkin's theory result just in such angular dependence. Therefore, the discussion presented below can be carried out without predetermining the value of the retardation coefficient χ_b.

Thus, assuming the velocity distribution to be known, we can calculate the effect of surface motion on the distribution of the surfactant whose nature, properties and concentration we know. Two limiting cases can be immediately identified. A redistribution of a surfactant can result in an additional retardation of the surface which is characterised by the retardation parameter χ_b. The first possibility corresponds to the condition

$$\chi_b \ll \chi_o, \tag{8.92}$$

the second to

$$\chi_b \approx \chi_o. \tag{8.93}$$

Only in the first case it can be assumed that the surface motion is given by Eq. (8.91). The dynamic adsorption layer and the diffusion boundary layer are interrelated through the boundary condition which expresses the compensation of surface divergence of surface flow of a surfactant I_s by the normal component of the flow density to the bulk phase or from the bulk to the surface (i.e., the analog of Eq. (8.10)). Note that this boundary condition differs only in the form of the constant surface velocity coefficient from that in the description of dynamic adsorption layers on a non-retarded surface.

The analysis of the second case leads to two limiting conditions which are realised also in the case of a strongly retarded surface. Under condition (8.70) the surface concentration variation along the surface is insignificant. In the opposite case at sufficiently large Γ_0 / c_0, the motion of the surface pushes the adsorption layer down to the bottom pole of the bubble so that everywhere condition (8.72) is fulfilled except at $\Theta \approx \pi$. This enables us to use the approximate boundary condition (8.74) for solving the convective diffusion equation.

The main distinction of the theory of a dynamic adsorption layer formed under weak and strong retardation arises when formulating the convective diffusion equation. At weak retardation the Hadamard-Rybczynski hydrodynamic velocity field is used while at strong retardation the Stokes' velocity field. Different formulas for the dependence of the diffusion layer thickness on Peclet numbers are obtained. The problem of convective diffusion in the neighbourhood of a spherical particle with an immobile surface at small Reynolds numbers and condition (8.74) is solved, so that the well-known expression for the density distribution of the diffusion flow along the surface can be used. As a result, Eq. (8.10) takes the form (Dukhin, 1982),

$$\frac{1}{a_b \sin\theta} \frac{\partial}{\partial\theta} \left[\sin\theta \frac{\eta v}{\chi_o} \sin\theta \Gamma(\theta) \right] = \frac{Dc_o}{1.15} \sqrt[3]{\frac{3v}{4Da_b^2}} \frac{\sin\theta}{\left(\theta - \dfrac{\sin^2\theta}{2}\right)^{1/3}}. \tag{8.94}$$

Its solution reads

$$\Gamma(\theta) = c_o \beta \frac{\phi(\theta)}{\sin^2 \theta}, \tag{8.95}$$

with

$$\beta \cong \frac{\chi_o a_b}{\eta Pe^{2/3}}; \quad \phi(\theta) = \int_0^\theta \frac{\sin^2 \theta d\theta}{\left(\theta - \frac{\sin 2\theta}{2}\right)^{1/3}}. \tag{8.96}$$

Now condition (8.72) can be defined more accurately as follows,

$$\Gamma_o / c_o \gg \beta \approx \frac{\chi_o}{\eta} \frac{a_b}{Pe^{2/3}}. \tag{8.97}$$

The same has to be done with condition (8.92). Clearly, the effect of a surfactant on surface motion is small, if the established surface tension gradient is small as compared with the viscous stress gradient calculated on the basis of Stokes' velocity distribution

$$\frac{d\gamma}{d\theta} \ll \eta \left(\frac{1}{a_b} \frac{\partial v_r}{\partial \theta} + \frac{\partial v_\theta}{\partial z} - \frac{v_\theta}{z} \right)_{z=a_b} = \frac{3}{2} \frac{\eta v}{a_b} \sin\theta \tag{8.98}$$

wherefrom it follows, using Gibbs' formula,

$$\frac{\partial \Gamma}{\partial \theta} \ll \frac{3}{2} \frac{\eta v}{RT} \sin\theta. \tag{8.99}$$

Under the conditions $(3/2)\eta u / RT \ll \Gamma_o$, Eq. (8.92) is stricter than

$$\Gamma(\theta) \ll \Gamma_o \tag{8.100}$$

and it turns out that the solution (8.96) is correct for adsorption values

$$\Gamma_o \ll \eta\eta v / RT \tag{8.101}$$

with $n = K_H \left(\dfrac{\eta Pe^{2/3}}{\chi_o a_b} \right) \gg 1.$

8.5.2. USE OF THE DORN EFFECT TO CHECK THE INCOMPLETE RETARDATION OF A BUOYANT BUBBLE SURFACE

It is well known (Samigin et al., 1964; Usui & Sasaki, 1978; Usui et al., 1980) that the potential difference measured in the column of buoyant bubbles arises due to the superposition of the electric fields of individual bubbles, which are caused by their dipole moments induced by tangential electric current. It was shown by Dukhin (1964a, 1964b) that three current components I_ζ, I_V, and I_S' have predominant significance while other components can be neglected. Like in Smoluchowski's theory, I_ζ is caused by the ion transport of the diffuse part of the equilibrium DL due to the tangential liquid flow, I_V is determined by the movement of both planes of the secondary DL (Dukhin & Derjaguin, 1958) due to the tangential liquid flow. The secondary double layer generates not only an electric field normal to the surface but also tangential and contributes to the tangential electric current. In addition, both sorts of DL ions diffuse from the rear stagnation pole of the bubble to the leading one along the surface under the effect of increasing surface concentration. The electric current originated by tangential components of electric potential gradients and ion concentration in the secondary DL is designated by I_S'.

It was shown by Dukhin (1983) that under the conditions of strong retardation the effect of an equilibrium DL is dominant for any surface activity of the surfactant. Therefore, Smoluchowski's formula is valid at any degree of surface activity.

A noticeable deviation of sedimentation potentials from Smoluchowski's formula takes place at large surface concentration variation along the bubble surface. Before considering experimental data, it has to be pointed out that the validity of Smoluchowski's formula for the description of the Dorn effect at large Peclet numbers applies only to solid spherical particles. In particular, the correctness of conclusions of some papers (Dukhin, 1964; Dukhin & Buikov, 1965; Derjaguin & Dukhin, 1967, 1971) is experimentally confirmed by Usui et al. (1980). Sedimentation potential for four sizes of glass balls appears to be the same. Since the radii of the particles under consideration are approximately 50, 150, 250, and 350 μm, the absence of any effect of Peclet and Reynolds numbers on the sedimentation potential could be demonstrated.

It follows from works of Usui et al. (1980) and Derjaguin & Dukhin (1967, 1971) that under the condition of complete retardation of the bubble surface, there are neither experimental nor theoretical reasons to expect an effect of bubbles size on sedimentation potential. Comparing these conclusions of the theory with sedimentation potentials found by systematic studies of Russian and Japanese scientists, it can be concluded that at Re<40 not a complete, but at least a strong retardation, of the bubble surface takes place. This conclusion was obtained by Usui et al (1980) only when using nonionic surfactants since these experiments were performed with bubbles of diameters a_b>300 μm.

In contrast to Usui et al. the (1980) the results of Sotskova et al. (1982) are obtained for small bubbles in the presence of nonionic surfactant, which is important for comparison with the theory for small Reynolds numbers. It is also instructive to consider alcohol concentration effects. In the interval of alcohol concentrations at which the bubble radius decreases, the sedimentation potential grows.

The possibility to realise the second condition (8.101) is considered by Usui et al. (1980), where the Stern potential was calculated together with the sedimentation potential from adsorption data of CTAB. In the region of high concentration 0.7-$2.5 \cdot 10^{-5}$ mole/cm^3 and adsorption, ζ and ψ_{st} coincide. At concentrations lower than $7 \cdot 10^{-8}$ mole/cm^3, differences between ζ and ψ_{st} become significant. The difference between ζ and Ψ_d decreases with decreasing surfactant concentration, reaching a value of the order of 100 mV.

These results show that at high concentrations (above $7 \cdot 10^{-8}$ mole/cm^3), addition of surfactant increases the degree of retardation of the bubble surface. Thus, under the condition $\chi_b > \chi_o$ the adsorption can considerably deviate from the mean value only in the vicinity of the leading pole of the bubble and the electrokinetic potential can be calculated from the equilibrium adsorption value. When the concentration of the surfactant decreases, retards to a lesser extent the motion of the surface. The condition $\chi_b < \chi_o$ is realised when the removal of surfactant to the rear of the bubble is possible, and adsorption is much lower than the equilibrium value over almost the whole bubble surface. This statement needs a confirmation if values of adsorption less than 10^{-10} mole/cm^2 are taken into account since then a deviation of the electrokinetic potential from Stern potential was observed (Sotskova et al. 1982). Substituting this value and the velocity of the buoyant bubble with a radius of 150 μm condition (8.98) is fulfilled.

Systematic studies of bubble hydrodynamics based on the Dorn effect were suggested by Dukhin (1983). A comprehensive study, comprising the measurement of adsorption on immobile surfaces with the calculation of the Stern potential and measurements of sedimentation potentials, should be performed with homologous series of ionic surfactant so that the condition (8.97) is fulfilled by higher homologues and the opposite condition (8.101) by the lower ones. With decreasing surface activity the condition (8.97) will be fulfilled at smaller adsorption values. This means that the lower the surface activity, the smaller should be the deviation of the electrokinetic potential from the Stern potential at respective values of adsorption. And finally, when condition (8.97) is not fulfilled the electrokinetic and Stern potentials must coincide over the whole concentration interval.

8.6. THEORY OF DYNAMIC ADSORPTION LAYER OF A BUBBLE AND RETARDATION OF ITS SURFACE AT LARGE REYNOLDS NUMBERS

In respect to large Reynolds numbers, still big difficulties appear with the non-linearity of the Navier-Stokes equation. Therefore, this section will be restricted to evaluations of conditions for different states of adsorption layer formation and retardation (Section 8.6.1.) and to the consideration of weak retardation conditions (Sections 8.6.2. and 8.6.3., cf. Dukhin 1965, 1981).

Evaluations are possible by using the concept of hydrodynamic and diffusion boundary layers, having a thickness δ_G and δ_D respectively, and independent of angle θ. The evaluations are simplified under the assumption of either a strong retardation of the surface (Section 8.3),

$$\left.\frac{\partial v}{\partial z}\right|_{z=a_b} \approx \frac{v}{\delta_G}, \tag{8.102}$$

or a weak retardation,

$$\text{div}_s\left[\Gamma(\theta)v_\theta(\theta)\right] \approx \frac{\Gamma(\theta)v_o}{a_b}. \tag{8.103}$$

At first conditions are considered under which the surfactant adsorption at the bubble surface is strongly retarded,

$$\left|\Gamma(\theta)-\Gamma_o\right| \ll \Gamma_o, v_o / v \ll 1. \tag{8.104}$$

An estimation of the right-hand side of Eq. (8.38) with regard to Eq. (8.102) results in

$$\left|\Gamma(\theta)-\Gamma_o\right|/\Gamma_o \approx \frac{\eta v}{RT\Gamma_o}\frac{a_b}{\delta_G} \ll 1, \tag{8.105}$$

which allows to approximate the bubble surface velocity by means of conditions (8.10) and (8.11),

$$D\frac{\left|c(a_b)-c_o\right|}{\delta_D} \approx D\frac{\left|\Gamma-\Gamma_o\right|}{\delta_D}\frac{c_o}{\Gamma_o} \approx \frac{\Gamma_o v}{\Gamma_o}. \tag{8.106}$$

$\delta_D = a_b / Pe^{1/3}$ if the bubble surface is almost completely retarded and $\delta_D = r_o / Pe^{1/2}$ in the opposite case of an almost free bubble surface (Levich, 1962). From (8.105) and (8.106) the second necessary condition of the considered state can be obtained,

$$\frac{v_0}{v} \approx \frac{\eta D c_o}{R\Gamma_o^2} \frac{a_b}{\delta_D} \frac{a_b}{\delta_G} \ll 1,$$

(8.107)

and the following estimate for χ_b at $Re \gg 1$ can be obtained (Dukhin & Derjaguin, 1961),

$$\chi_b \approx \frac{R\Gamma_o^2}{Dc_o} \frac{\delta_D}{a_b} \frac{\delta_G}{a_b}.$$

(8.108)

Now conditions for the formation of the second state of the dynamic adsorption layer formation of nonionic surfactants is formulated under conditions where the surface concentration slightly deviates from the equilibrium state Γ_o and the bubble surface is weakly retarded,

$$(\Gamma(\theta) - \Gamma_o)/\Gamma_o \ll 1 \text{ at } v_0/v \approx 1.$$

(8.109)

The conditions of a slight deviation of an adsorption layer from equilibrium at $Re \gg 1$ are derived in the same way as at $Re \ll 1$ which yields Eq. (8.71). The second necessary condition is that the viscous stresses on the surface of a rising bubble must be much smaller than the characteristic value of a strong surface retardation,

$$\eta \left[\frac{\partial v_\theta}{\partial z} - \frac{v_\theta}{a_b} \right]_{z=a_b} = \frac{1}{a_b} RT \frac{\partial \Gamma}{\partial \theta} \ll \eta \frac{v}{\delta_G}.$$

(8.110)

After rearrangements we get

$$\frac{\eta D c_o}{R\Gamma_o^2} \frac{a_b}{\delta_D} \frac{a_b}{\delta_G} \gg 1.$$

(8.111)

Note that the left-hand sides in inequalities (8.107) and (8.111) are identical which indicates that the regions of realisation of the first and second conditions of dynamic adsorption layer formation do not overlap.

Now conditions of formation of the third state of dynamic adsorption layers is considered. In this case the adsorbed surfactant is almost completely dislocated to the lower pole and the main part of the bubble surface (except for a narrow rear zone) is weakly retarded,

$$|\Gamma(\theta) - \Gamma_o|/\Gamma_o \approx 1 \text{ at } v_0/v \ll 1.$$

(8.112)

The derivation of the condition of strong variation of surface concentration along the bubble surface at $Re \gg 1$ is carried out in the same way as at $Re \ll 1$ and leads to condition (8.72). The second necessary condition is (8.110) which can be rewritten in the from,

$$\frac{\eta v}{RT(\Gamma(\pi)-\Gamma(0))}\frac{a_b}{\delta_G} \gg 1. \tag{8.113}$$

Comparing condition (8.71) with Eq. (8.72) and (8.113) with (8.105) it becomes clear that regions of realisation of the different states do not overlap. These conditions are presented in Fig. 8.4 in a graphic form.

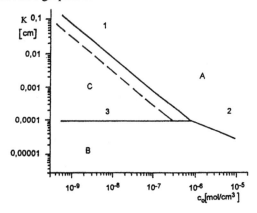

Fig. 8.4. Conditions of realisation of different states of dynamic adsorption layer formation of nonionic surfactant. Estimates are given for bubbles with a radius of 0.05 cm. The regions of parameters c_o and Γ_o / c_o are as follows:
A - slight deviation of surface concentration from equilibrium and strong surface retardation
B - slight deviation of surface concentration from equilibrium and weak surface retardation
C - almost a complete displacement of adsorbed surfactant to the rear stagnation pole and a weak retardation of the main part of the surface

The regions of the three states of dynamic adsorption layer formation are separated by curves 1, 2 and 3 which are given by the following respective equations:

$$\frac{\eta_\alpha v}{RT\Gamma_o}\frac{a_b}{\delta_G} = 1, \tag{8.114}$$

$$\frac{\eta_\alpha}{\chi_b} = 1, \tag{8.115}$$

$$\frac{\Gamma_o}{c_o \delta_D} = 1. \tag{8.116}$$

Regions A, B, and C are separated from each other not by lines but rather by wide bands as there is a change form "much greater than" into the condition "much less than", and vice versa.

When the boundary between regions B and C, and B and A is crossed, one of the properties is retained while the other changes. In contrast, when crossing the border between regions C and A, two properties, degree of retardation and deviation of surface concentration from equilibrium, change (cf. Eqs (8.109) and (8.112)). It can be expected that in a wide region between A and C, a region D is located in which the conditions $v_o / v \ll 1, |\Gamma(\theta) - \Gamma_o| / \Gamma_o \approx 1$ are fulfilled. It was shown that this region exists at $Re \ll 1$ (cf. Section 8.3).

8.6.2 THE THEORY OF A DYNAMIC ADSORPTION LAYER OF A BUBBLE WITH $RE \gg 1$ AND WEAK SURFACE RETARDATION

At present no quantitative theory of drop motion in liquids at large Reynolds numbers exists. For bubbles, however, the situation is much simpler because gas viscosities are negligible with respect to liquid viscosities. For $Re < 500 \div 800$, the bubble retains its spherical shape and the velocity field is that for an ideal liquid, (Levich, 1962) provided that no surfactant is present on the bubble surface,

$$v_z^{(o)} = -v(1 - a_b^3 / z^3)\cos\theta, \quad v_\theta^{(o)} = v(1 + a_b^3 / 2z^3)\sin\theta. \tag{8.117}$$

As a second approximate it is necessary to set the tangential stress on the bubble surface equal to zero. The resulting boundary condition is not satisfied by Eq. (8.117),

$$\eta\left(\frac{1}{z}\frac{\partial v_z}{\partial \theta} + \frac{\partial v_\theta}{\partial z} - \frac{v_\theta}{z}\right)_{z=a_b} = 0. \tag{8.118}$$

The real solution can be written in perturbation form $\bar{v}(z,\theta) = \bar{v}^{(o)}(z,\theta) + \bar{v}^{(1)}(z,\theta)$, in which

$$v^{(1)}(z,\theta) \ll v^{(0)}(z,\theta) \tag{8.119}$$

everywhere within the hydrodynamic boundary layer $\delta_G \approx a_b / \sqrt{Pe}$. Within this boundary layer

$x = (z - a_b) \ll a_b$, and Eq. (8.117) simplifies to

$$v_\theta^{(o)} = v_o \sin\theta, \quad v_z^{(o)} = -3v_o \frac{x}{a_b}\cos\theta, \tag{8.120}$$

where

$$v_o = \frac{3}{2}v. \tag{8.121}$$

Based on Eqs (8.119) to (8.121) it may be shown (Dukhin 1965, 1981) that $v_\theta^{(1)}$ satisfies the relation,

$$\frac{3v}{a_b}v_0^{(1)}\cos\theta + \frac{3}{2}\frac{v}{a_b}\sin\theta\frac{\partial v_\theta^{(1)}}{\partial\theta} - \frac{3v}{a_b}y\cos\theta\frac{\partial v_\theta^{(1)}}{\partial y} = v\frac{\partial^2 v_\theta^{(1)}}{\partial y^2}. \tag{8.122}$$

For the first time $v_\theta^{(1)}$ was considered by Levich (1962). The derivation contains some errors which were later corrected independently by Moor (1963) and Dukhin (1965). Transforming to variables $t = p_1(\theta)$, $\psi = x\sin^2\theta$, $f = v_\theta^{(1)}\sin\theta$, Eq. (8.122) turns into the heat conduction equation

$$\frac{\partial f}{\partial t} = K_1\frac{\partial^2 f}{\partial\psi^2}, \tag{8.123}$$

in which

$$K_1 = 2a_b v/3v. \tag{8.124}$$

The boundary condition for Eq. (8.123) is obtained by substituting $v_\theta = v_\theta^{(0)} + v_\theta^{(1)}$ into (8.118) and transforming to the new variables. The result is

$$\frac{\partial f}{\partial\psi}\bigg|_{\psi=0} = \frac{3v}{a_b}. \tag{8.125}$$

As $\psi \to \infty$, i.e. $x \to \infty$ it is necessary that $f \to 0$. Using a source function

$$\frac{1}{2}\frac{\exp[-\psi^2/4K_1(t-z)]}{(\pi K_1(t-z))^{1/2}}$$

and recognising that Eq. (8.125) is equivalent to a non-steady heat source of strength $K_1\dfrac{\partial c(t)}{\partial\psi}$, then for an initially homogeneous field the solution to Eq. (8.123) is

$$f(t,\psi) = \sqrt{\frac{K_1}{\pi}}\frac{3v}{a_b}\int_0^t\frac{\exp(-\psi^2/4K_1(t-z))}{(t-z)^{1/2}}dz, \tag{8.126}$$

whence

$$v_\theta^{(1)}(\theta,x) = -\frac{2v}{\sin\theta}\sqrt{\frac{3}{2\pi\,\text{Re}}}\int_0^{p_1(\theta)}\frac{\exp(-x^2\sin^4\theta/4K_1(p_1(\theta)-z)}{(p_1(\theta)-z)^{1/2}}dz,$$ (8.127)

and on the bubble surface

$$v_\theta^{(1)}(\theta,0) = -\left(\frac{8}{\pi\,\text{Re}}\right)^{1/2}\frac{v\sin\theta\sqrt{2+\cos\theta}}{1+\cos\theta}.$$ (8.128)

To derive a convective diffusion equation, it is important to simplify the expression for $v_\theta(\theta,x)$ inside the boundary layer

$$v_\theta(\theta,x) = v_o\sin\theta + \frac{\partial v_\theta(\theta,0)}{\partial x}x.$$ (8.129)

Using Eq. (8.118) and taking into account Eq. (8.119) and Re»1 yields

$$\frac{\partial v_\theta(\theta,0)}{\partial x} = \frac{v_\theta(\theta,0)}{a_b} = \frac{v_\theta^{(0)}(\theta,0)+v_\theta^{(1)}(\theta,0)}{a_b} \approx \frac{v_o}{a_b}\sin\theta.$$ (8.130)

According to (8.129) and (8.130), $v_\theta(\theta,x)$ changes only slightly within the boundary layer and has the same angular dependence as for Re<1; Pe»1, so that it is possible to reduce the convective diffusion equation to the form (8.16). This has to be done carefully because the coefficient on the right hand side of (8.16) has a different interpretation. Furthermore, at Re»1 it is justified to neglect surface diffusion with respect to convection in the boundary condition at Re«1 and Pe»1. Thus, if v_o is independent of n, the convective diffusion equation and the boundary condition (8.58) at Re«1 and Pe»1 as well as at Re»1 are completely identical. This implies that the adsorption fields given by Eqs (8.69) and (8.79) are equally valid if Re»1. On the other hand, if Re»1, the velocity v in the Peclet number can no longer be expressed by Eq. (8.24).

According to Eq. (8.117), $n = v/v_o = 2/3$. For weakly retarded bubbles $\eta' = \chi_b = 0$. This simplifies Eqs (8.24) and (8.25) and, finally, n=2 is obtained at Re < 1. The range of validity of Eqs (8.69) and (8.79) expressing the adsorption distribution changes from Re»1 to Re«1, the diffusion layer thickness δ_D is smaller in the former case.

8.6.3. WEAK SURFACE RETARDATION

The Eq. (8.123) allows to characterise a weak surface retardation. To do so, the boundary condition (8.125) has to be supplemented by a term reflecting the effect of surface tension gradients on surface motion. This yields the following boundary condition,

$$\frac{\partial f}{\partial \psi} = 3\frac{v}{a_b} + \frac{RTK_H}{a_b \eta} \Phi(t), \quad \Phi(t) = \frac{1}{\sin\theta} \frac{\partial c(a_b, \theta)}{\partial\theta}.$$ (8.131)

The solution of Eq. (8.123) together with this boundary condition was obtained by Dukhin & Derjaguin (1961) and Rulyov & Leshov (1980). From the solution it follows

$$v_0 = v_0^{(1)}(\theta) + v_0^{(2)}(\theta)$$ (8.132)

with

$$v_0^{(2)}(\theta) = -\frac{4RT\Gamma_o}{\sqrt{\pi \, Re}\,\sin\theta} \int_0^{t(\theta)} \frac{\Phi(\theta')\sin^3\theta'}{(t(\theta) - t(\theta'))^{1/2}} d\theta',$$ (8.133)

where $v_\theta^{(1)}(\theta)$ is expressed by Eq. (8.128). The function $v_\theta^{(2)}(\theta)$ characterises small variations of the velocity distribution along the bubble surface caused by surface tension gradients. For the limiting cases (8.72) and (8.71), numerical calculations of the integral are performed by Dukhin (1965) and Rulyov & Leshov (1980). To avoid the bulky results of these integrations, only an asymptotic solutions of the angular dependence is given here,

$$v_0^{(2)} = \begin{cases} -\dfrac{2\sqrt{2}RTc_o \sqrt{Dv}}{3\pi v \eta} \dfrac{I_1(\theta)}{\sin\theta} & \text{under Eq. (8.72)} \\[4mm] -\dfrac{2RT\Gamma_o^2 \sqrt{v}}{\sqrt{3D\pi\eta c_o a_b}} \dfrac{I_2(\theta)}{\sin\theta} & \text{under Eq. (8.71)} \end{cases}$$ (8.134)

with

$$I_1 = 2\pi / (1 + \cos\theta) \text{ at } \theta \to \pi,$$ (8.135)

$$I_2 = -\ln(1 + \cos\theta) \text{ at } \theta \to \pi.$$ (8.136)

At weak retardation of almost the whole surface, in the neighbourhood of $\theta = \pi$ the ratio $v_0^{(2)}(\theta) / v_0^{(o)}(\theta)$ grows very fast, so that a strong retardation can exist close to the point $\theta = \pi$. The transition zone between the strongly and the weakly retarded regions is very narrow because of the fast increase of $v_0^{(2)}(\theta)$ at $(\pi - \theta) \ll \pi$ (cf. Eqs (8.133) and (8.135)). Neglecting this transition zone the model of weakly retarded surfaces results in a stagnant cap model (Savic 1953). In addition, the model allows calculation of the concentration at which a stagnant cap appears and a decreased rising velocity of bubbles is observed. This decrease is caused by the growth of the hydrodynamic resistance due to the widening of the turbulent zone. In absence of any surfactant, the turbulent zone is located at $\theta > \theta_0$ and caused by the viscous correction $v_0^{(1)}(\theta)$ in Eq. (8.128) to the potential velocity distribution. A substantial extension of the turbulent zone and a decrease of the bubble rising velocity sets in only when the adsorption term $v_0^{(2)}(\theta)$ at the point $\theta = \theta_0$ becomes comparable with the viscous term.

Equating the mentioned quantities, from Eqs (8.128) and (8.133) the following rough estimate is obtained for the surfactant concentration, at which the rising velocity begins to decrease (Rulyov & Leshov 1980),

$$c_h = \frac{3\sqrt{2\pi}v^2\eta}{RT(Re)^{2/3}\sqrt{Dv}}, \tag{8.137}$$

$$c_1 = \frac{\sqrt{3\pi Dv}\eta}{RTK_H^2(Re)^{1/2}\ln(2Re)}. \tag{8.138}$$

c_1 and c_h are the lower critical surfactant concentration of the onset of the velocity decrease at low and high surface activities. Calculated data from these dependencies are in satisfactory agreement with experimental data of Okasaki (1964).

8.7. THE REAR STAGNANT REGION OF A BUOYANT BUBBLE

At low surfactant concentration the surface of a bubble is not uniformly stagnant. Usually the main portion of the surface can be weakly stagnant, and only a narrow truly stagnant region can exist in the vicinity of the rear bubble pole (Savic 1953, Harper 1982). Results of papers describing the structure of this region in analytical form are to be considered now. As already pointed out in the introduction, results of such kind can be obtained for bubbles floating up at $Re \ll 1$, whereas the results presented in Sections 8.4 to 8.6.3 can be used for flowing bubbles close to the potential flow, i.e. $Re \gg 1$. The existence of the stagnant cup was confirmed experimentally by Savic (1953), Garner & Scelland (1956), Elzinga & Banchero (1961), Griffith (1962), Horton et al. (1965), Huang & Kintner (1969) and Beitel & Hedeger(1971).

8.7.1. SPECIAL FEATURES OF THE PROCESS OF DYNAMIC ADSORPTION LAYER FORMATION IN THE REAR STAGNANT REGION

It is likely that the unrestricted growth of adsorption at $\theta \to \pi$ given by Eqs. (8.69) and (8.79) in both limiting cases was considered for the first time by Derjaguin et al. (1959). The necessity of the growth of $\Gamma(\theta)$ at $\theta \to \pi$ is physically obvious since the adsorption layer is compressed. The increase of $\Gamma(\theta)$ is naturally restricted by parameters which were not taken into account in deriving Eq. (8.85).

At $\theta \to \pi$ the process of dynamic adsorption layer formation is complicated by three factors: the expansion of the diffusion boundary layer, the necessity to consider surface diffusion, and the retardation of the surface taking place at $\theta \to \pi$ even when the complete remaining surface is essentially not stagnant. The appearance of all the three factors is qualitatively different at high and low surface activity, respectively. The special features of the process of adsorption layer formation at $\theta \to \pi$ are considered by Dukhin (1965) for each case separately.

It was shown by Harper (1974, 1988) that the solution representing the distribution of the surface concentration diverges in the neighbourhood of the rear stagnant point (RSP) as

$$\Gamma \approx \Gamma_o B \left[\frac{3}{16}(\pi - \theta)^4 \right]^{\alpha}.$$

(8.139)

The constant α $(-1/2 < \alpha < 0)$ can be determined from the relation

$$\Lambda = \sqrt{\frac{4\eta RD}{u}} \frac{c_o}{\Gamma_o} = \frac{-4\tilde{\Gamma}(-\alpha)}{\tilde{\Gamma}\left[-\frac{1}{2} - \alpha\right]}$$

(8.140)

where $\tilde{\Gamma}$ is Γ-function and $B \approx - \Lambda / (A + 2\sqrt{\pi})$. From the solution derived by Harper (1974) it follows that at a weak stagnation in the neighbourhood of the RSP, the surfactant adsorption can be calculated by the relation,

$$\Gamma_{RSP} = \Gamma_o B(Pe)^{-2\alpha} \left[\frac{3}{4} \right]^{\alpha} \tilde{\Gamma}(2\alpha + 1)$$

(8.141)

where $\Gamma_{RSP} \gg \Gamma_o$ as Pe is large and α is negative.

8.7.2. STRUCTURE OF THE REAR STAGNANT REGION OF A BUBBLE

Harper's results (1974, 1988) permit calculation of the surfactant concentration at $Re \ll 1$ and evaluation of the conditions at which the surface of the bubble in the vicinity of this point is weakly stagnant.

Since the surfactant is not uniformly distributed over the surface of a buoyant bubble, the stagnating effect of the surfactant appears to a greater extent in the neighbourhood of the RSP. Therefore, the surface motion model proposed for the first time by Savic (1953) is very

310

valuable. Within the framework of this model it is assumed that some region in the vicinity of the RSP is completely stagnant due to the adsorbed surfactant and the complete remaining surface is almost free from surfactant and weakly stagnant. Unfortunately, all theoretical work employing this model (Savic 1953, Harper 1973, 1974, 1982, 1988, Heet et al. 1991) is restricted by $Rc \ll 1$. The structure of the strongly stagnant region in the vicinity of the RSP was studied numerically by Davis & Acrivos (1966). An analytical description of dynamic adsorption layers at bubble surfaces is proposed by Harper (1973, 1982) which is asymptotically suitable for small dimensions of the stagnant cap. The mathematical apparatus of the theory is simplified in the case of small dimensions of the stagnant cap region (Harper 1973, 1982) and the results can be presented more clearly. Therefore, the present analysis is restricted to these works.

It is assumed that the surface of the bubble is completely free at $\theta < \pi - \psi$ and completely stagnant at $\theta < \psi$. In addition, it is assumed that in the former region the adsorption is much lower than the equilibrium value and in the latter much higher than Γ_0. This allows to assume that adsorption of surfactant at the free surface takes place in the surfactant free region, and desorption of surfactant from the surface to the bulk of the solution with $c_0 \approx 0$ in the stagnant region. The size of the strongly stagnant region is determined by the balance of the total adsorption flow to the bubble surface and the total desorption flow (Fig. 8.5).

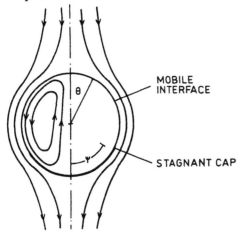

Fig. 8.5.   Stagnant cup model

The total adsorption flow to a weakly stagnant bubble surface was calculated by Levich (1962),

$$J_{ad} = 5.79 \sqrt{\frac{Pe}{n}} a_b Dc_0. \tag{8.142}$$

The function $\Gamma(\theta)$ and total desorption flow can be calculated by using the distribution of viscous stresses in the strongly stagnant region near the lower pole (Harper 1973),

$$\frac{RT}{a_b} \frac{\partial \Gamma}{\partial \theta} = \eta \frac{\partial v_\theta}{\partial z}\bigg|_{z=a_b} = \eta \frac{v}{a_b} \frac{4m}{\pi \sqrt{1-m^2}}, \tag{8.143}$$

with $m = (\pi - \theta)/\theta^*$, $s = (z - a_b)/(a_b \theta^*)$ being inner coordinates within the stagnant cap. This method of estimation of $\Gamma(\theta)$ in the case of a strong stagnation was proposed for the first time by Dukhin & Derjaguin (1961). On the basis of variable transformations

$$x = \int_m^1 \frac{m^2}{(1-m^2)^{1/4}} dm, \tag{8.144}$$

$$y^2 = \left[\frac{2}{81\pi^2} \frac{a_b^2 v^2 \theta^{*4}}{D^2} \right]^{1/3} \frac{m^2 s^2}{\sqrt{1-m^2}}, \tag{8.145}$$

the equation of convective diffusion in the neighbourhood of the RSP can be reduced to the following form,

$$\frac{\partial c}{\partial x} = \frac{1}{y} \frac{\partial^2 c}{\partial y^2}, \tag{8.146}$$

which can be solved analytically. Thus, the total desorption flow of surfactant from the surface of stagnant cap can be calculated (Harper 1973),

$$J_{des} \approx 1.99 a_b Dc_0 \left[\frac{a_b v}{D} \right]^{1/3} \frac{\eta v}{RT\Gamma_0} \psi^{8/3}. \tag{8.147}$$

Equating the total desorption and adsorption flows for $Re \ll 1$ results in

$$\psi = 1.761 \left[\frac{a_b v}{D} \right]^{1/16} \left[\frac{RT\Gamma_0}{\eta v} \right]^{3/8}, \tag{8.148}$$

$$\Gamma_{RSP} = 2.242 \Gamma_0 \left[\frac{a_b v}{D} \right]^{1/16} \left[\frac{RT\Gamma_0}{\eta v} \right]^{5/8}. \tag{8.149}$$

It is of great interest to know the conditions under which it is advisable to use the model of bubble surface motion proposed by Savic (1953). The surfactant adsorption at the main portion of the bubble strongly deviates from its equilibrium value when the condition (8.71) holds. The condition for a small stagnant cap size $\psi \ll 1$ with regard to Eq. (8.148) can be written in the following form,

$$\frac{\eta_\alpha V}{RTT_o} \gg 1.22 \, Pe^{1/6}. \tag{8.150}$$

8.7.3. THE REAR STAGNANT CAP AND BUBBLE BUOYANT VELOCITY AT SMALL Re

The creeping flow past a bubble with a stagnant cap has been investigated by Savic (1953), Davis & Acrivos (1966) and Harper (1973, 1982), Holbrook & Levan (1983). In each case the difficulty came about in dealing with the mixed boundary conditions of the stagnant cap. The formulation led to an infinite set of algebraic equations for the coefficients of a series solution. Savic (1953) truncated the series after the sixth term, while Davis & Acrivos (1966) used 150 terms. Harper (1973, 1982) studied the case of small cap angles and carried out an asymptotic analysis using oblate spheroidal coordinates. Sadhal & Johnson (1983) generalized the problem to include both drops and bubbles by allowing internal circulation within the drop.

As a purely hydrodynamic problem, the velocity field due to a stagnant cap at the rear of a moving drop was solved exactly in terms of an infinite series of Gegenbauer polynomials with constants depending on the cap angle ϕ. From this series, an analytical solution for the drag $F(\phi)$ exerted on the drop can be obtained, from which the terminal velocity was computed once the external force on the drop is resolved,

$$F(\phi) = 4\pi\eta VRa_b \left\{ \frac{\eta}{4\pi(\eta+\eta')} \left[2\phi + \sin\phi - \sin 2\phi - \frac{1}{3}\sin 3\phi \right] + \frac{2\eta + 3\eta'}{2\eta + 2\eta'} \right\}. \tag{8.151}$$

For the limiting case of $\phi=0$ (no surfactant), equation (8.151) is transformed into the Hadamard-Rybczynski equation. For $\phi=\pi$ (completely stagnant interface), a solid sphere behaviour results.

If the drop viscosity becomes infinitely large ($\eta'\to\infty$), a solid sphere drag is also obtained. The drag force for the special case of a bubble is easily obtained from (8.151) by letting $\eta'\to 0$. This yields

$$F(\phi)_{\text{bubble}} = 4\pi\eta UR\left\{\frac{1}{4\pi}\left[2\phi + \sin\phi - \sin 2\phi - \frac{1}{3}\sin 3\phi\right] + 1\right\}. \tag{8.152}$$

The cap angle is determined by the adsorbed amount of surfactant necessary to cause a surface pressure that balances the compression due to the viscous shear on a cap of angle ϕ.

Sadhal & Johnson (1983) give a dimensionless equation for the difference in shear stresses exerted by the surrounding phases on the interface in the cap region

$$\left(\tau_{z\theta(s)}^{(2)} - k\tau_{z\theta(s)}^{(1)}\right) = h(\theta,\phi)/\lambda, \tag{8.153}$$

where the tangential stress is scaled by $\eta V/a_b$, k is the ratio of the droplet to the continuous phase viscosity, and $h(\theta,\phi)$ is a very complicated function.

The balance of Marangoni and viscous stresses (8.153), reformulated in terms of Γ, is integrated to obtain the surfactant distribution and yields Γ as a function of ϕ and the dimensionless Marangoni number Ma. The surfactant distribution can be integrated over the cap region to obtain the total amount on the surface, M. The variable M is also computed independently from the surfactant conservation equations and equating the two expressions yields ϕ. Once ϕ is specified, the drag coefficient and terminal velocity can be calculated.

The above procedure was first introduced by Griffith (1962) whose study is incomplete since he did not use the proper hydrodynamic solution, and later by Sadhal & Johnson (1983) in their exact solution of the problem. Each of these authors assumed that the surface pressure exerted by the compression of the surfactant in the stagnant cap may be represented by a linear isotherm.

8.8. TOTAL AMOUNT OF SURFACTANT AT MOBILE BUBBLE SURFACES

8.8.1. BUBBLE FRACTIONATION AND DYNAMIC ADSORPTION LAYER

The problem of total amount of surfactant on a mobile bubble surface is important for bubble fractionation. In bubble fractionation (Clarke & Wilson 1983) surface active material is transferred to the upper section of a liquid column through adsorption on rising bubbles followed by the release at the top of the column as the bubbles burst or accumulate in foam.

In calculating surfactant transfer to foam, it is usually assumed that the value of surface concentration of surfactant on the surface of a buoyant bubble is equal to Γ_o. On the other hand we know that the equilibrium value of surface concentration cannot be established in each case within the rising time, and the total amount of adsorbed surfactant differs from the equilibrium value. It was shown in the preceding section that the adsorption layer of a rising bubble is strongly deformed in a wide range of system parameters. Therefore, the structure of the dynamic adsorption layer must be taken into account in order to calculate surfactant transfer into foam.

In the present section, a parameter is calculated

$$K_b = \frac{1}{4\pi a_b^2 \Gamma_o} \oint \Gamma(\theta) ds, \qquad (8.154)$$

which considers only one surfactant species in a surfactant solution. The parameter K_b takes into account the effect of non-equilibrium factors on the process of surfactant transfer to foam.

Results presented in this chapter are, of course, preliminary. In order to obtain reliable predictions and recommendations, which are of interest for technology and mathematical modelling of surfactant transfer to foam it is important to take into account the following factors:

1. real state of solution and adsorption layer;

2. presence of several surfactants in solution;

3. bubble deformation and real structure of flow at arbitrary Re;

4. polydispersity of bubbles;

5. coagulation of bubbles during buoyant motion;

6. bubble volume fraction dependence of the process of surfactant transfer to foam;

7. transfer processes in foams.

8.8.2. ESTIMATION OF TIME OF STEADY-STATE ESTABLISHMENT OF THE DYNAMIC ADSORPTION LAYER OF BUOYANT BUBBLES

As noted above, the effect of kinetic factors on surfactant transfer to foam can be complicated by a time t_b of bubble rising, which can be insufficient for a steady-state of the dynamic adsorption layer to be reached in the solution column. If the dynamic adsorption layer is far from its steady-state, the required parameter K_b can be expressed as

$$K_b \approx \frac{J_{ad} t_b}{4\pi a_b^2 \Gamma_o} = \frac{t_b}{\tau_{ad}}, \qquad (8.155)$$

where J_{ad} is total adsorption flow to the bubble surface, and

$$\tau_{ad} = \frac{4\pi a_b^2 \Gamma_o}{J_{ad}} \qquad (8.156)$$

is the characteristic adsorption time. At the initial stage of adsorption when the total surfactant adsorption on the bubble surface is much less than in the steady-state, the flux J_{ad} to the bubble surface at strong retardation is equal to (Levich 1962),

$$J_{ad} = 2\pi a_b Dc_o Pe^{1/3}, \qquad (8.157)$$

and at weak retardation (Levich 1962),

$$J_{ad} = 7.09 a_b Dc_o \sqrt{Pe}. \qquad (8.158)$$

The value of τ_{ad} can be estimated for these two limiting cases as follows:

$$\tau_{ad} = \frac{2a_b \Gamma_o}{Dc_o} Pe^{-1/3} \text{ at } Re \ll 1 \qquad (8.159)$$

$$\tau_{ad} = \frac{1.77 a_b \Gamma_o}{Dc_o} Pe^{-1/2} \text{ at } Re \gg 1 \qquad (8.160)$$

The rising time t_b results simply from the height of cell l and buoyant velocity v,

$$t_b = \frac{1}{v}. \qquad (8.161)$$

It is convenient to introduce a characteristic length

$$l_a = v\tau_{ad} \qquad (8.162)$$

along which the adsorption layer of buoyant bubble approaches equilibrium. According to (8.161) this characteristic length is given by

$$l_a = \frac{2\Gamma_o}{c_o} Pr^{2/3} Re^{2/3}, \text{ at } Re \ll 1 \text{ and } l_a = \frac{1.77\Gamma_o}{c_o} \sqrt{Pr\, Re} \text{ at } Re \gg 1. \qquad (8.163)$$

At $Re < 1$, even at very high surface activity ($\Gamma_o / c_o \sim 10^{-2}$ cm) $l_a \approx 1$ cm, since $Pr^{2/3} \approx 10^2$. This means that l_a is much less than the height of the liquid column l, usually of the order of $10 - 10^2$ cm.

l_a grows with buoyant velocity v and Re. Even at Re ~ 10^3 when the bubble behaviour is complicated due to the variation of its form by surface oscillations etc., l_a attains a values of the order of 10 cm, again at very high surface activity ($\Gamma_o / c_o \sim 10^{-2}$ cm).

Hence, the state of the adsorption layer can be considered as stationary in the transition of bubbles to the foam layer. The situation can change if the liquid column is thin (less than 1 - 10 cm).

8.8.3.   EVALUATION OF THE TOTAL AMOUNT OF SURFACTANT ON MOBILE BUBBLE SURFACES

It can be shown that the total amount of surfactant adsorbed at the bubble surface under stationary conditions is equal or even exceeds the respective equilibrium value $4\pi a_b^2 \Gamma_o$. Using Levich's model of a diffusion layer of constant thickness (Levich 1962), the surfactant flux density to the bubble surface can be evaluated by

$$j = D \frac{c_o - c(a_b, \theta)}{\delta_D}.$$
(8.164)

Under stationary conditions, the total surfactant flux to the bubble surface is equal to zero,

$$\oint j \, dS = 0.$$
(8.165)

Substituting (8.164) into (8.165) leads to

$$\oint c(a_b, \theta) dS = 4\pi a_b^2 c_o ; \oint \Gamma(\theta) dS = 4\pi a_b^2 \Gamma_o.$$
(8.166)

Therefore, in this model the total amount of surfactant adsorbed on the bubble surface under stationary conditions does not depend on conditions of dynamic adsorption layer formation, and the parameter K_b is equal to unity.

The total amount of surfactant adsorbed to a moving bubble proves to be greater than that at rest if we take into consideration that the thickness of the diffusion layer grows monotonically along the meridian from the upper pole to the lower. As a result, the diffusion layer thickness δ_A averaged over the part of surface S_A adsorbing surfactant is less than the thickness of diffusion layer δ_D averaged over the part of surface S_D desorbing surfactant

$$\bar{\delta}_A < \bar{\delta}_D.$$
(8.167)

The condition for a stationary state of the dynamic adsorption layer (8.165) can be presented in the form

$$\frac{D}{\overline{\delta}_A} \int_{S_A} (c_o - c(a_b, \theta)) dS = \frac{D}{\overline{\delta}_D} \int_{S_D} (c_o - c(a_b, \theta)) dS. \tag{8.168}$$

Now, accepted simplifications of the stagnant cap theory are used. At the adsorbing surface, c can be neglected as compared with c_o and at the desorbing surface c_o can be neglected as compared with c

$$\int_{S_D} c(a_b, \theta) dS \approx \frac{\overline{\delta}_D}{\overline{\delta}_A} \int_{S_A} c_o dS \tag{8.169}$$

or

$$\int_{S_D} \Gamma(\theta) dS \approx \frac{\overline{\delta}_D}{\overline{\delta}_A} \int_{S_A} \Gamma(\theta) dS \approx \frac{\overline{\delta}_D}{\overline{\delta}_A} \int_{S_A} \Gamma_o dS, \tag{8.170}$$

where $S = S_A + S_D$ is the total bubble surface. Here, it is additionally taken into account that the stagnant cap constitutes a small portion of the bubble surface $S_D \ll S$.

Thus,

$$\int_S \Gamma(\theta) dS > \int_{S_D} \Gamma(\theta) dS > \Gamma_o S \tag{8.171}$$

i.e. the adsorption layer in dynamic state contains as a whole more surfactant than a bubble at rest. Note that a similar result was obtained also under the conditions of strong retardation of the whole bubble surface. As the result of a strict solution of the convective diffusion equation, it was shown (Section 8.3) that the relative increment of adsorption averaged over the bubble surface amounts to approximately 17%.

The stagnant cap theory permits a quantitative evaluation of the variation of the total amount of surfactant adsorbed at the bubble as a function of its buoyant velocity. The discussion of these results are avoided since they are restricted to Re « 1 and experimental data about the mobility of the bubble surface at small Reynolds numbers do not exist.

8.8.4. EXPERIMENTAL INVESTIGATION OF THE TIME DEPENDENCE OF BUBBLE BUOYANCY

Both in theory, and experiment, attention is paid to the stationary state of adsorption layer and, respectively, to stationary bubble rising velocity. Pioneering experiments by Loglio et al. (1989) demonstrate the possibility to study relaxation of the adsorption layer velocity. The time dependence of the rising velocity of bubbles of 2 - 3 mm in diameter in aqueous surfactant solutions was studied. For this purpose the rising velocity was measured at different height inside a Pyrex glass column of 140 cm height. As one would expect, the velocity decreases since the amount of adsorbed surfactant is the higher the longer is the path length. For the same reason the velocity decreases with increasing surfactant concentration. These results are presented in Fig. 8.6 as a function of the ratio between rising time in solution of a given concentration and the time of rising in pure water. One can conclude that a path of the order of several meters is required for the stationary state of adsorption layer of bubbles of the size of 2 - 3 mm. This result does not strongly contradict the estimate according to formula (8.163) which ignores the desorption flux, and thus gives the time to establish an approximately stationary state. It is well known that a relaxation process slows down as it approaches a stationary state. In the present case the net flux is determined by the balance of adsorption and desorption fluxes. It decreases as the stagnant cup grows and consequently the desorption increases too. The decrease of the net flux can lead to an increase of the characteristic length. It is not excluded that an impurity flux to the bubble surface is controlled by the retarded adsorption kinetics, which can also cause an increase of the characteristic length.

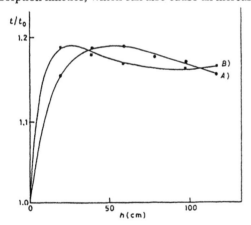

Fig. 8.6.   Variation of the rising time ratio t/t_0 vs. the rising distances for solutions of $DC_{12}PO$ in distilled water at concentrations: A) 2 mg/dm^3 ; B) 4 mg/dm^3. Diameter of the bubble = 0.36 cm, according to Loglio et al. (1989)

In sea and tap water (both containing impurities) the bubble velocity decreases along its path too. However, the contamination level is not sufficient to suppress the mobility of the bubble surface completely. The non-monotonous concentration and height dependence of the rising velocity cannot be interpreted by the presence of these impurities.

This work by Loglio et al. (1989) demonstrates the possibility of experimental investigations of the relaxation of adsorption layers on a mobile bubble surface which stimulate the development of the DAL theory for large Reynolds numbers. In experimental aspect the investigation of the height dependence of the rising velocity for the bubbles of different dimensions is very important. Suzin & Ross (1984) have also observed the height dependence of the bubble rising velocity.

8.9. SUMMARY

The coupling of the transport of momentum with the mass transport practically excludes any analytical solution in the field of physico-chemical hydrodynamics of bubbles and drops. However, a large number of effective approximate analytical methods have been developed which make solutions possible. Most important is the fact, that the calculus of these methods allows to characterise different states of dynamic adsorption layers quantitatively: weak retardation of the motion of bubble surfaces, almost complete retardation of bubble surface motion, transient state at a bubble surface between an almost completely retarded and an almost completely free bubble area.

With increasing surfactant concentration, the dynamic adsorption layer changes from the state of weak retardation to the transient state, characterised by the appearance of a stagnant cap and its growth with further surfactant concentration increase. In the process of the growth of the stagnant cap, weakly and strongly retarded parts of the bubble surface coexist.

The theory of weak retardation of surface motion allows relations to be obtained to give an estimate of the minimum surfactant concentration for the appearance of a stagnant cap which exerts and effects the buoyant velocity. The theory of strong retardation yields a maximum surfactant concentration which separates the transient state from a complete retardation of the bubble surface. Thereby, the transition between the theory of limiting states of the dynamic adsorption layer and the theory of the transient state is obtained, which is important for two reasons. First of all, the theories were developed by different teams of scientists independently. Secondly, it allows to conclude the appropriateness of the approximate methods employed which gives a complete picture of different states of the dynamic adsorption layer. This is possible without huge efforts necessary for numerical solutions.

Unlike the limiting cases of the dynamic adsorption layer theories, using well-known hydrodynamic solutions, the theory of the transient state is very complex. The hydrodynamic equations have to be solved with very complicated boundary conditions taking into consideration the region of the stagnant cap as well as a part of the bubble surface, completely free from adsorbed molecules. For this reason, the theory of limiting cases is developed for small as well as large Reynolds numbers, while the theory of the transient state was developed only for small Reynolds numbers. The theories of limiting cases were more or less developed from solutions of the convective diffusion, while the transient case of the dynamic adsorption layer is to a large extend a hydrodynamic problem.

When passing from a weakly to a strongly retarded bubble surface, the diffusion layer and its thickness, do not change significantly. Boundary conditions used for dynamic adsorption layers show frequently more substantial changes. Evidently, adsorption and desorption kinetics and the formation of a dynamic adsorption layer are strongly interrelated and represent a single process. The established rules of the formation of a dynamic adsorption layer simultaneously characterise the specific nature of the processes of convective diffusion and adsorption/desorption kinetics under the effect of the hydrodynamic field of the moving bubble.

Although considerable success in the development of the dynamic adsorption layer theory has been reached, there has been less progress experimentally. This is not marked for the transient state, where the theoretical advances are most impressive. It turns out that experimental works devoted to the stagnant cap theory are more or less of empirical interest as they are restricted to small Reynolds numbers. At small, and even intermediate, Reynolds numbers the bubble surface can initially behave immobile and the formation of a stagnant cap is almost impossible.

The essence of the problem is that on one hand the phenomenon of a stagnant cap appears only at large Reynolds numbers and on the other, the theory of a stagnant cap was developed only for small Re. The development of a stagnant cap theory for large Re should be possible nowadays because the solution of the very complex hydrodynamic part has been presented already in other studies (Rivkind et al. 1971, 1976).

There are two reasons in favour of the theory of the dynamic adsorption layer at weakly and strongly retarded bubble surfaces. At first, the theory has been developed also for large Re, and secondly it was developed for small Re also for the case of surface retardation controlled by surfactant concentration as well as by other retardation factors.

The state of experiments in general seems to be underdeveloped as compared with the level of broad theoretical studies of dynamic adsorption layers on moving bubbles and drops. Most of all the lack of systematic studies, which are obviously necessary for a further development of this area of physico-chemical hydrodynamics. This is clearly an important area for the optimisation of many technological processes. Only experiment can prove the existence of different states of a dynamic adsorption layer proposed by the theories. Two questions remain to be answered: "does the motion of the bubble surface in absence of surfactants arise at small or intermediate Reynolds numbers" and "is it possible to attain a low surfactant concentration at which its effect on the surface motion remains small?"

The measurements of buoyant bubble velocity is not suitable for solving these problems and attention must be paid to other experimental techniques. Ii is clear that investigations of surfactant transfer in foams and of sedimentation potential measurements deserve more attention.

8.10. REFERENCES

Beitel, A. and Heideger, W.J., Chem. Eng. Sci., 26(1971)711

Bond, W., Phil. Mag., 4(1927)889

Boussinesq, M.J., Comp. Rend., 56(1913)983, 1035, 1124

Chabra, R.P. and Kee, D. De, Transport Processes in Bubbles, Drops and Particles, Hemisphere, New York, 1990

Clarke, A.N. and Wilson, P.J., Foam Flotation, N.Y. and Basel: Marcel Decker, 1983

Clift, R., Grace, J.R. and Weber, M.E., Bubbles, Drops and Particles, N.Y., etc.: Acad. Press., 1978, 380 p.

Davis, R.E., Acrivos, A., Chem. Engen. Sci., 21(1966)681

Derjaguin, B.V. and Dukhin, S.S., in Issledovania v Oblasti Poverkhnostnykh sil, V. 3,(B.V. Derjaguin, Ed.), Nauka, 1967; Research in Surface Forces, Vol 3, Consultants Bureau, New York and London, 1971

Derjaguin, B.V. and Dukhin, S.S., Trans. Inst. Mine. Metal., 70(1960)221

Derjaguin, B.V., Dukhin, S.S. and Lisichenko, V.A., Zh. Phys. Chim., 33(1959)2280

Derjaguin, B.V., Dukhin, S.S., and Lisichenko, V.A., Zh. Phys., Chim., 34(1960)524

Dukhin, S.S., Thesis, Moscow, Institute of Physical Chemistry, 1965

Dukhin, S.S., in "Modern Theory of Capillarity" (F.G. Govard and A.I. Rusanov, Eds.), Berlin, Akademic Verlag, 1981, 83

Dukhin, S.S., Kolloidn. Zh., 26(1964)36

Dukhin, S.S., in "Research in Surface Forces" (B.V. Derjaguin, Ed.), Vol. 2, N.Y., London; Consultant Bureau, 1966

Dukhin, S.S., Kolloidn. Zh., 44(1982)896

Dukhin, S.S., Kolloidn. Zh., 45(1983)22

Dukhin, S.S. and Buykov, M.V., Zh. Phys. Chim., 39(1965)913

Dukhin, S.S. and Derjaguin, B.V., Zh. Phys. Chim., 35(1961)1246

Dukhin, S.S. and Derjaguin, B.V., Zh. Phys. Chim., 35(1961)1453

Dukhin, S.S. and Derjaguin, B.V., Kolloidn. Zh., 20(1958)705

Elsinga, E.R. and Banchero, J.T., AICHE J., 7(1961)394

Frumkin, A.N. and Levich, V.G., Zh. Phys. Chim., 21(1947)1183

Hadamard, Comp. Rend., 152(1911)1735

Hamielec, A.E., Jonson, A.I., Canad. J. Chem. Eng., 40(1962)40.

Harper, J.F., J. Fluid Mech., 58(1973)58

Harper, J.F., Q. J. Mech. Appl. Math., 27(1974)87

Harper, J.F., Appl. Sci. Res., 38(1982)343

Harper, J.F., Q. J. Mech. Appl. Math., 41(1988)203

He, Z., Maldarelli, C. and Dagan, Z., J. Colloid Interface Sci., 146(1991)442

Holbrook, J.A. and Levan, M.D., Chem Eng. Commun., 20(1983)273

Horton, Y.J., Frish, T.R. and Kintner, R.G., Can. J. Chem. Eng.,(1965)143

Huang, W.S. and Kintner, R.C., AIChE J., 15(1969)735

Garner, F.H. and Skelland, A.H.P., Eng. Des. Equip., 48(1956)51

Gorodetskaya, A.B., Zh. Phys. Chim., 23(1949)7

Grifith, R.M., Chem. Eng. Sci., 17(1962)1057

Lebedev, A.A., ZhRPhHO, Fiz. Otd., 48(1916)3

Levich, V.G., Physico-Chemical Hydrodynamics, Prentice-Hall, Englewood Cliffs, N.Y., 1962

Listovnichy, A.V., Kolloidn. Zh., 47(1985)512

Loglio, G., Degli Innocenti, N., Tesei, U. and Cini, R., Nuovo Cimento, 12(1989)289

Moor, D.W., J. Fluid Mech., 16(1963)161

Okazaki, S., Bull. Chem. Soc. Japan, 37(1964)144

Quintana, G.C., in Transport Processes in Bubbles, Drops and Particles, Hemisphere, New
    York, 1990

Rivkind, V.Ya. and Riskin, G.M., Inzh.-Phys. Zh., 20(1971)1027

Rivkind, V.Ya., Riskin, G.M., Izv. AN SSSR, Mech. Zhidk. Gaza, 1(1976)8

Rybczynski, Bull. de Cracovie (A), (1911)40

Rulyov, N.N. and Leshchov, E.G., Kolloidn. Zh., 42(1980)521

Sadhal, S.S. and Johnson, R.E., J. Fluid Mech., 126(1983)237

Samigin, V.D. and Derjaguin, B.V., Kolloidn. Zh., 26(1964)493

Savic, P., Nat. Res. Counc. Can., Div. Mech. Engng. Rep. MT-22, 1953

Savill, D.A., Chem. Eng. Sci., 5(1973)251

Sotskova, T.Z., Bazhenov, Yu.F. and Kulski, L.A., Kolloidn. Zh., 44(1982)989; 45(1983)108

Suzin, Y. and Ross, S., J. Colloid Interface Sci., 103(1984)578

Usui, S. and Sasaki, H., J. Colloid Interface Sci., 65(1978)36

Usui, S., Sasaki, H. and Matsukava, H, J. Colloid Interface Sci., 81(1980)80

CHAPTER 9

9. **DYNAMIC ADSORPTION LAYERS OF SURFACTANTS AT THE SURFACE OF BUOYANT BUBBLES. KINETIC - CONTROLLED SURFACTANT TRANSPORT TO AND FROM BUBBLE SURFACES**

The rate of the exchange process of surfactant molecules between the surface of a bubble (drop) and the bulk solution is determined not only by convective diffusion but in the general case also by the kinetics of the adsorption step itself. Details of the physical model of the adsorption process are given in chapter 2 and 4. A method which takes into account the effect of adsorption kinetics on the formation of the dynamic adsorption layer was developed by Levich (1962). Using this method, attempts were made to generalize the theory of the dynamic adsorption layer of bouyant bubbles (Dukhin 1965).

9.1. DYNAMIC ADSORPTION LAYERS OF NONIONIC SURFACTANTS

$P(\Gamma)$ denotes the number of molecules desorbing from a unit surface in unit time and $Q[c(a_b,\theta),\Gamma(\theta)]$ the number of molecules adsorbing in unit time. Note that in the expression for Q, the value of the concentration at the bubble surface is used, i.e. in the sublayer adjacent to the bubble surface. It should also be emphasized that in this case no simple relation between $c(a_b,\theta)$ and $\Gamma(\theta)$ exsists since the adsorption is in non-equilibrium. It is well known that at low surface coverages, the flow Q does not depend on the adsorption $\Gamma(\theta)$ and the functions P and $Q[c(a_b,\theta)]$ can expressed by linear relations,

$$P(\Gamma) = k_{des}\Gamma(\theta), \tag{9.1}$$

$$Q[c(a_b,\theta)] = k_{ad}c(a_b,\theta). \tag{9.2}$$

In the equilibrium state $P(\Gamma_o) = Q(c_o)$, i.e.

$$k_{des}\Gamma_o = k_{ad}c_o, \; K_H = \Gamma_o/c_o = k_{ad}/k_{des}. \tag{9.3}$$

The total adsorption rate at a unit surface is

$$j_n = -P(\Gamma) + Q[c(a_b,\theta)] = -k_{des}\Gamma(\theta) + k_{ad}c(a_b,\theta). \tag{9.4}$$

Expressing j_n by means of Eq. (8.10), in the case of a small retardation of the surface, and neglecting surface diffusion, Eq. (9.4) can be rearranged to

$$\frac{1}{a_b \sin\theta} \frac{\partial}{\partial\theta}\left(\sin^2\theta v_0\Gamma(\theta)\right) = -k_{des}\Gamma(\theta) + k_{ad}c(a_b,\theta). \tag{9.5}$$

This relation contains two unknown functions $\Gamma(\theta)$ and $c(a_b,\theta)$. A second equation relating $\Gamma(\theta)$ and $c(a_b,\theta)$ is obtained if the convective diffusion equation is considered together with the boundary condition (8.58). As shown in Chapter 8, this can be rearranged to give the integral equation (8.85). However, it should be noted that the adsorption cannot reach equilibrium. Therefore, the substitution $\Gamma(\theta) \rightarrow K_H c(a,\theta)$ cannot be made in the left-hand side of Eq. (8.85). It must be rewritten in its original form,

$$\frac{\Gamma(\theta)}{K_H}\sin^2\theta = m\left[c_0\sqrt{t} - \frac{1}{2}\int_0^t \frac{c(a_b,t')}{\sqrt{t-t'}}dt'\right], \tag{9.6}$$

where $t(\theta)$ is expressed by Eq. (8.17). Thus, Eq. (9.5) takes into account the effect of adsorption-desorption kinetics on the adsorption layer formation and Eq. (9.6) the effect of convective diffusion. In the general case, a common solution of Eqs (9.5) and (9.6) seems to be impossible.

The bubble surface restoration time τ_m and the adsorption time τ_a can be conventionally expressed by the following relations

$$\tau_m = a_b / v_0 \text{ and } \tau_a = \frac{1}{k_{des}}. \tag{9.7}$$

The ratio of these characteristic times is,

$$v = \tau_m / \tau_a \tag{9.8}$$

When the characteristic adsorption time is much smaller than the bubble surface restoration time,

$$v \gg 1, \tag{9.9}$$

it can be expected that the adsorption at the bubble surface is close to equilibrium. It can be shown that, under condition (9.9), the rate of the adsorption-desorption processes has no effect on the concentration distribution since it is completely determined by convective diffusion. Clearly, this is possible in the case where the equilibrium condition $\Gamma(\theta) = K_H c(a_b,0)$ in Eq. (9.5) holds,

$$K_H c(a_b,0) - \Gamma(\theta) \ll \Gamma(\theta) = K_H c(a_b,0). \tag{9.10}$$

The left-hand side of this inequality can be expressed also in terms of $c(a_b,\theta)$ using Eq. (9.5). From Eqs (9.3), (9.8) and (9.10) the following equation results,

$$\frac{1}{\sin\theta} \frac{\partial}{\partial\theta} \left(\sin^2 \theta v_o c(a_b,\theta) \right) \ll v c(a_b,\theta), \tag{9.11}$$

where $c(a_b,\theta)$ is the solution of Eq. (8.85), as obtained for the two limiting cases (8.71) and (8.72). Substituting $c(a_b,\theta)$ with respect to Eqs (8.78) and (8.69) it becomes clear that condition (9.10) is fulfilled at any values of θ, if $v \gg 1$. Eq. (9.11) makes it possible to neglect the left-hand side in Eq. (9.5) compared with each term on the right-hand side. This means that in condition (9.9) only the diffusion processes have to be considered and the calculation of the concentration field as well as of the distribution of adsorption can be performed under adsorption equilibrium condition (8.4). Note that condition (9.10), used in deriving the relation for $c(a_b,\theta)$, imposes restrictions on the accuracy of the obtained results. At low surface activity, the variation of concentration along the surface, according to Eq. (8.69) at $m \gg 1$, can prove to be less than the term $\Gamma(\theta) - K_H c(a_b,\theta)$ of the approximate solution of Eq. (9.6). Thus, at $m \gg 1$, the following condition of Eq. (8.85) must be fulfilled in addition to Eq. (9.9),

$$v \gg m. \tag{9.12}$$

According to Eqs (8.85) and (9.5) $c_o - c(a_b,\theta) \gg K\Gamma(\theta) - c(a_b,\theta)$. Since $m \gg 1$, in essence Eq. (9.12) defines condition (9.9) more accurately.

Now the condition will be analysed when convective diffusion has no effect on the formation of dynamic adsorption layer. If in Eq. (9.5) $c(a_b,\theta)$ can be replaced by c_o, Eq. (8.85) is no longer needed. Writing the right-hand side of Eq. (9.5) as $-k_{des}\Gamma(\theta) + k_{ad}c_o + k_{ad}(c(a_b,\theta) - c_o)$, we conclude that given

$$\left| c(a_b,\theta) - c_o \right| \ll \frac{k_{des}}{k_{ad}} \left| \Gamma_o - \Gamma(\theta) \right|, \tag{9.13}$$

the rate of surfactant exchange between the drop surface and the bulk is limited by the adsorption-desorption process. Given condition (9.13) the form of Eq. (9.5) becomes simpler,

$$\frac{1}{a_b \sin\theta} \frac{\partial}{\partial\theta} \left(\sin^2\theta v_o \Gamma(\theta) \right) = -k_{des}\Gamma(\theta) + k_{ad}c_o, \tag{9.14}$$

and can easily be solved,

$$\Gamma(\theta) = \frac{v\Gamma_o}{\sin^2\theta \tan^v(\theta/2)} \int_0^\theta \sin\theta \tan^v(\theta/2) d\theta = \frac{v\Gamma_o(1+x^2)}{x^{2+v}} \int_0^x \frac{x^{v+1}}{(1+x^2)^2} dx \tag{9.15}$$

with

$$x = \tan(\theta/2). \tag{9.16}$$

Deriving Eq. (9.15), the integration constant was omitted which provides the finite value of $\Gamma(0)$. If condition (9.9), $\tau_a \ll \tau_m$, holds, it can be expected that the adsorption layer is close to equilibrium, i.e.

$$|\Gamma_o - \Gamma(\theta)| \ll \Gamma_o. \tag{9.17}$$

Indeed, it is shown by Dukhin (1965) that given condition (9.9) the Eq. (9.15) is changed into the following simplified form,

$$\Gamma(\theta) = \Gamma_o - 2\frac{\Gamma_o}{v}\cos\theta. \tag{9.18}$$

Thus, both a physically reasonable result which corresponds to condition (9.10), and complete quantitative agreement with Frumkin & Levich theory is obtained using exactly the same equation (see Levich (1962), Eq. 74.3). Thus, Eq. (9.15) can be considered as a generalization of Frumkin & Levich's equation which is independent of restriction (9.9). If condition (9.9) is not fulfilled, a substantial deviation of $\Gamma(\theta)$ from equilibrium appears and, in the general case, the angular dependence proves to be very complex. It is shown by Dukhin (1965) that the angular dependence $\Gamma(\theta)$ becomes simpler at greater adsorption times:

$$\Gamma(\theta) = \frac{v\Gamma_o}{2} / \cos^2(\theta/2), \tag{9.19}$$

if

$$v \ll 1. \tag{9.20}$$

As one would expect, the value of surface concentration $\Gamma(\theta)$ becomes much less than the equilibrium value when the adsorption rate decreases. Because convective diffusion was neglected in deriving Eq. (9.15), it may be used at any value of the Peclet number. As the velocity distribution on a weakly retarded surface is given by Eq. (8.14), which is true at Re«1 and Re»1, Eq. (9.15) refers to these limiting cases.

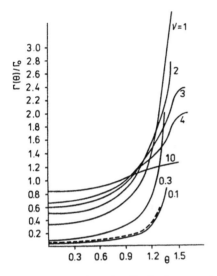

Fig. 9.1.   Adsorption distribution over the surface of bubble given by Eq. (9.15) for different values of $v = 0.1; 0.3; 1; 2; 3; 4; 10$

Examples of the distribution of adsorption calculated by Eq. (9.15) are presented in Fig. 9.1 for different values of v. The dotted line represents the results of calculations using the approximate Eq. (9.19).

To find the concentration distribution at the surface in a non-equilibrium state and to check condition (9.13) used in deriving Eq. (9.15), the latter is substituted into the left-hand side of Eq. (8.85). In this case Eq. (8.85) turns ino the Abel integral equation and its solution can be obtained immediately (Dukhin 1965),

$$c(a_b,\theta) = c_o\left(1 - \frac{2v}{\pi m}\Phi(\theta,v)\right)$$

(9.21)

with

$$\Phi(\theta,v) = \int_0^\theta \frac{\sin\theta'd\theta'}{\chi_1^{1/2}(\theta,\theta')} - \frac{v}{2}\int_0^\theta \frac{\int_0^{\theta'}\sin\theta''\tan^v(\theta''/2)d\theta''}{\tan^{v+1}(\theta'/2)\cos^2(\theta'/2)\chi_1^{1/2}(\theta,\theta')}d\theta'.$$

(9.22)

The condition (9.13) can be rearranged to give the following form

$$\frac{2v}{\pi m}|\Phi(\theta,v)|\ll\left|\frac{\Gamma(\theta)}{\Gamma_o}-1\right|, \tag{9.23}$$

where $\Gamma(\theta)$ is expressed in the general case by Eq. (9.15).

When the neighbourhood of the point $\theta = \pi$ is excluded and $v\ll1$ is assumed, condition (9.23) can be easily analysed for the limiting cases. In accordance with Eq. (9.19) the right-hand side of inequality (9.23) is close to unity and condition (9.23) can be roughly approximated by

$$v\ll m \text{ at } v\ll1. \tag{9.24}$$

In the case correponding to condition (9.9), the right-hand side of inequality (9.23) is of the order of v^{-1} due to Eq. (9.18), and conditions (9.23) can roughly be approximated by

$$1\ll v\ll m^{1/2}. \tag{9.25}$$

Another rough approximation has shown that condition (9.24) is also suitable at $v\approx1$. Therefore, condition (9.24) can be generalized to some extent,

$$v\ll m \text{ at } v\leq1. \tag{9.26}$$

9.2. DYNAMIC ADSORPTION LAYERS OF IONIC SURFACTANTS

The aim of this section is to consider the dynamic adsorption layer structure of ionic surfactant on the surface of rising bubbles. Results obtained in the previous section cannot be transferred directly to this case. The theory describing dynamic adsorption layers of ionic surfactant in general should take into account the effect of electrostatic retardation of the adsorption kinetics of surfactant ions (Chapter 7). The structure of the dynamic adsorption layer of nonionic surfactants was analysed in the precedings section in the case when the adsorption process is kinetic controlled. In this case, it was assumed that the kinetic coefficients of adsorption and desorption do not depend on the surface coverage. On the other hand, the electrostatic barrier strongly depends on Γ_o, and therefore, the results of Section 9.1. cannot be used for the present case..

To obtain the results in analytical form, two limiting cases should be considered:

a - the adsorption and desorption of surfactant ions is controlled by the overcoming of the diffuse layer;

b - the electrostatic retardation has no substantial effect on adsorption kinetics.

In the second case, the mathematical descriptions of the adsorption layer of ionic and nonionic surfactants qualitatively do not differ. In this section, the distribution of adsorption of ionic surfactants over the weakly retarded bubble surface at $Re \ll 1$, $Re \gg 1$ is analysed. In Section 9.3., the structure of the stagnant cup of rising bubbles at $Re \ll 1$ is discussed. Conditions required for the formation of different regimes of dynamic adsorption layers at rising bubbles will be considered in Section 9.4.

9.2.1. EQUATIONS AND BOUNDARY CONDITIONS DESCRIBING THE FORMATION OF THE DYNAMIC ADSORPTION LAYER OF A BUBBLE IN PRESENCE OF IONIC SURFACTANTS

The surface electric current is approximately zero since electric currents of the anions and cations are of a convective nature and roughly compensate each other. The electric current density within the diffuse layer of ionic surfactants obeys the equation

$$\text{div} I = 0, \tag{9.27}$$

and the boundary conditions

$$I\big|_{r \to \infty} = 0; \ I_{no} \approx 0, \tag{9.28}$$

where I_{no} is the normal component of electric current density on the outer boundary of the DL. Neglecting the electric current leads to results described by Eqs (7.17) to (7.37).

As the diffusion and the electromigration lateral flows in the monolayer are much less than the convective flow at $Pe \gg 1$, the first boundary condition at the bubble surface has the form

$$j = \text{div}_s \left[\Gamma(\theta) \cdot v(\theta) \right]. \tag{9.29}$$

If diffusion is the controlling process of adsorption of surface-active anions the boundary condition is,

$$j = D \frac{\partial c}{\partial x}\bigg|_{x=0}. \tag{9.30}$$

The limiting stage is the overcoming of the electric double layer, with electrostatic retardation of adsorption, the value of j is given by Eq. (7.36).

9.2.2. DISTRIBUTION OF ADSORPTION OF IONIC SURFACTANTS OVER WEAKLY RETARDED BUBBLE SURFACES IN THE CASE WHERE THE ADSORBED SURFACTANT IS ALMOST COMPLETELY MOVED TO THE REAR STAGNATION POLE

Under the conditions considered,

$$\Gamma \ll \Gamma_o, \tag{9.31}$$

$$v_\theta \approx \frac{v}{n} \sin\theta, \tag{9.32}$$

where $n = v/v_o$.

Substituting (7.36) and (9.32) into (9.29) leads to

$$\frac{Dc_o}{K(\psi_o)} \left[\frac{\Gamma}{\Gamma_o}\right]^{1-\frac{2z}{z^+}} = \frac{3}{2} \frac{v}{a_b \sin\theta} \frac{d}{d\theta} \left[\sin^2\theta \, \Gamma\right]. \tag{9.33}$$

The solution of this equation has the form,

$$\frac{\Gamma}{\Gamma_o} = \left[3 \frac{z}{z^+} \frac{a_b Dc_o}{vK(\psi_o)\Gamma_o}\right]^{\frac{z^+}{2z}} \frac{1}{\sin^2\theta} \left[\int_0^\theta [\sin\theta]^{\frac{4z}{z^+}-1} d\theta\right]^{\frac{z^+}{2z}}, \tag{9.34}$$

which is valid near the leading pole of bubble but deviates near the rear stagnation pole.

Now the condition under which, electrostatic retardation of adsorption kinetics of surfactant anions appears at the main part of the surface, is determined. Substituting the estimate for Γ from Eq. (9.34) into (7.26) and (7.29) yields the following condition,

$$\left[\frac{c_{el}}{c_o}\right]^{1-\frac{z^+}{2z}} \left[\frac{1}{\kappa\delta_D}\right]^{1-\frac{z^+}{z}} \ll 1. \tag{9.35}$$

The first factor in the left-hand side is greater than unity and the second one is less than unity. Hence, the realization of conditions for the dynamic adsorption layer formation is possible both with electrostatic retardation of the adsorption and without, depending on the system parameters. Note that condition (9.35) can be fulfilled only for multiple-charged anions. Conditions considered in the present section correspond to the situation where the dynamic adsorption layer of ionic surfactant is far from equilibrium. As on the outer boundary of the DL, the ion concentrations close to bulk values, the concentration of surfactant anions on the inner boundary and the value of adsorption are much smaller than the equilibrium values.

9.3 STRUCTURE OF REAR STAGNANT CUP OF A BUBBLE RISING IN SOLUTION OF IONIC SURFACTANT AT Re<<1

When electrostatic barrier has no effect on adsorption, the result presented in Chapter 8 can be used,

$$J_{ad1} = 5.8 a_b Dc_o \sqrt{\frac{Pe}{n}}. \tag{9.36}$$

In cases where electrostatic retardation of adsorption kinetics arises and the controlling step of the adsorption process is the overcoming of the electric double layer, the total adsorption flow is,

$$J_{ad2} = 2\pi a_b^2 D \int_0^{\pi-\theta^*} \frac{\partial c}{\partial r}\bigg|_{r=a_b} \sin\theta \cdot d\theta = 2\pi a_b^2 \int_0^{\pi-\theta^*} div_s\left[\Gamma \frac{v}{n}\sin\theta\right]\sin\theta d\theta = \frac{2\pi va_b}{n}\left[\sin^2 \theta \Gamma\right]_{\pi-\theta^*}. \tag{9.37}$$

Substituting the value of Γ calculated in the previous section into Eq. (9.37), at $\theta^* \ll \pi$ gives

$$J_{ad2} \approx \frac{2\pi va_b\Gamma_o}{n}\left[\frac{2}{\beta}\frac{z}{z^+}\frac{a_b Dc_o}{v\Gamma K(\psi_o)}\right]^{\frac{z^+}{2z}} B, \tag{9.38}$$

with

$$B = \left[\int_0^{\pi}(\sin\theta)^{\frac{4z}{z^+}-1}d\theta\right]^{\frac{z^+}{2z}}. \tag{9.39}$$

Eq. (9.38) should be used in the case when condition (7.29) is fulfilled on the main part of the bubble surface. Then, the adsorption flow of surfactant anions calculated in accordance with (9.38) must be smaller than the flow calculated from Eq. (9.36). Otherwise the inverse inequality must be fulfilled. Using this restriction, we can present the flow of surfactant anions to the surface of the bubble in the form of the well-known equation,

$$J_{ad} = min(J_{ad1}, J_{ad2}) = B'a_b Dc_o \sqrt{Pe}, \tag{9.40}$$

where

$$B' = \min\left[\frac{5.8}{\sqrt{n}}, \frac{\pi}{n}B\frac{z^+}{z}\left[2n\frac{z^2}{(z^+)^3}(z-z^+/2)\right]^{\frac{z^+}{2z}}\left(\frac{c_o^+}{c_o}\right)^{1-\frac{z^+}{2z}}\left(\frac{1}{\kappa\delta_D}\right)^{1-\frac{z^+}{z}}\right]. \tag{9.41}$$

An estimate of the total desorption flow from the surface of a strongly retarded region in the neighbourhood of the rear pole of the bubble is derived as follows. When electrostatic retardation of adsorption-desorption kinetics does not exists, the results of Chapter 8 [Eq. (8.145)] can be applied. For ionic surfactant, the equation for surface tension variation somewhat differs from that for non-ionic surfactant. With regard to these differences, the following estimate of desorption flow results,

$$J_{desl} \approx 1.99\left(1+2\frac{z}{z^+}\right)^{-1}a_bDc_o\left[\frac{a_bv}{D}\right]^{1/3}\frac{\eta v}{RT\Gamma_o}(\theta^*)^{8/3}. \tag{9.42}$$

Equating it to the total adsorption flow, an estimate of the strongly retarded region located at the rear pole of the bubble is obtained,

$$\theta^* \approx 0.840Pe^{1/16}\left[B'\left(1+2\frac{z}{z^+}\right)\frac{RT\Gamma_o}{\eta v}\right]^{3/8}. \tag{9.43}$$

The calculation of the total desorption flow in the case of electrostatic retardation of desorption kinetics follows from Eq. (7.36). The density of the surfactant anions flux within the stagnant cap can be estimated by

$$\dot{j}_{des} \approx -\frac{Dc_o}{K(\psi_{Sto})}\left[\frac{\Gamma(\theta)}{\Gamma_o}\right]^2, \tag{9.44}$$

since within this region the condition

$$\frac{\Gamma(\theta)}{\Gamma_o} \gg 1 \tag{9.45}$$

holds. Using the equation for viscous stresses derived by Harper (1973), the surface concentration distribution of ionic surfactant in the neighbourhood of the rear stagnant point and the total desorption flux, respectively, are

$$\Gamma = \frac{4}{\pi}\left[1 + 2\frac{z_s}{z^+}\right]^{-1}\frac{\eta v}{RT}\theta^\bullet\sqrt{1 - m^2},$$ (9.46)

$$J_{des2} = \frac{8}{\pi}\frac{a_b^2}{K(\psi_{Sto})}Dc_0\theta^{\bullet 4}\left[\left(1 + 2\frac{z}{z^+}\right)^{-1}\frac{\eta v}{RT\Gamma_0}\right]^2.$$ (9.47)

Equating the desorption flux to the total adsorption flux results in the following expressions for $\theta^\bullet$:

$$\theta^\bullet \approx \left[\frac{\pi}{8}B'\frac{K(\psi_{Sto})}{a_b}\sqrt{Pe}\right]^{1/4}\sqrt{\left(1 + 2\frac{z}{z^+}\right)\frac{RT\Gamma_0}{\eta v}}.$$ (9.48)

Numerical estimates show that at $Re \ll 1$ also for ionic surfactants the rising of the bubbles also becomes strongly retarded even at very low surfactant concentrations, of the order of 10^{-8} mol/l, which are difficult to control in experiments.

9.4. CONDITIONS OF REALIZATION OF REGIMES OF IONIC SURFACTANT DYNAMIC ADSORPTION LAYER FORMATION

Now, the conditions for the realization of different regimes of dynamic adsorption layer formation of ionic surfactant in the case $Re \gg 1$ are considered. The first regime corresponds to the situation where the surfactant adsorption at the bubble surface deviates only slightly from its equilibrium value and the bubble surface is strongly retarded. Proceeding in the same way as in Section 8.6 yields

$$|\Gamma - \Gamma_0| \approx \left[1 + \frac{2z}{z^+}\right]^{-1}\frac{\eta v}{RT}\frac{a_b}{\delta_G}.$$ (9.49)

Therefore, the first necessary condition of realization of the regime under consideration has the same form as in the case of a non-ionic surfactant [Eq. (8.103)]. To derive the second condition, the bubble surface velocity $v_0(\theta)$ has to be estimated. In the absence of electrostatic retardation of surfactant anion adsorption kinetics, the estimate derived in Section 8.6. is valid and the condition is identical to (8.105).

At a given electrostatic retardation of adsorption-desorption kinetics of the surfactants, its approximate anion flux to the bubble surface is

$$j \approx \frac{Dc_o}{K(\psi_{Sto})}\left[1-\left(\frac{\Gamma}{\Gamma_o}\right)^{\frac{2z}{z^+}+1}\right] \approx \frac{Dc_o}{K(\psi_{Sto})}\frac{\eta v}{RT\Gamma_o}\frac{a_b}{\delta_G}.$$

(9.50)

Substituting j into the boundary condition (8.10) gives

$$\frac{v_o}{v} \approx \frac{\eta Dc_o}{RT\Gamma_o^2}\frac{a_b}{K(\psi_{Sto})}\frac{a_b}{\delta_G},$$

(9.51)

when neglecting the surface diffusion.

Comparing (9.51) with (8.107) it is easy to verify that under the condition (7.29) the bubble surface velocity decreases. This can be explained in the following way. For a completely retarded surface, $\Delta\Gamma$ is required to be of the same order of magnitude as in the absence of electrostatic retardation. However, a substantially smaller deformation of the adsorption layer at the bubble and surface velocity v_o are required to create such adsorption gradient because the supply of surfactant anions from the bulk solution is retarded.

The second condition of realization of the regime can be obtained from Eqs (8.105), (7.29), and (9.51) in the following form:

$$\frac{\eta Dc_o}{RT\Gamma_o^2}\frac{a_b}{\max\left[K(\psi_{Sto}),\delta_D\right]}\frac{a_b}{\delta_G} \ll 1.$$

(9.52)

An estimate of the surface retardation coefficient has the following form:

$$\chi_b/\eta \approx \frac{RT\Gamma_o^2}{Dc_o}\frac{\max\left[K(\psi_{Sto}),\delta_D\right]}{a_b}\frac{\delta_G}{a_b}.$$

(9.53)

The conditions for the second and third regimes of formation of a dynamic adsorption layer of ionic surfactant will be presented without derivation. They are similar to the equivalent conditions derived in Section 8.6. The second regime, at which the value of surfactant adsorption only slightly deviates from equilibrium Γ_o and the bubble surface is only slightly retarded, is realized under conditions

$$\frac{\max\left[K(\psi_{Sto}),\delta_D\right]}{\delta_D}\frac{\Gamma_o}{c_o\delta_D} \ll 1,$$

(9.54)

$$\frac{\eta Dc_o}{RT\Gamma_o^2} \frac{a_b}{\max\left[K(\psi_{Sto}),\delta_D\right]} \frac{a_b}{\delta_G} \gg 1.$$

(9.55)

The third regime, where the adsorbed surfactant is almost completely moved to the rear stagnation pole and the main part of the bubble (expect for a narrow rear zone) is slightly retarded, is realized given the condition (8.111) and

$$\frac{\max\left[K(\psi_{Sto}),\delta_D\right]}{\delta_D} \frac{\Gamma_o}{c_o\delta_D} \gg 1.$$

(9.56)

The conditions derived in the present section are presented graphically in Figs 9.2.-9.4. The notations used in the figures are as follows:

A- a slight deviation of adsorption from equilibrium and strong surface retardation;

B- slight deviation of adsorption from equilibrium and slight surface retardation;

C- almost complete removal of adsorbed surfactant to the rear pole and slight retardation of the main part of the surface.

The regions of the dynamic adsorption layer regimes for ionic surfactant are separated by curves 1,2, and 3, which are defined by the following equations, respectively:

$$\frac{\eta v}{RT\Gamma_o} \frac{a_b}{\delta_G} = 1,$$

(9.57)

$$\frac{\eta}{\chi_b} = 1,$$

(9.58)

$$\frac{\max\left[K(\psi_{Sto}),\delta_D\right]}{\delta_D} \frac{\Gamma_o}{c_o\delta_D} = 1.$$

(9.59)

Dotted curves in the plots corresponds to the condition

$$\frac{K(\psi_{Sto})}{\delta_D} = 1.$$

(9.60)

Electrostatic retardation of surfactant anion adsorption is realized in the region above the dotted lines.

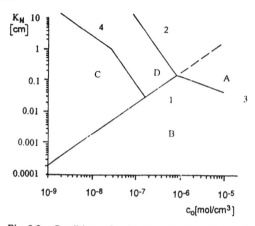

Fig. 9.2. Conditions of realization of ionic surfactant dynamic adsorption layer formation. Bubble radius a_b = 0.05 cm, $c^+ = 0$, $z^+ = 1$, $z = 1$

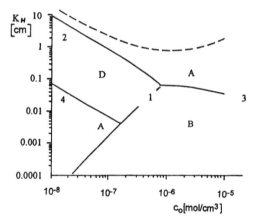

Fig. 9.3. Conditions required for the formation of ionic surfactant dynamic adsorption layer. Bubble radius $a_b = 0.05$ cm, $c = 10^{-3}$ M, $z^+ = 1$, $z = 1$

Some conclusions can be drawn from the results of the calculations, presented in Figs 9.2 to 9.5. It follows from Figs 9.2, 9.3 that electrolyte suppresses the electrostatic barrier of adsorption. The first regime of formation of a dynamic adsorption layer of surfactant is practically never realized for multiple-charged surfactant anions. This situation can be qualitatively explained in the following way. The degree of deformation of an adsorption layer of a rising bubble depends on two factors. The first factor is the convective pulling down of the adsorption layer which generates the above-mentioned deformation. The second factor is the diffusion supply (or take-off) of surfactant from the bulk of solution, which supports the establishement of equilibrium of the adsorption layer. For multiple-charged ions, the second factor is suppressed by the electrostatic barrier. Hence, the region of the first regime shrinks for respective surfactant.

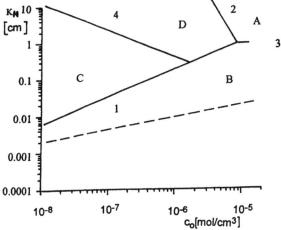

Fig. 9.4. Conditions required for the formation of ionic surfactant dynamic adsorption layer. Bubble radius $a_b = 0.05$ cm, $c^+ = 0$, $z^+ = 1$, $z = 2$.

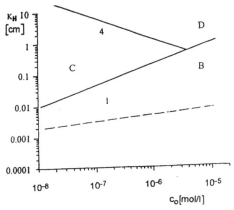

Fig. 9.5. Conditions required for the formation of ionic surfactant dynamic adsorption layer. Bubble radius $a_b = 0.05$ cm, $c^+ = 0$, $z^+ = 1$, $z = 3$.

9.5. THE SIZE OF THE STAGNANT CAP OF THE BUBBLE (DROPLET) USING SURFACTANTS WITH A SLOW RATE DESORPTION

When a spherical fluid droplet translates through a liquid medium containing surfactants, with a desorption rate which is much slower than the convective rate, surfactant collects at the trailing pole in a stagnant cap of angle ϕ which reduces the terminal velocity. The cap angle was obtained by computing the surfactant distribution in the cup region (Sadhal & Johnson 1983).

When the surfactant density becomes large, the finite size of the adsorbed molecules gives rise to strong repulsions between surfactant molecules and generates surface pressures that vary much more strongly than linearly with the surface concentration. The concentration of surfactant in the stagnant cap will depend principally on the amount adsorbed and the degree of compression exerted by the viscous forces. If these forces are large enough, they can compress the surfactant to a density high enough so that gaseous expression does not apply.

Since the use of gaseous constitution equation underestimates the surface pressure, Sadhal and Johnson's result underestimates the cap angle and consequently the drag coefficient. He et al. [1991] obtained a more realistic value for the cap angle by allowing for nonlinear interactions. In their study, Frumkin's equation of state (Frumkin & Levich, 1947, Chapter 2)

$$\gamma_0 - \gamma = -RT\Gamma_\infty \ln(1 - \Gamma/\Gamma_\infty).$$ (9.61)

was used to formulate an expression for a nonlinear surface pressure. We shall also adopt this approach.

From the Frumkin equation (9.61) and the Marangoni stress balance, a differential equation for the surfactant distribution may be obtained

$$\frac{Ma}{(1-\Gamma)}\frac{\partial \Gamma}{\partial \theta} = h(\theta, \Psi)/\lambda.$$ (9.62)

where Ma identifies the Marangoni number, and is defined as $Ma = RT\Gamma_\infty/(\eta v)$; h is the tangential component of the gradient of liquid velocity distribution [Sadhal & Johnson, 1983]; λ is the drag coefficient. The surfactant distribution $\Gamma(\theta)$ is obtained by integrating equation (9.62) from an angular position θ in the cap to Ψ. From the second integration the total (non-dimensional) amount on the surface, $M(\Psi)$ is obtained. Thus,

$$\Gamma(\theta) = 1 - \exp\left[\frac{1}{Ma\lambda}\int_\theta^\Psi h(\theta', \Psi)d\theta'\right]$$ (9.63)

$$M(\Psi)/2\pi = \int_0^\Psi \sin\theta\left\{1 - \exp\left[\frac{1}{Ma\lambda}\int_\theta^\Psi h(\theta', \Psi)d\theta'\right]\right\}d\theta.$$ (9.64)

In obtaining (9.63), the condition $\Gamma(\Psi) = 0$ has been used.

The second relation for the total amount adsorbed is obtained by multiplying the Eq. (9.5) by θ and integrating from 0 to π. The net diffusive flux is equal to zero,

$$\int_0^\pi \sin\theta\frac{\partial c}{\partial r}\bigg|_{r=0} d\theta = 0,$$ (9.65)

while the generalised equation (9.5) results in an equation for $M(\Psi)$,

$$M(\Psi)/2\pi = \int_0^\pi \sin\theta\left[kc_s(\theta)\left(1 - \Gamma(\theta)/\Gamma_\infty\right)\right]d\theta. \tag{9.66}$$

The Eq. (9.5) is generalized by inserting a multiplier $\left(1 - \Gamma(\theta)/\Gamma_\infty\right)$ in the expression for the adsorption flux (9.2).

As was shown in Section 9.1, the retarded desorption corresponds to the weak flux in the diffusion layer and the small difference of the sublayer and bulk concentrations. Substituting $C_s(\theta) = 1$ into Eq. (9.66) leads to the very simple result

$$M(\Psi)/2\pi = \frac{2k}{(1+k)}. \tag{9.67}$$

Combining equations (9.64) and (9.67) results in the following implicit equation for the cap angle Ψ:

$$\frac{1}{(1+k)} = \frac{1}{2}\int_0^\Psi \sin\theta\left\{1 - \exp\left[\frac{1}{\text{Ma}\lambda}\int_0^\Psi h(\theta',\Psi)d\theta'\right]\right\}d\theta \tag{9.68}$$

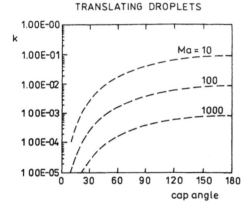

Fig. 9.6. Cap angle Ψ as a function of non-dimensional bulk concentration k, calculated for a bubble by Eq. (9.68) for large Marangoni numbers Ma: 10; 100; 1000; according to He et al. (1991)

As the inner integral in equation (9.68) can be evaluated analytically, solutions for Ψ as a function of k and Ma may be obtained by fixing Ma and Ψ and numerically evaluating the outer integral in order to solve for k.

Consider first the case in which Ma is very large (Ma>10), and therefore the characteristic linear surface pressure forces are much larger than the compressive viscous shear forces. As a result of this disparity, the adsorbed monolayer cannot be compressed very much by the viscous forces, and the cap angle increases dramatically with small increase in the bulk concentration. This trend is observed as expected in Fig. 9.6, which represents the numerical solution of equation (9.68) for k(Ψ) with Ma equal to 10, 100, and 1000.

When the Marangoni number is small, the viscous compression forces far exceed the linear surface pressure, and the surfactant is compressed considerably so that nonlinear repulsion provides the added surface pressure necessary to balance the viscous action. As a result of the significant compression, relatively high bulk concentrations of surfactant (compared to the large Marangoni regime) are required to achieve large cap angles. These results are shown in Fig. 9.7. again for k=0.

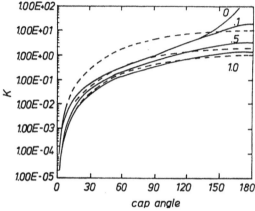

Fig. 9.7. Cap angle Ψ as function of nondimensional bulk concentration k, calculated for a bubble by Eq. (9.68) for small Marangoni numbers Ma: 0; 0.1; 0.5; 1; according to He et al. (1991)

For small Ma, we do not expect the linear gaseous equation to give accurate results because of the strong viscous compression. For this case, k(Ψ,Ma) is given by Sadhal and Johnson (1983) in the form which coincides with the leading term in the expansion of the nonlinear result for large Ma. This equation is plotted as the dashed line in Fig. 9.7 and it shows that the use of the linear equation of state allows the monolayer to be more compressible.

The asymptotic behavior of the cap angle, Ψ(Ma,k) as Ma→0, if a gaseous equation of state is used, is Ψ→0 and all the surfactant is compressed into the rear pole. However, when the nonlinear Frumkin equation is used, when Ma→0 and the viscous compressive stresses become infinite, surfactant is compressed to the extent that Γ→1 and the surface pressure forces become singular in order to balance the infinite viscous forces.

9.6. SUMMARY

As the dynamic adsorption layer deviates farther from equilibrium, the drop in the surface concentration increases along the surface and leads to the formation of a stagnant cup. The deviation from equilibrium is caused by compressive viscous shear forces and surface convection of adsorbed molecules: it grows as the bubble dimensions and velocity increase and it diminishes as the exchange of adsorbed molecules between the solution bulk and surface increases. A retardation of the adsorption rate contributes to the deviation of the adsorption layer from the equilibrium state. A set of equations is derived, taking into account the influence of convective diffusion and transfer mechanisms on the structure and steady state of the adsorption layers when movement to the surface is slightly inhibited, i.e. at low surfactant concentrations. One of the two mechanisms predominates. The model was derived and solved using the angle dependence of the surface concentration at high Reynolds numbers and any surface activity of the surfactant. The lower the adsorption rate and the greater the bubble diameter the greater is the surface area of the bubble free of adsorbed molecules. These molecules are compressed at the rear stagnation point.

With increasing surfactant concentration the extension of the stagnant cup grows and an accurate description becomes important. An efficient analytical theory derived for small Reynolds numbers was elaborated by He et al. (1991). It is impossible to apply this theory directly to experimental data obtained from rising bubbles because at small Re, even in the initial stages the bubble surface is strongly retarded by impurities (even using twice-distilled water). This theory, with high Re, can be used to describe the dynamic adsorption layer of a drop in a viscous liquid. The absence of an analytical theory, for the hydrodynamic field around a bubble with a stagnation cup, hampers this generalisation. For high Re values only Dukhin's (1965) theory is available, based as it is on a weak retardation of the bubble surface and the existence of a small cup.

Electrostatic retardation of the transport of ionic surfactants through the diffuse part of the DL also contributes to the deviation of the adsorbed layer from equilibrium, when the valency of the ion is sufficiently high and the background electrolyte concentration low. Under these conditions, the transport is controlled by electrostatic retardation and convective diffusion and adsorption-desorption mechanisms can be neglected. This leads to a quantitative theory of the DAL at large Re and weak surface retardation. At strong surface retardation, a variation in surface concentration along the bubble surface is perhaps controlled by parameters such as surfactant bulk concentration and adsorption energy.

The condition for the stagnant cup formation, the surface concentration variation along the bubble surface, and the adsorption saturation near the rear stagnation point is determined in terms of Marangoni number. If the compressive viscous shear force exceeds the characteristic linear surface pressure force, i.e. at small Marangoni numbers, the adsorption layer is compressed considerably and no saturation near the rear stagnation point can result.

9.7. REFERENCES

Dukhin, S.S., Thesis, Institute of Physical Chemistry, Academy of Sciences of USSR, Moscow, (1965)

Harper, J.F., J. Fluid Mech., 58(1973)539

Levich, V.G., Physicochemical Hydrodynamics, Prentice-Hall, Englewood Cliffs, N.Y., 1962

He, Z., Maldareli, C., and Dagan, Z., 146(1991)442

Sadhal, S.S. and Jonson, R.E., J. Fluid Mech., 126(1983)126

CHAPTER 10

10. DYNAMIC ADSORPTION LAYER IN MICROFLOTATION

Only recently the flotation theory has been reduced to the study of attachment of particles to the surface of a bubble. This approach made it possible to analyse the problem of selectivity of large-size particles ($\approx 20-40\mu m$ or more), which is of major importance for ore dressing.

The flotation of small particles represents an independent scientific problem inasmuch as the transition from coarse to fine grinding may be accompanied by qualitative changes in the mechanism of an elementary flotation act considered as the interaction of a particle with a bubble.

The traditional treatment of flotation in which the emphasis is on the formation of the bubble-particle aggregate and on the physical chemistry of flotation reagents is insufficient for solution of a number of technological problems of flotation, in particular of flotation technology of small particles (less than $20-40\mu m$ in size). As applied to water purification, the behaviour of such small particles is of interest in the elementary flotation act. Since flotation of small particles by small bubbles is a qualitatively new process, it is quite natural to use a special term: microflotation (Clarke & Wilson 1983).

Unlike the original flotation whose elementary act is complicated by an inertia impact and the accompanying deformation of bubble surface, microflotation is completely a colloid chemical process and it can be described in terms of modern colloid chemistry as orthokinetic heterocoagulation (Derjaguin & Dukhin, 1960).

In general, in the elementary flotation act, one may distinguish a stage at which the particle approaches a bubble and a stage involving attachment of the particle onto the bubble surface. A more detailed examination shows that the number of stages is greater and that the division of the process into stages is conventional.

Passing from large to small particles, the mechanism of the elementary flotation act changes qualitatively, both in the stages of approach of attachment, as suggested by Derjaguin and Dukhin (1959 and 1960) almost three decades ago.

An analysis of DAL effect on microflotation is complicated as it has an effects both on the transport stage and the stage of attachment. It is advisable to consider first the regime at which attachment of particles is provided. Then the DAL effect manifests itself at the transport stage. It is natural to begin considerations with the simpler case. It is necessary to consider first the mechanism of particle transport to the surface of a bubble under microflotation conditions (Section 10.1). Sections 10.2 - 10.4 are devoted to questions connected with the effect of DAL on the transport stage of microflotation. Once the mechanisms of particle attachment to a bubble surface have been considered in Section 10.5, the joint effect of DAL on all stages of the elementary microflotation act is considered in Sections 10.6 and 10.16.

10.1   MECHANISM OF TRANSFER OF SMALL PARTICLES TO BUBBLE SURFACE

10.1.1.   SPECIFIC FEATURES OF THE MECHANISM OF TRANSFER OF SMALL PARTICLES TO THE BUBBLE SURFACE

The process of approach of a particle to a bubble surface undergoes qualitative changes when the distance between them diminishes compared with the particle size. At large distances this process is determined by two factors: the forces of inertia and the long-range hydrodynamic interaction.

A sufficiently large particle moves linearly under the effect of the forces of inertia until it collides with the bubble surface, which takes place if the target distance $b < a_b + a_p$ (Fig. 10.1), where a_b is the radius of the bubble.

The liquid flow envelops the bubble surface, and the particles are entrained to a greater or a lesser extent by the liquid. The smaller the particles and the less different their density relative to that of the medium, the weaker are the inertia forces acting upon them and the more closely the particle trajectory coincides with the liquid streamlines. Thus, at the same target distance fairly large particles move almost linearly (Fig. 10.1, line 1), while fairly small particles move essentially along the corresponding liquid flow line (line 2). The trajectories of particles of intermediate size are distributed within lines 1 and 2; as the size of particles decreases, the trajectories shift from line 1 to line 2 and the probability of collision decreases.

The deviation of the trajectory of a small particle from the straight path to the bubble surface at a distance of the order of the bubble size is caused by long-range hydrodynamic interactions. The bubble causes a curving of liquid streamlines and thereby bends the trajectory of small particles, i.e. acts on them hydrodynamically due to the liquid velocity field. In the case of large particles the forces of inertia considerably exceed the long-range hydrodynamic interaction (LRHI) which can neglected. In the opposite case of small particles the forces of inertia are small compared with the LRHI (Derjaguin and Dukhin, 1959).

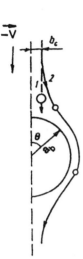

Fig. 10.1.   The influence of the inertia of particles on their trajectory in the vicinity of the floating bubble. Trajectories of large (inertia, line 1) and small (inertia-free, line 2) particles at the same target distance b

Thus, the process of approach of large particles to the bubble is controlled by forces of inertia, while in the case of small particles this process occurs in an inertia-free manner and is strongly hindered by the LRHI. In addition, the hydrodynamic interaction at distances comparable to the particle radius has to be taken into account; this lets the particle's trajectory deviate from the liquid flow line and is called the short-range hydrodynamic interaction (SRHI). Using Taylor's solution of the hydrodynamic problem involving the squeezing out of liquid from the gap between the approaching spherical particles and the flat surface, Derjaguin and Dukhin (1960) assumed that the SRHI may prevent particles from coming into contact with the bubble.

According to Taylor (1924) at a gap thickness h, much smaller than a_b, the hydrodynamic resistance of the film against the thinning process is given by

$$F_\eta \approx v\eta a_b^2 / h, \tag{10.1}$$

where η is the liquid viscosity and v is the velocity with which the particles approach a certain surface area of the bubble. The bubble surface can be considered to be flat because the bubble radius is much greater than the particle radius. If a constant pressing force F_σ is applied to the particle, we obtain according to Eq. (10.1)

$$v \approx F_\sigma h / a_b^2 \eta. \tag{10.2}$$

It may be inferred that a complete removal of the liquid from the gap requires an infinitely long time:

$$t \approx -\int_h^0 \frac{\eta a_b^2 dx}{F_\sigma x} \approx -\frac{a_b^2 \eta}{F_\sigma} \ln x \Big|_h^0.$$ (10.3)

The questions arise about the nature of forces pressing the particle against the bubble surface about the effect of surface attraction forces which increase with decreasing distance more rapidly than the film's resistance, and the thinning of the film at a thickness greater than the effective range of surface forces. In the area above the equatorial plane the liquid flow lines approach the bubble surface, which means that the radial component of the liquid velocity is directed towards the bubble surface. Since the motion of the particle towards the surface is obstructed within the zone of the SRHI, the radial velocity of the liquid is higher than that of the particle. Thus, at a small gap thickness and high viscous resistance, the radial velocity of the liquid is even much greater. The radial flow of liquid envelops the particle whose approach to the bubble has been retarded and presses it against the latter. As a first approximation this hydrodynamic force can be estimated from the Stokes formula by substituting the particle radius and the difference in the local values between the velocity of the liquid and the particle.

The important distinction is that in the case of large particles, thinning of a liquid interlayer is accomplished through an impact, while in the case of small particles it is due to the effect of hydrodynamic pressing forces.

With very large particles the liquid interlayer thinning process is complicated by the deformation of the bubble surface by an inertia impact of the particle. It was shown by Derjaguin et al. (1977) and Dukhin & Rulyov (1977) that in the inertia-free deposition of small particles on a bubble surface its deformation under the influence of the hydrodynamic pressing force is insignificant. This third important feature facilitates the development of a quantitative kinetic theory of flotation of small particles.

10.1.2. QUANTITATIVE THEORY OF FLOTATION OF SMALL SPHERICAL PARTICLES

The process of approach of particles to the bubble surface can be described quantitatively by taking into account both the LRHI and the SRHI. For estimating the flotation efficiency, we introduce a dimensionless parameter of the collision efficiency,

$$E = b_{cr}^2 / a_b^2,$$ (10.4)

where a_b is the bubble radius and b_{cr} is the maximum radius of the cylinder of flow around the bubble encompassing all particles deposited on the bubble surface (Fig. 10.2). The particles moving along the streamline at a target distance $b < b_{cr}$ are deposited on the bubble surface

346

(Fig. 10.2, as indicated by a dashed line). Otherwise the particle is carried off by the flow. From Fig. 10.2 it is evident that the calculation is essentially reduced to the so-called "grazing trajectory" (continuous curve) and, correspondingly, the target distance. A similar approach has long been used in the science of aerosols (Langmuir and Blodgett, 1945).

Figure 10.2 Continuous lines illustrate the concept of the grazing trajectory of particles, dashed lines indicate the trajectories of the particle at $b < b_{cr}$ and $b > b_{cr}$.

The bubble surface moves together with liquid and at Reynolds numbers

$$Re = \frac{2a_b v}{v} \gg 80,$$ (10.5)

the flow of liquid around the rising bubble is a potential flow if the motion of its surface is not retarded by surfactants (Levich 1962). Here v is the buoying velocity of the bubble and v is the kinematic viscosity of the medium. At a potential distribution of liquid flow, according Levich (1962), the velocity is given by

$$v = \frac{g a_b^2 \rho}{9 \eta}.$$ (10.6)

If the quadratic dependence of the bubble velocity on its radius is taken into account, it is easy to see that the Reynolds number changes very rapidly with the radius. Re is equal to unity at $a_b = 90$ μm.

In order to understand the mechanism of inertia deposition of particles on the rising bubble, the particle inertia path l is introduced, which is defined as the distance the particle is able to pass in the presence of viscous resistance of the liquid of an initial velocity v_∞,

$$l = \frac{2}{9} \frac{V_\infty a_p^2 \rho_p}{\eta},$$

where ρ_p is the density of the particle.

Since the bubble surface is impermeable to liquid, the normal component of the liquid velocity on the surface is zero. As the distance from the bubble surface increases, the normal component of the liquid velocity also increases. The thickness of the liquid layer, in which the normal component of the liquid velocity decreases due to the effect of the bubble, is of the order of the bubble radius. The particle crosses this liquid layer due to the inertia path whereby deposition of a particle depends on the dimensionless Stokes parameter,

$$St = l / a_b = \frac{2}{9} \frac{\rho_p V a_p^2}{a_b \eta}. \tag{10.7}$$

When $St > 1$, the inertia deposition is obviously possible, yet calculations have shown that it can also take place at $St < 1$ as long as St is not too small. This conclusion becomes apparent if it is considered that in a layer of thickness a_b the particle moves toward the surface not only due to inertia but also together with the liquid. The motion component of the particle normal to the bubble surface becomes zero at the surface of the bubble. Inertia deposition proves to be impossible if St is smaller than some critical value St_{cr}. In the case of a potential flow regime and negligible particle size, Levin obtained (1961)

$$St_{cr} = \frac{1}{12}. \tag{10.8}$$

Substituting this value into Eqs (10.6) and (10.7) yields an expression for the critical particle radius below which the inertia forces cannot cause the particle to approach the bubble,

$$a_p^{cr} = \frac{9}{\sqrt{48}} \sqrt{\frac{v\eta}{g\rho_p a_b}}, \tag{10.9}$$

Inertia deposition of particles on the bubble surface occurs when $a_p > a_p^{cr}$, but its intensity decreases with decreasing a_p. This finding is in agreement with the formula for the collision efficiency, which had been first derived by Langmuir & Blodgett (1945) for the coagulation of aerosols. Langmuir's expression was later extended by Derjaguin & Dukhin (1960) to the case of an elementary flotation act. Still later, Eq. (10.10) was confirmed by Fonda & Herne (1966) with an accuracy of 10%.

$$E = \frac{St^2}{(St + 0.2)^2}.$$ (10.10)

This relationship was experimentally verified by Samygin et al. (1977) for the range of values St =0.07-3.5, where also the efficiency of particle capture of several particle sizes by a single bubble of a strictly fixed size was measured.

In the process involving inertialess approach of particles to the bubble surface, their size also plays an important role. It is in the equatorial plane that the closest approach of the streamline to the bubble surface is attained. In Fig. 10.3 the broken line (curve 1) represents the liquid streamline whose distance from the bubble surface in the equatorial plane is equal to the particle radius. Some authors erroneously believe that this liquid streamline is limiting for the particles of that radius. The error consists in that the SRHI is disregarded. Under the influence of the SRHI the particle is displaced from liquid streamline 1 so that its trajectory (curve 2) in the equatorial plane is shifted from the surface by a separation greater than its radius. Therefore, no contact with the surface occurs and, correspondingly, $b(a_p)$ is not a critical target distance.

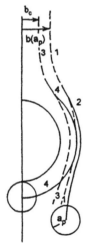

Figure 10.3   The influence of the finite dimension of particles in inertia-free flotation on their trajectory in the vicinity of a floating bubble. The liquid flow lines corresponding to target distances $b(a_p)$ and b_{cr} are indicated by dashed lines. The continuous lines are characteristic of the deviation of the trajectory of particles from the liquid flow lines under the influence of short-range hydrodynamic interaction

Due to the SRHI the distance from the particle to the surface in the equatorial plane is greater than the distance from the surface to the liquid streamline with which the trajectory of the particle coincides at large distances from the bubble. It may thus be concluded that $b_{cr} < b(a_p)$. The grazing liquid streamline (curve 3) is characterised by the fact that the particle trajectory (curve 4) which branches off under the influence of the SRHI, lies in the equatorial plane at a distance a_p from the bubble surface.

The value of b_{cr} decreases, first due to the deflection of the liquid streamlines under the influence of the LRHI and, second, due to the deflection of the particle trajectory from the liquid streamline under the influence of the SRHI. Therefore, the collision efficiency is expressed as the product of two factors, E_0 and f, each of which is smaller than unity. The first represents the influence of the LRHI and the second, the SRHI.

In the case of the potential regime of flow (Sutherland, 1948)

$$E_{0P} = 3a_p / a_b,$$ (10.11)

where the subscript P corresponds to the potential regime of flow around the bubble. With Re < 1, the liquid flow becomes viscous. In this case the motion of the bubble surface is usually completely retarded by the adsorption layer of surfactants so that the distribution of velocities is described by Stokes' formula. Moreover, according to the observation of Okazaki (1964), at Re < 40 the bubble rises like a solid sphere even in the absence of surfactant.

In the Stokes regime an expression similar to Eq. (10.11) is given by

$$E_0 = \frac{3}{2}\left(\frac{a_p}{a_b}\right)^2.$$ (10.12)

Eq. (10.12) was derived by Derjaguin & Dukhin (1959) by analogy to the mechanics of aerosols. Unfortunately, these results were not mentioned later and remained almost unknown. Eq. (10.12), which has been rendered more accurately by taking into account the gravitational component, was derived by Reay & Ratcliff (1973).

The problem of the hydrodynamic pressing force which contributes to thinning of the liquid interlayer in the course of the SRHI was solved by Goren & O'Neil (1971) for the Stokes regime and by Dukhin & Rulyov (1977) for the potential regime. It has been established that the pressing force is two or three times larger than the value calculated by Stokes' formula. The hydrodynamic pressing force is not sufficient to ensure contact between particle and bubble because it is of a finite value while the resistance increases indefinitely as the liquid interlayer grows thinner (cf. Eq. (10.1)).

It has been shown by Derjaguin & Dukhin (1960) that the inertia-free flotation is facilitated by the effect of surface forces. Two cases may be encountered here. It is known from the work of Derjaguin & Zorin (1955) that interphase films may lose their stability and spontaneously disintegrate as the critical thickness h_{cr} is attained in the course of thinning. Substituting h_{cr} instead of zero in the upper limit of the integral in Eq. (10.3), we obtain $\ln(h/h_{cr})$ instead of infinity. Thus, the inertia-free flotation becomes possible in the presence of flotation reagents providing $h_{cr} \neq 0$.

The molecular force of interaction of a spherical particle with a flat surface at fairly small h is expressed by (Mahanta & Ninham 1976),

$$F = A_H a_p / 6h^2,$$

(10.13)

where A_H is the Hamaker constant. Since the attractive force (Derjaguin 1976, Derjaguin et al. 1976) increases more rapidly with decreasing h than the resistance of the viscous interlayer to thinning (Derjaguin and Dukhin, 1960), the contact causes the rupture of the film. The influence of the SRHI on the particle capture efficiency at $h_{cr} \neq 0$ was calculated by Derjaguin (1976), Derjaguin et al. (1976), and Dukhin (1976) with neglecting the long-range effect of molecular forces. The formulas derived by using the Stokes and potential distribution of velocities in the liquid enveloping the bubble have the form

$$E_{cS} = E_{0S} f_S,$$

(10.14)

$$E_{cP} = E_{0P} f_P,$$

(10.15)

where f_p and f_s are the functions which reflect the influence of the SRHI on the elementary act of flotation and depend on the dimensionless parameter $H_{cr} = h_{cr}/a_p$. At all values under consideration these functions are smaller than unity; they decrease with H_{cr} and become zero at $H_{cr} = 0$. This confirms with the aforementioned representation of the mechanism based on the influence of the SRHI on the particle deposition process. Index "c" denotes the capture efficiency. Under the action of thin film rupture or molecular attraction forces a particle-bubble collision leads to particle capture.

As H_{cr} decreases from 10^{-1} to 10^{-3}, f_s decreases from 0.5 to 0.15. Thus, the inclusion of the SRHI is important not only in considering the problem of the possibility of flotation; this effect reduces the number of contacts of particles with the bubble by several times. Rulyov (1978) has developed the SRHI theory without considering the phenomenological parameter, h_{cr}, by directly taking into account the dependence of molecular forces on h and obtained

$$E_{cS} = E_{0S} f_S(W_S), \quad W_S = \frac{A_H a_b^2}{27 v_S \pi \eta a_p^4},$$

(10.16)

$$E_{cP} = E_{0P}f_P(W_P), \ W_p = \frac{4A_Ha_b}{27v_p\pi\eta a_p^3},$$ (10.17)

where v_S and v_p are the bubble buoying velocities in the Stokes and potential regimes, respectively. Numerical analysis of the universal functions f_S and f_p allowed Eqs (10.16) and (10.17) to be approximated in the following way (in cgs units):

$$E_{cS} \approx 0.11\frac{a_p^{1,4}A_H^{1/6}}{a_b^2},$$ (10.18)

$$E_{cP} \approx 1.1\frac{a_p^{0,8}}{a_b}A_H^{1/15}.$$ (10.19)

Although the absolute value of E_c only weakly depends on the Hamaker constant A_H, which varies within the range of 10^{-14} to 10^{-12} erg, collision cannot occur when $A_H < 0$, i.e., if the dispersion forces are directed away from the bubble surface. However the collision can occur due to the non zero value of h_{cr}. Thus, molecular forces act essentially on the yes-no principle, depending on the sign.

A profound effect of the LRHI in inertia-free flotation was convincingly corroborated by Collins & Jameson (1977) and Reay & Ratcliff (1975). The most important manifestations of the LRHI revealed experimentally are rapid decreases in E_{col} as the particle size decreases and the bubble size increases as well as abrupt changes in the character of the dependence of E_{col} on a_b in passing over from the viscous to the potential flow regime. It should be emphasised that Eq. (10.18) is in quantitative agreement with the data of comprehensive studies of Collins & Jameson (1977) and Reay & Ratcliff (1975) concerning the dependence on both the particle and bubble radius. The dependence of E_{col} on a_b as characterised by the exponent 1.5 is in better agreement with Eq. (10.18) than with Eq. (10.12). These results may be evidence for the influence of the SRHI on the elementary flotation act.

10.2. EFFECT OF DYNAMIC ADSORPTION LAYER ON THE TRANSPORT STAGE OF THE ELEMENTARY FLOTATION ACT

10.2.1. INFLUENCE OF SURFACE RETARDATION BY DAL ON TRANSPORT STAGE. QUALITATIVE CONSIDERATIONS

Experimental studies have shown that bubbles behave as solid spheres not only at Re < 1, but also at $1 < Re < 40$, and even in distilled water (Okazaki, 1964). Based on the method described by Hamielec et al. (1962, 1963) and using functions describing the hydrodynamic field under transient conditions, formulas for E_{col} were obtained in the form of a product where the first cofactor E_0 characterises LRHI and the second SHRI. In the interval $5 < Re < 100$ these formulas can be approximated by expressions derived by Rulyov (1978),

$$E_{cP} \approx \frac{a_p^{0.8}}{b_b},$$ (10.20)

$$E_{cS} \approx \frac{a_p^{1.4}}{a_b^2}$$ (10.21)

which are very similar to Eqs. (10.18) and (10.19). Comparing expressions (10.18) and (10.19) with (10.20) and (10.21) we can conclude that the dependence of the efficiency of collisions over the whole interval of Reynolds numbers upon the degree of mobility is determined only by the state of the bubble surface and remains approximately constant.

Thus, collision efficiency at given bubble size can strongly change with the degree of mobility of the surface, i.e. the effect of DAL on collision efficiency is very strong. It is important to characterise this effect not only for limiting states of DAL, as researchers usually restrict their studies. We will try to characterise the effect of DAL on the transport stage of microflotation without restrictions in well-known limiting cases.

At first the effect of sedimentation on collision efficiency is taken into account since it can strongly decrease the role of the hydrodynamic field of bubble and DAL. This consideration proves to be more obvious when the method of collision efficiency calculation is used as proposed by Dukhin & Derjaguin (1958).

10.2.2. EQUATION FOR PARTICLE FLUX ON BUBBLE SURFACE

In case of non-inertia flotation and without regard to SHRI, the velocity of particles is well-known at any distance from the bubble:

$$\vec{v} = \vec{v}(z,\theta) + \vec{v}_g,$$ (10.22)

so that it is advisable to derive an expression for the radial component of the particle flux density,

$$J_z = n(z,\theta)\, v_z(z,\theta),$$ (10.23)

where $n(z,\theta)$ is the particle number concentration next to a bubble surface. Integrating the flux density over that part of the surface on which sedimentation takes place we obtain the number of particles colliding with the bubble in unit time N which relates to E as

$$N = 4\pi a_b^2 E v, \qquad (10.24)$$

Earlier Dukhin & Derjaguin (1958) have proved a theorem according to which the particle concentration remains constant if the velocity field is solenoidal, i.e. the condition div $v = 0$ is met. From this it follows that

$$E = \frac{1}{\pi a_b^2 v n_0} \int_0^{\theta_{\varepsilon}} n_0 (v_z + v_g) 2\pi a_b^2 \sin\theta d\theta = \int_0^{\theta_{\varepsilon}} \left(\frac{v_z}{v} + \frac{v_{qz}}{v} \right) \sin\theta d\theta, \qquad (10.25)$$

where θ_c characterises the boundary of the region of particle deposition at the bubble surface and n_0 is the number of particles beyond the bubble. To consider the effect of a finite particle size, integration over a concentric sphere of radius $a_b + a_p$ has to be performed. From this condition θ_c can be derived according to

$$V_z (a_b + a_p, \theta_c) = 0. \qquad (10.26)$$

Substituting v_z for a potential velocity field, Sutherland's formula is obtained. When we substitute Stokes' velocity field into Eq. (10.25) and use Eq. (10.22), we obtain

$$E_s = \frac{3}{2} \frac{a_p^2}{a_b^2} + \frac{\Delta\rho}{\rho} \frac{a_p^2}{a_b^2}, \qquad (10.27)$$

where $\Delta\rho = \rho_p - \rho$. It is important that Eq. (10.25) is applicable not only under Stokes and potential fields but also at any solenoidal hydrodynamic fields which ensures its wide application.

This method of calculation was proposed almost a quarter of the century later by Weber (1981) and Weber et al. (1983a, 1983b) who were probably not aware of the work of Dukhin & Derjaguin (1958). The usefulness of the method was well demonstrated, although the discussion was restricted by consideration of systems in which the particle density differs only slightly from water density (bacterium, latices, emulsions, coal) and allows to ignore the role of sedimentation. The more general approach was demonstrated in the recently appeared work of Nguen Van & Kmet, (1992) who again erroneously cite Weber as the author of the method.

10.2.3. APPLICATION OF EQUATION OF PARTICLE FLUX TO ESTIMATION OF ROLE OF SURFACE RETARDATION AND SEDIMENTATION

To analyse the effect of surface retardation on the radial velocity of liquid at the bubble surface, we will write the expression for velocity divergence

$$\frac{1}{z\sin\theta}\frac{\partial}{\partial\theta}(v_\theta\sin\theta)+\frac{\partial v_z}{\partial z}=0 \tag{10.28}$$

Integrating from a_b to z under the assumption that at $z-a_b \ll a_b$ the tangential velocity changes insignificantly, the radial velocity decreases inversely with the retardation coefficient,

$$v_z(z,\theta)\Big|_{z=a_b+a_p} \cong \bar{v}\frac{\eta}{\chi_b}\frac{(z-a_b)}{a_b}\Big|_{z=a_b+a_p}\cos\theta = \bar{v}\frac{\eta}{\chi_b}\frac{a_p}{a_b}\cos\theta. \tag{10.29}$$

At moderate values of the retardation coefficient, deposition of particles happens under the effect of the radial component of liquid velocity. At strong retardation, deposition of particles happens due to sedimentation if the density of the particle differs remarkably from the density of the medium.

This important statement can be approximated by the ratio of sedimentation velocity of a particle to the velocity of its inertialess movement together with liquid according to Eq. (10.29) that yields dimensionless values

$$\Lambda = \frac{\Delta\rho}{\rho}\frac{a_p}{a_b}\frac{\chi_b}{\eta}. \tag{10.30}$$

Large Λ-values correspond to the predomination of sedimentation; at small Λ-values the influence of a residual mobility on collision can predominate, even at the strongly retarded surface of a bubble. If the normal component of liquid velocity is proportional to $\cos\theta$ the angular dependence diminishes in the derivation of Eq. (10.30) because the normal component of the sedimentation velocity is proportional to $\cos\theta$ too.

The ratio v_p / v is proportional to $\Delta\rho / \rho$, which changes in the range $0 - 0.2$ for emulsions up to $2 - 7$ for minerals. Taking into account usually strongly retarded bubble surfaces, we can conclude that sometimes sedimentation can predominate. For emulsions the sedimentation influence is restricted and the surface retardation can play a significant role.

Eq. (10.29) corresponds only to one of the components of the two-term formula describing the normal component of the velocity, according to

$$v_z(\theta, a_p) = 0 \text{ at } \chi_b \to \infty,$$
(10.31)

whereas in reality this does not take place. In deriving Eq. (10.29) the change of the tangential velocity as a function of distance to the surface was ignored. If we do take it into account, then a second component of velocity appears which corresponds to the limiting case of a completely retarded surface, i.e. the solid particle. Thus, when χ_b increases, the radial velocity component can decrease approximately a_b / a_p times. In other words, an increase of χ_b up to values $\chi_b / \eta \approx a_b / a_p$ results in a decrease of collision efficiency; its further increase does not change E significantly.

At larger Reynolds numbers, hydrodynamic and diffusion fields are characterised by the appearance of boundary layers which enable one to estimate a local value of the retardation coefficient, as given by Eq. (8.108). A joint application of Eq. (10.31) and (8.109) allows to describe the effect of DAL on the radial component of particle velocity at the instant it approaches the bubble surface. Since in the non-retarded state the collision efficiency E_0 is described by Sutherland's formula (10.11), the effect of DAL can be approximated by the expression,

$$E \approx E_0 \frac{\eta}{\chi_b}.$$
(10.32)

The accuracy and applicability of this formula is restricted by many factors. First, the real angular dependence of the tangential and normal velocity component is very complex. Second, calculation of the collision efficiency by Eq. (10.25) presupposes the knowledge of the angle θ at which the particle motion changes direction. Third, formula for χ_b (8.108) was obtained for the case of strong surface retardation. Thus, the transition to the limit given by Sutherland's formula at $\chi_b \to \eta$ cannot be expected. More exactly, this transition formally exists but Eq. (8.108) becomes less and less accurate with decreasing χ_b. Finally, the approximate Eq. (10.29) is sensible only if the surface as a whole is characterised by one retardation coefficient. This means that a condition similar to Eq. (8.71) must be fulfilled.

As in the case of moderate Reynolds numbers, there must be a two-term formula instead of Eq. (10.32), which represents only one term of the corrected formula. The second term represents the collision efficiency at $\chi_b \to \infty$. This problem can be solved in the same way as the second term in Eq. (10.27) was obtained.

A theory of the hydrodynamic boundary layer of a solid spherical particle exists (Hamielec et al., 1963) and agrees with experimental data by Carner & Crafton (1954). The distribution of the radial velocity component is known and θ_1 can be calculated. Then it is sufficient to calculate the integral in Eq. (10.25).

Eqs. (10.25) and (10.32) in the general case must be supplemented with components taking into account the contribution of particle sedimentation on a bubble surface.

Although Eq. (10.32) is an approximation of low accuracy, it is of definite value for predicting very strong effects of bubble radius, the degree of surface activity and surfactant concentration. Depending on these parameters E can vary by two orders of magnitude which is more essential than errors even of the order of 100%, which may have arisen in deriving Eq. (10.32).

10.2.4.   *DAL INFLUENCE ON TRANSPORT STAGE OF MICROFLOTATION AND LEVEL OF WATER CONTAMINATION*

There are two regimes of DAL influence on the transport stage of microflotation. At sufficiently high level of water contamination the transport of a particle to the retarded surface of a bubble is the same as that of a solid sphere. Taking into account the results of Section 10.2.3 the condition of this regime is,

$$\frac{\chi_b}{\eta} \geq \frac{a_b}{a_p}. \qquad (10.33)$$

There is no specificity of the transport stage of microflotation at the condition (10.33) in comparison with a solid sphere. However let us emphasise that the surface retardation is caused by the dynamic state of the adsorption layer, i.e. by the change of the surface concentration and surface tension along the bubble surface.

However, at very high surfactant concentration the adsorption layer becomes saturated making a bubble surface immobile, even in the absence of a surface tension gradient.

At smaller surfactant concentrations and correspondingly smaller values of the surface retardation coefficient

$$\frac{\chi_b}{\eta} < \frac{a_b}{a_p} \qquad (10.34)$$

the residual mobility of a bubble surface leads to the increase of particles deposition on a bubble in comparison with a solid sphere. Taking into account that a_b / a_p is very large, for example, of order of 10 or even of 100, we can conclude that the residual mobility of a bubble surface can intensify the particle transport also at strong surface retardation. In other words, the strong retardation of a bubble surface does not always lead to a particle capture by a bubble similar to the capture by a solid sphere.

We emphasise this statement because in many papers the identity of buoyant bubbles and solid spheres is considered as the ground for neglecting the specificity of the particle capture by a bubble. To discriminate the "solid body" regime and regime of the residual surface mobility let us introduce the condition

$$\frac{\chi_b}{\eta} = \frac{a_b}{a_p}, \tag{10.35}$$

which determines the boundary between the two regimes given by the conditions (10.33) and (10.34).

Due to rapid decrease of the retardation coefficient with increasing bubble dimension, a residual surface mobility remains for big bubbles. The transport stage for sufficiently small bubbles and solid spheres is identical. The characteristic bubble value which describes the boundary between the two ranges of bubble dimensions can be determined from Eq. (10.35) after substitution of χ_b as a function of bubble radius, according to Eq. (8.106),

$$\frac{2}{3} \frac{RT}{\eta} K_H^2 \frac{c_o}{D} \frac{\delta_D}{a_b} \frac{\delta}{a_b} = \frac{a_b}{a_p}. \tag{10.36}$$

The derivation corresponds to the condition of uniform retardation, i.e. Eq. (8.71). Thus, Eq. (10.36) is reasonable to compare with Eq. (8.106) and with the results of the analysis of Eq. (8.106), represented in Fig. 8.3. As it is seen from Fig. 8.3 the strong retardation under condition (8.71) is possible at very high surfactant concentration ($10^{-3} - 10^{-2}$ M) only. Comparing the r.h. sides of Eqs. (10.36) and (8.106) we can conclude that neglecting the influence of the residual surface mobility on microflotation is possible at very high surfactant concentration only, i.e. $10^{-3} - 10^{-2}$ M multiplied by large values of a_b / a_p. However this conclusion corresponds to the big bubbles because the data of Fig. 8.3 were calculated for $a_b = 0.05$ cm. To connect the boundary value of the bubble dimension with the surfactant concentration let us substitute $K_H = \delta_D$ into Eq. (10.36) and express δ_D by Eq. (8.152),

$$\frac{2}{3} \frac{RT}{\eta v} c_o = \frac{Re^{1.7}}{a_p a_b}. \tag{10.37}$$

Surfactants contamination in water usually does not exceed 10^{-4} M. According to Eq. (10.37), the residual mobility influences the microflotation at high Reynolds numbers.

These conclusions preserve at not too large K_H values, for example $K_H < 10^{-3}$ cm. Thus at not very high surface activity the residual surface mobility influences the microflotation at high

Reynolds number. Naturally these general conclusions can be quantified at different concentration according Eq. to (10.37) and different K_H-values according to Eq. (10.36).

At high Reynolds numbers the residual surface mobility can influence the flotation even at high surface activity ($K_H > 10^{-3}$ cm, see Section 10.2.5). However its influence can be neglected with respect to the intermediate range of Reynolds number because χ_b increases proportionally with K_H^2, according to Eq. (10.36).

When the rear stagnant cap and the angle Ψ increase, the collision efficiency decreases. Thus, at uniform surface retardation, i.e. under condition (8.71) and during the rear stagnant cap formation, the mechanisms of DAL effect on the transport stage differ qualitatively. In the former case, with increasing surfactant concentration, the normal component of velocity and, respectively, the flow of particles uniformly decrease over the leading surface. In the latter case, the area admissible for sedimentation of particles decreases.

Using the formulas for the hydrodynamic field of a bubble carrying a rear stagnant cap (cf. Section 8.7), we can calculate the effect of the cap on collision efficiency. It is unlikely that such work is of interest since the theory described in Section 8.7 is restricted to small Reynolds numbers. Thus, we cannot expect agreement between this theory and reality since the leading surface must be either completely or strongly retarded, according to experimental data by Okazaki (1964).

The consideration made is of semi-quantitative nature, and is justified both at moderate and large Reynolds numbers. At moderate Reynolds numbers, a refinement of the theory seems to be premature since the notion of the incomplete retardation of the surface is a hypothesis which needs an experimental check. At large Reynolds numbers, a quantitative consideration of the effect of DAL on the elementary flotation act will prove to be possible generally only after the quantitative theory of DAL has been developed. The given evaluations confirm that the effect of DAL on the transport stage of microflotation is high at large Reynolds numbers and, possibly, also at moderate values.

In conclusion, the development of comprehensive experimental and theoretical studies of DAL and rheology of bubble surfaces is needed. Having shown the importance to calculate the collision efficiency for very strong and very weak retardation of the bubble surface, the necessity of a detailed development of the theory of DAL of a bubble at moderate and high Reynolds numbers becomes quite obvious.

10.2.5.   *FREE SURFACE ZONE NEAR FRONT STAGNANT POINT*

At high surface activity due to the residual surface mobility, the zone of low surface concentration can appear near the front stagnation point. In other words the angle ψ characterising the dimension of the stagnant cup can be slightly less than π and

$$\pi - \psi \ll \pi. \tag{10.38}$$

For the sake of simplicity we introduce the special term front stagnation free zone (f.s.f.z.).

When the bubble is sufficiently big and the Marangoni number (see Section 8.3) is sufficiently small, the viscous compression forces far exceed the linear surface pressure, which decreases the cup. The decrease of the cup angle at bubble velocity increase can be described by Eq. (8.41) at small Reynolds numbers. It turns out that a f.s.f.z. appears at small Reynolds number in sufficiently clean water only.

In contaminated water an f.s.f.z. can appear at sufficient large Reynolds numbers only. Due to the absence of a quantitative theory of the stagnant cup at high Reynolds numbers one can only estimate the condition of the existence of an f.s.f.z..

The Marangoni stress balance, given by Eq.(8.38), can be used as the basis for this estimation. Eq. (8.38) derived for small Re, must be generalised taking into account that the vorticity is proportional to $\sqrt{Re}$ (Clint et al., 1978),

$$\frac{1}{a_b}\frac{\partial \gamma}{\partial \theta} \approx \eta \frac{v}{a_b} \sqrt{Re}. \tag{10.39}$$

Integrating both sides of this equation between $\theta = 0$ and $\theta = \theta^*$, the l.h.s. can be estimated as $\Delta\gamma.$, the r.h.s. as $\eta v \, Re^{1/2} a_b$. The existence of an f.s.f.z. means that $\gamma(0) = \gamma_w$, where γ_w is the surface tension of pure water.

The angle θ^* corresponds to that part of the adsorption layer which is in equilibrium with the bulk solution. It means that $\gamma_w - \gamma(\theta^*)$ equals the decrease of the equilibrium tension in the contaminated water. The value changes in the range 1-10 mN/m in distilled, tap or river water. The approximate result of the integration on the r.h.s. of Eq. (10.39) can be represented by introducing an multiplier a_b

$$\gamma_w - \gamma(\theta^*) \approx \eta v \sqrt{Re}. \tag{10.40}$$

The r.h.s. of Eq. (10.40) depends very strongly on bubble size. It can exceed the l.h.s. of the equation at high Reynolds number such that the viscous stress compresses the adsorption layer to such a degree that the f.s.f.z. appears. On the other hand, the l.h.s. can predominate at sufficiently high surface tension changes and not very big bubbles, for example at Re = 100, and a f.s.f.z is absent. In the intermediate range of Reynolds number the appearance of an f.s.f.z. in a contaminated water is probably impossible.

These conclusions can change if micellisation in surfactant solutions is taken into account. If the surfactant concentration exceeds the critical concentration of micellisation (CMC) the surface tension does not change with surfactant concentration. It means that the Marangoni-Gibbs effect and the surface retardation of a bubble also disappear.

Both CMC and surface activity increase with higher homologs. As it was shown in Section 10.2.4. the residual mobility of a bubble influences the microflotation at high Reynolds number and at $K_H < 10^{-3}$ cm. At $K_H > 10^{-3}$ cm a residual mobility in a contaminated water can be provided by micellisation. Indeed this high surface activity is possible for sufficiently high homologs, corresponding to rather low values of the CMC, for example 10^{-4} - 10^{-3} M. Thus the mobility of a bubble surface can preserve at $K_H > 10^{-3}$ - 10^{-2} cm and $c_o < 10^{-4}$ - 10^{-3} M. However, this conclusion relates to sufficiently big bubbles and the formation of an f.s.f.z. only.

The conclusion cannot be generalised with respect to other surface active species. For example, proteins can remarkably decrease the surface tension at very small concentrations (Izmailova et al., 1988). However, these data relate to equilibrium conditions. Thus, it is not excluded that under dynamic conditions, i.e. during the adsorption on a mobile bubble surface, the increase of surface concentration is retarded by slow adsorption kinetics.

10.2.6.   PARTICLE DEPOSITION ON THE FREE ZONE NEAR THE FRONT STAGNANT POINT. CONTAMINATED WATER

Particles can be deposited on the f.s.f.z. and on the stagnant cup. Let us estimate that part of collision efficiency connected with the deposition on the f.s.f.z. It means that the coordinates $(\pi - \Psi, a_b + a_p)$ must be substituted into the equation of the stream function for the potential flow. Note that the substitution of the coordinates $(\pi / 2, a_b + a_p)$ yields Sutherland's equation (10.11). The substitution of the coordinates $(\pi - \Psi, a_b + a_p)$ leads to the equation,

$$E_p(\pi - \psi) = E_0 \sin^2(\pi - \psi) \cong 3\frac{a_p}{a_b}(\pi - \psi)^2. \tag{10.41}$$

The exactness of Eq. (10.41) is restricted because the stagnant cup can deform the potential flow. However, it is well known that in the theory of the hydrodynamic boundary layer of a solid sphere a potential flow can be used for the description of the hydrodynamic field outside the boundary layer. This approach is verified only near the front stagnant point. Clift et al. (1978) stressed that the pressure distribution follows a potential flow up to about 30° from the front stagnation point. In the present case of a free mobile surface near the front stagnation point the potential distribution can be used at any small distance to the surface and verifies a sufficient exactness of Eq. (10.41).

At large Re the particles sediment only on a section of the leading part of the bubble surface because the normal component of the liquid velocity changes its sign at $\theta_i < \pi/2$. It means that the hydrodynamic flow can prevent the sedimentation near the equator. As a result we can use the second term of Eq. (10.27) to estimate the collision efficiency

$$E \approx 3\frac{a_p}{a_b}(\pi - \psi)^2 + \frac{\Delta\rho}{\rho}\frac{a_p^2}{a_b^2}. \tag{10.42}$$

There are two small parameters in Eq. (10.42), $(\pi - \Psi)^2$ and a_p/a_b. The first or the second term can predominate in Eq. (10.42). The bubble buoyancy cannot be sensitive to the existence of f.s.f.z., not even at very small angle $\pi - \Psi$. In Section 8.7.3 this was proven for small Re. The coincidence of the measured bubble velocity with that calculated for a solid sphere is not the ground to calculate the collision efficiency neglecting the mobility of the bubble surface. Its influence can predominate according Eq. (10.2.23).

10.2.7.   BUBBLE BUOYANCY FOR A WIDE RANGE OF REYNOLDS NUMBERS AND DIFFERENT DEGREES OF WATER CONTAMINATION, AND THE FREE SURFACE ZONE NEAR THE FRONT STAGNANT POINT

In the preceding sections the possibility of the strong influence of the free surface zone near the front stagnant point on the transport stage of microflotation was emphasised. Eq. (10.40) enables us to estimate the critical condition of the appearance of an f.s.f.z. The parameters v and Re relate to the bubble with a completely retarded surface and can be described by equations derived for solid spheres.

At first the equations for v and Re are presented in a convenient form. Afterwards the r.h.s. of Eq. (10.40) is expressed as a function of Re only.

In the range $2 < Re < 750$ the empirical relation between Reynolds number and Froude numberreads (Peebles & Garber 1953, Grassman 1961, Hosler 1964),

$$Re = 47Fr^{3/2} \tag{10.43}$$

The definition of Froude number

$$Fr = v^2/2ga_b,$$ (10.44)

and Eqs. (10.43) and (10.44) together with the definition of Reynolds number Eq. (8.37) are the set of equations which allow to exclude a_b and Fr, which leads to a dependence of bubble rising velocity on Reynolds number,

$$v = m\,Re^{5/9},$$ (10.45)

with $m = (gv)^{1/3}(47)^{-2/9} \approx 1$. A comparison of Eq. (10.45) with the measured value of buoyant velocities in non-purified water is demonstrated in Table 10.1 and Fig. 10.4., where experimental and theoretically determined velocities from different authors are presented.

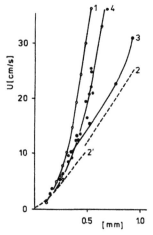

Fig. 10.4.   Measured bubble velocities in water; tube of diameter 75 mm and of length 2 m, T = 18°C; 1 - equation of Levich, 2 - experiments by Allen, 2' - experimental curve of Shabalin, 3 - experiments by Luchsinger, 4 - experiments by Gorodetzkaja, according to Levich (1959)

Table 10.1   Measured and calculated bubble velocities at different bubble diameters

a_b, cm	v_{exp}, cm/s	Re	v_{theor}, cm/s	$\Delta\gamma$
0.025	5	25	≈5	0.25
0.05	10	100	≈10	1
0.1	22	440	≈20	5

The velocities of bubbles of diameters between 0.1 and 2 mm are of the same order as those obtained from Eq. (10.45). Measured velocities are also presented in the second column of Table 10.1, taken from Schulze (1992), from which the Reynolds numbers are calculated (third column). The substitution of Reynolds numbers into Eq. (10.45) yields theoretical velocity values (column 4). The comparison of values of the second and fourth columns confirms the usefulness of Eq. (10.45).

The substitution of values for v and Re into Eq. (10.40) yields values for $\Delta\gamma = \gamma_w - \gamma$ which are given in the last column.

If we assume

$$\Delta\gamma_{exp} < \Delta\gamma, \tag{10.46}$$

a free zone near the front stagnant point exists for bubbles of a dimension exceeding the diameters given in the first column. The critical $\Delta\gamma$ values (column 5) can be compared with experimental data for different water samples.

The large scatter is caused by differences in place and time of the measurements. According to Fainerman (private communication) the initial value of the surface tension of bidistilled water is approximately 72.7 mN/m at 20^oC. Its value decreases by 0.02 mN/m during 8 minutes. The initial value relates to a freshly formed water-air interface and can be characterised by γ_w. Thus $\Delta\gamma = \gamma_w - \gamma_{eq}$ can be estimated to be 0.02 mN/m . For water from river Kalmius (Donetsk, Ukraine) Fainerman obtained $\Delta\gamma \sim 0.4$ mN/m, and from the see of Azow $\Delta\gamma = 0.1$ mN/m. Parallel to these measurements the content of ionic surfactants was determined by another method and found to be smaller than the surfactant concentration calculated from the surface tension. This is reasonable because the ionic surfactants are only part of the impurities. If we compare the total content of organic compounds determined by the well known oxidation method it usually exceeds by 10 times the content determined from the surface tension by means of Gibbs equation.

We can conclude that the residual surface mobility cannot be neglected even in the case of very contaminated water from the river Kalmius. However, it is valid for bubbles of larger dimension, approximately 1 mm. Concerning sea water or tap water, residual surface mobility can appear even in the intermediate range of Reynolds numbers.

However, it has to be emphasised that the given estimation is of low accuracy. There are several sources of errors which have to be avoided in future when reliable information about the residual mobility is needed for technological developments.

At first the surface tension gradient is estimated by the ratio of the surface tension difference between γ_w and $\gamma(\Gamma_o)$ over a characteristic length, equal to the bubble radius. This uncertainty can be bypassed after the elaboration of a stagnant cap theory for high Reynolds number. In the r.h.s. of Eq. (10.40) the viscous stress is estimated from the boundary layer thickness. However, the systematic elaboration of bubble (droplet) hydrodynamics (Protodiaconov & Ulanov, 1983) uses the same approximation up to $Re < 100$.

The boundary layer thickness is given by the well known estimation

$$\delta_G = a_b / Re^{1/2}, \tag{10.47}$$

which agrees with the vorticity estimation used in Eq(10.40). According to experimental investigations and numerical solutions (cf. Clint et al., 1978) qualitatively different hydrodynamic regimes exist at different Reynolds numbers. At $7 < Re < 20$ a so-called unseparated flow is observed. Flow separation is indicated by a change in the sign of the vorticity and first occurs at the rear stagnation point, approximately at $Re = 20$.

If Re increases beyond 20 the separation ring moves forward so that the attached re-circulating wake widens and lengthens. The separation angle measured in degrees from the front stagnation point is well approximated by $\theta_s = 180 - 42.5 \, (\ln(Re/20))^{0.48}$ at $20 < Re < 400$. The steady wake region appears at $20 < Re < 130$. The onset of wake instability corresponds to $130 < Re < 400$.

This qualitative discussion cannot be reflected by Eq. (10.47) which is a very primitive approximation especially with respect to intermediate Reynolds number $1 < Re < 100$.

To increase the accuracy of Eq. (10.40) and correspondingly condition (10.46), the theories of flow pattern around the bubble at intermediate Re can be used (Luttrel et al., 1988; Yoon & Luttrel, 1989; Yoon, 1991; Nguen Van & Kmet, 1992). At $Re \sim 1$ the r.h.s. of Eq. (10.40) is reasonable because it agrees with Stokes law. Thus, Eq. (10.40) can be characterised as an interpolation for the intermediate range of Re.

At microflotation the process of deposition of disperse particles and molecular contaminants on a bubble surface proceed in parallel. If the rates of these processes are commensurable in the process of microflotation the level of impurities in the bulk decreases. It means that their adsorption on bubbles surfaces decreases too, leading to the increase of the residual mobility.

Let us compare the fluxes of the disperse particles and molecular impurities on the bubble surface assuming that the hydrodynamic field is described by the Stokes equation. It enables to use Eq. (8.152) for the description of the impurity flux. This equation can be transformed into an analogue of the capture efficiency,

$$E_D = 2.5(\frac{D}{a_b v_{St}})^{2/3}, \tag{10.48}$$

where v_{St} is the rising velocity of a bubble according to Stokes' equation.

The comparison of both fluxes can be characterised by the ratio E_D and E according to Eq. (10.12)

$$E_D/E = \frac{5}{3} \frac{(D/\alpha)^{2/3}}{a_p^2}, \tag{10.49}$$

where $\alpha = 2g/9\eta$.

Let us emphasise that this ratio does not depend on bubble radius which simplifies the description of the common recovery of disperse particles and molecular impurities. As seen from Eq. (10.49) the fluxes can be comparable. Thus, we can conclude that the role of the residual mobility can increase if the recovery of molecular surface active impurities is taken into account.

10.2.8   ROLE OF R.S.C. IN TRANSPORT STAGE AT DIFFERENT PARTICLE ATTACHMENT
          MECHANISMS

In transient state the DAL has a slight effect on the transport stage if the rear stagnant cap covers a smaller part of the surface. If the rear stagnant cap is not too small and characterised by the angle φ (cf. Section 8.6) essentially less than $\pi/2$, the possibility of its effect depends substantially on the mechanism of fixation of particles on a bubble surface (see Appendix 10D).

10.3.   INVESTIGATION OF MICROFLOTATION KINETICS AS A METHOD OF DAL STUDIES

As it was pointed out in Chapter 8, the experimental verification of the theory of DAL of a bubble has so far been based on investigations of the effect of concentration and surface activity of a surfactant on velocity of buoyant bubbles of different size. As follows from the theory and from experiments, this effect is not very appreciable, it decreases sharply with the decrease of bubble size and of Reynolds number, and at Re < 40 it becomes unnoticeable at all. It was shown in preceding sections that the DAL structure determining the degree of retardation of surface motion has a strong effect on the deposition of small particles on a bubble surface. In this case it is important that papers by Reay & Ratcliff (1973, 1975), Collins & Jameson (1977), and Anfruns & Kitchener (1976, 1977) have demonstrated the possibility of experimental determination of collision efficiency with rather high accuracy. Thus, investigations of deposition of small spherical particles on buoyant bubbles can be an efficient method for the experimental verification of theories of the DAL, its retardation of bubble surfaces, and the hydrodynamic field of bubble changing under the effect of DAL.

The first example of such a kind are experiments by Anfruns & Kitchener (1977). As a result of the comparison of experimental data with theory, they concluded that under conditions of their experiments ($Re = 40$) the bubble surfaces were completely retarded. In addition, they have substantiated in a similar way the applicability of Hamielec's theory (1962, 1963) for the description of the hydrodynamic field of a bubble. In conclusion, problems of the theory of DAL can be solved within the framework of the proposed methods.

10.3.1. HYPOTHESIS OF INCOMPLETE RETARDATION OF BUBBLE SURFACES AT $Re < 40$

It was pointed out in Chapter 8 repeatedly, that it is impossible to discriminate between a complete and a strong retardation by measuring the buoyant bubble velocity. Since the retardation of bubble surface is derived up to now from the value of its buoyancy velocity only, conclusions in the literature about a complete retardation of the bubble surface at $Re < 1$ cannot be considered as really proven.

The initial degree of retardation (before addition of surfactant) can be checked by varying the surfactant concentration and observing the variation of collision efficiency or flotation kinetics. If the surface is initially completely retarded, the addition of surfactant cannot affect the degree of retardation and flotation kinetics remain unchanged. If the surface was not completely retarded, the addition of surfactant leads to larger retardation coefficients and measurements of collision efficiency as a function of surfactant concentration allows to check for the retardation coefficient given by Eq. (10.32). Finally, if this formula is confirmed, the initial degree of surface retardation can be judged by the value of collision efficiency before addition of surfactant (see Section 8.5).

When deposition of emulsion drops is investigated, sedimentation can be neglected so that the first term in Eq. (10.27) dominates even at strong retardation. Retardation of the surface becomes less efficient with increasing bubble dimensions and, respectively, Reynolds number. Therefore, the experimental verification of the hypothesis of incomplete retardation of the surface at intermediate Reynolds numbers is of interest. A maximum removal of impurities from the water used is important in such experiments.

Maximum water purification can create the conditions for new results with respect to the mobility of a bubble surface because in experiments performed earlier the water was not particularly purified. As sufficiently clean, distilled or bidistilled water was used for measurements of a bubble velocity. However, even during the distillation process transport of impurities by evaporated water to distillate can happen. The application of deflegmators hampers this transport and the content of organic compounds in bidistillate decreases. According to Fainerman (private communication) the decrease of surface tension caused by traces of impurities is of the order of 0.02 mN/m after the application of deflegmators. Thus, such highly purified water in bubble investigations is recommended. As a method of water purification special flotation is often used. However, some secondary processes restrict the degree of water purification by microflotation. Attention is paid to these processes and the possibility of enhanced water purification in Appendix 10D. The analysis of the problem shows that a total remove of impurities is impossible. However, a decrease by 10 to 100 times of the initial low concentration of admixtures seems possible. The removal of these contaminants can reveal incomplete retardation of surface motion at Re < 40. It has to be pointed out once more that this purification does not lead to an increase of the velocity of bubble rising. More probable is an increase of collision efficiency due to a decreased retardation coefficient which still exceeds many times the viscosity of pure water. To perform such complex measurements, the strong contamination of bubble surfaces by organic impurities in prepared dilute suspensions has to be avoided.

10.3.2. *INVESTIGATION OF THE STATE OF SURFACE OF BUBBLES RISING IN A THIN LAYER OF LIQUID*

It is assumed that in the process of bubble formation at the tip of a capillary, its growing surface is slightly retarded by impurities. If this is true, the bubble surface adsorbs a small amount of impurities from the liquid also on a rather small segment of its path (of course, if the water was specially purified to a possibly low level of the content of impurities). In experiments which have revealed no influence of added surfactant on the rising velocity (Okazaki, 1964), the velocity was calculated from the time of bubble rising for a sufficiently large distance, in order to increase the measurement accuracy. No data were found about the velocity of rising bubbles during the initial part of its path of the order of few bubble diameters. Probably it caused also by the fact that transient hydrodynamic conditions for a bubble have not been studied sufficiently. The bubble velocity changes along the initial part of its path not only due to the accumulation of contaminants and immobilisation of the surface but also due to the transformation of its hydrodynamic field. Hence, studies under simpler steady-state conditions were performed, such as the measurement of the velocity averaged over sufficiently long segments of a bubble trajectory. Anyhow, there is no need to measure absolute velocity and to compare it with theory in order to determine incomplete retardation of a bubble surface.

It is sufficient to ensure that addition of surfactant decreases the velocity at a given sufficiently small distance from the capillary tip. Thus, investigations in a thin layer of liquid, practically in a film of a few millimetres thick seem reasonable. Even at such a small thickness, decimicron latex particles (possibly with a modified surface for the improvement of their optical properties) can be observed for a sufficiently long time, taking into consideration their small sedimentation and diffusion rate. Determination of the number of particles on a bubble surface as a function of surfactant concentration would provide very valuable information.

10.3.3.   INVESTIGATION OF DAL UNDER CONDITION OF LARGE REYNOLDS NUMBERS

Reasonable investigations under these conditions are restricted by the state of the DAL theory which has been developed so far only for conditions of very strong and weak surface retardation (cf. Section 8.6). Collision efficiency has been derived only for potential flow conditions (Sutherland, 1948). With increasing surfactant concentration up to c_{cr}^l (Eqs 8.135 - 8.136), a beginning decrease of bubble velocity may be expected. A respective rear stagnant cap results in a decrease of collision efficiency only when attachment of the particle is accomplished not due to the of instability of the water interlayer at some thickness h_{cr} but under the effect of attraction forces (Appendix 10B).

With increasing surfactant concentration and rear stagnant cap, respectively, the collision efficiency should decrease to a limiting small value corresponding to strong retardation of the whole surface. A formula of collision efficiency is proposed by Nguen & Kmet (1992) (Section 10.16). There is also a relation for the concentration c_{cr}^h Eq.(8.57) at which the bubble surface becomes strongly retarded. At surfactant concentration $c > c_{cr}^h$, the bubble velocity remains invariable. Thus, bubble velocity and collision efficiency decrease with increasing concentration from c_{cr}^l to c_{cr}^h. Then the velocity decrease stops and the collision efficiency continues to decrease due to enhanced surface retardation. Similar results are expected for surfactants of high surface activity (condition (8.72)). A similar picture occurs under condition (8.71), i.e. a joint decrease of bubble velocity and collision efficiency should be initially observed at increasing surfactant concentration. The velocity decrease stops with the further increase of surfactant concentration while the collision efficiency continues to decrease. The discussed difference in regularities can be revealed from simultaneous measurements of the concentration dependence of bubble velocity and collision efficiency. Such measurements are based on the idea that the concentration at which the velocity decrease stops and the decrease of collision efficiency persists is controlled by different conditions.

With the constraint (8.72) that this takes place at $c = c_{cr}^h$, and with the constraint (8.71), the limiting concentration corresponds to a rather large χ_b compared with η. It is of interest to carry out such a measurement at several dimensions of large bubbles. Since c_{cr}^h and χ_b are known functions of the bubble radius, verification of their relation is possible.

In conclusion of this section and in accordance with the results obtained in the preceding sections it can be recalled that in studies of DAL by microflotation methods it is necessary to provide attachment of particles on bubble surfaces. The results are then controlled only by the transport stage of the elementary flotation act.

10.4. DYNAMIC ADSORPTION LAYER AND OPTIMISATION OF TRANSPORT STAGE OF FLOTATION

It was shown in Section 10.1, that the collision efficiency depends on the ratio between the radii of particle and bubble, and on the degree of surface retardation.

At high degrees of surface retardation, a decrease of the bubble size very strongly increases the collision efficiency. The situation is more complex when the surfactant content is not very high and the surface of sufficiently large bubbles is slightly retarded so that the collision efficiency is given by Sutherland's formula. The advantage exists only in the case of substantially reduced bubble diameters. Let a_b^I be the radius of a large bubble for which Sutherland's formula (10.11) is valid at a surfactant content available in the suspension, and a_b^{II} be the radius of a bubble whose surface is retarded to the extent necessary for the use of Eq. (10.12). If the collision efficiencies in the two cases are denoted by E_1 and E_2, respectively, we obtain

$$\frac{E_1}{E_2} = 2 \frac{\left(a_b^{II}\right)^2}{a_b^I a_p}.$$ (10.50)

Advantageous is a strong decrease of the bubble dimension down to values less than

$$a_b^{II} = \sqrt{\frac{a_b^I a_p}{2}}.$$ (10.51)

If a_b^{II} exceeds this value, a decrease of bubble size will only decrease the collision efficiency. Large bubbles with the size of the order of $a_b^I \approx 0.5\,\text{mm}$ are obtained in usual pneumatic dispersions formed in flotation machines. It is much more difficult to obtain bubbles 10 to 50 times smaller, which can be achieved by using completely different methods of bubble generation, electroflotation and air-dissolved flotation. Since a decrease of bubble dimensions is connected with substantial complication and a rise in price of the technology, it is necessary to forecast sufficiently accurately the required size of bubbles as a function of surfactant concentration and its surface retardation. Therefore, the approximate Eq. (10.11) needs refinement which is possible only with further investigations of the DAL of bubbles. Instead of Eq. (10.32) a more general formula must be used taking into consideration a possibly incomplete retardation of the bubble surface, short range hydrodynamic interaction, and the sedimentation term. Such more detailed and exact consideration is beyond the scope of the present section which is related to DAL rather than to microflotation problems. Such considerations are so far not developed to the degree of DAL theories.

However taking into consideration the importance of optimisation for bubble diameter, some additional information on the question touched upon is presented in Appendix 10C.

10.5. SPECIFIC FEATURES OF THE MECHANISM INVOLVING ATTACHMENT OF SMALL PARTICLES ON THE SURFACE OF A BUBBLE

10.5.1. GENERAL CONSIDERATION

The probability of formation of stable particle-bubble aggregates is determined by the probabilities of its attachment and retention on the bubble. The detachment is affected either by gravity or by inertia. These forces are proportional to the volume of particles, i.e. as the cube of the linear dimension of a particle; hence, they are very big for large particles and very small for fine particles. This trivial fact results in radical consequences when analysing the role played by the size of particles in the mechanism of the elementary act of flotation (Derjaguin & Dukhin 1960, 1979).

If the size of a particle is about 100 μm, the detachment forces are a million times greater than those arising when the particle size is about 1 μm. Therefore, in the case of large particles only one form of attachment is possible, namely, by forming a three-phase wetting perimeter which will be able to resist strong detachment forces at high values of the contact angle. Such a flotation will be called contact flotation. In the case of fine particles, along with contact flotation, contactless flotation is also possible. This conclusion is the first fundamental feature of the flotation of small particles.(Derjaguin et al. 1984).

Contactless flotation happens if two conditions are satisfied: a potential well must be present, this well must be sufficiently deep. Under these conditions there is no necessity to use the reagents which destroy the hydrophilic layer. Thus the contactless flotation can be collectorless. However, the addition of another type of reagents, i.e. cationic surfactants, can become necessary (see third peculiarity of micro flotation).

The discussed specificity of microflotation can be characterised easier by using the notion of contact angle. In distinction from flotation of big particles the flotation of small particles is possible at very low values of contact angle.

The basis for the introduction of the notion of contactless flotation was the analogy with the well known phenomena of colloid particle coagulation in the secondary energetic minimum. Due to the predomination of the attractive molecular forces at large distances the particles can form aggregates in which some distance between particles is preserved. Thus, there is no direct contact between particles in this type of aggregation. However the notion of "contact" is not so simple. It is sufficient to point to the fact that a water monolayer remains on the hydrophobic surface. Thus the term "contactless" is maybe not suitable.

If the interaction energy between water and particle molecules exceeds that of water molecules with each other, i.e. $A_{wp} > A_w$, the composite Hamaker constant is negative and the molecular component of the disjoining pressure is positive. Consequently, the molecular forces stabilise the wetting film. If A_p is the Hamaker constant of the particle substance, we have $A_{wp} \approx \sqrt{A_w A_p}$. For example for quartz $A_p = 5.47 \cdot 10^{-20}$ J, i.e. it exceeds the Hamaker constant of water $A_w = 5.13 \cdot 10^{-20}$ J. The Hamaker constants of most minerals exceed that of quartz, so that molecular forces stabilise wetting films in flotation. The exceptions are coal and talc, for which $A_{wp} < A_p$. It is worth mentioning that for them natural flotation is possible.

The composite Hamaker constant can change its sign due to surfactant adsorption at the bubble and/or particle surfaces (see Appendix 10D).

Note, that the bubble surface can be covered by an adsorption layer even without the special addition of surfactant due to contaminations in natural, tap, and industrial waters. Thus, the conclusion that wetting films are stable due to the stabilising effect of molecular forces is at least doubtful for natural waters containing traces of surface active compounds which can destabilise the wetting film and can provide contactless flotation. Collectorless microflotation has been observed for example by Goldman et al. (1974). However, they did not perform any colloid-chemical investigations.

Surface forces also include electrostatic interaction forces arising from the overlap of the double layers (DL) of a particle and a bubble, which usually have equal charges (Huddleston & Smith 1975), i.e., the electrostatic component of the disjoining pressure of an interlayer between them (Derjaguin 1934), which may be positive. In the case of large particles, the positive disjoining pressure of the double layer is overcome by an inertia impact on the bubble surface. The small particles do not undergo such an impact; the approach occurs in an inertialess way and can be hampered by electrostatic repulsion (second peculiarity).

The general regularities involved in the influence of both components of the disjoining pressure on the attachment of small particles may be established by considering the familiar dependence of the interaction energy on the shortest distance between the surfaces of the particle and the bubble, which are usually charged to different potentials (Derjaguin, 1954). This dependence has been derived by Derjaguin in the heterocoagulation theory which was repeatedly used in the interpretation of flotation processes (Derjaguin & Dukhin, 1960). When the bubble and the particle are charged to the same potential the general theory of stability of lyophobic colloids is applicable, as developed by Derjaguin (1937), Derjaguin & Landau (1945), and Verwey & Overbeek (1948).

Fig. 10.5. presents the free energy of interaction W of a spherical particle and a bubble, which are equally charged, as a function of distance h. In this case,

$$W = \int_{y}^{\infty} F_\sigma(y)dy.$$

(10.52)

F_σ is the interaction force defined by the familiar expression

$$F_\sigma(h) = 2\pi a_b \int_{h}^{\infty} \Pi(z)dz,$$

(10.53)

where Π is the Derjaguin disjoining pressure, and z is the thickness of a plane-parallel interlayer between the plate-like particle of the same substance and the air phase. Eqs (10.52) and (10.53) have been derived under the assumption that the radius of the spherical particle a_p is much smaller than that of the bubble a_b. At large and small separations molecular attraction forces predominate, while at intermediate distances electrostatic repulsion may prevail. Under flotation conditions it is valid for the particular case of talc and coal.

The bubble and the particle differ from each other not only in the potential of their surfaces but also in their behaviour, which may drastically change the arising interaction. The simplest situation is where the potential of the particle surface and the charge density of the bubble surface do not depend on the interlayer thickness. In the particular case where this change is zero, Frumkin & Gorodetskaja (1938) and Langmuir (1938) showed that electrostatic repulsion results, which is exactly equal to the repulsion of two identical particles at the double thickness of the interlayer.

Heterocoagulation theory predicts that the inertialess flotation cannot take place if the bubble and the particle posses identical potentials ($\psi_b = \psi_p$), if the absolute values are reasonably large and the ionic strength low (Fig. 10.5). The same theory enables one to calculate at what values of the potential (or charge) of one of the interacting objects the electrostatic repulsion decreases in order for the barrier of repulsion forces to disappear.

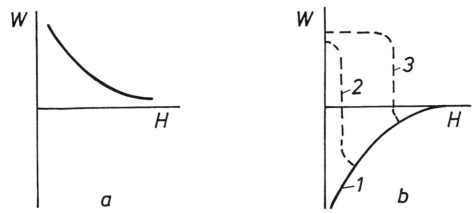

Fig. 10.5. Characteristic curves of the total contribution of molecular attraction forces and electrostatic forces conditioned by the overlap of the diffuse parts of the double layers into the energy W of bubble particle interaction at various distances and the surface charges of the same sign (a); and in the case of recharging the bubble (b) (curve 1). In (b) dotted lines characterise the contribution to the energy of interaction of non electrostatic repulsion forces when their effective radius is smaller (curve 2) or greater (curve 3) than the thickness of the double layer.

The negative potential of a surface can be decreased by the adsorption of surface-active cations, which may also lead to coagulation. On the other hand, when electrolytes are added, the thickness of the diffuse part of the double layer decreases, resulting in a decrease of the effective range of electrostatic repulsion forces.

From the technological point of view it is impractical to introduce electrolytes in order to ensure floatability. A more economical method for controlling the electrostatic component of the disjoining pressure and, correspondingly, the floatability consists in using ionic surfactants in certain concentrations; when adsorbed by a bubble, ionic surfactants may recharge it. Very small surfactant concentrations, however, will not provide floatability if this is hampered not only by the electrostatic barrier but also by non electrostatic factors which can interfere with the approach of particle and bubble. Such a factor may be the presence of a polymolecular hydrate layer on the particle surface, the possible existence of which on lyophilic surface was demonstrated in several studies (Derjaguin & Churaev 1970, 1974, Derjaguin 1976). This layer would interfere with the thinning of wetting films due to the structural component of the disjoining pressure.

In Fig. 10.5b the barrier due to the structural component of the disjoining pressure is schematically represented for the case where its range of action is either smaller or greater than the thickness of the electric double layer. One of the ways of ensuring enhanced floatability in the presence of the structural component consists in the use of surfactants. Adsorption of surfactants causes hydrophobisation of the particle surface; it either destroys the structural component of the disjoining pressure or changes its sign.

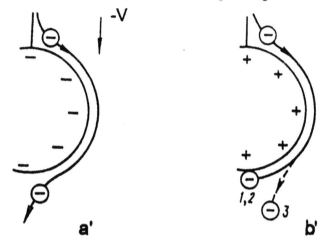

Fig. 10.6.   Illustrations of the impracticability of flotation at sufficiently great charges of the same sign of particle and bubble (a), the practicability of flotation in the case of bubble recharging of the bubble charge (b, curve 1). Curves 2 and 1 in (b) illustrate the possibility of flotation at the effective radius of the non electrostatic repulsion forces, which is smaller than the thickness of the double layer (2) and its impracticability in the opposite case (3).

Another way of overcoming the difficulty of flotation of dispersions when the non-electrostatic stability factor is operative was indicated by Derjaguin & Dukhin (1960) and involves stimulation of electrostatic attraction. For this purpose it is necessary to charge strongly a bubble by adsorption of a considerable amount of surfactants, yielding the bubble charge opposite to that of the particle. As a result, the interaction of double layers will give rise to attraction forces. If the thickness of the DL exceeds that of the hydrate layer, attraction forces between the opposite charges of the particle and the bubble act beyond the effective range of repulsion forces. Thus, a bubble particle aggregate may be formed, yet a gap of a thickness comparable with the extension of the barrier of non electrostatic repulsion forces may be preserved between the surface of the bubble and the particle. This is the third peculiarity of microflotation because this type of the stabilisation of the bubble-particle aggregate is questionable in case of big particle due to great detaching force.

When the range of structural repulsion increases to the same thickness as the DL, the potential well becomes more shallow. It will be very expedient to choose surfactants that would not only enhance electrostatic attraction but simultaneously decrease repulsion caused by a change in the water structure.

Thus, if the electrostatic attraction forces prevail over the non-electrostatic repulsion forces (caused by the water structure), floatability is quite possible (Figs. 10.5b and 10.6b, curves 2). If the effective ranges of these forces are comparable, flotation becomes impossible (Figs. 10.5b and 10.6b, curves 3) in the presence of the barrier of non-electrostatic repulsion forces.

The physico-chemical sense of "hydrophilicity" or "hydrophobicity" has been changed after the introduction of notion of structural forces in wetting phenomena (Churaev 1974, Derjaguin & Churaev 1984). It complicates the understanding and the analysis of literature, in particular for the paper by Yoon & Luttrel (1989, 1991). Initially these terms were related to small or large values of the contact angle. Now the notion of a hydrophobic surface can be understood as the presence of hydrophobic (attractive) structural forces. Correspondingly, the presence of hydrophilic (repulsive) structural force characterises the surface as hydrophilic. Meanwhile, contact angle values depend on the properties of both surfaces and, in addition, on other components (molecular and electrostatic) of the disjoining pressure.

The capture of particles at a bubble surface is determined by the collision efficiency and the probability of attachment. If the transport and attachment stages are independent processes the capture efficiency can be expressed by $E_c = E E_a$.

10.5.2. PRESSING FORCE AND COLLECTORLESS MICRO FLOTATION

The hydrodynamic pressing force can exceed the force barrier of the disjoining pressure, thereby allowing the possibility of flotation without the use of reagents. This problem may be examined when considering the motion of a particle along the symmetry axis of a bubble from the incoming liquid flow (Derjaguin, Rulyev et all 1977). Under this condition, the hydrodynamic pressing force is maximum: hence one can obtain the necessary and sufficient condition for the deposition of particles on the surface of a rising bubble.

Along the bubble axis the tangential flow velocity is equal to zero. Thus, the duration of the deposition process can be indefinitely long. This means, in turn, that the viscous resistance of the interphase film can be neglected in the balance of acting forces (Derjaguin et al. 1977),

$$F_\Sigma(h, a_b) = F_v + F_h + F_e + F_g < 0, \qquad (10.54)$$

where F_h is the hydrodynamic pressing force, F_v is the molecular attraction force, and F_e and F_g are the electrostatic and gravitational components of forces, respectively.

Eq. (10.54) indicates that for all values of h, the particle is subjected to a force directed towards the surface (front pole) of the bubble; otherwise the deposition cannot occur. The same

376

expression imposes limitations on parameter values at which the disjoining pressure can be overcome. The limitation is represented graphically in Figs 10.7 and 10.8 as a curve which characterises the dependence of the particle radius on the electrolyte concentration and straddles the region close to the origin, where flotation is impossible.

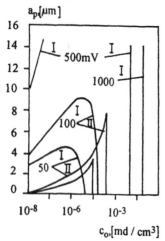

Fig. 10.7.   Regions of floatability in the case of Stokes flow (shaded): I - lower boundary of flotation in primary minimum; II - upper boundary of flotation in the secondary minimum. In terms of cgs units $\varepsilon = 80, g = 10^3$

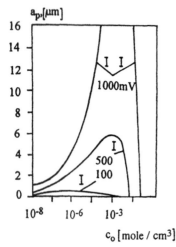

Fig. 10.8.   Regions of floatability in the case of potential flow. In terms of cgs units $A_H = 5 \cdot 10^{-14}, \varepsilon = 80, a_b = 2,5 \cdot 10^{-2}, \eta = 10^{-2}$ and $v_p = 15$

The region of floatability due to hydrodynamic pressing forces were calculated for the Stokes (Fig. 10.7) and potential (Fig. 10.8) regimes and for several values of the product of surface potentials of the bubble, ψ_b and the particle ψ_a. As could be expected, flotation can occur even below the heterocoagulation threshold if particles are not too small. In the potential regime the hydrodynamic pressing force is much greater than in the Stokes regime, so that flotation can occur at greater electrostatic repulsion forces or with smaller particle sizes.

Flotation without overcoming the force barrier is hampered by detachment of particles in the vicinity of the rear pole of the bubble. The component of the liquid flow velocity which is normal to the bubble surface induces a force pressing the particle against the surface at the front pole and a detaching force at the rear pole. Consequently, the detaching force differs from the pressing force only in sign. Thus, the possibility of fixing particles in the secondary energy minimum (Verwey & Overbeek, 1948) may also be analysed with the aid of Eq. (10.54). Since the detaching force decreases with particle size, the flotation due to attachment of particles in the secondary minimum becomes possible for sufficiently small particles. In the potential regime, when the detaching force is stronger, attachment in the secondary minimum becomes possible only for particles small enough that thermal motion does not interfere.

So far a quantitative description of inertia-free flotation is given only for limiting cases; namely, at sufficiently small and large values of the potential product $\psi_r \psi_a$. At small values the electrostatic interaction does not complicate flotation so that Eqs (10.20) and (10.21) can be used, whereas at large values flotation cannot occur.

As the concentration of the electrolyte increases, the double layer contracts and the parameter $\psi_r \psi_a$ decreases. Hence, in light of the data presented in Fig. 10.8, the possibility of flotation of small particles in the absence of a reagent would be expected at a sufficiently high electrolyte concentration if, of course, the non-electrostatic stability factors are absent.

The necessary condition of microflotation under the action of the pressing force (10.54) was used later by Okada et al. (1990) for the interpretation of their results obtained by direct observation of the elementary act of microflotation.

A different representations of the general necessary condition for the microflotation (10.54) is possible. The representation in papers by Derjaguin et al. (1977) and Okada et al. (1990a) are oversimplified and needs generalisation by taking into account recent developments in heterocoagulation theory. First, it relates to the description of electrostatic interaction. Both representations disregard the transition from electrostatic repulsion to attraction at a distance h_1 according to Eq. (10D.3). In the first paper it is justified by assuming the equal potentials of bubble and particle. In the second paper there is no justification because the measured values for bubbles and latex particles are different. Okada et al. (1990) use the first term of the equation by Hogg et al. (HHF) (1966) which was rewritten by Schulze (1984). First, the neglecting of this second term leads to an overestimation of the barrier height. Second, the use of the HHF equation is based on a linear approximation and needs justification too. Recently

Overbeek (1990) developed non-linear formulations for the calculation of the double layer interaction of divers particles. Kihira et al. (1992) used Overbeek results and established an applicability limit of the HHF approximation. It is not valid if the Stern potential difference is to small and the individual values exceed 25 mV. On the other hand it works well if the ratio of the potential values is sufficiently high even though one of the two potentials is high.

The literature analysis in Appendix 10D showed that even at high electrolyte concentration ($10^{-2} - 10^{-1}$ M) the absolute values of Stern potentials of bubbles and particles cannot be small and their difference very large. It means that Overbeek's equations have to be used for the estimation of the electrostatic barrier in microflotation and it can be lower than predicted by the HHF approximation. For example in the investigations by Yoon & Luttrel (1991) on the role of hydrophobic interaction in microflotation Ψ_b equals -20 mV and $\Psi_p = -45$ mV. The HHF approximation leads to an overestimation of electrostatic repulsion by almost two orders of magnitude according to Fig. 3 of Kihira et al. (1992).

On the contrary, the HHF approximation can be used for the estimation of the barrier disappearance due to the decrease of the Stern potential caused by cation adsorption.

The results presented in Figs. 10.7 and 10.8 have to be modified and generalised. The negative complex Hamaker constant and the difference of the Stern potentials of the bubble and the particle have to be incorporated into the calculations. The force of hydrophobic attraction has to be added into the general necessary condition of microflotation.

Both short range (Israelachvili & Pashley, 1984) and long range hydrophobic forces may be represented in the form $\dfrac{a_b a_p}{a_b + a_p} K_s e^{-h/\lambda}$, where λ is decay length determining the radius of action of the forces, and K_s controls the order of magnitude of the attraction forces. For short-range forces experimentally obtained values of K_s vary (at $h < 5$ nm) within the range $(1.8 \div 4.8)10^3$ N $\cdot$ cm^{-2} and λ values change from 1 to 2.5 nm, respectively (Cleasson & Christenson 1988, Rabinovich & Derjaguin 1988). For long range hydrophobic forces λ-values change from 15 to 50 nm and K_s is negligible compared to short range forces. In their estimation of the role of the hydrophobic interaction Yoon and Luttrel (1991) take into account long range hydrophobic interaction. The calculation leads to h = 13nm because the experimental data correspond to a very low ionic strength and a large barrier distance from the surface. However, the ionic strength in industrial flotation varies in the range $10^{-2} - 10^{-1}$ M. Thus, it is reasonable to incorporate stronger short range hydrophobic forces into the necessary condition of microflotation.

Similar calculations are necessary for the estimation of cationic surfactant adsorption needed for the decrease of the energy barrier. This leads to a decrease of the electrostatic repulsive force and simultaneously makes the water-air interface hydrophilic. In that respect the adsorption of the multivalent inorganic ions (Somasundaran et al, 1991, 1992, 1993) may be more effective.

10.5.3. EXPERIMENTAL CORROBORATIONS

The interpretation of the attachment process during flotation of small particles in terms of the heterocoagulation theory, as suggested by Derjaguin & Dukhin (1960), has been considered in a number of review articles on the theory of flotation (Joy & Robinson 1964, Usui 1972, Rao 1974) and was further confirmed in many studies (Derjaguin & Shukakidse 1961, Jaycock & Ottewil 1963, Rubin & Lackay 1968, Devivo & Karger 1970, Collins & Jameson 1977).

Ottewil et al. (1963) detected the maximum of floatability at the isoelectric point by varying the electrokinetic potential of silver iodide particles by adsorption of a cation-active surfactant. Furthermore, it has been established that the flotation rate is high within a narrow pH range and very low outside this range. In the former case the pH values correspond to very small ζ potentials of the particle, i.e. in the vicinity of their isoelectric point, (Jaycock & Ottewil 1963, Rubin & Lackay 1968, Devivo & Karger 1970). Addition of aluminium hydroxide extends the range of pH values which promotes flotation.

The latter authors interpret their data as proof of the decisive influence of the electrostatic component of the disjoining pressure on the process of particle attachment to the bubble. Collins & Jameson (1977) considered another interpretation of these results. The non-appearance of the barrier of electrostatic repulsion forces can bring about a rapid coagulation of particles at the isoelectric point. The resulting aggregates may deposit on the bubble surface more rapidly than the individual particles. In order to prevent this possibility, Collins & Jameson (1977) measured the change in the distribution of spherical polystyrene particles of 4 to 20 μm diameter in the course of flotation in the presence of different electrolyte concentrations. The addition of the salts altered the electrophoretic mobility, i.e. it caused a change in the values of the electrokinetic potentials. For each of the eight fractions that were studied, flotation rate increased monotonically as the ζ-value of particles and bubbles decreased. These experiments enabled the authors to separate the influence of the size and charge of particles on flotation. Hence, an unambiguous proof was provided that the elimination of the disjoining pressure is the necessary condition for microflotation.

Thus, the flotation of fine particles can be controlled with the aid of ionic surfactants and the floatability is ensured even in the case of non-electrostatic stability factors. As a corroboration of this, let us consider research into the floatability of quartz. Laskowski & Kitchener (1969) have shown that the surface not only of pure but even of methylated quartz exhibits hydrophilic areas that contribute to stability of slurries and hinder flotation (Dibbs et al. 1972). The floatability of quartz was ensured within the range of concentration of dodecylamine chloride, in which the sign of charges of the bubble and particle surfaces are opposite (Dibbs et al. 1972).

Systematic investigations by Schulze & Cichos (1972a, 1972b) have shown that the interphase water films separating the water-air and water-quartz interfaces become unstable under the influence of adsorption of trivalent cations or cation-active surfactants if their thicknesses are less than a critical thickness, h_{cr}, ranging between 300 and 450Å. Recharging the bubble surface by the adsorption of cation surfactants causes rupture of the film at thicknesses of about 150Å if the electrolyte concentration is high and at thicknesses of about 1500Å if the electrolyte concentration is low, which is in agreement with the electrostatic nature of attraction. Schulze also suggested that the floatability would be aided by recharging the bubble surface due to adsorption of surfactants at the lowest possible electrolyte concentrations, which leads to thickening of the DL and, accordingly, to increases in h_{cr}.

The suggestions of Schulze are in agreement with the experimental data of Goddard et al. (1977), who induced flotation of quartz particles by a bubble whose surface was recharged by amines. A sharp enhancement in floatability corresponded to an abrupt increase in the adsorption of amines. The floatability is usually associated with the hydrophobisation of the particle surface, but the cause may be quite different when the bubble surface charge is modified by surfactants. In experiments by Goddard and co-workers the ζ-potential of quartz varied very little when surfactants were added; hence, it might be assumed that the surface state of the particles also changed insignificantly (Bleier et al. 1977).

In recharging the bubble surface, the depth of the potential well formed beyond the limits of the barrier of non-electrostatic repulsion forces is insufficient to ensure the contactless flotation of large particles. Therefore, in the presence of the non-electrostatic component of the disjoining pressure recharging can cause the contactless flotation only if the particles are sufficiently small.

Anfruns & Kitchener (1976) examined the efficiency of capture of several fractions of hydrophobised (methylated) quartz by individual bubbles of 0.5-1.1 mm in diameter.

Their investigations corroborated the relationship which accounts only for the LRHI and which was derived on the basis of Hamielec's hydrodynamic field. The authors arrive at the conclusion that the disjoining pressure of the electric double layer does not complicate the flotation act, at least not under the experimental conditions. This conclusion, however, cannot be considered to be well-established because another interpretation of these interesting experimental data may be offered. The theory of the SRHI and of the influence of DL on flotation has been developed for spherical particles, whereas in the research discussed by

Anfruns & Kitchener (1976) non-spherical particles of crushed Brazilian quartz were used (Fig. 10.9). Therefore, it is necessary to take into consideration surface roughness which determines the face of such particles. The resistance to thinning of the interfacial film can be drastically reduced if the angles between the faces are sufficiently sharp so that "intrusion" of the liquid interlayer by a sharpened section of the particle surface may take place. Such geometric conditions of the elementary flotation act also drastically facilitates motion against the disjoining pressure of the DL.

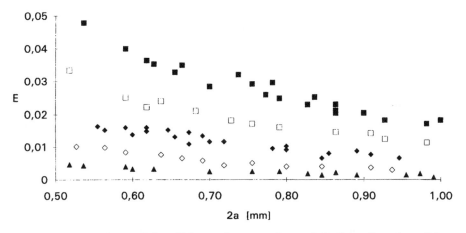

Fig. 10.9. Dependence of the efficiency of capture of several fractions of quartz particles on their dimensions d and bubble dimensions. d = 40.5µm (■), 31.4µm (□), 27.6µm (◆), 18µm (✧), 12µm (Δ), according to Anfruns & Kitchener (1976)

Anfruns & Kitchener (1977) determined the efficiency of the capture of glass spheres with diameters between 25 µm and 40 µm, and quartz particles with diameters between 8 µm and 46 µm, i.e. the intermediate range of particles, by a single bubble of 600 µm in diameter. The particle surfaces were methylated so that a probability of particle attachment near 1 can be assumed. The observed capture efficiency E_c did not exceed 20% of the predicted value for pure water. The capture could be intensified in 1 M KCl; however, E_c turned out to be only half the predicted value.

This result was interpreted by the manifestation of short range hydrodynamic interactions. The role of electrostatic interaction in flotation is confirmed by investigations of the influence of added sodium dodecylsulfate (SDS). Due to SDS adsorption the charge of bubbles increases und the collision efficiency decreases. When following the capture rate it is observed, that at SDS concentrations above 10^{-3} M particle deposition completely stops while the contact angle continues to be large. Thus, electrostatic repulsion prevents the contact between particles and bubbles.

The micro-relief of the surface cannot strongly affect the LRHI; therefore, Eqs (10.11) and (10.12) and Kitchener's relationship (Anfruns & Kitchener, 1976), which have been derived for spherical particles, also retain their significance as approximate formulas for non-spherical but isometric particles.

The foregoing discussion of the theory of flotation of small particles involving the influence of the SRHI and surface forces shows that the elementary flotation act may remarkably differ for particles with smooth and rough surfaces; thus they must be investigated separately.

The role played by the SRHI and the DL in flotation may not only be decreased but also considerably increased by the deviation of the particle shape from spherical. The plane-parallel liquid interlayer between the bubble and a plate-like particle undergoes thinning even more slowly than in the case of a spherical particle, so that in this case the role of the SRHI is considerably enhanced. This is to be expected for a particle in the form of a thin plate. If the plate is not thin, then it may be expected that its shifting along the bubble surface should be accompanied by rolling. In the latter motion orientations of the particle to the bubble surface occur at which the viscous resistance of the interphase film and the disjoining pressure of the DL due to the influence of the edges are overcome so rapidly that the SRHI slightly affects the value of E.

Qualitatively different and more complicated situations are observed in flotation of colloidal sulphur of particle diameters between 0.4 μm and 1.8 μm during the recharge of bubble surfaces due to addition of lignosulfonates. Recharging was established by adsorption of different radicals of the surfactant. As one can see from Fig. 10.10., flotation is intensified at opposite signs of surface charge of bubbles and particles. At higher surfactant concentrations the surface charge of bubbles and particles have the same sign, which leads to the retardation of flotation (Sotskova, 1981; 1982; 1983).

An extreme dependence on concentration of added ionic surfactant was observed by Somasundaran & Chari (1983) by investigating the flotation of quartz and alumina. Flotation decrease at high surfactant concentrations is also explained by the appearance of the same sign due to recharge processes. Calculated interaction energies correlate well with the observed flotation recovery.

The papers by Okada et al. (1990a, 1990b) can be estimated as an important novelty in the experimental investigation of the elementary microflotation act and the role of hydrodynamic and surface forces. The mean diameter of bubbles fed into the cell was 33μm. Three kinds of uniform latex particles of different diameter (0.9, 2.9, and 6.4μm) were used as model particles. In order to change the surface charges of bubbles and particles three types of surfactants (cationic, anionic and nonionic) and AlCl$_3$ were used. The electrokinetic potentials of bubbles and particles were determined by microelectrophoresis. The quartz glass cell of rectangular shape was attached to the microscope stage which was capable to move vertically with the same velocity as that of the rising bubbles in the cell. The particle trajectory and the particle attachment to the bubble surface was visualised by a video system mounted to the microscope. The quantitative characterisation of the particle trajectory was possible at h values exceeding 1 μm. This limit of the visual method allows to discriminate between particle trajectories inside and outside of the grazing trajectory.

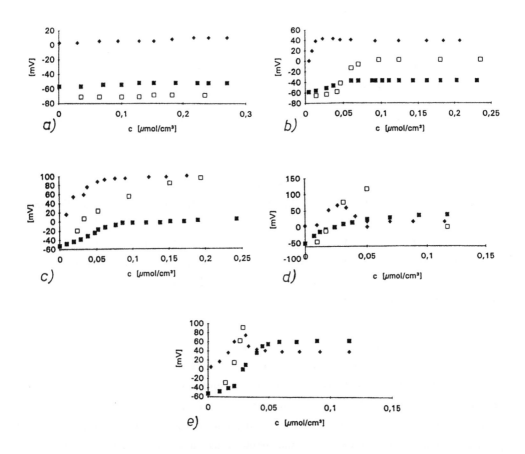

Fig. 10.10. Dependence of electrokinetic potentials of colloidal sulphur (■) and bubbles (□) and flotation recovery (◆) on concentration of ammonium salts with radicals CH_3 (a), C_5H_{11} (b), C_8H_{17} (c), $C_{10}H_{21}$ (d), $C_{12}H_{25}$ (e), according to Sotskova (1982, 1983)

If the horizontal distance of a particle trajectory from the vertical axis $\theta = 0$ at large distance from the bubble surface exceeds a definite critical value the particles approaching the bubble surface do not attach to the bubble surface but departed from it. At smaller initial horizontal distance from the vertical axis $\theta = 0$ the particles attach at the front part of the bubble surface, move further along the bubble surface until $\theta = 180°$ and left there being attached.

No attachment was observed at sufficiently high charges of bubbles and particles. On the contrary at smaller values ($\varsigma_b = -25\text{mV}$, $\varsigma_p = -4\text{mV}$) particles attach to the bubble.

In addition dissolved air flotation experiments were carried out to obtain the flotation efficiency under the same experimental conditions as the zeta potential measurement. The values of particle capture efficiency were a maximum when the absolute values of zeta potentials of the bubbles and particles were at a minimum.

10.6. INFLUENCE OF DYNAMIC ADSORPTION LAYER ON ATTACHMENT OF SMALL PARTICLES ON BUBBLE SURFACE

If flotation is not complicated by an energy barrier and h_{cr} is not too small so that a attachment of particles can happen without the need of overcoming the barrier (barrierless microflotation), no question about the dynamic nature of adsorption layer exists.

More often a situation exists when particles and bubbles carry charges of the same sign and the electrostatic barrier is localised so that the secondary potential well is not sufficiently deep to fix small particles, or is practically absent. At the same time a particle attachment in the near potential well is possible by overcoming the barrier. Several cases may exist depending on the bubble and particle size. If the bubble size is sufficiently large, the hydrodynamic pressing force can provide the overcoming of the barrier for particles (Section 10.5). This contribution to the attachment of particles is not efficient for very small size particles or at strongly retarded bubble surfaces which always applies at Re < 40.

At any of these conditions, the addition of cationic surfactant for changing the sign of the bubble surface charge provides attachment of particles. When the bubble surface is mobile and the surfactant's surface activity is high, i.e. under constraint (8.72), adsorption is strongly decreased at the leading bubble surface. The application of highly surface active surfactants is of interest, which is connected with an increasing role of DAL. It opens the possibility of particles attachment at the lower part of the bubble surface at given sufficiently deep potential wells. Indeed, their low dynamic adsorption on the upper half of the bubble is insignificant since attachment takes place on the rear stagnation cup. Adsorption is increased here resulting in a deepening of the potential well.

10.7. INFLUENCE OF DYNAMIC ADSORPTION LAYER ON SMALL PARTICLE DETACHMENT

If flotation is not complicated by an energy barrier and too small h_{cr}, the deposition of particles is accompanied by a break of the film and formation of a three phase contact line so that the possibility of particle detachment can be practically excluded.

After the disjoining pressure of the double layer under the effect of hydrodynamic pressing forces (see Figs. 10.7 and 10.8) has been overcome, a subsequent detachment of the particle displaced to the rear pole of a bubble is hindered only at a sufficiently deep potential energy minimum. Even when the bubble is charged opposite to the particle, this minimum may be insufficiently deep if the α-film is thick enough (see Appendix 10D). The question about the depth of the primary minimum is very complex (cf. Derjaguin & Kudryavtseva 1964, Martynov & Muller 1972, Overbeek 1977).

The possibility of detachment increases with bubble size and its rising velocity. Surface retardation at $Re > 40$ due to the presence of surfactant can appear. At low surface activity the whole surface is retarded almost uniformly which prevents particles from detachment. At high surface activity an increase of surfactant concentration only yields a larger rear stagnant cap, while the rest of the surface is not very strongly retarded.

However, one can assume that detachment of small particles from the stagnant cap is unlikely. The bubble surface of the rear stagnant cap is strongly retarded and the normal component of liquid velocity is much lower than that at the upper not retarded part. This can prevent detachment of sufficiently small particles from the stagnant cap.

Agreement between experimental velocities of rising bubbles (Levich, 1962) in purified water and velocities calculated from Eq. (10.6) for a non-retarded surface means that at $Re > 40$ the portion of the lower half of the bubble should be mobile. This mobility is accompanied by a sharp increase in the normal velocity component and leads to detachment of particles. Thus, the non-contact flotation is hampered when passing to the regime at $Re > 40$. It is worth mentioning that the addition of ionic surfactant is necessary even when hydrodynamic pressing forces provide the overcoming of the disjoining pressure, except the case of a deep primary well.

When passing from Stokes field to the potential field of a bubble, hydrodynamic detachment forces grow a_b / a_p times. The radial velocity of the liquid flow at a distance of maximum approach of the particle centre to the bubble surface increases by the same factor. This difference between Stokes and potential conditions somewhat decreases with respect to particle detachment when bubble surface retardation is taken into account. On a completely retarded part of the bubble surface, the radial component of liquid velocity in a distance z from the bubble surface follows a quadratic law at $z \ll a_b$, in equivalence to Stokes' law.

The velocity dependence for the partly retarded section follows a linear law with a coefficient comparable to the ratio between the local retardation coefficient and the viscosity of water. The higher the degree of bubble surface retardation, the less is the possibility of detachment of particles in non-contact flotation. Therefore, the question about the degree of surface retardation is of great interest.

The data presented in Fig. 8.2 show that at $Re \geq 40$ the bubble velocity is the higher the lower the concentration of added surfactant is. At $Re \leq 40$ bubbles rise like solid spheres, even in highly purified water.

Hydrodynamic resistance, determining the rising velocity of bubbles, comprises the resistance of the shape, which depends on the position of the hydrodynamic flow separation line on the spherical bubble surface and the viscous resistance, which is a function of the degree of retardation of the whole surface. Since the contribution of viscous resistance is not small at $Re < 40$, the whole lower part of the bubble is strongly retarded. Otherwise the bubble velocity would strongly differ from the velocity of a solid sphere. It follows that at $1 < Re < 40$ the detachment force increases insignificantly compared with the detachment force under the Stokes regime.

10.8. PERFECTION OF MICROFLOTATION BY GOVERNING DYNAMIC ADSORPTION LAYER

After the main stages of the elementary act have been considered, preliminary considerations about microflotation control are discussed. Control of microflotation processes is based first of all on selection of optimal bubble size and optimal reagent to manipulate electrostatic interaction. The control is efficient if all stages of the elementary flotation act are optimised: long and short range interaction, attachment process detachment prevention.

If flotation is not complicated by an energy barrier and h_{cr} is not very small so that the approach of particles leads to a break of the film and a formation of three-phase contact (fulfilled for small particles), microflotation is perfected by controlling the transport stage. This question was considered in Section 10.4 and the control of microflotation reduces to the choice of an optimal bubble size.

The control of the DAL becomes important when an electrostatic barrier has to be removed by the addition of cationic surfactants. Here again two cases have to be distinguished.

Through a recharge of the bubble surface not only the electrostatic barrier is removed, but also the near potential well becomes deeper, so that a subsequent detachment of particles becomes impossible. In this case microflotation perfection reduces to the control of the transport stage and the bubble charge. If the primary energy minimum turns out to be not deep enough, also after bubble recharging, additional efforts are needed to prevent particle detachment.

Two techniques decrease the possibility of particle detachment. One consists of the use of reagents which decrease the non-electrostatic component of the disjoining pressure. The second technique consists of choosing bubble size and hydrodynamic regime so as to decrease detachment forces.

The term "small particles" in microflotation covers a very wide range of sizes and also the values of different factors vary strongly having an effect on transport, attachment process and detachment of particles. A general analysis of the elementary microflotation act without a unified theory is impossible. Such theory needs to take into account the role of all these factors within the whole range of particle sizes, and in addition a unified DAL theory. To bypass this difficulty two regions of particle sizes are discussed separately in subsequent sections: micron/submicron particles, and decimicron particles (see Appendix 10E, 10F).

The problem of DAL can be neglected with respect to small bubbles which are recommended for microflotation. However, the problem of DAL structure deserves attention. As it was explained in Section 10.4 the application of big bubbles with a mobile surface can have advantages in the case of decimicron particles. This conclusion can be generalised if the aggregation of micron or submicron particles is provided.

10.9. EFFECT OF PARTICLE AGGREGATION ON ELEMENTARY MICROFLOTATION ACT AND DYNAMIC ADSORPTION LAYER

The addition of surfactant to decrease the electrokinetic potential of bubbles can simultaneously result in a decrease of the electrokinetic potential of particles and, respectively, in their aggregation. The effect of aggregation on the elementary flotation act was considered by Dukhin et al. (1979) under the assumption of stiff, spherical aggregates with small radius a^* compared with a_b. Under these simplifying assumptions, the efficiency of the collision of aggregates with a bubble E^* can be described by Eqs (10.11) and (10.12), when replacing the particle radius a_p by the aggregate radius a^*. The following estimates for the intensity of the process of deposition of particles on a bubble as a result of aggregation under Stokes and potential regimes are obtained,

$$\frac{E_S^*}{E_S} \approx \left(\frac{a^*}{a_p}\right)^2, \frac{E_P^*}{E_P} \approx \frac{a^*}{a_p}. \tag{10.55}$$

An increase of the aggregate sizes plays a positive role for flotation only in the stage of aggregate-bubble system formation while the probability of detachment of an aggregate from the bubble increases. Indeed, the greater the size of aggregates or particles, the greater is the radial velocity component in the centre of an aggregate which is oriented to the lowest surface of the bubble along the external normal. Formulas for hydrodynamic detaching forces and the maximum radius of a bubble able to flotate aggregates of a given radius are given by Dukhin et al. (1979). The derivation of these formulas is based on the solution of the hydrodynamic problem of two spheres rising along a common axis in a Stokes regime.

At a given bubble size, an optimal size of aggregates exists which provides most effective flotation. Aggregates of larger size detach from the bubble surface while aggregates of smaller size are deposited too slowly on the bubble surface. An optimal size of bubbles for flotating aggregates of a given size exists too. Unfortunately, it is difficult to determine this optimum reliably. First, it is difficult to estimate the number of aggregates contacting a bubble and withstand the detachment forces. This number depends on the deformation of aggregates under the effect of hydrodynamic pressing forces and under the effect of tangential liquid flow. Second, the detachment of an aggregate is possible by breaking coagulation links within the aggregate.

Since in the present section the inertia-free transport of particles is considered, we can restrict ourselves to aggregates of a sufficiently small size, approximately tens of microns. At small degrees of packing in an aggregate, its sedimentation rate can be so small that their transport from below to the rear stagnant cap is possible against gravity.

Intensification of aggregate transport to the surface of rising bubbles is currently of great practical interest since the development of flotation technology of small particles involves flocculation (Babenkov 1977, Veister & Mints 1980, Dobias, 1993) and the transport of flocks is more efficient.

It can be assumed that the use of not too small bubbles with a stagnant cap over the major part allows to fulfil simultaneously the conditions of a sufficiently intensive transport and of preventing particles from detachment. The transport is achieved by a considerable rising velocity of bubbles. Secondary flow developing near the stagnant cap is more intense the greater the bubble velocity is, and the transport of particles from below to the rear stagnant cap is intensified. Prevention from detachment is given by a strong retardation of the surface within the rear stagnant cap. Unfortunately, lack of a theory of a stagnant cap at large Reynolds numbers does not allow to substantiate quantitatively the application of this mechanism.

Since Brownian coagulation proceeds very slowly for particles of the order of a micron and larger, the main role in this process is played by the orthokinetic or gradient coagulation caused by an inhomogeneous hydrodynamic field. Under microflotation conditions, these heterogeneities can be caused both by the motion of individual bubbles and by macroscopic convective flows (cf. Section 2.4 in Derjaguin et al. 1986).

The hydrodynamic field of a bubble, and the degree of its heterogeneity strongly change depending on the size of the stagnant cap. By varying the reagent regime and the stagnant cap, the formation of aggregates and microflotation kinetics can be affected.

10.10. COLLISION EFFICIENCY, BUBBLE VELOCITY AND MICROFLOTATION KINETICS

Usually flotation is carried out at not very small volume fractions φ of bubbles which results in a substantial deviation of the liquid velocity distribution in the neighbourhood of bubbles as compared with the case of individual bubble considered above. As a result of considering a system of bubbles, the following equation was obtained (Bogdanov & Kiselwater, 1952)

$$n(t) = n_0 e^{-Kt}, \tag{10.56}$$

where n(t) is the number of particles at time t, and n_0 is the initial number of particles.

The following expression for K was obtained by Bogdanov et al. (1980) for monodisperse systems,

$$K = 3qE / 4a_b, \tag{10.57}$$

where q is the bubbling velocity being equal to the volume of gas blown per unit time through a unit cross section of the flotation chamber. It is possible to generalise this formula for polydisperse systems and for particles and bubbles with a normal size distribution. Eq. (10.57) can be used after the bubble size and particle size have been replaced by the mean statistical values (Derjaguin et al, 1986). Thus the kinetic theory of flotation of small particles enables the most important technological parameter K to be related to measured flotation system characteristics.

10.11. BUBBLE COALESCENCE AND DYNAMIC ADSORPTION LAYER

An extremely high bubbling velocity results in capture of bubbles in a foam which is subjected to coalescence. However, investigations by Rulyov (1985) show that enlargement of bubbles by coalescence takes place also at small bubbling velocities. Integral distribution functions of bubble diameters in an electroflotation cell, obtained by microphotography, are shown in Fig. 10.11.

The curve (■) corresponds to the region close to the electrodes (less than 2 cm), and curve (□) to the region 2 to 8 cm away from the electrodes. As one can see, bubbles located near the source are approximately 25% smaller than in the remaining volume of the flotation cell. Calculations by Rulyov (1985) show that in the leading part of the cell 60% of bubbles are the coalescence product of two, and 40% of three initial bubbles. This results in a decrease of extraction intensity by a factor 1.5 - 2 when the height of flotated liquid layer increases from 2 to 8 cm. It was also shown by Rulyov (1985) that in gradient coalescence the rate of disappearance of initial bubbles can be presented by

$$\frac{dN_D}{dt} \sim \frac{q^{5/2}}{8a_b^3} N_D,$$ (10.58)

where N_D is the number of bubbles of radius a_b.

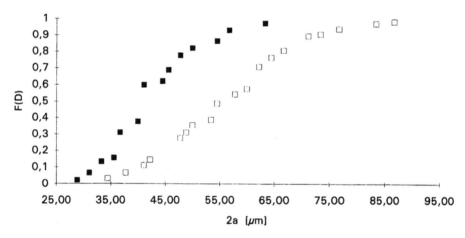

Fig. 10.11.   Function of integral distribution of bubbles diameters in an electroflotation cell at a height of 2 cm (■) and 8 cm (□) from the cell bottom, according to Derjaguin et al. (1986)

At steady-state bubbling conditions, it is difficult to provide a large number of small bubbles. The smaller the bubbles are, the faster they coalesce. Since microflotation intensity depends very strongly on the bubble radius (experimentally confirmed by Bogdanov et al., (1980)) it is necessary to minimise the negative coalescence effect. One of the methods of stabilising bubbles consists in the use of respective reagents. This is not always acceptable, especially in water purification. Another method consists in distribution of bubble sources and its intensity over the volume of the flotation cell. This decreases the number of small bubbles in the vicinity of their source and reduces the negative effect of their coalescence.

Unfortunately, generation of small bubbles (~20 μm) and maintaining their high concentration is one of the most difficult problems of microflotation which still remains to be solved. Since the surface of small bubbles is strongly retarded, their transport to the surface of large bubbles can be described, as a first approximation, by the equation derived for the sedimentation of solid spheres on a bubble surface, i.e. Eqs. (10.31) and (10.27). It has to be taken into account that small bubbles do not sediment on the surface of a large bubble. On the contrary, they rise from it so that the coefficient before the second term in Eq. (10.27) becomes negative, which yields

$$E_{os} = a_p^2/2a_b^2. \tag{10.59}$$

At a free surface of a large bubble ($Re > 100$ and very small Stokes numbers) the collision efficiency sharply increases and tends to the value given by Sutherland's formula. At a uniform retardation of the surface of large bubbles the collision efficiency decreases inversely proportional to the value of the retardation coefficient. Thus, the coalescence proceeds most intensively in the presence of large bubbles and the effect of a DAL is stronger. The formation of a DAL at higher surfactant concentration suppresses coalescence. The degree of surface activity plays a major role. For very strongly surface active substances reflected by Eq. (8.72), it becomes impossible to prevent motion over the whole surface of rather large bubbles which contributes to the transport of small bubbles on the upper bubble part. DAL and surface retardation influences the fixing stage in the same direction. To provide heterocoagulation of particles and bubbles, the bubble surface should carry a high opposite charge caused by the adsorption of cationic surfactant, so that foam films in microflotation can be stable. For large bubbles and high surface activity of cationic surfactants, i.e. under constraint (8.72), adsorption in the neighbourhood of the front pole of bubbles is insignificant, so that an electrostatic barrier is much lower and can be overcome by hydrodynamic pressing forces (see Section 10.5). During the approach of a small and a large bubble to a distance of the order of the diffusion layer thickness, non-equilibrium attraction forces arise (see Section 12.5) which allow to overcome the electrostatic barrier.

The formation of bubble-bubble aggregates under microflotation conditions is complicated by the presence of dispersed particles on their surface. This question will be considered in the next paragraph.

Gradient coagulation (cf. Van de Ven 1989) of small bubbles with the use of rather high amounts of cationic surfactant under microflotation conditions can be slowed down substantially due to electrostatic repulsion forces. High stresses develop in the zone of bubble contact in a shear flow, so that the adsorption and the double layers of bubbles can deviate from their equilibrium states. Thus, the traditional approach based on the theory of equilibrium surface forces is insufficient. This will be discussed in more detail in Section 12.5.

The non-equilibrium state of an adsorption layer has to be taken into account when selecting the foaming agent. If the bubble surface is mobile and the foaming agent is of high surface activity (condition 8.72), it can form a rear stagnant cap so that leading surface of the bubble is not protected against coalescence.

The appearance of bubble surface retardation under production conditions is often discussed in the literature (for example, Leya 1982, Gaudin 1957). In such cases surfactants with high surface activity as foaming agent are used to decrease their consumption. However, complete and partial but strong retardation cannot be differentiated through measurements of rising velocities of bubbles. Therefore, it cannot be excluded that the bubble surface is partly mobile which imposes a limitation on the surface activity of foaming agent. The smaller the retardation coefficient, the lower should be the surface activity of foaming agents and the higher the applied amounts.

10.12. TWO-STAGE FLOTATION OF MICRON AND SUBMICRON PARTICLES AND DYNAMIC ADSORPTION LAYER

As noted above, along with aggregation, there is one more possibility to accelerate flotation by using small bubbles exceeding the size of particles by not more than one order of magnitude. This means that decimicron bubbles should be used for micron or submicron size particles. On the other hand, separation of small bubbles from the liquid after floating is also an intolerably slow process. Therefore, repeated flotation can be used to separate microbubbles. Bubbles of a size one order of magnitude larger than the size of bubbles in the first stage can be used in the second stage.

Eq. (10.57) for the kinetic microflotation constant was presented in Section 10.10. The use of bubbles with $a_{b1} \sim 10$ μm instead of bubbles with radius $a_{b2} \sim 100$ μm makes it possible to decrease the time of flotation extraction of particles by one order of magnitude. The second stage proceeds much faster than the first one and has a small effect on the total rate of the technological process. The constant of the first stage can be estimated to be

$$K_1 = B \frac{a_p^2}{a_{b1}}, \tag{10.60}$$

and the constant of the second stage

$$K_2 = B\frac{(10a_{pl})^2}{10\cdot10a_{pl}}, \qquad K_2/K_1 \approx 10,$$ (10.61)

since $a_{bl} \sim 10a_{pl}$, $a_{b2} \sim 10a_{bl}$.

Using the following values: $\varphi = 0.03$, $\alpha = v/a_b^2 \approx 2\cdot10^4$, $a_p \approx 3\cdot10^{-4}$ cm, $a_{bl} = 10a_p$, the value of K_1 can be estimated to $K_1 \approx 10^{-2}$ s. A time of flotation extraction of particles of the order of 200 s is acceptable.

The interaction of larger bubbles, used at the second stage of the process, with microbubbles (decimicron bubbles), should be discussed here. Let us first assume that coalescence of bubbles is a negligible process, i.e. foam films are highly stable. Fixing of a microbubble to the surface of a larger one can be caused by the ability of particles to be fixed to the surface of microbubbles as well as a larger bubble, so that the particle acts as a bridge between them. If fixing of a particle happens as a result of coagulation in a sufficiently deep potential well, this link between bubbles is rather strong. The different aspects of two stage microflotation and the manifestation of DAL are discussed in Appendices 10G, 10H, 10I, 10K, and 10L). Some examples of industrial applications of two-stage microflotation are given in Appendix 10J.

10.13. SELECTION AND APPLICATION OF CATIONIC SURFACTANT IN MICROFLOTATION AND DYNAMIC ADSORPTION LAYER

Addition of cationic surfactant (at negative particle charge) leads to three effects which enhance flotation: no difficulties to overcome the disjoining pressure during particle deposition; attractive forces are provided which prevent detachment, hydrodynamic detachment forces are decreased due to a retardation of the motion of the lower bubble surface.

Application of ionic surfactant provides neutralisation of the bubble charge which can be roughly estimated from the surface potential Ψ. From the Gouy-Chapman theory (see Chapter 2) it follows that

$$\sigma = F\Gamma' = 4Fc_{el}\kappa^{-1}\sinh\left(\frac{\tilde{\Psi}}{2}\right)$$ (10.62)

where Γ' is the adsorption of cationic surfactant required for neutralising the bubble charge at an electrolyte concentration c_{el}. The addition of surfactant should provide a sufficiently high adsorption Γ' so that the following relation is fulfilled,

$$\Gamma > \Gamma' \approx c_{el}\kappa^{-1}.$$ (10.63)

In the most important case, when the electrolyte concentration is deci/centi-normal,

$$\kappa^{-1} < \delta_D.$$ (10.64)

Eq. (10.63) can be fulfilled at low and high surface activity of surfactant. It is not recommended to use a surfactant of very high surface activity (cf. condition (8.72)) in the absence of a rear stagnant cap since the adsorption layer is carried away to the rear pole of the bubble. Addition of such surfactant does not enhance particle deposition on the leading surface of bubbles and does not prevent their detachment from the rear surface. The surfactant molecules are concentrated in the close neighbourhood of the rear pole of the bubble and detachment of particles can occur from any part of the surface in the vicinity of which the normal velocity components are directed to the liquid.

The use of surfactant of very low surface activity is unfavourable since it is associated with excessive amounts of reagent. The higher the ratio Γ_o/c_o, the lower is the concentration c_o at which one and the same required adsorption value Γ can be provided, all within the limits of condition (8.72). The retardation coefficient increases with Γ_o/c_o even at equivalent decrease of c_o which is also important to prevent particle detachment. Therefore, we propose to use a surfactant with a surface activity at which

$$\Gamma_o / c_o \approx \delta_D,$$ (10.65)

when deposition on the surface from above happens at Re > 40. To satisfy condition (10.63), the surfactant concentration needs to be equal to

$$c_o = c_{el}\kappa^{-1}/\delta_D.$$ (10.66)

Using the following values $c_{el} \approx 10^{-2}$ M/l; $\kappa^{-1} \approx 3$ nm, $\delta_D \approx 1$ μm, we obtain a low surfactant concentration $c_o \approx 3 \cdot 10^{-5}$M. In practice it is important to use still lower surfactant concentrations.

Surfactant moving along the bubble surface is proportional to surface velocity and adsorption. If the velocity is decreased χ_b/η times, the surfactant concentration can be estimated by

$$c_o \sim \frac{c_{el}\kappa^{-1}}{\delta_D} \frac{\eta}{\chi_b}.$$ (10.67)

It has to be taken into account that the amount of reagent necessary to provide an equilibrium concentration of c_o is not simply identical to the bulk concentration added. At very high surface activity a substantial part of surfactant is adsorbed at the bubble surfaces and dispersed particles. This determines the limits of applicability of estimate (10.67) which can be generalised when the specific surface of bubbles and particles and the adsorbability of surfactant is taken into account.

The possibility of decreasing the required bulk concentration of surfactant is not only important for saving expensive reagent. A possible application of intensive microflotation conditions for preparing drinking water requires very low residual surfactant concentration. When microflotation is used in closed production water supply cycles, the residual surfactant concentration is allowed to be higher.

The minimum amount of cationic surfactants can also be estimated from Eq. (10.67) for large bubbles (Re > 40) with a strongly retarded surface.

10.14. NEGATIVE EFFECT OF INERTIA FORCES ON FLOTATION OF SMALL PARTICLES. GENERALISATION OF SUTHERLAND'S FORMULA. EXTENSION OF LIMITS OF APPLICABILITY OF MICROFLOTATION THEORY

When passing from small particles to somewhat larger particles, it is necessary to take into account the specific effect of inertia forces. In general this is a rather difficult problem. However, as long as the particles are not too large and their trajectories only slightly deviate from the respective liquid stream lines an approximation of the problem is possible which shows that the effect of inertia forces is negative. Relatively small deviations of particles from liquid stream lines result in significant effects.

The ratio between the values of inertia force and viscous resistance of the medium is characterised by the dimensionless Stokes number (10.7).

The less Stokes' number is, the smaller is the effect of inertia forces on the particle trajectory since the viscous resistance of the medium inhibits displacement of the particle from the respective liquid stream line.

Levin (1961) has shown that inertia deposition of particles below a critical size, which corresponds to a critical Stokes number $St_c = 1/12$, is impossible. Regarding a finite size of particles, the collision is characterised by Sutherland's formula (10.11). Comparison of the results obtained from Sutherland's relation and by Levin enables to conclude that in the region of small $St \leq St_c$ the approximation of the material point, accepted by Levin and useful at fairly big St, becomes unsuitable for $St < St_c$. Thus, the conditions of small Stokes numbers were studied by Dukhin (1982; 1983b) for particles of finite size. Under these conditions inertia forces retard microflotation.

396

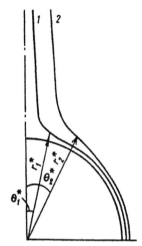

Fig. 10.12. Liquid stream-lines of a potential flow around a sphere calculated for different values of the radius of collision (the point of inflection is shown which separates the near and the distant parts of stream-lines): θ^*, z^* are polar coordinates of points of inflection with a small (1) and a large (2) radius of collision.

There is a point of inflection on each liquid stream line, which divides the line into two parts (Fig. 10.12). One of these branches is called the near and the other the distant part. On the distant part of the stream line, inertia forces shift the direction of particle motion to the bubble surface promoting deposition, a positive effect of inertia forces.

Displacement of particles along the near part is similar to displacement along the circumference so that inertia forces appear as centrifugal forces inhibiting deposition. The primary effect of inertia forces on the near part of the trajectory at $St < St_c$ was determined by Dukhin (1983). A displacement of the particle trajectory with respect to stream-line 1 which is the grazing trajectory is shown in Fig. 10.14. After displacement, particles move along the stream-line 1 away from the bubble and do not touch its surface.

If we take into account the negative effect of inertia forces on particle capture, it turns out that the grazing trajectory (Fig. 10.13) corresponds to values of b_a^2 smaller than those in Sutherland's theory and the point of tangency moves from the equator towards the front pole.

It is worth noticing that negative effects of inertia forces appear at subcritical values of Stokes numbers when a positive effect is practically absent (cf. Section 10.1). The inertia-free approach of a particle and a bubble is caused by the radial particle velocity when its centre is located at a distance from the bubble surface approximately equal to a_p. When the particle radius tends to zero, this velocity also tends to zero and deposition depends on the finite size of the particle.

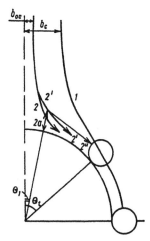

Fig. 10.13.   Diagram of grazing trajectories of particles taking into account inertia forces and SHRI (near hydrodynamic interaction): 1 - grazing trajectory in terms of Sutherland; 2 - liquid stream-line coinciding with grazing trajectory; 2' - trajectory branching out from stream-line 2 under the effect of inertia force; 2" - trajectory branching out from trajectory 2' under the effect of SHRI; θ_t collision angle; θ_1 - angle characterising the boundary of the part of trajectory controlled by SHRI.

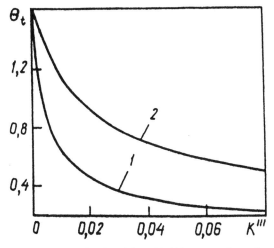

Fig. 10.14.   Angle of particle and bubble contact θ_t as a function of K''' at $E_0 = 0.01(1)$ and $E_0 = 0.05(2)$

According to this approach the consideration of a finite velocity is equivalent to the consideration of a finite size of particles in the interception effect discussed by Sutherland. Thus, it is necessary to take into consideration low radial velocity in the vicinity of the equator: first because the angular dependence (cf. Eq. 8.117) of the velocity on equator vanishes and, second, as a result of decreasing velocity with decreasing the particle size. Even at small centrifugal force and small Stokes numbers deposition in the neighbourhood of the equator is prevented. Particle deposition at a smaller angle θ is possible since the radial component of the velocity increases and the tangential component decreases, and consequently the centrifugal force.

This process is characterised by some boundary values of angle θ_t. At $\theta > \theta_t$ centrifugal forces prevent deposition. The collision efficiency E_{St} decreases under the effect of centrifugal force as it was calculated by Dukhin et al. (1981) and Dukhin (1982)

$$E_{St} = E_0 \mu\left(\frac{a_p}{a_b}, St\right)$$

(10.68)

with

$$\mu\left(\frac{a_p}{a_b}, St\right) < 1,$$

(10.69)

$$\mu = \sin^2 \theta_t \left[1 - \frac{2\cos\theta_t}{\sin^2 \theta_t}(1 - \cos\theta_t)^2(2 + \cos\theta_t)\right]$$

(10.70)

The collision angle (Fig. 10.14) characterises the position of the point of tangency of the grazing trajectory and results from the following equation,

$$\cos\theta_t = \sqrt{1 + \beta^2} - \beta,$$

(10.71)

where

$$\beta = 2E_0 / 9K''', \quad K''' = \frac{\Delta\rho}{\rho_p} St.$$

(10.72)

Considering the negative effects of inertia forces, most important is the calculation of the point of tangency θ_t where the radial component of the particle velocity at the time of contact with bubble surface becomes zero,

$$v_z\big|_{\theta=\theta_t, z=a_b+a_p} = 0$$

(10.73)

The use of a successive approximation method, i.e. as a first approximation, the identification of the particle velocity $v_p(z, \theta)$ with the local liquid velocity $v(z, \theta)$ seems to be not justified for calculating the point of tangency and it is recommended to avoid. The description of a part of the grazing trajectory of a particle was based on this approximation (Dukhin 1982) and requires generalisation. It is also necessary because θ_t decreases continuously (cf. Eq. (10.65)) with increasing Stokes numbers so that the limits of applicability of Eqs (10.68)-(10.72) have to be discussed.

In a more rigorous method by Dukhin (1983) the description of particle deposition on bubble surfaces taking into account both negative and positive effects of inertia forces is reduced to the solution of a Fuchs type equation. The treatment has shown that at sufficiently small subcritical Stokes numbers, Eqs (10.68)-(10.72) hold true. Positive effects of inertia forces become dominant with growing St, so that the relation between E_{St} and a_p has an extreme. Passage through a minimum means that at values of a_p greater than those in the minimum the increase of positive effects dominates over the negative effects with increasing St while negative effects dominate at subcritical Stokes numbers. Thus, the concept of a critical Stokes number introduced by Levin (1961) in considering deposition based on the particles' material preserves its importance also when passing to finite size particles but it requires new contents.

While neglecting the finite size of a particle at subcritical Stokes numbers excludes inertia forces at all, the situation changes with the consideration of a finite particle size. Inertia forces become essential at subcritical but not too small Stokes numbers. This effect can turn out to be negative. Thus, a critical Stokes number separates the regions of positive and negative effects of inertia forces on particle deposition.

The negative effect of the centrifugal force can be summarised by the negative effect of SRHI, which is an essential deviation from Sutherland's formula. A common action of these factors appear if the limit trajectory ends not at the equator but at $\theta = \theta_t$. Results of such common action are shown in Fig. 10.15 for a fixed bubble radius $a_b = 0.04$cm and for a number of critical film thicknesses $H_c = h_c / a_b$.

The effect of the specific density of a particle and its radius on the combined effect of centrifugal forces is shown in Fig. 10.16. It can be expected that Sutherland's formula describes the transport stage of the elementary act also at Stokes number close to the critical one. It becomes clear from the results in Fig. 10.16, that it is applicable only at θ_t close to 90°. This condition is not fulfilled over a wide range of θ_t and $\Delta\rho$.

The condition of applicability of Sutherland's formula obtained from Eq. (10.71) results in

$$\beta = 2E_0 / 9K''' \gg 1. \tag{10.74}$$

If the condition is not fulfilled, Sutherland's formula needs a refinement given by Eq. (10.68), which can be called the generalised Sutherland formula.

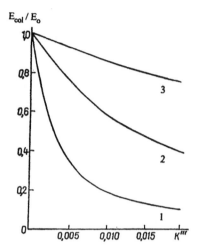

Fig. 10.15. Relative collision efficiency E / E_0 characterising the negative influence of inertial forces K''' at different E_0-values; $E_0 = 0.01(1)$, $=0.02(2)$, $=0.05(3)$

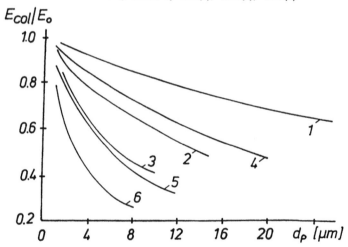

Fig. 10.16. Relative collision efficiency E / E_0 as a function of particle radius calculated for different bubble radii $a_b = 0.04$cm (1, 2 ,3) and 0.06cm (4, 5 ,6), and density $\Delta\rho = 3$ (3, 6); 1.5 (2, 5); 0.5 (1, 4); according to Dukhin (1982, 1983)

At insignificant density differences between particle and medium this condition is not fulfilled over some range of ρ and ρ' values which can be easily characterised by collision efficiency calculations based on the results of Dukhin (1982, 1983). The fulfilment of condition

$$St(a_b, a_p, \Delta\rho) < St_c \qquad (10.75)$$

is a confirmation that the elementary flotation act can be considered as an inertia-free process only when the finite particle size is neglected. At the same time Sutherland's formula does not allow to neglect the finite particle size and the possible effect of inertia forces on flotation of particles of subcritical size ($a_p < a_c$). Prior to Dukhin's works (1982, 1983) the limits of applicability of the microflotation theory was indefinite and corresponded to a radius much less than a_c. The generalisation by Dukhin (1982, 1983) of the transport stage of elementary microflotation act in terms of a quantitative description of the negative effect of inertia forces allow to extend the limits of the applicability of the flotation theory, and condition (10.75) describes this limit. This does not mean that the generalised Sutherland equation (10.68) can be used at St values arbitrarily close to St_c. Deriving Eq. (10.68), the positive effect of inertia forces on the remote part of the trajectory was ignored and only the negative action of inertia forces on the near part of the trajectory was taken into account which is true only at

$$St \ll St_c. \tag{10.76}$$

Only in this region of Stokes numbers the use of the generalised Sutherland formula is justified. Under a more severe constraint,

$$St \leq St_c \ll 1, \tag{10.77}$$

in other words, far from the bubble surface the effect of inertia forces is substantial and a more rigorous theory of the transport stage of microflotation has to be used (Dukhin, 1983).

The negative effect of inertia forces increases with the bubble size and particle density. With decreasing bubble size centrifugal forces can become equal to gravitational forces since the ratio $v^2(a_b)/a_b$ decreases with a_b. Hence, gravitational forces should be taken into account along with centrifugal forces at decreasing bubble size. Their ratio does not depend on the particle density. Thus, a critical bubble radius a_{bc} can be introduced independently of the particle density when the negative effect of inertia forces is completely compensated by the effect of gravitational forces,

$$\frac{v^2(a_b)}{a_b} = g. \tag{10.78}$$

If v is expressed by Levich's formula (10.6), we obtain

$$a_{bcr} = \left(\frac{g^2 \eta^2}{g \rho^2} \right)^{1/3} \approx 400 \mu m. \tag{10.79}$$

At $\frac{\Delta\rho}{\rho} \leq 0.1$ the negative effect of inertia becomes insignificant at any size of bubbles and particles. It is substantial at $\frac{\Delta\rho}{\rho} \approx 0.5$ and $a_b = 0.06mm$ for particles with $a_p > 20\mu m$, at $\frac{\Delta\rho}{\rho} \approx 2-3$ and $a_b = 0.03mm$ for particles with $a_p > 10\mu m$.

This theory enables to characterise quantitatively the conditions under which flotation proceeds inertia-free. It is simultaneously a generalisation of the theory of small particle flotation which not only allows to describe inertia-free flotation, but also flotation complicated by a negative effect of inertia forces. If flotation proceeds practically inertia-free at $a_p < 3-10\mu m$, the negative effect of inertia its quantitatively described over the range of bubble size from 10 to 30 μm.

Unfortunately, up to now all quantitative investigations of collision efficiency have been performed under conditions which do not allow to check the validity of the present theory. In experiments by Anfruns & Kitchener (1977), bubble surfaces were retarded to a large degree which strongly decreases the negative effect of inertia forces.

A direct method for detection of negative inertia force effects on collision efficiency at $St < St_c$ consists in performing experiments with particles of same size but with different and strongly varied density.

The calculation of E cannot be restricted to the consideration only of LRHI and the microflotation systems have to be classified in general terms as follows.

A strong effect of SRHI can be expected for particles with a smooth surface. With respect to inertia forces, two limiting cases are of interest. In oil emulsions (oil-water type) the effect of inertia forces can be neglected due to small $\Delta\rho$ and all relations derived in Section 10.1 which take into account only LRHI and SRHI can be used. Drops of hardening metal melts are spherical particles with a smooth surface and their flotation is retarded by the effect of SRHI and of inertia forces. Such systems, for example mercury emulsions, are most convenient for studying the total effect of SRHI and inertia forces. For flotation technology it is also of interest that for these systems the flotation velocities have the lowest attainable values.

For particles with an uneven, rough surface, the effect of SRHI can be weak and two interesting limiting cases are possible. At $\frac{\Delta\rho}{\rho} \ll 1$ the effect of inertia forces is negligible. This regime is most convenient for testing the theory of E_0 since the experimentally measured E will be close to E_0. At $\dfrac{\Delta\rho}{\rho}\!\gg\!1$ we can observe the effect of inertia forces in the absence of a remarkable effect of SRHI.

The process of particle deposition on a bubble is hampered by the action of three factors: LRHI, SRHI and inertia forces, which can be described by separate relations only. Therefore we will try to present their total effect as a product of the respective functions. Such an approach is not quite rigorous since these factors depend on each other. The accuracy of such an approach can be evaluated by comparing the results of more rigorous calculations of Dukhin et al. (1986). Thus, E can be presented as the product

$$E = E_P \, \mu(\theta_t), \qquad\qquad\qquad (10.80)$$

where E_P is expressed by Eq. (10.5) and describes the effect of SRHI and LRHI on collision efficiency; $\mu(\theta_t)$ is expressed by Eq. (10.70) and describes the negative effect of inertia forces. Curves calculated from Eq. (10.80) almost coincide with respective curves in Fig. 10.16 and differences between them are of the order of several per cent.

10.15. INFLUENCE OF DYNAMIC ADSORPTION LAYER ON INERTIA FORCES IN MICROFLOTATION

Dynamic adsorption layers (DAL) cause bubble surface retardation and thus decrease the negative effect of inertia forces on microflotation. It is important to keep in mind the substantial qualitative distinctions of the DAL effect and the inertia forces at high (condition (8.72)) and at low (condition (8.71)) surface activity of surfactants.

10.15.1   LOW SURFACE ACTIVITY

Uniform surface retardation allows to introduce a common retardation coefficient for the whole surface in the form of Eq. (8.108). As a first approximation the tangential velocity can be presented by $v(a_b)\sin\theta$ and the centrifugal force by $\dfrac{v^2(a_b)\sin^2\theta\,\Delta\rho}{a_b}$. Taking into account a quadratic dependence of the centrifugal force on the retardation coefficient, we can conclude that negative inertia force effects can be suppressed very easily by small amounts of surfactant. At the same time the radial component of the velocity adjacent to the bubble surface leading to particle deposition decreases, which is described by Sutherland's formula. According to Eq. (10.29), the normal component of velocity depends linearly on the tangential component and, therefore, it depends also linearly on the retardation coefficient. If centrifugal forces essentially affect the collision efficiency, its role can be substantially decreased with the introduction of surfactant. Collision efficiency increases but not to the extent expected from Sutherland's formula.

Because retardation has an effect on both the normal and the tangential components of velocity near a bubble surface and as long as centrifugal forces dominate, a non-monotone dependence of collision efficiency on surfactant concentration can be expected. Initial collision efficiency increases with increasing surfactant concentration as the decrease of negative inertia force effects dominates over the decrease of the radial velocity component. However, in the region of rather high surfactant concentrations where the surface is retarded already to such extent that negative inertia forces effects are negligible, the collision efficiency monotonically decreases with increasing surfactant concentration and retardation coefficient. Thus, this function passes through a maximum which corresponds to a compensation of the negative effects of centrifugal force and the normal component of velocity. This can be characterised quantitatively by generalising Eqs. (10.68)-(10.72)

$$\beta(\chi_b) = \frac{2E_0\eta}{9K'''\chi_b}, \quad E_{St} = \left(E_0\frac{\eta}{\chi_b}\right) \cdot \mu\{\theta_t[\beta(\chi_b)]\}, \tag{10.81}$$

and these equations preserve their importance. The first cofactor accounts for the monotonic decrease of depositing particles at each point of the surface with increasing concentration and retardation coefficient. The second cofactor accounts for the monotonic expansion of the part of surface on which particle deposition takes place when inertia forces decrease. The effect of the second cofactor initially dominates the concentration dependence of collision efficiency. Collision efficiency first grows with surfactant concentration. When the second cofactor approaches unity, i.e. at a considerable decrease of inertia forces, the concentration dependence of the first cofactor becomes dominant so that the collision efficiency decreases with further concentration increase. Even such semi-quantitative analysis is difficult for highly surface active surfactants because of the absence of a quantitative theory of the rear stagnant cap and the hydrodynamic field of bubbles at large Reynolds numbers.

10.15.2.   HIGH SURFACE ACTIVITY

The extension of the rear stagnant cap takes place at increasing surfactant concentration, adsorption and retardation of the surface which also results in a decrease of inertia forces. Any variation of surface concentration and its gradient at any point of the surface results in a change of the velocity gradient. This is a local effect.

Extension of the rear stagnant cap far from the front pole of a bubble exerts an effect on velocity distribution even in the neighbourhood of the front pole. This is a non-local effect.

As already mentioned in Section 10.2., extension of the rear stagnant cap results in an earlier separation of the hydrodynamic flow, i.e. θ_s decreases and the interception angle θ_i also decreases.

It is well known that even an insignificant expansion of the rear stagnant cap leads to a substantial increase of energy dissipation and to a substantial decrease of surface velocity. Nevertheless, as long as the secondary flow embraces a very small part of the bubble surface, its direct effect on the primary flow is possibly insignificant.

Though velocity drop leads not only to a decrease of inertia forces, but also to a decrease of radial liquid velocity at any point close to the surface. The first effect is quadratic and dominates over the second which is linear.

10.16. Effect of Hydrodynamic Boundary Layer on Elementary Act of Microflotation and Dynamic Adsorption Layer of Bubbles

Non-trivial effects of dynamic adsorption layers of bubble on microflotation are observed at large Reynolds numbers when a complete retardation of the bubble surface is observed. In this most interesting case the elementary act of microflotation is complicated by the formation of a hydrodynamic boundary layer.

10.16.1. Hydrodynamic Boundary Layer at a Slightly Retarded Bubble Surface

The effect of hydrodynamic boundary layers on particle transport (of Section 8) to slightly retarded bubble surfaces is insignificant. This is shown in Appendix 10M where the work of Mileva [1990] is discussed.

10.16.2. Hydrodynamic Boundary Layer near Strongly Retarded Bubble Surface and Sutherland's Formula

In contrast to free bubble surfaces, the role of boundary layers for the strong or complete retardation of a bubble surface is of substantial interest. Levich (1962), who introduced the notion of the hydrodynamic potential field of bubbles in the absence of surfactant pointed out that the velocity drop across the boundary layer is small. For completely or strongly retarded surfaces, the velocity distribution beyond the hydrodynamic boundary layer is a potential one, i.e. on its outer boundary the velocity is of the order of magnitude of v and on the bubble surface it vanishes. In distinction from the preceding case at strong retardation of the bubble surface, the tangential velocity change across the section of the boundary layer is equal to v. Unexpectedly we come to the conclusion that under special conditions modified Sutherland's formula (10.42) can serve as a very rough approximation also at a strongly retarded bubble surface because the potential flow is valid at $\psi < 30°$ (Clint at al. 1978).

Indeed, if a sufficiently large particle of radius a_p substantially larger than the thickness of the boundary layer moves in a potential velocity field the trajectory of its inertialess motion is described exactly in the same way as by Sutherland's formula. However, different and very difficult refinements and restrictions are necessary. First, the thickness of the hydrodynamic boundary layer is of the order of 1 - 3 μm. Therefore, particles should be substantially larger so that the inertialess approximation can fail. However, there is an interesting number of systems with particle of density only slightly differing from the density of water (emulsions, latices, biological cells). Dimensions less than 10 μm are characteristic for particles of such systems. Coarsely dispersed systems of this kind are encountered and Sutherland's formula can serve as a first approximate. At least two complications can arise. When a particle moving along the stream-lines of a potential flow it partly sinks into the boundary layer which makes it rotating. Second, even outside the boundary layer the velocity field of particles deviates from the liquid velocity field. In view of special inhomogeneities of hydrodynamic velocity, a force predicted by Faxen (1922) arises which was addressed by Mileva (1990). It increases with the square of particle radius and becomes remarkable at $a_p \gg \delta_G$. It should be added that with increasing particle size and Stokes number, it becomes more difficult to approximate the Basset integral (Basset 1961) in the general equation of non-stationary motion of particles in heterogeneous hydrodynamic flows. The effect of hydrodynamic boundary layers on the motion of particles with a size $a_p \gtrsim \delta_G$ is without doubt.

10.16.3. LIFT-FORCES

The role of lift-forces F_l (Safman, 1965) in flotation has been raised by Mileva (1990) but it is considered under conditions most unfavourable for F_l. Schulze (1992) draws attention to a possible neglecting of F_l as shown by Clift et al. (1978). Some remarks addressing the part of Mileva's work (1990) concerning lift-forces are given in Appendix 10M.

10.16.4. *Unsuitability of Traditional Methods for Describing Particle Transport through Hydrodynamic Boundary Layer of almost Completely Retarded Bubble Surface*

Since water can often be strongly contaminated by organic matter and bubble surfaces can become almost completely immobilised by an adsorption layer, a development of the theory of particle transport to a bubble surface through a hydrodynamic boundary layer is very important.

In a paper by Nguen Van & Kmet (1992), a difficult numerical integration of the Navier-Stokes equation was carried out in order to obtain information necessary for calculating collision efficiency. As usual, it is very difficult to evaluate the reliability of numerical calculations. Usually some judgement can be drawn when comparison with results from analytical formulas for known limiting cases or with numerical calculations based on substantially different methods are performed. Unfortunately such data are not available.

Collision efficiency was calculated by the method proposed for the first time by Dukhin & Derjaguin (1958). To calculate the integral in Eq. (10.25) it is necessary to know the distribution of the radial velocity of particles whose centre are located at a distance equal to their radius from the bubble surface. The latter is presented as superposition of the rate of particle sedimentation on a bubble surface and radial components of liquid velocity calculated for the position of particle centres. Such an approximation is possibly true for moderate Reynolds numbers until the boundary hydrodynamic layer arises. At a particle size commensurable with the hydrodynamic layer thickness, the differential of the radial liquid velocity at a distance equal to the particle diameter is a double liquid velocity which corresponds to the position of the particle centre. Such a situation radically differs from the situation at Reynolds numbers of the order of unity and less when the velocity in the hydrodynamic field of a bubble varies at a distance of the order $a_b \gg a_p$. At a distance of the order of the particle diameter it varies by less than about 10%. Just for such conditions the identification of particle velocity and liquid local velocity was proposed and seems to be sufficiently exact. In situations of commensurability of the size of particle and hydrodynamic boundary layer thickness at strongly retarded surface such identification leads to an error and nothing is known about its magnitude.

Unfortunately Nguen Van & Kmet (1992) even do not touch this principal difficulty. It is necessary to stress this fact at the evaluating the comparison of the formula obtained by these authors with their experiment.

This difficulty manifests itself not so remarkably at incomplete surface retardation since the radial component of liquid velocity appears and is associated with a certain bubble surface mobility which slightly varies along the cross-section of the boundary layer. Therefore the use of Eq. (10.25) in Section 10.2. is justified.

10.16.5. EXPERIMENTAL AND THEORETICAL INVESTIGATIONS OF COLLISION EFFICIENCY IN A WIDE RANGE OF REYNOLDS NUMBERS AT STRONGLY RETARDED BUBBLE SURFACE

The work by Nguen Van and Kmet (1992) is the reason for a number of open questions. First, the authors solved the Navier-Stokes equation at boundary conditions of liquid sticking to an immobile surface (see Section 10.16.3). Second, a special purification of water from contamination was not carried out. Third, the velocity of the motion of liquid with respect to a bubble fixed at a syringe needle was one-third of that at a free rising bubble with a non-retarded surface. The flow velocity in the cell varied between 1 and 10 cm/s and was 30% lower than in the case of a solid sphere (approximately 15 cm/s). If a velocity of 35 cm/s would be created (which corresponds to $Re = 200$) and measured by Okazaky (Fig. 8.2), the bubble surface possibly would manifest its mobility. Particle suspension was prepared in distilled water. It is not excluded that a high volume fraction of particles used in the experiment (1 gram of given fraction per 1 ml of distilled water), introduces impurities which could be adsorbed in such an amount to retard the bubble surface. We point out that the conditions could contribute to almost complete retardation of the bubble surface despite the fact that both large bubbles and distilled water were used. Thus, the results of the work cannot be generalised since under different conditions the residual mobility of big bubble surface can appear.

A comparison was carried out by Nguen Van and Kmet (1992) between the theoretical results and the experimental data as a function of particle radius and two dimensionless parameters: Reynolds number and Galileo number. Thus, four parameters were varied in the experiment: bubble radius, flow velocity, particle radius and density. The experiment as a whole has confirmed the theoretical dependence for all of the four parameters as predicted by the authors. Because of the importance of these results we will discuss some points which need further clarification.

There is no doubt that retardation of the surface contributes to an agreement between theory and experiment, the more the density of particles differ from the density of water. In such a situation any inaccuracy in describing the hydrodynamic field only slightly affects the collision efficiency which to a large measure depends on particle sedimentation. Second, it is difficult to see a discrepancy between theory and experiment with the logarithmic scale of graphical presentation of the results. Third, it is difficult to understand that "the accuracy of the measurement of collision radius was 0.01 mm" in connection with particles of the size of 30 µm used. The diameter of the top needle nozzle was still larger. Fourth, the grazing trajectory cannot be visually observed so that the accuracy cannot be identified. However, all of these remarks apply only to the accuracy of the experiments which was not sufficiently discussed in the work.

The non-trivial part of the study consists in calculations of the hydrodynamic field and its effect on collision efficiency. To check this result, the collision efficiency should be studied at a minimum difference between particle and water densities. Unfortunately this was not done.

10.17. SUMMARY

Although individual elements of microflotation theory have been published earlier (for example, Sutherland (1948)), they originate from the work of Derjaguin & Dukhin (1960). The most essential point in this work is not the characterisation of important features of microflotation as a separate field of colloid science, but the demarcation between microflotation and traditional flotation as the method of mineral processing. This radical demarcation made clear that flotation and microflotation belong to two different parts of surface and colloid science. Scientific problems of traditional flotation of large particles in which attention is mainly paid to thermodynamics of formation of bubble-particle aggregates and to physical chemistry of flotation reagents belong mainly to surface science and especially to surface chemistry. Scientific problems of microflotation belong to colloid science and especially to the sections of surface forces, colloid stability, and colloid hydrodynamics (Van de Ven, 1988) and thin films (Schulze, 1975, Ivanov, 1980).

Justification for a separate microflotation theory has created favourable conditions for the development of a new and independent section of surface and colloid science, which can be called Derjaguin-Dukhin microflotation theory. After more than thirty years, it can be evaluated to what extent this theory is correct and useful; it can also be pointed out which further stages of development and unsolved problems exist.

FUNDAMENTAL OF MICROFLOTATION THEORY

The most important distinction of an elementary flotation act of small particles from a similar act of large particles consists in the fact that transport of particles to a bubble surface controls the flotation kinetics rather than the sticking stage. Therefore the main reserve of intensification of water purification and selective flotation is connected with the increase of the number of collisions of particles with a collective of rising bubbles.

As the size of particles and bubbles decreases, the inertia forces affecting the process of their approach to each other, the formation of a bubble-particle aggregate, and the destruction of this aggregate decrease drastically. As the inertia forces decrease, the influence of viscous forces on these processes increases. These forces especially interfere with the process of thinning out of the interlayer between the surfaces of a particle and a bubble and long-range surface forces are of growing importance. Therefore, the flotation of small-sized particles obeys the general principles which have been intensively studied during the post-war decades in connection with the general problem of the stability of colloids controlled by surface forces.

Extension of these principles to the flotation of small-sized particles require, first, the inclusion of the hydrodynamic factor because particles deposit on a bubble from the stream of liquid flowing around it, and second, the consideration of mobility of the bubble surface if the level of the impurities which retard the surface movement is not very high. The second factor is manifest in flotation, and the specific feature of this process is connected with it.

Many of the aspects of microflotation theory were formulated in an early work by Derjaguin & Dukhin (1960). They were complemented and detailed later in another original paper by Derjaguin & Dukhin (1979), in a review (Derjaguin et al., 1984) and in monographs (Derjaguin et al. 1986, Dukhin et al, 1986). Some aspects of microflotation theory relate to non-equilibrium surface forces and will be considered separately in Chapter 12.

A number of consecutive stages can be distinguished in an elementary flotation act: first, the LRHI between a bubble and a particle, in which the particle moves along the streamlines of liquid flowing around the bubble; second, the SRHI which arises after the bubble has approached the particle to a distance of the order of the particle size.

In this case the movement of a particle to the surface slows down because thinning of the liquid interlayer involves hydrodynamic resistance which is much greater than that during the motion in a free liquid. On the other hand, since the particle slows down, the local radial velocity of liquid is higher than the velocity of the particle. Therefore, the liquid flows around the particle and presses it against the surface, contributing to the thinning out of the liquid interlayer. The final stage of thinning of the interlayer is controlled by surface forces; in this case, either the film is ruptured at a critical thickness h_{cr} of the interlayer, or it thins out gradually.

Both possibilities have been taken into account in the theory of flotation of weakly stable dispersions of spherical particles with a smooth surface. The sign of Hamaker's constant is important here. The bubble-particle aggregate is not formed in the absence of attraction forces or at $h_{cr} = 0$.

Under the influence of the LHRI the efficiency of collision drops as the particle size decreases and the bubble radius increases. This drop is characterised by a linear dependence of the collision efficiency on the ratio between the radii of particles and bubbles in the hydrodynamic potential mode and by a quadratic dependence in the Stokes mode. The LRHI causes the rate of flotation of small-sized particles to decrease by several orders of magnitude and the SRHI by several times. Aggregation of small-sized particles leads to an increase in the efficiency of collision with a bubble, but in this case the detaching forces increase.

The effect of gravity on the collision efficiency is small if the bubble surface is mobile and substantial at stagnant bubble surface.

In the absence or weakness of non-electrostatic stability factors and at excess electrolyte concentration above a threshold, possibilities of contactless flotation should be checked. The particle attachment can take place through heterocoagulation in the secondary minimum of the total-interactions energy curve, if its depth is deep enough. High electrolyte concentration, suppressing electrostatic repulsion, weakness of non-electrostatic factors of stability and the presence of attraction forces caused by molecular attraction or action of the adsorbed surfactant are favourable prerequisites for the appearance of a sufficiently deep potential well.

If the signs of charges of the bubble and particle coincide, the disjoining pressure of electric double layers (at rather low electrolyte concentration) may inhibit flotation. The electrostatic force barrier can be overcome by the hydrodynamic pressing force, so that particles are fixed on bubbles in the near potential well (coagulation in the near well). This possibility increases as the bubble velocity and the particle size increase. On the contrary, the possibility of coagulation in the far well (without overcoming the barrier) decreases as the size of bubbles and particles increases because the detaching hydrodynamic force may increase while the depth of the far potential well may be insufficient for retaining the particles.

The greater the equilibrium thickness of the liquid interlayer in a contactless flotation, the higher is the probability of detachment of particles from the bubble and the more important is the decrease of the bubble size, which reduces the detaching forces.

Realisation of microflotation is ensured by opposite signs of the charges of the particle and the bubble and at low electrolyte concentration; the effect of electrostatic attraction forces is extended over large distances and the depth of the potential well increases. Since particles and bubbles usually carry negative charges, it is expedient to use cation-active surfactants, which are predominantly adsorbed at bubbles in order to ensure contactless flotation.

VERIFICATION OF MICROFLOTATION THEORY

Main points of the theory were confirmed either in later theoretical works of other investigators or, which is especially important, experimentally or even by microflotation practice (see Sections 10.1. and 10.5.).

412

The theory of the transport stage of elementary microflotation at strong surface retardation is confirmed by works of Reay & Ratcliff (1975), Collins & Jameson (1977) and Anfruns & Kitchener (1976, 1977). Numerous confirmations of the possibility of contactless and collectorless microflotation, of the importance of overcoming or removing electrostatic barrier at microflotation and of the possibility of flotation even of hydrophilic particles through adsorption of cationic surfactant on bubble surface are presented in Section 10.5.

Note that the experience of flotation in water purification, which has so far been developed on an empirical basis, is in good agreement with the hydrodynamic theory outlined above. Small-sized bubbles are commonly used in order to ensure the reasonable rate of flotation to remove dispersed impurities, which corresponds to an increase in the efficiency of capture of particles by decreases in bubble size. Instead of flotation with large-size bubbles arising from mechanical dispersion of air, water purification is carried out by using electroflotation, microflotation, flotation by bubbles nucleated in supersaturated solutions of air in water, that is, by the methods producing small-radius bubbles. In passing over to small-sized bubbles (electroflotation, microflotation), intensification of the flotation can be attributed to the weakening of the hydrodynamic forces responsible for detachment of particles from the bubble.

DISCUSSION AND SPECIFICATION OF MICROFLOTATION THEORY

Main propositions of the Derjaguin-Dukhin microflotation theory (1961) are confirmed both with respect to the transport stage and to the attachment of small particles on bubble surfaces. This is demonstrated in Sections 10.1 and 10.5, in a monograph by Schulze (1984) and in reviews by Schulze (1989, 1991, 1993). At the same time, it should be pointed out that some propositions have been substantially changed in the process of development during the period of 30 years. Some of them are not so important as they seemed to be earlier; others are still not understood.

The proposition of importance in flotation of small particles to use small bubbles was confirmed in practice. Soon it became clear that application only of decimicron bubbles does not form a closed process and that two-stage flotation is necessary (Section 10.12). In some cases a following flotation of small bubbles is substantial or their capture by large bubbles which usually arise in much smaller amount in parallel with small ones.

With respect to short range hydrodynamic interactions its influence is not so general as assumed earlier. SHRI manifests itself in emulsion and bubble coalescence. Simultaneously, it is not so important for particles of rough surfaces.

The role of diffusiophoresis was overestimated in the original work of Derjaguin & Dukhin (1961) despite the fact that its effect was later confirmed by more detailed calculations (Dukhin et al., 1985) and by experiment (Derjaguin & Samigin, 1982). The rate of diffusiophoresis is rather high and it can have a substantial effect on flotation at low electrolyte concentration, for example below 10^{-3} M. When the electrolyte concentration increases, the rate of diffusiophoresis quickly decreases. Under industrial microflotation conditions the salt content is high and diffusiophoresis is negligible.

The possibility of collectorless microflotation (Derjaguin et al., 1984, 1986) was treated later by some investigators in the sense that flotation of any small particles can be carried out without reagents. This arouses astonishment since it was especially pointed out that even with application of cationic surfactants small hydrophilic particles do not flotate when the effective range of structural repulsive forces exceeds the effective range of electrostatic attractive forces. As applied to these systems, application of flotation reagents is required which are used in flotation of large particles. Refusal to use cationic surfactants is possible only when two conditions are met: the dimensions of bubbles and small particles should not be too small so that hydrodynamic pressing forces (Section 10.5) can surmount an electrostatic barrier, the depth of the first energetic minimum must be sufficiently large even at increasing repulsive forces due to increasing surface charge (accompanied by a contraction of the mobile surface of bubble) which prevent particle detachment. Only the development of the DAL theory at large Reynolds numbers allows one to define more accurately the restriction concerning the increase of surface concentration of ionic surfactant up to saturation and, respectively, of the charge along stagnant cap.

Water can contain impurities which immobilise the surface even of larger bubbles. In this situation to which insufficient attention was paid so far (Derjaguin et al., 1960- 1986) the peculiarity of the transport stage disappears and the problem of DAL loses its importance. At the same time it is important to underline that the loss of interest in the problem of DAL as applied to microflotation has no grounds and is fraught with serious losses for microflotation technology.

DYNAMIC ADSORPTION LAYER AND MICROFLOTATION IN CONTAMINATED WATER. RESIDUAL MOBILITY OF BUBBLE SURFACE AND COLLISION EFFICIENCY

Since surface tension of tap water and even distilled water is lower than expected for real clean water, it is commonly supposed that under conditions of industrial flotation the bubble surface is completely retarded due to water contamination. Justification is based on the fact that the bubble rising velocity is close to that calculated for solid spheres. The latter conclusion is not unambiguous. Small residual mobility of a bubble can exist without any effect on the rising velocity but strongly increases collision efficiency. Formulas are derived for a rough estimation of the effect of residual mobility of bubble surface on collision efficiency for limiting cases of a high surface activity, Eq. (10.42); and low surface activity, Eq. (10.32). Their refinement is complicated because of the lack of quantitative theory of DAL at high Reynolds numbers.

DEGREE AND NATURE OF WATER CONTAMINATION AND COLLISION EFFICIENCY

The nature and degree of contamination of waste water varies over a wide range and only a qualitative classification of the situations can be offered. The most important factor is a very strong increase of surface retardation with decreasing bubble size. Therefore the problem of residual surface mobility can be related to bubbles characterised by large Reynolds numbers and less commonly to those with Reynolds numbers in the intermediate range. It is shown in Sections 10.2.4 to 10.2.6 that in two limiting cases of low and high surface activity the residual mobility can substantially increase the collision efficiency in the region of high Reynolds numbers and at surfactant concentrations not exceeding 1-10 mg/l. At high surface activity this is encouraged by the disappearance of the effect of surfactant concentration on surface tension above CMC.

Soluble substances exist which can immobilise the surface even of large bubbles present in water at extremely low concentrations. The problem of the effect of residual mobility of a bubble surface loses its meaning if these impurities are contained in water. However, their influence on surface mobility can be hampered at retarded adsorption kinetics. At a given surface tension decrease due to impurities a critical bubble dimension exists. For bubbles exceeding the critical size a residual surface mobility is present. Eq. (10.40) interrelates the critical bubble size with the surface tension drop. On the basis of Eq. (10.45) it was shown that residual mobility is important even for highly contaminated river water at high Reynolds numbers (cf. Section 10.2.7).

DYNAMIC ADSORPTION LAYER, RESIDUAL MOBILITY OF BUBBLE SURFACE AND MICROFLOTATION TECHNOLOGY

As shown in Section 10.12, effective microflotation is possible as a two-stage process. At the first stage the problem of residual mobility does not arise because bubbles of decimicron dimension are used. At the second stage the use of bubbles of millimetre size is technologically effective if their surface is not completely retarded. Methods to obtain intermediate-size bubbles are insufficiently developed. Residual mobility of bubble surfaces can cause a substantial increase in collision efficiency. At the same time it has an effect on the distribution of the flotation reagents along the bubble surface and thus on the possibility of attachment and detachment of particles. A decreased input of cationic flotation reagents made possible through the change to higher surface active homologs is limited by the need of a residual mobility.

Residual mobility of a bubble surface can substantially intensify microflotation of emulsions since sedimentation of emulsion droplets on a bubble surface is very slow. On the contrary, in microflotation of suspensions it can be insignificant since the rate of sedimentation of dispersed particles of high density on bubble surface is one or two orders of magnitude higher than of emulsions. Aggregated suspensions represent the intermediate case since the density of aggregates is lower than that of original particles but higher than for emulsions.

To model the variety of processes effecting the second stage of microflotation, the development of the theory of DAL at high Reynolds numbers is necessary.

PROCESSES LEADING TO THE INCREASE OF BUBBLE SURFACE MOBILITY AND TO INTENSIFICATION OF MICROFLOTATION.

Two factors which can sharply increase bubble surface mobility and intensify the second stage of microflotation are usually not taken into account and even not discussed.

Experiments by Loglio et al. (1989) have shown, that the stationary rising velocity of large bubbles in sea and tap water is reached after a path of the order of several meters. This means that on the initial section of about one meter, the surface of sufficiently large bubbles cannot be completely retarded. Thus, we can conclude that a high initial collision efficiency corresponding to an originally clean surface can be preserved on a sufficiently long path of rising as long as the process of accumulation of impurities on bubble surface is not too fast. At not too high water contamination, the microflotation process can largely proceed on the "initial" path of rising where the initial collision efficiency can exceed the final one by one or two orders of magnitude.

Thus, in microflotation the processes of deposition of disperse particles and adsorption of molecular contaminants on the bubble surface proceed in parallel. It has been estimated in Section 10.2.7. that the rates of these processes are commensurable even when particles are aggregated which accelerates particle recovery. This means that in microflotation the level of impurities immobilising the bubble surfaces decreases and that the residual mobility of the bubble surface and also the role of DAL in microflotation increases.

Let us point out two more aspects of this problem, which contribute to the preservation of a noticeable mobility of the bubble surface. It is expected that in the development of microflotation technology, increasingly deeper purification from dispersed particles and therefore from molecular contaminations can be attained under the condition of a remarkable residual mobility.

Another aspect is connected with a joint consideration of two processes affecting a residual mobility. At the beginning a high initial collision efficiency is retained since the process of adsorption of molecular impurities on the bubble surface is not too fast. Then the decrease of residual mobility slows down due to drop of concentration of molecular impurities. Thus, a joint action of two factors considered above separately can provide a prolonged preservation of surface mobility.

A possibly substantial intensification of the microflotation process follows from the findings. If we provide a maximum intensive microflotation process, it simultaneously intensifies the purification of the system from molecular impurities. This leads to an increase of the residual mobility and, correspondingly, to an increase of the collision efficiency. Unfortunately this way of intensification is restricted by the existence of an optimal volume fraction of bubbles (Derjaguin & Dukhin, 1986).

The coupling of the processes of microflotation of dispersed particles and molecular impurities presents big problems in optimising the technology. This points to the need of mathematical modelling and respective development of the theory of nonstationary DAL.

EFFECT OF MICROFLOTATION THEORY ON THE THEORY OF FLOTATION

Little attention was paid to the transport stage of flotation until the theory of microflotation was developed. Only one possibility was considered for the approach of particle and bubble surface and of the thinning of the liquid interlayer (colliding process, Section 11.1). In connection with the well-known experiments modelling the transport stage in microflotation (Schulze & Gottschalk, 1981), attention was drawn by Schulze & Dukhin (1982) to the possibility to realise a detailed study of the transport process in microflotation (sliding process). It is the process which later attracted attention of investigators of the elementary flotation act. Sliding processes and how they are strongly affected by dynamic adsorption layers are discussed in Chapter 11.

10.18 REFERENCES

Anfruns, J.P. and Kitchener, J.A., The Absolute Rate of Capture of Single Particles by Single Bubbles, Flotation, A. M. Gaudin Memorial, Vol.2, Fuerstenau, M.G. (ed.), American Institute of Chemistry, New York, (1976)

Anfruns, J.P. and Kitchener, J.A., Trans. Inst Mining and Mat., 86(1977)C9

Babenkov, E.D., Water Purification by Coagulation, Nauka, Moskow, (1977)

Basset , A.B., A Treatise on Hydrodynamics V.II, Chapter 5, Deigton Bell, Cambridge 1888; Dover Publ., NY (1961)

Blake, T.D. and Kitchener, J.A., J. Chem Soc., Faraday Trans., 68(1972)1435

Bleier, A., Goddard, E.D. and Kulkarni, R.D., J. Colloid Interface Sci., 59(1977)490

Bogdanov, O.S. and Kiselwater, B.W., in Proceedings of a scientific session of the Institute Mechan. Metallurgizdat, Moskow, (1952)51

Botsaris G.D.and .Plazman Yu.M, J. Dispersion Sci. Technol., 3(1982)67

Brodskaya E.N.and Rusanov A.I. , Kolloidn. Zh., 48(1986)3

Churaev, N.V., "Properties of the Wetting Films of Liquid", in Surface Forces in Thin Films and Stability of Colloids, Derjaguin, B.V. (Ed.), Nauka, Moskow, (1974)81

Churaev N.V., Kolloidn. Zh., 46(1984)302

Churaev N.V., Colloids and Surfaces, 79(1993)25

Claesson P.M. and Christenson, H.K. J. Phys. Chem., 92(1988)1650

Clarke, A.N. and Wilson,J., FoamFlotation, Marcel Decker, N.Y and Basel, (1983)

Clift, R., Grace, J.R. and Weber, M.E., Bubbles, Drops and Particles, N.Y., etc.: Acad. Press., 1978, pp 380

Coffin Vernau, L., Column Flotation of Gaspe., 14th Int. Miner. Process. Technol. Toronto, (1982)

Colic, M.,.Fnerstenau, W. ,Kallay, W and Matijevie, E., Colloids and Surf. 59(1991)169

Collins, C.L. and Jameson, G.L., Chem. Eng. Sci., 32(1977)239

Derjaguin, B.V., Kolloid-Z., 69(1934)155

Derjaguin, B.V., Izv. Akad. Nauk SSSR, Ser. Khim. 5(1937)1153

Derjaguin B.V.Zh. Fiz. Khim. 14(1940)137

Derjaguin, B.V. and Landau, L.D., Zh. Eksp. Teor. Fiz., 15(1945)663

Derjaguin, B.V., Kolloidn. Zh.. 16(1954)425

Derjaguin, B.V. and Zorin, Z.M., Zh. Fiz. Khim., 29(1955)1755

Derjaguin B.V.and Zorin, Z.M., Zh. Fiz. Khim, 29(1955)1010

Derjaguin, B.V. and Dukhin, S.S., Izv. Akad. Nauk SSSR, Otdel.Metall.Topl., 1(1959)82

Derjaguin, B.V. and Dukhin, S.S., Trans. Inst. Mining Met., 70(1960)221

Derjaguin, B.V. and Shukakidze N.D., Trans.Inst.Mine Metal, 70(1961)569

Derjaguin, B.V. and Kudryavtseva, N.M., Kolloidn. Zh., 26(1964)61

Derjaguin, B.V. and Churaev, N.V., Croat. Chem. Acta, 50(1970)187

Derjaguin, B.V. and Churaev, N.V., J. Colloid Interface Sci., 49(1974)249

418

Derjaguin, B.V., Chem. Scr., 9(1976)97

Derjaguin, B.V., Dukhin, S.S., Rulyov, N.N. and Semenov, V.P., Kolloidn. Zh., 38(1976)258

Derjaguin, B.V., Dukhin, S.S. and Rulyov, N.N., Kolloidn. Zh., 38(1976)251

Derjaguin, B.V., Dukhin, S.S. and Rulyov, N.N., Kolloidn. Zh., 39(1977)1051

Derjaguin, B.V., Rulev, N.I. and Dukhin, S.S., Kolloidn. Zh., 39(1977)680

Derjaguin, B.V., Dukhin, S.S., Kinetic Theory of the Flotation of Small Particles , Proc. 13. Int.Miner.Process Cong.Warsawa, 2(1979), 21-62

Derjaguin, B.V. and Churaev, N.V., Wetting Films, Nauka, Moscow, (1984)

Derjaguin, B.V., Dukhin, S.S. and Rulyov, N.N., Kinetic Theory of Flotation of Small Particles, Surface and Colloid Science (E. Matijevic, Ed.), Vol 13, Wiley Interscience, New York, (1984) 71-113

Derjaguin B.V. and Churaev N.V., J. Colloid Interface Sci., 103(1985)542

Derjaguin, B.V., Dukhin, S.S. and Rulyov, N.N., Microflotation, Nauka, Moscow, (in Russian), (1986)

Derjaguin B.V., Churaev N.V., and Muller V.M., Surface Forces, Consultants Bureau-Plenum, New York, 1987

Derjaguin B.V. and Churaev N.V., Colloids Surfaces, 41(1989)223

Devivo, D.G. and Karger, B.L., Sep. Sci., 5(1970)145

Dibbs, N.F., Sireis, L.L. and Bredin, R., Department of Energy, Mine and Resources, Ottawa, Research Rep. R 248(1972)

Dobias, B. (Ed.), "Coagulation and Floculation", Marcel Decker, New York, (1993)

Dukhin, S.S. and Derjaguin, B.V., Kolloidn. Zh., 20(1958)326

Dukhin S.S. and Semenikhin, N.M., Kolloidn Zh. 32(1970)366

Dukhin, S.S., Abh. Akad. Wiss. DDR, 1(1976)561

Dukhin, S.S., Kolloidn. Zh., 44(1982a)896

Dukhin,S.S., Kolloidn. Zh., 44(1982b)431

Dukhin, S.S., Kolloidn. Zh., 45(1983)207

Dukhin, S.S. and Rulyov, N.N., Kolloidn. Zh., 39(1977)270

Dukhin, S.S., Rulyov, N.N. and Semenov, V.P., Kolloidn. Zh., 41(1979)263

Dukhin, S.S., Rulyov, N.N., Leshchov, E.S., and Yeremova, Yu.Ya., Chim. i Technol.Vodi, 3(1981)387

Dukhin, S.S., Listovnichiy, A.V. and Zholkovski, E.K., Kolloidn. Zh., 47(1985)240

Dukhin, S.S., Rulyov, N.N. and Dimitrov, D.S., Coagulation and Dynamics of Thin Films, Naukova Dumka, Kiev, 1986, (in Russian)

Dukhin, S.S., Adv. Colloid Interface Sci., 44(1993)1

Fainerman V.B., private communication.

Fonda, A. and Herne, H., Nat. Coal. Board, M. R. E., Rep. 2068, in Aerosol Science, C.N. Davies (Ed.), Academic Press, New York, (1966)393

Fowkes, F.M, J. Adhes. Sci. Technol., 4(1990)669

Frumkin, A.N., Zhurn. Fiz. Chim., 12(1938)337

Frumkin, A.N. and Gorodetskaja, A.V., Acta Physicochim. USSR, 9(1938)327

Gaudin, A.M., Flotation, Mc Graw Hill, N.Y., London, (1957)

Golman et al., Flotazionie Sistemi, Procesi i Apparati, Institut Fiziki Zemli AN SSSR, Moskow, (1974), (in Russian)

Goren, S.L. and O'Neill, M.E., Chem. Eng. Sci., 25(1971)325

Grassman, P., Physikalishce Grundlagen der Chemie-Ingenieur Technic, Verlag Sauerlander, Aarau,(1961), p. 707, 748, 759

Hamielec, A.E. and Johnson, A.I., Can. J. Chem. Eng., 40(1962)41

Hamielec, A.E., Storey, S.H. and Whitehead, J.M., Can. J. Chem. Eng., 41(1963)216

Hesleiter,P. Kallay, N.and .Matijevie, E., Langmuir, 7(1991)1

Hogg R, Healy T.W. and Fuerstenau D.W., J. Chem. Soc. Faraday Trans., 1, 62(1966)1638

Hosler, A., Wasser-Abwasser, 105(1964)764

Hornsby, D. and Leya, J., Surface and Colloid Science (E. Matijevic, Ed.), Wiley, N.Y., 1982, V. 13, p. 234

Huddleston, R.W. and Smith, A.L., 10th Int. Conf. Soc. Chem. Ind., Brunel Univ., 1975

Iljin V.V., Chrjapa V.M., and Churaev N.V., Fiz. Mnogochastich. Sist, 18(1991)50

Israelachvili J.N. and Pashley R.M., J. Colloid Interface Sci., 98(1984)500

Iunzo, S. and Isao, M., Application of air-dissolued flotation for seperation, 14th Int. Mineral Process. Congr. World wide Ind. Appl. Miner. Process Technol. Toronto, (1982

Ivanov , I.B., Pure Appl.Chem., 52(1980)1241

Izmailova, V.N., Yampolskaya, G.P. and Summ, B.D., Surface Phenomenian Protein Systems, Chimiya, Moscow, 1988

Jaycock, M.J. and Ottewil, R.N., Trans. IMM, 72(1963)497

Joy, A.S. and Robinson, A.I., in Recent Progress in Surface Science, Vol 2, Academic Press, New York, (1964)169

Kihira Hiroshi and Matijevie E., Adv. Colloid Interface Sci., 42(1992)1

Kihira Hiroshi, Ryde N, and Matijevie E., Colloids Surfaces 64(1992)317

Kijlstra, S..von Leenvan, H. P., and Lyklema H. Langmuir,(1993)

Klassen, V.I. and Mokrousov, V.A., Vedenie v teoriju flotacii, Gosgortechizdat, Moscow 1959

Klassen, V.I., Vodnie Resursi, 6(1973)99

Kovalenko, V.S., Skripnik, V.N. and Jakovleva, E.D., Sudovie energetickie ustanovki, N9(1979)165

Langmuir, J. and Blodgett, K., Mathematical Investigation of Water Droplet Trajectories, Gen. Elec. Comp. Rep., July, 1945

Laskowski, J. and Kitchener, J., J. Colloid Interface Sci., 29(1969)670

Levich, V.G., Physicochemical Hydrodynamics, Prentice-Hall, Englewood Cliffs, New Jersey, (1962)

Levin, L.I., Research into the Physics of Coarsely Disperse Aerosols, Isd. Akad. Nauk SSSR, Moscow, (1961)267, (in Russian)

Li, C. and Somasundaran, P., J. Colloid Interface Sci., 146(1991)215

Li, C. and Somasundaran, P., J. Colloid Interface Sci., 148(1992)587

Li, C. and Somasundaran, P., Colloids & Surfaces, 81(1993)13

Loglio, G., Degli-Innocenti, N., Tesei, U. and Cini, R., Il Nuovo Cimento, 12(1989)289

Lopis, J., in Modern Aspects of Electrochemistry. J.O.M.Bockris and B.E.Convay Eds.,
    Plenum Press, New York, 1971

Luttrel, G.H., Adel, G.T. and Yoon, R.H., Proc. 16 Int. Miner. Process Cong., K.S.E.Foresberg
    (Ed.), Stockholm, Elsevier, Amsterdam, Vol. 2, (1988)1791

Lyklema, J. Dukhin, S.S., and Shilov, V.N., J. Electroanal. Chem. Interfacial Electrochem.
    143(1983)1

Mahanty, J. and Ninham, B.V., Dispersion Forces, Academic Press, New York, (1976)

Mamakov, A.A., Modern State and Perspectives of Electroflotation, Shtiintsa, Kishinev,
    (1975), (in Russian)

Martinov, G.A. and Muller, V., in Surface Forces in Thin Films and Disperse Systems,
    Derjaguin, B.V. (Ed.), Nauka, Moskow, (1972)7

Matov, B.M., Electroflotation, Karta Moldovanska, (1971), (in Russian)

Matsnev, A.I., Purification of effluant by flotation , Budivelnik, Kiev, (1976), (in Russian)

Moor, D.W., J. Fluid Mech., 16(1963)161

Nebera, V.P., Rebrikov, D.N. and Kuzmin, V.I., Flotation of secondery mined lead-zine ores in
    column machines 14th Int. Miner Process Conep. World Wide , Ind. Appl. Miner.
    Process Technol. Toronto, (1982)

Nguen Van, A. and Kmet, S., Int. J. Miner.Process, 35(1992)205

Okada Keniji , Akagi Yasuhary , Kogure Masahiko, and Yoshioka Naoya , Canad. J. Chem.
    Eng., 68(1990)393

Okada Keniji , Akagi Yasuhary , Kogure Masahiko, and Yoshioka Naoya , Canad. J. Chem.
    Eng., 68(1990)614

Okazaki, S., Bull. Chem. Soc. Japan, 37(1964)144

Overbeek, J.Th.G., J. Colloid Interface Sci., 58(1977)408

Overbeek J.T.G., Colloids Surfaces, 51(1990)61

Pashley R.M. and Israelachvili J.N., Colloids Surfaces, 2(1981)169

Peebles, F.N. and Garber,H.J., Chem.Eng.Progress, 49(1953) 88/97

Petrarz, A.A., Tsvetnie Metali, 10(1981)109

Protodiaconov, I.O. and Ulanov, S.B. Hydrodynamics and Mass Transport in Disperse Systems
Liquid-liquid, Nauka, Leningrad, (1986), (in Russian)

Rabinovich Ya.I. and Derjaguin B.V., Colloids Surfaces, 30(1988)243

Randles J.E.B., Phys. Chem Liquids, 2(1977)107

Rao, S.R., Minerals Sci.Eng., 6(1974)45

Reay, D. and Ratcliff, G.A., Can. J. Chem. Eng., 51(1973)178

Reay, D. and Ratcliff, G.A., Can. J. Chem. Eng., 53(1975)481

Rubin, A.J. and Lackay, S.C., J. Amer. Water Works Assoc., 10(1968)1156

Rulyov, N.N., Kolloidn. Zh., 40(1978)898

Rulyov, N.N., Chimija, Technologija Vodi, 7(1985)9

Samygin, V.D., Chertilin, B.S. and Enbaed, I.A., Kolloidn. Zh., 42(1980)898

Samygin, V.D., Chertilin, B.S. and Nebera, V.P., Kolloidn. Zh., 39(1977)1101

Schulze, H.J., Physico-Chemical Elementary Process in Flotation, Analysis from Point of View of Colloid Science, Amsterdam, Elsevier, (1984)348

Schulze, H.J., Mineral Processing and Metallurgy Review, 5(1989)43

Schulze, H.J. and Cichos, C., Z. Phys. Chem., 251(1972)252

Schulze, H.J. and Cichos, C., Mitt. Forschungsinstitut für Aufbereitung, Freiberg, (1972)7

Schulze, H.J. and Dukhin, S.S., Kolloidn. Zh., 44(1982)1011

Schulze, H.J. and Gottshalk, Kolloidn. Zh., 43(1981)934

Schulze, H.J., Adv. Colloid Interface Sci., 40(1992)283

Schulze, H.J., Colloid and Polym. Sci., 253(1975)790

Schulze, H.J. in Dobias Ed."Coagulation and Floculation", Marcel Decker, New York,(1993)

Somasundaran, P. and Char. K., Colloids & Surfaces, 8(1983)121

Skrilev, L.D. et al., Zh. Prikl. Chimii, 50(1977)1410

Skvarld J. and Kmet,S. Colloid and Surfaces, 79(1993)89

Sotskova, T.Z., Gutovskaja, V.V., Golik, G.AS., Losinskij, A.M and Kulskij, L.A., Chimija i Technologija Vodi, 3(1981)396

Sotskova, T.Z., Bazhenov, Yu.F. and Kulskij, L.A., Kolloidn. Zh., 44(1982)989

Sotskova, T.Z., Poberezhnij, V.Ya., Bazhenov, Yu.F. and Kulskij, L.A., Kolloidn. Zh., 45 (1983) 108

Stahov, E.A., Purification of Waste Water from Oil, Nedra, Leningrad, (1983), (in Russian)

Sutherland, K.L., J. Phys. Chem., 58(1948)394

Taylor, P., Proc. Rog. Soc. A, 108(1924)11

Townsen R.M.and Rice R.M., J. Chem. Phys., 94(1991)2207

Usui, S., in Progress in Surface and Membrane Science, Vol 5, Academic Press, New York, (1972)233

Van de Ven, T.G.N., Colloidal Hydrodynamics, Academic Press, London, (1989)

Van Oss, C.J., Chandhury, M.K. and Good, R.J. Adv. ColloidInterface Sci., 28(1987)35

Verwey, E.J.W. and Overbeek, J.Th.G., Theory of the Stability of Lyophobic Colloids, Elsevier, Amsterdam, 1948

Weber, M.E., J. Separ. Technol., 2(1981)29

Weber, M.E., Blanchard, D.C. and Syzdeck, L.D., 28(1983)101

Weber, M.E. and Paddock, D., J.Colloid Interface Sci., 94(1983)328

Veitser, Y. and Mints, D.M., Polymer Floculants in Water Purification,Stroiizdat, Moscow, (1980)

Xu Z. and Yooh R.H. J. Colloid Interface Sci., 143(1990)427

Xu Z.and Yoon, R.H. J. Colloid Interface Sci., 132(1989)532

Yoon, R.H. and Luttrel, G.H., Preprints of 17 Int. Mineral Process Cong., Dresden, Vol. 2, (1991)17

Yoon, R.H. and Luttrel, G.H., Miner. Process Extr. Metall. Rev., 5(1991)101

Zorin, Z.M., Romanov, V.P. and Churaev, N.V., Colloid Polymer Sci., 257(1979)986

patent 310683 USSR, MKI B 03 D 1/24

patent 4226706 USA, MKI B 03 D 1/24

CHAPTER 11

11. DYNAMIC ADSORPTION LAYER IN FLOTATION

Depending on the extent of inertia force effects the hydrodynamic stage of the elementary act of flotation proceeds very different. Four quantitatively different conditions of particle deposition on bubbles can be differentiated depending on the Stokes number St (Dukhin et al., 1986):

1. St«0.1:   Inertia forces have practically no effect on the motion of particles and can be considered as inertia-free (see Section 10.14).

2. St ≤ 0.1:   As shown in Section 10.14, inertia forces can impede particle deposition on a bubble.

3. 0.1 < St < 1:   An inelastic inertia impact of particles on a bubble surface is characteristic and, as will be shown below, a major portion of kinetic energy of the particles get lost both during the approach to the bubble and at the impact itself, when a liquid interlayer of liquid is formed between the surfaces of a particle and a bubble.

4. St>3:   The trajectory of a particle deviates very slightly from a straight line and the energy of the particle as it approaches the bubble and on collision changes so little that the impact can be considered as being quasi-elastic.

A characteristic feature of the latter two cases is the availability of inertia impact of a particle on a bubble surface. As a result, a thin liquid layer is formed between them and the dynamics of its thinning and the dissipation of its energy to a large extent determine the likelihood of particle attachment to the bubble or its jump back from the surface (Spedden & Hannan 1948, Whelan & Brown 1956; Schulze & Gottschalk 1981). Particles are repelled from the bubble surface if the film has no time, during collision, to flow out to the critical value at which its spontaneous break-through and the formation of a three-phase wetting perimeter takes place. It was found experimentally that attachment of a particle on a bubble can happen not only at the first but also at a repeated collision - once the particle has lost a considerable part of its kinetic energy (Stechemesser 1989; Bergelt et al. 1992). In Sections 11.1 and 11.2 dynamics of thinning of a thin liquid film formed at an inertia impact of a spherical particle on a bubble surface is considered and energy losses of the particle at such impact are calculated.

11.1. QUASI-ELASTIC COLLISION

If the initial kinetic energy of a particle (in the system of coordinates of the rising bubble) is sufficiently large, the dissipation of particle energy to overcome viscous forces in the course of impact can be neglected. In this case processes of film formation and deformation and its thinning can be considered as being independent which enables us to calculate the finite thickness of the film after an inertia impact and to determine the possibility of its break-through. It was shown by Dukhin et al. (1986) that the rate of deformation of a locally flat bubble surface is small compared with the curvature and velocity of a spherical particle falling normally on it under the condition

$$h \gg a_p \sqrt{3De/2\alpha},$$ (11.1)

where $De = v\eta/\gamma$ is the Derjaguin number. The quantity α reflects the degree of retardation of the bubble surface by the surfactant adsorption layer. Its value is 1 when the surface is completely retarded and 4 when the surface is free. With decreasing h the outflow of liquid from the gap becomes difficult due to viscosity. At a fixed particle velocity, a positive pressure and the capillary pressure compensating it grow with decreasing h. The section of the surface which is approached by the particle distorts to a greater extent and the speed of its displacement increases. Clearly, the quasi-stationary state of the liquid surface (the state in which the normal velocity is much smaller than the velocity of the particle) is possible only when the capillary pressure can compensate the positive hydrodynamic pressure. Since the curvature of the liquid surface cannot be greater than a_p (except for the cases where a dimple is formed, cf. Derjaguin & Kusakov 1939), beginning at $h = h_o$ the velocity of the liquid surface becomes equal to the particle velocity which then is completely determined by the forces of the film surface tension. The equation of motion takes the form

$$m\frac{d^2x}{dt^2} = -\frac{dW_\gamma}{dx},$$ (11.2)

where m is the mass of the particle, x is coordinate of particle centre normal to the surface (Fig. 11.1), W_γ is free energy of the liquid surface. If we assume that the shape of the surface at the place of impact is a sphere, we obtain

$$W_\gamma = \pi\gamma\left(a_p + h - x\right)^2 + \text{const}, \tag{11.3}$$

which after substitution into (11.2) and integration with the initial conditions

$$t = 0, \; x = a_p, \; \frac{dx}{dt} = -v, \; h = h_o \tag{11.4}$$

gives the expression

$$a_p + h - x = \frac{v}{\omega}\sin\omega t + h_o\cos\omega t, \tag{11.5}$$

where

$$\omega = \sqrt{\frac{2\pi\gamma}{m}} = \sqrt{\frac{3\gamma}{2\rho a_p^3}}. \tag{11.6}$$

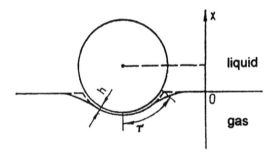

Fig. 11.1.   Diagram illustrating the impact of a spherical particle on a bubble surface with the formation of a liquid interlayer.

At a quasi-elastic impact the film thinning velocity is much smaller than the particle velocity. Thus dh/dt is very much less than unity and can be neglected in deriving Eq. (11.5). It is easy to obtain the well-known formula of Evans (1954) for the film deformation time from Eq. (11.6),

$$T_o = \sqrt{\pi m/2\gamma} = \sqrt{2\pi^2\rho a_p^3/3\gamma}. \tag{11.7}$$

It was shown by Dimitrov & Ivanov (1978) that a dimple is not formed at the impact of a particle on liquid surface if the following condition is fulfilled,

$$a_p^2 g\rho/3\gamma \ll 1,$$ (11.8)

which is always true for particles of flotation size. This allows us to take advantage of a so-called plane-parallel approximation (Ivanov et al. 1979) which assumes that the film in the gap is uniform in thickness and follows the shape of the particle surface (see dashed line in Fig. 11.1). Then the film thinning velocity can be obtained from Reynolds formula (1886). It is convenient to present it in the form

$$dh/dt \approx -2\alpha h^3 p_\gamma/3\eta l^2,$$ (11.9)

where l is the radius of the film, p_γ is the capillary pressure, in the present case given by

$$p_\gamma \cong 2\gamma/a_p,$$ (11.10)

$$l = \sqrt{\frac{a_p}{2}(a_p + h - x)}.$$ (11.11)

After substitution into (11.9) with regard to (11.5), we obtain

$$\frac{dh}{dt} = -\frac{2\alpha\omega\gamma}{3va_p^2\eta}\frac{h^3}{\left(\sin\omega t + \frac{\omega h_o}{v}\cos\omega t\right)}.$$ (11.12)

Integration of Eq. (11.12) with the initial condition $h(t = 0) = h_o$, yields

$$h_T = \frac{h_o}{\sqrt{1 + \frac{4\alpha\gamma h_o^2}{3va_p^2\eta\sqrt{1+A}}\ln\left(\frac{\sqrt{1+A}+1}{\sqrt{1+A}-1}\right)}},$$ (11.13)

where h_o and h_T are the thickness of the film at the beginning and at the end of impact (when the particle takes up position $x = a_p$), respectively, and A is given by

$$A = (\omega h_o/v)^2.$$ (11.14)

As mentioned above, h_o corresponds to the thickness of a film at which the liquid surface can no more be in a quasi-stationary state and its velocity starts to approach the particle velocity. Therefore, the inequality should be fulfilled at $h \geq h_o$,

$$|F_\gamma| \geq |F_n|,$$

(11.15)

where $F_\gamma = p_\gamma \pi^2$ is the capillary force acting on the deformed liquid surface, F_n is the liquid interlayer resistance against thinning which is equal to the integral of hydrodynamic pressure over the film area expressed by

$$F_\gamma = -3\pi\eta vl^4/2\alpha h^3.$$

(11.16)

Thus on the basis of Eqs (11.10), (11.11), (11.15), and (11.16) we obtain for $x = a_p$ and $h = h_o$,

$$h_o \geq a_p \sqrt{3De/2\alpha}.$$

(11.17)

Using (11.17), one can show that the second term in the root in Eq. (11.13) is much smaller than unity. Moreover, a quasi-elastic impact is obtained in the case A<0.01. Therefore Eq. (11.13), after some simplifications, takes the form

$$h_T \approx \frac{a_p \sqrt{De}}{2\sqrt{\dfrac{\alpha}{3}\ln\dfrac{4}{A}}}.$$

(11.18)

A remarkable feature of Eq. (11.18) is, that the finite film thickness h_T shows only a slight dependence on parameter A. This allows us to replace the approximate inequality in Eq. (11.17) by an exact equality. Then, after substituting Eq. (11.17) into Eq. (11.14), we obtain

$$A = \frac{9\eta}{4\alpha a_p v\rho} = \frac{a_p}{2\alpha a_b}\frac{1}{St}.$$

(11.19)

It then follows from Eq. (11.18), that $h_T \approx a_p \sqrt{v\eta/\gamma}$. The increase of h_T with increasing v is determined by the decrease in the duration of the initial stage of film deformation (when the initial radius of the film l is still small and the thinning velocity is maximum), as shown in Eq. (11.19). Besides, according to Eq. (11.11), $l = \sqrt{2a_p h_o}$, which gives $h_o \approx a_p$.

Using real values, it is easy to make sure that at a quasi-elastic impact (0.005<A<0.01) and at a non-retarded bubble surface ($\alpha=4$) the values of h_T at $2a_p = 100$ μm they are approximately 1μm. For a completely retarded bubble surface ($\alpha = 1$) the values are doubled. The critical thickness of wetting films commonly amounts to 1 to 5 nm, sometimes 10-50 nm and in rare cases it amounts to 100 nm. Thus, when the inertia impact of the spherical particles on the bubble surfaces is in most cases large the liquid interlayer does not have enough time to thin down to the critical thickness. Since the energy loss in this case is small, the particles are thrown away from the bubble surface by capillary forces. Clearly, the results obtained pertain to particles which are non-spherical and smooth.

11.2. INELASTIC COLLISION

It follows from Eq. (11.19) that a quasi-elastic collision (A<0.01) is only possible over a rather narrow range of values of a_p and ρ which are of interest for flotation. In most cases an inelastic inertia impact takes place. The majority of the energy is lost not only during the syneresis of the liquid interlayer but also during the approach to the bubble. As before, we consider a head-on collision of a particle with a bubble where the radial component of the liquid velocity along the trajectory of the particle is determined by a potential flow.

After a particle has approached the bubble surface, the velocity of the liquid drops to zero and a viscous resistance acts on the particle

$$F_\eta = \frac{1}{2}\Psi(\mathrm{Re})\pi a_p^2 \rho_0 u^2 \eta, \tag{11.20}$$

where $u = v_p - v$ is the velocity of the particle relative to that of the liquid, $\mathrm{Re} = 2a_p u / v$. According to Brauer (1971), the function $\Psi(\mathrm{Re})$ can be expressed by the empirical relation

$$\Psi(\mathrm{Re}) = \frac{24}{\mathrm{Re}} + \frac{4}{\sqrt{\mathrm{Re}}} + 0.4, \tag{11.21}$$

which is valid in the region $0.5 <\mathrm{Re}<10^3$. The loss of the kinetic energy of the particle can be estimated by

$$\Delta W = -\int_{\infty}^{a_p+a_b} F_\eta dr. \tag{11.22}$$

The energy of the particle at the instant of liquid interlayer formation W_1 is given by

$$W_1 = K_1 W_0; \ K_1 = 1 - \frac{\Delta W}{W_0}; \ W_0 = \frac{2v^2 \pi \rho a_p^3}{3}, \tag{11.23}$$

where W_0 is the initial kinetic energy of the particle in the coordinate system of the bubble centre, K_1 is the coefficient of viscous energy dissipation during the approach to the bubble. From Eqs (11.20), (11.22), and (11.23) this coefficient reads

428

$$K_1 = 1 - \frac{\text{Re}\,\Psi(\text{Re})}{12\text{St}} \frac{\int\limits_{1+\frac{a_p}{a_b}}^{\infty} \bar{u}^2 d\bar{r}}{}. \tag{11.24}$$

If, as a first approximation, the particle velocity is assumed to be constant when the energy losses are small, the integral in Eq. (11.24) my be evaluated for the case of potential flow. The result can be refined by successive approximation methods and finally takes the form

$$K_1 = 3\sum_{n=0}^{\infty} \frac{(-2)^n}{(2n+3)!!(K^*)^n}, \tag{11.25}$$

where quantity K^* is related to Stokes number St by

$$K^* = \frac{24}{\text{Re}\,\Psi(\text{Re})}\text{St}. \tag{11.26}$$

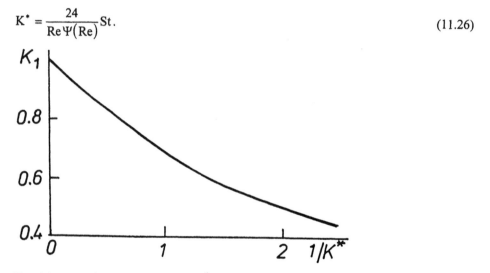

Fig. 11.2. K_1 plotted as a function of $1/K^*$.

Values of K^* calculated from Eq. (11.25) are presented in Fig. 11.2. At $K^*>2/5$ a particle loses no more than 50% of its kinetic energy. The loss grows with decreasing K^*.

Let W_1 be the energy of a particle at the instant when its centre has the coordinate $x = a_p$ (see Fig. 11.1). The following motion of the particle must obey the law of conservation of energy

$$W_k + W_y + W_\eta = \text{const}, \tag{11.27}$$

where W_k is the kinetic energy, W_η is work done by capillary forces to overcome the viscous stresses in the liquid interlayer syneresis given by

$$W_\eta = \pi \int_{h_o}^{h} p_\gamma l^2 dh. \tag{11.28}$$

Transformation of variables $\tilde{z} = (a_p - x)/h_o$, $\tilde{t} = tv_1/h_o$ in Eqs (11.3), (11.9)-(11.11), (11.28), with $v_1 = \sqrt{2W_1/m}$, and differentiating Eq. (11.27) with respect to $\tilde{t}$ after substitution of Eq. (11.3) into (11.27), we obtain the following system of equations describing the kinetics of film deformation and thinning,

$$\ddot{\tilde{z}}\tilde{z} + A(\dot{\tilde{z}} - \dot{\tilde{h}})(\tilde{z} + \tilde{h}) = 0, \tag{11.29}$$

$$\dot{\tilde{h}} = -B\tilde{h}^3/(\tilde{z} + \tilde{h}), \tag{11.30}$$

with

$$B = \left(\frac{h_o}{2a_p}\right)^2 \frac{8\alpha}{3De}; \quad A = \frac{3h_o^2\gamma}{2v_1^2\rho a_p^3}; \quad \tilde{h} = \frac{h}{h_o}. \tag{11.31}$$

The functions $\tilde{z}$ and $\tilde{h}$ must meet the initial conditions

$$\tilde{z} = 0, \; \dot{\tilde{z}} = \tilde{h} = 1, \; \dot{\tilde{h}} = -B \text{ at } \tilde{t} = 0. \tag{11.32}$$

From Eqs (11.28) and (11.30) it follows that the viscous energy dissipated by the particle during the complete cycle of film deformation is determined by

$$\Delta\tilde{W}_\eta = 4AB \int_0^{\tilde{t}} \tilde{h}^3 d\tilde{t}, \tag{11.33}$$

where $\Delta\tilde{W}_\eta = \Delta W_\eta/W_1$. Introducing the coefficient of viscous loss in the film, and using

$$K_2 = 1 - \Delta\tilde{W}_\eta, \tag{11.34}$$

the energy of the particle after collision reads

$$W_2 = K_2 W_1 = K_1 K_2 W_0. \tag{11.35}$$

430

It is seen from the plots of K_2 as a function of parameters A and B (calculated numerically from Eqs (11.29)-(11.35)) that energy losses in a liquid interlayer begin to appear at A>0.01. When the values of A are small, the impact can be considered to be quasi-elastic. It is also seen from Fig. 11.3 that in the region B>0.2 K_2 depends only slightly on B. This allows us to use (as for h_o) the simplified Eq. (11.17) which after substitution into (11.31) results in B≈1, and expression (11.19) for A. As an example, the relations $\tilde{z}(\tilde{t})$ and $\tilde{h}(\tilde{t})$ calculated for B=1, A=0.1, α=4 (free bubble surface) are presented in Fig. 11.4. It is clear that the main thinning of the film (about 50%) takes place during the early stage of the impact, when the spreading of the film is still small. The relation $K_2(A)$ is presented in Fig. 11.5a showing that energy losses in the film increase with A and K_1 and that K_2 decreases as the density and particle diameter are reduced, i.e. energy losses during the approach towards the bubble and in the liquid interlayer increase as the inertia of the particle is reduced.

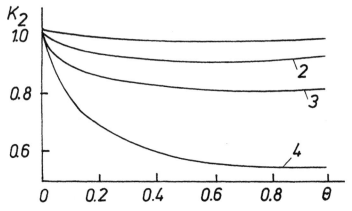

Fig. 11.3. K_2 plotted as a function of B at A=10^{-3} (1), 10^{-2} (2), 3·10^{-2} (3), 10^{-1} (4).

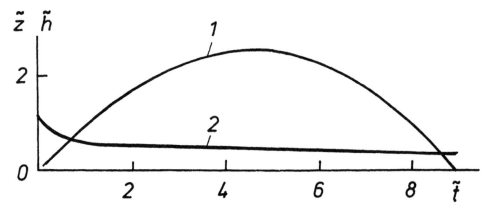

Fig. 11.4. $\tilde{z}$ (1) and $\tilde{h}$ (2) plotted as functions of $\tilde{t}$.

The relation between h_T / h_o and A at B=1 has been calculated (Fig. 11.5b). It shows that the relative thinning of film during an inelastic impact (A≥0.01 with $h_T \approx h_o / 3$) depends even less on A than in the case of a quasi-elastic impact (A<0.01 with $h_T \approx h_o / 4$).

It follows from (11.17) that $h_o \approx a_p K_1^{1/4}$ and that $A \approx 1 / a_p$. Therefore, we can achieve a substantial decrease of the finite thickness of the film (more than by an order of magnitude) by decreasing the particle size decreases and we enter the region of inelastic impact (A≥0.01). In this case, the final energy of the particle substantially decreases and, therefore, the probability of the particle jumping back decreases while the probability of its detaching during subsequent impact increases.

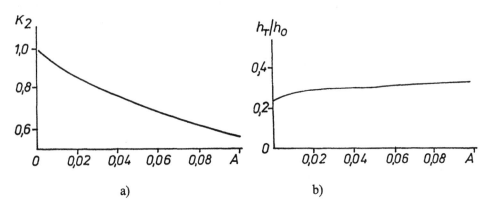

Fig. 11.5. K_2 (a) and h_T / h_o (b) plotted as functions of A at B=1.

11.3.   PREVENTION OF PARTICLE DEPOSITION ON BUBBLE SURFACE AT $\Theta > \Theta_T$ UNDER THE EFFECT OF CENTRIFUGAL FORCE

The mechanism of the centrifugal force effects on particle deposition (Section 10.14) qualitatively changes with increasing particle size and Stokes numbers, respectively. At a point on the surface, the centrifugal force exceeds the hydrodynamic pressing force (for a particle of a subcritical size it can be approximately estimated by $6\pi\eta a_p v_r$) and particle deposition is impossible. For larger particles, the value of the radial velocity at the point of impact on the surface can be extremely large. Therefore, the centrifugal force does not prevent the impact, although it will prevent the particle from being captured. It was calculated in Section 11.1 that a film thins insignificantly during the impact and does not reach the break-through value h_{cr}. It has to be taken into account that the final film thickness h_T is proportional to the particle size. Thus, the impact may not occur if the Stokes number is sufficiently large. In addition, h_T is proportional to the root of the radial component of particle velocity at the instant of impact, which is smaller closer to the equator. The radial particle velocity close to the equator is much smaller than at the upper pole. If it is close to zero an impact near to the equator is prevented by the action of centrifugal forces. At smaller angles (close to Θ_t) an impact is possible, but after an impact the subsequent motion of the particle is characterized by a smaller velocity close to v_r and thus the next impact on the surface is prevented by the action of centrifugal forces.

The deposition at $\Theta > \Theta_t$ is impossible, not only at subcritical values of the Stokes number but also at its supercritical values under conditions of inertia impact.

Let us now explain how the centrifugal force varies with increasing St. The description of the tangential motion of a particle as inertia-free (cf. Section 10.14) assumes small particles and very small St numbers (less than the critical value). When St grows, the tangential velocity v_Θ deviates more and more from the surface velocity. This results in a variation of the centrifugal force and also affects the value of angle Θ_t. It should be pointed out that the following equation (Dukhin, 1982, 1983),

$$K'\frac{d\tilde{v}_{p\Theta}}{d\tau} + \tilde{v}_{p\Theta} - \tilde{v}_\Theta = -K''\left(\tilde{v}_r \frac{\partial \tilde{v}_\Theta}{\partial \tilde{r}} + \frac{\tilde{v}_\Theta}{\tilde{r}}\frac{\partial \tilde{v}_\Theta}{\partial \tilde{r}} + \frac{\tilde{v}_r \tilde{v}_\Theta}{\tilde{r}} \right), \tag{11.36}$$

is more accurate than those in Section 10.14. Here $K' = K(1+\dfrac{\rho}{2\rho_p})$, $K'' = 3K\dfrac{\rho}{\rho_p}$, $K = St$, ρ and ρ_p are the densities of water and particle. Restrict ourselves to considering only a short jump-back. Under this condition, equations describing the hydrodynamic liquid field can be simplified and allow us to change Eq. (11.36) into one containing only one unknown function $\Theta(\tau)$. The velocity and acceleration of a particle in terms of this function can be expressed by

$$v_{p\Theta}(\tau) = r(\tau)\frac{d\Theta}{d\tau}, \tag{11.37}$$

$$\frac{dv_p}{d\tau} = r(\tau)\frac{d^2\Theta}{d\tau^2} + \frac{dr}{d\tau}\frac{d\Theta}{d\tau}. \tag{11.38}$$

Under the restriction

$$\frac{dr}{d\tau}\frac{d\Theta}{d\tau} \ll r(\tau)\frac{d^2\Theta}{d\tau^2} \approx \frac{d^2\Theta}{d\tau^2}, \tag{11.39}$$

and taking into account

$$\Theta < \Theta_t, \tag{11.40}$$

and the approximations

$$\sin\Theta \approx \Theta, \quad \cos\Theta \approx 1, \tag{11.41}$$

we can write

$$v_{p\Theta}(\Theta)\big|_{\bar{r}\approx 1} = \frac{d\Theta}{d\tau}; \quad \frac{dv_p}{d\tau}(\Theta)\big|_{\bar{r}\approx 1} = \frac{d^2\Theta}{d\tau^2}. \tag{11.42}$$

Eq. (11.36) can be transformed into a linear second-order equation with constant coefficients

$$K'\frac{d^2\Theta}{d\tau^2} + \frac{d\Theta}{d\tau} - \frac{3}{2}\Theta = \frac{9}{4}K''\Theta. \tag{11.43}$$

Condition (11.40) allow us to neglect the first and the third term on the right-hand side of Eq. (11.36) and second can be reduced to the term on the right-hand side of Eq. (11.43). Thus, the solution of Eq. (11.43) can be presented as

$$\Theta(\tau) = A'e^{S\tau}, \tag{11.44}$$

with

$$S = -1 + \sqrt{1 + 6K'\left(1 - \frac{3}{2}K''\right)}\bigg/2K'. \tag{11.45}$$

$\Theta(\tau)$ is a monotonically increasing function, when the second negative root of the characteristic equation is not taken into account. Differentiating (11.44) and using Eqs (11.37) and Eq. (10.72), we obtain

$$v_{p\Theta}(\Theta) = A'Se^{S\theta} = S\Theta(\tau) = \frac{2}{3}S\frac{3}{2}\Theta = \frac{2}{3}Sv_\Theta(\Theta). \tag{11.46}$$

If this expression is inserted into expression for the centrifugal force instead of $\tilde{v}_\Theta$, a new expression for β, instead of Eq. (10.74), is obtained,

$$\beta = \frac{E_0}{2K'''S^2} = \frac{4E_0K'^2}{2K'''\left[-1 + \sqrt{1 + 6K'\left(1 - \frac{3}{2}K''\right)}\right]^2}, \tag{11.47}$$

where $K''' = K \dfrac{\Delta\rho}{\rho_p}$ and $\Delta\rho = \rho_p - \rho$.

This new expression for β can be substituten into Eq. (10.80). Let us investigate how the angle Θ_t varies with increasing size of the particle interacting with a bubble. The Stokes number is given by

$$St = \frac{1}{9}\frac{\rho'}{\rho}\frac{2va_b}{v}\frac{a_p^2}{a_b^2} = A''E_0^2,$$
(11.48)

and thus we obtain

$$E_0 = \sqrt{9St}\,(\rho'/\rho\,Re)^{-\frac{1}{2}}.$$
(11.49)

The Reynolds number for a particle Re_p of supercritical size, deposited on the surface of a sufficiently large bubble (for which a potential distribution of the liquid velocity field is valid), is much larger than unity. In this case, the hydrodynamic resistance is expressed by a resistance coefficient. In aerosol mechanics a technique is used (Fuks, 1961) in which the non-linearity from the resistance term is displaced by the inertia term. As a result, a factor appears in the Stokes number which, taking into account Eq. (11.20), can be reduced to $\left(1 + Re_p^{2/3}/6\right)^{-1}$. This allows us to find the upper and the lower limits of the effect by introducing K^* instead of K''' into Eq. (10.47) and the factor X in the third term,

$$K^* = K'X \text{ with } X = \begin{cases} 1 \\ \left(1 + \sqrt{Re_p}/6\right)^{-1}. \end{cases}$$
(11.50)

Let us emphasise that the non-linearity in the particle drag coefficient is caused by the particle movement relative to the liquid. Its manifestation in the condition of particle rebound is weak and can be neglected. Let us estimate the effect for the onset of the recoil of the particle movement away from the surface. In this moment the particle changes its direction and the velocity is zero. In the same coordinate system the liquid velocity is equal to $\dfrac{3a_p}{2a_b}v_b$ if the length of the recoil path is comparable with the particle radius. Thus the relative velocity can be estimated by the liquid velocity. The maximum Reynolds number occurs under potential flow and for large bubbles. For example, in experiments of Ralston & Wolfe (1992) bubble of

2 mm diameter had a rising velocity of 30 cm/s which corresponds to $Re_b = 600$. The Reynolds number of a particle can be expressed by its radius, the bubble radius and Re_b

$$Re_p = Re_b \frac{a_p}{a_b} \frac{v_p}{v_b} = Re_b \frac{3}{2} \left(\frac{a_p}{a_b} \right)^2.$$ (11.51)

The substitution of data from experiments by Hewitt et al. (1994), $a_p = 30\mu m$, $a_b = 1mm$ and $Re_b = 600$, we get $Re_p = 2$, $X = 0.98$. However, for $a_p = 100\mu m$ we obtain $Re_p = 20$ and $x = 0.6$. As the equation (11.47) is not valid at

$$St > 0.5 \left(1 - \frac{\rho}{4\rho'} \right),$$ (11.52)

we cannot consider particle of diameter $a_p \geq 100\mu m$. At the onset of rebound the drag coefficient is controlled by the resistance of the liquid interlayer between particle and bubble against to thinning and has to be described by the Taylor equation. Thus we have to omit X and the particle rebound is short at St<0.5. Along the entire path of the particle the drag coefficient can be described by a more exact and general Taylor equation. A new theory taking this into account is under preparation.

The dependence of Θ_t on St or E_0, for the case of a bubble, with radius $a_b=0.07$ cm (which corresponds to $Re = 400$) is shown in Fig. 11.6.

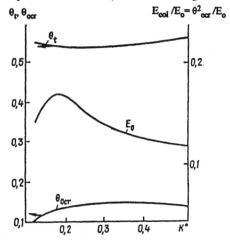

Fig. 11.6.   Results of the calculation of angles Θ_t and Θ_{0cr} as a function of St at X=1 and the decrease of collision efficiency due to a combined action of reflection and centrifugal forces as compared with Sutherland's formula for E_0.

11.4. PARTICLE REFLECTION FROM A BUBBLE SURFACE

The determination of the particle velocity after its inelastic collision with a bubble is based on the calculation of the energy losses spent in overcoming the viscous resistance during the approach and thinning of the film formed between the bubble and the particle (cf. Section 11.2). This may be estimated by taking the ratio between the particle velocity after collision v_0 and its initial velocity (equal to bubble velocity),

$$v_0/v = \sqrt{K_1 K_2}.$$ (11.53)

and using a value of $K_1=0.6$, we obtain

$$v_0/v = 0.6.$$ (11.54)

As in Section 11.1, we consider the reflection of a particle from a flat boundary at normal sedimentation. This assumption can be used for a bubble when we are interested in impacts close to the pole, at $\Theta<\Theta_t$. Thus, we can simplify the expression for the normal flow of liquid setting cosine to unity and assuming that over whole section of the surface at $\Theta<\Theta_t$ the length of recoil is characterized by a constant value. In dimensionless form, the equation for calculating the inertia path change due to the opposite motion of the liquid which has a velocity distribution expressed by a linear relationship (differs from a linear second-order differential equation with constant coefficients only due to a variation of Reynolds number for a retarded particle). It reads

$$K'\left(\frac{d^2\tilde{r}}{d\tau^2} + \frac{d\tilde{r}}{d\tau} + 3(\tilde{r}-1)\right) = 0.$$ (11.55)

Taking into account the approximation (11.54), we can solve Eq. (11.55) by assigning a constant coefficient at the second-order derivative (minimum and the maximum values, respectively) and by satisfying the initial conditions

$$\left.(\tilde{r}-1)\right|_{\tau=0} = \frac{E_0}{3},$$ (11.56)

$$\left.\frac{d\tilde{r}}{d\tau}\right|_{\tau=0} = \frac{v_0}{v}.$$ (11.57)

The solution has the form

$$\left(\tilde{r}(\tau)-1\right)=\left[\left(\frac{V_o}{v\omega_1}+\frac{E_0\sigma_1}{3\omega_1}\right)\sin\omega_1\tau+\frac{E_0}{3}\cos\omega_1\tau\right]e^{-\sigma_1\tau},\tag{11.58}$$

with

$$\sigma_1=\left(2K'\right)^{-1}\text{ and }\omega_1=\sqrt{12K'-1}/2K'.\tag{11.59}$$

For a supercritical value of $K'=1/12$ the roots of the characteristic equation, corresponding to Eq. (11.55), are imaginary. The solution (11.58) describes a damped oscillation. Once the particle has lost velocity, the opposite motion of liquid brings it back to the surface. Therefore, when the distance of the particle from the surface is maximum at time τ_o its velocity approaches zero. Based on (11.58), we obtain

$$\dot{r}(\tau)=\left\{\frac{V_o}{v}\cos\omega_1\tau-\left[\frac{\sigma_1}{\omega_1}\frac{V_o}{v}+\frac{E_0}{3\omega_1}\left(\sigma_1^2+\omega_1^2\right)\right]\sin\omega_1\tau\right\}e^{-\sigma_1\tau},\tag{11.60}$$

from which follows

$$\tau_o=\frac{1}{\omega_1}\text{arctg}\frac{V_o/v}{\sigma_1V_o/v+\left(\omega_1+\sigma_1^2/\omega_1\right)E_0/3}.\tag{11.61}$$

Substituting this value into Eq. (11.59), gives a very long equation for the length of the inertia path. This equation can be simplified when the path is sufficiently short and $E_0/3$ can be neglected,

$$1=\frac{V_o}{v}a_b\frac{\sin\omega_1\tau_0}{\omega_1}e^{-\sigma_1\tau_0}=a_b\frac{V_o}{v}\left(\omega_1^2+\sigma_1^2\right)^{-\frac{1}{2}}\exp\left(-\frac{\text{arctg}\left(\omega_1/\sigma_1\right)}{\omega_1/\sigma_1}\right)$$

$$=a_b\frac{V_o}{v}\left(\frac{K'}{3}\right)^{\frac{1}{2}}\exp\left(-\frac{\text{arctg}\sqrt{12K'-1}}{\sqrt{12K'-1}}\right).\tag{11.62}$$

In the other limiting case when the length of recoil is commensurable with the radius of the particle, the third term in (11.55) can be considered as a constant quantity equal to E_0. This allows us to decrease the order of the equation and to obtain a simpler solution,

$$\frac{d\tilde{r}}{d\tau}=\left(\frac{V_o}{v}+\frac{E_0}{3}\right)e^{-\tau/K'}-\frac{E_0}{3},\tag{11.63}$$

$$\tilde{r}(\tau)-1-\frac{E_0}{3}=K'\left(1-e^{-\tau/K'}\right)\left(\frac{v_o}{v}+\frac{E_0}{3}\right)-\frac{E_0}{3}\tau.$$

(11.64)

For the length of recoil we get

$$l=a_b K'\left[\frac{v_o}{v}-\frac{E_0}{3}\ln\left(1+\frac{v_o}{v}\frac{3}{E_0}\right)\right]$$

(11.65)

and the limits of applicability of this result is

$$-\frac{v_o}{v}\le\frac{1}{K'}\frac{E_0}{3}.$$

(11.66)

Thus we can define the conditions for which the methods used for the calculation of the grazing trajectory hold.

11.5. PREVENTION OF PARTICLE DEPOSITION ON BUBBLE SURFACE AT ANGLES $\theta<\theta_{ocr}$ DUE TO JOINT ACTION OF PARTICLE REFLECTION FROM A BUBBLE SURFACE AND CENTRIFUGAL FORCES

Inertia forces manifest themselves in at least in three effects: 1 - shift of particles trajectories away from the liquid stream-lines from the bubble (Section 10.1); 2 - deformation of the bubble surface by a particle at the instant of impact (cf. Section 11.1); 3 - prevention of particle deposition on some parts of the bubble surface due to centrifugal forces (cf. Section 11.3). In the present section these effects are considered simultaneously (Dukhin et al. 1986), although some restrictions are imposed on the properties of the bubbles and the particles.

For this investigation, the bubbles are assumed to be spherical and their diameters are taken to be of the order of millimetre having a non-retarded surface (this is necessary for the use the theory developed in Section 11.3, which enables to describe the hydrodynamic field of bubble by a potential one). In addition, only spherical particles with smooth surface are considered (this makes it possible to use the results of the consideration of film thinning at the inertia impact of a particle on bubble surface from Section 11.2 and experimental data). Upper and lower limits of particle size are assumed, the necessity of which will be explained later.

To trace the final location of a particle, it is important to know how the conditions of an inertia impact change as a function of its localisation on the bubble surface. This dependence is clear in experimental studies by Schulze & Gottschalk (1981). Their observations have shown that two zones exist on a bubble surface. In the first zone adjacent to the bubble pole with an angle

$0 < \Theta < \Theta_{cr}$ (where $\Theta = 0$ corresponds to the bubble pole), impact of particles in this region is followed by a recoil; in the second zone no recoil is observed at any angle $\Theta > \Theta_{cr}$ and the particle slips along the surface. At $\Theta < \Theta_{cr}$ the radial velocity of a particle during its approach is much higher than the radial velocity of the liquid. This means that particle deposition occurs due to the inertia path to the surface. An decrease of the radial velocity with increasing angle Θ results in a decrease of the kinetic energy of the particle on impact at the surface. At $\Theta > \Theta_{cr}$ the recoil of the colliding particle becomes small and is not observed. Larger values of Θ result in larger tangential surface velocities, so that mainly the tangential motion of the particles, slipping is observed at $\Theta > \Theta_{cr}$. An additional reason for the decrease in the radial velocity of the particles and of the decrease of recoil at rather large Θ consists in the fact that an increase of centrifugal forces corresponds to an increase of the tangential velocity of particles (cf. Section 10.14).

Two mechanisms are possible during deposition in the zone $\Theta < \Theta_t$. Firstly after recoil the particle is carried by a tangential flow to the zone $\Theta > \Theta_t$ where deposition is impossible. Clearly, the extent to which this occurs increases as the distance from the impact to the pole and the recoil distances increase. The second possibility is that a subsequent impact occurs at $\Theta < \Theta_t$ which culminates in a film break through.

Let us clarify the conditions for the first variant. Eq. (11.60) characterises the normal component of the motion of the reflected particle, its withdrawal from the surface and return to it. At a small recoil length, Eq. (11.46) reflects the tangential displacement of a particle. Using Eq. (11.58) it is possible to calculate the time T_0 between the beginning of the recoil and the second collision of the particle with the surface. From Eq. (11.44) we can calculate the tangential displacement of a particle in this time, expressed in terms of angular units. We denote the angle at which the recoil occurs by Θ_0, and the angle of the second impact by Θ_1. The situation at which $\Theta_1 = \Theta_t$ is most interesting . We denote the corresponding value of Θ_0 by Θ_{0cr}. Clearly, the functional relationship $\Theta_1(\Theta_0)$ is such that Θ_1 increases with increasing Θ_0. If $\Theta_0 > \Theta_{0cr}$, then $\Theta_1 > \Theta_t$, i.e. a reflected particle is carried to the zone where the centrifugal forces are strong and hence it does not reach the bubble surface. Thus, collisions are inefficient not only at $\Theta > \Theta_t$ but also in the range $\Theta_{0cr} < \Theta < \Theta_t$.

The time taken for the particle to return to the bubble surface T_0, corresponds to a repeated fulfilment of condition (11.56). Comparing Eqs (11.56) and (11.58), we obtain a transcendental equation for determining T_0. If the recoil length is sufficiently large that

$$\beta = \frac{\exp\left(\dfrac{\pi\sigma_1}{\omega}\right)}{\left(\dfrac{v_0}{v}\dfrac{3}{E_0} + 1\right)} \ll 1, \qquad (11.67)$$

440

the return time is expressed approximately by

$$T_o \cong (\pi - \varepsilon)\omega^{-1},$$

(11.68)

where

$$\varepsilon = \beta/(1+\beta).$$

(11.69)

In the other limiting case, when the recoil length is commensurable with the particle radius and the Eqs (11.65) and (11.66) are valid, we have

$$T_o \cong K'\left(1 + \frac{V_o}{v}\frac{3}{E_0}\right) \quad at \quad \frac{V_o}{v}\frac{3}{E_0} \geq \frac{1}{K'},$$

(11.70)

$$T_o \cong 2K'\frac{V_o}{v}\frac{3}{E_0} \quad at \quad \frac{V_o}{v}\frac{3}{E_0} \ll 1.$$

(11.71)

If we define $\Theta|_{\tau=0} = \Theta_0$, from Eq. (11.44) we obtain

$$\Theta = \Theta_0 e^{ST_o}.$$

(11.72)

Together with Eqs (11.45), (11.59) and (11.68) we finally obtain

$$\Theta_{0cr} = \Theta_t e^{-ST_o} = \Theta_t \exp\left(-\pi \frac{-1 + \sqrt{1 + 6K'\left(1 - \frac{3}{2}K''\right)}}{\sqrt{12K'-1}}\right),$$

(11.73)

from which the effect of the process on the effectiveness of capture can be estimated.

11.6. ESTIMATION OF COLLISION EFFICIENCY

Unlike the collision effectiveness which is expressed by Langmuir's formula (cf. Section 10.1), the capture efficiency E_c is determined by the grazing trajectory which characterises the possibility of particle attachment. Results obtained above make it possible to estimate this quantity since collisions on the main part of the surface at $\Theta > \Theta_{0cr}$ do not result in capture. All of the simplifications used cause an increase of E_c. The cross-section of the stream tube from which a capture can be carried out is less than $\left(\Theta_{0cr}^{real}\right)^2$ since the stream tube expands when approaching the surface. Here Θ_{0cr}^{real} expresses the real value of Θ_{0cr} as distinct from the calculated value Θ_{0cr}^{calc}. Introducing a number of simplifications substantially overestimates the value of Θ_{0cr}. Centrifugal forces also act at $\Theta < \Theta_t$ and its account should result in a decrease of Θ_{0cr}. A continuous particle deposition is slowed down by short-range hydrodynamic interactions which also results in a decrease of Θ_{0cr}. Thus,

$$E_c < \left(\Theta_{0cr}^{real}\right)^2 < \left(\Theta_{0cr}^{calc}\right)^2. \tag{11.74}$$

Results obtained using the relationship

$$E_c^{calc} = \Theta_t^2 \left(\Theta_{0cr}/\Theta_t\right)^2, \tag{11.75}$$

where the first term is calculated from Eq. (10.12) and the second from Eq. (11.73) are presented in Fig. 11.7.

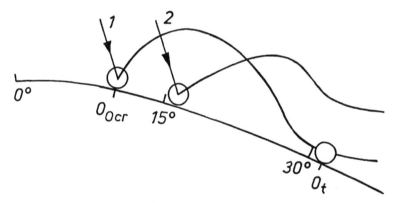

Fig. 11.7.   Illustration of the mechanism of deposition prevention in the zone $\Theta_{0cr} < \Theta < \Theta_t$ due to the joint action of inertia reflection of a particle from the bubble surface and centrifugal forces: 1 - grazing trajectory at a single reflection; 2 - impossibility of deposition at $\Theta > \Theta_{0cr}$

The theory developed substantially changes the notion of kinetics of flotation of smooth particles since the calculated efficiency is decreased by approximately one order of magnitude when compared with Sutherland's formula. However, the theory needs further development because the final Eq. (11.46) underestimates the value of $v_{p\theta}$ and the centrifugal force due to

$$\beta = \frac{E_0}{2St} \frac{\rho_p}{\Delta\rho} y^{-2}, \tag{11.76}$$

where $\dfrac{2S}{3} < y < 1$. A detailed explanation is given in Appendix 10D.

11.7. KINETICS OF EXTENSION OF THREE-PHASE CONTACT (t.p.c.)

This important problem was not discussed in the microflotation theory of Derjaguin & Dukhin. Scheludko and co-workers (Scheludko et al., 1970) discussed the propriety for flotation.

The formation of a primary hole during the rupture of a thin film is a very rapid process and its kinetics is not important for flotation. The extension of the t.p.c. requires a considerably longer time. Thus, this microprocess can play a decisive role in flotation of particles of normal size. The initial state of the spontaneous hole formation in a thin film is a state far from equilibrium. The resulting relaxation process terminates in the formation of a final equilibrium t.p.c. and an equilibrium contact angle. Clearly, the t.p.c. expands during this process. The driving force of this process is the difference between the interfacial energies which can be approximated by the difference between the instant value of the dynamic receding contact angle θ_R^* and the equilibrium value $\theta_{R\infty}$.

A kinetic equation of the t.p.c. expansion can be presented by the rate of increase of radius r of a cylindrical hole in thin film $\dfrac{dr}{dt}$. Such process in a wetting film can in general be discussed similar to the expansion of holes in free foam films (Scheludko et al., 1970)

$$\frac{dr}{dt} = a_{tpc}\left(\cos\theta_R^* - \cos\theta_{R\infty}\right),$$
(11.77)

where a_{tpc} is the instantaneous mobility coefficient of the t.p.c. Estimates show that the initial rate is of the order of 100 km/h. Such a fast flow in a liquid interlayer gives rise to high viscous resistance. Experiments made by Tschaliovska (1988) have shown that a_{tpc} depends in a very complicated manner on surface properties and the instantaneous radius r(t). Thus, only a very crude estimates of t_{tpc}, the time of the t.p.c., can be given at present (Schulze 1989). Scheludko estimates the radius of the t.p.c. which is necessary to form a stable bubble-particle aggregate to be proportional to the square of the particle radius,

$$r_{tpc} \approx a_p^2.$$
(11.78)

Thereby, it was assumed that the detachment force is caused by gravity.

The dependence of $\dfrac{dr}{dt}$ on r discussed above and the strong effect of pH have been studied in works by Stechemesser et al. (1980), Hopf & Geidel (1987), and Hopf & Stechemesser (1988). The important role of surfactant adsorption-desorption processes was also pointed out by Hopf

& Stechemesser (1988). The role of the DAL in flotation seems to be most important for the spontaneous expansion of the t.p.c.

Another important feature of kinematics of the t.p.c. expansion consists in the fact that its initial rate is very high and then slows down tens and hundreds times long before the equilibrium is attained. The initial rate corresponds to small radii r, so that the smaller r, the higher is the expansion rate. This peculiarity explains why this microprocess can be ignored in microflotation and why its role is the higher the bigger the particles are to be flotated.

If we assume that the characteristic radius of particles in microflotation is about 5 µm and in flotation 50 µm and that the rate of expansion of the t.p.c. does not depend on the instantaneous radius, we can conclude that in microflotation this process proceeds 100 times faster than in flotation.

Actually, the difference is even larger and can amount up to three or four orders of magnitude since the rate of expansion is very high at small r (microflotation conditions) and decreases by a factor of 10 or 100 at greater r (flotation conditions).

Flotation of large particle is often observed, which is made possible by a sufficiently high rate of expansion of the t.p.c. This fact together with the need of about three orders of magnitude lower expansion rate of the t.p.c. for small particles let us conclude that this process cannot control microflotation kinetics.

Thus, the t.p.c. in microflotation is a fast process so that it cannot be the limiting stage. In contrast, in flotation the t.p.c. expansion is a slow process which can be the limiting stage of flotation.

Scheludko's (1970) statement that slow t.p.c expansion can control flotation kinetics may be important for large particles. Note that one of the decisive differences between flotation and microflotation also consists in the electrostatic barrier. In Section 10.5 it was shown that the weight and the hydrodynamic pressing force of micron particles, in contrast to large particles, are not sufficient to surmount the electrostatic barrier. Thus, electrostatic repulsion can exclude microflotation while it does not affect flotation, and the t.p.c. expansion rate can control flotation kinetics but has probably no influence on microflotation.

This means that there are no contradictions between the kinetic theory of microflotation by Derjaguin & Dukhin, in which the limiting stage is the overcoming of the electrostatic repulsion between bubble and particle, and the capillary theory of flotation by Scheludko in which a special attention is paid to t_{tpc} and the contact angle. Both models are identical in the definition of thermodynamic condition of stability of bubble-particle aggregates. While

Derjaguin & Dukhin formulate this condition on the basis of surface forces and the disjoining pressure, Scheludko uses the contact angle, which can be expressed through the disjoining pressure (Appendix 10D).

With respect to kinetic aspects of flotation and microflotation, a general model could be developed on the basis of both notions.

For both flotation and microflotation cationic surfactants can be recommended. In microflotation the electrostatic barrier can be removed by cationics (cf. Section 10.5). In flotation the adsorption of cationics provides large contact angles (see Appendix 10D).

11.8. ATTACHMENT BY COLLISION.

11.8.1. STAGES OF ATTACHMENT BY COLLISION BEFORE (BCS) AND AFTER CONTACT (ACS)

Theories by Rulyov & Dukhin (1986) and Dukhin & Schulze (1987) (see Sections 11.1-11.5) demonstrate the importance of two distinct collision process drawn either by inertial forces or by inertiafree microflotation (Sections 10.1-10.4). Some of these separate processes have been studied theoretically and experimentally also by other researchers. For example, Schulze et al. (1989) considered inertial collision theoretically and Stechemesser et al. (1980), and Hopf et al. (1988, 1987) studied it experimentally. These experiments have confirmed the theoretical model of inertial collision and that the two models by Rulyov & Dukhin (1986) and Schulze et al. (1989) are essentially identical. Substantial progress was also achieved in theoretical and experimental investigations of the process of extension of t.p.c. (Section 10.8). Note that other parts of the theories by Rulyov & Dukhin (1986) and Dukhin & Schulze (1981) also obtained recognition by specialists. Thus, for example, the general classification of collision process regimes proposed by Rulyov & Dukhin (1986) was used in a review by Schulze (1992); the system of equations describing the effect of inertia forces (Dukhin & Schulze 1987, cf. Sections 11.3-11.7) was reproduced in a recent paper by Van Nguen (1993). The quality of prediction by the theory of attachment by collision has been increased by the investigations of Schulze & Birzer (1987) on the critical thickness h_{cr} as a function of interface forces (such as surface tension γ and advancing contact angle θ_A), which can be controlled by additives. The advanced contact angle is used to characterise the degree of particle hydrophobicity. Schulze & Birzer (1987) have found a correlation between h_{cr} and θ_A. The former parameter is essentially experimentally inaccessible, whilst θ_A can be easily measured,

$$h_{cr} = 23.3 \left[\gamma (1 - \cos\theta_A) \right]^{0.16}.$$ (11.79)

Thereby, prerequisites for the description of collision by attachment, based on a combination of the two theories by Rulyov & Dukhin (1986) and Dukhin & Schulze (1987), were developed to some extent.

Let us formulate the main elements of this theory, restricting the derivation to a spherical particle with a smooth surface, and next to particles with a well-defined surface roughness. The theory presented in Section 11.2 allows us to calculate h_T as a function of bubble and particle sizes, particle density, and degree of surface retardation. The first necessary condition for attachment by collision is

$$h_T \leq h_{cr}, \tag{11.80}$$

When the particle returns, two processes take place simultaneously; widening of the t.p.c. and velocity increase during the particle recoil. The first process stabilises the bubble-particle aggregate, the second destabilise it. The larger the particle and the bubble (ans so the larger the Stokes number), the greater the return velocity. At some critical Stokes number, the extending t.p.c. and resulting capillary force cannot prevent the separation of the particle from the bubble surface. At smaller St values, $St < St_{ACS}$, the kinetic energy of a rejected particle is insufficient to do the work against the capillary force caused by the t.p.c. expansion, and detachment is impossible.

The theory described in Sections 11.1-11.7 is based on investigations discussed by Schulze (1989, 1992, 1993) and refers to BCS. The theory of ACS in the attachment process by collision has not been elaborated yet because of difficulties in the further improvement of the t.p.c. expansion theory.

Thus, only a general qualitative characterisation and classification of different regimes is possible. Two conditions control the final result of the particle-bubble collision. The first condition relates to the BCS stage, the second to the ACS stage. The condition concerning the ACS stage at $St < St_{ACS}$ was formulated above. The condition for the BCS stage can be formulated in a similar way because h_T decreases with the Stokes number and satisfies condition (11.80). Thus, attachment by first collision occurs if

$$St < min(St_{BCS}, St_{ACS}), \tag{11.81}$$

and reflection occurs if

$$St > min(St_{BCS}, St_{ACS}). \tag{11.82}$$

Let now introduce the effect of particle roughness and shape irregularity. They differ in the scale of surface inhomogeneity. Roughness relates to small scale inhomogeneity. If large irregularities are absent it is possible to define a linear dimension which relates to a mean roughness. After averaging a macroscopically smooth and homogeneous analogue of a rough surface is obtained. (Note that notion of not too rough surface preserves irs physical sense for particles with macroscopically smooth (rough) surface.)

The notion of the critical thickness of the wetting film, h_{cr}, can preserve its physical meaning if the micro-geometry of a rough surface satisfies some conditions. The necessary condition of a smooth spherical particle attachment Eq. (11.80) can be used for a spherical macroscopically smooth surface, too.

Roughness also affects the ACS. The shape of irregularities on a particle surface and their distribution have an effect on the t.p.c. extension.

11.8.2 ATTACHMENT BY REPEATED COLLISION

Assuming that attachment is controlled by BCS, it is possible to estimate the minimum h_{cr} value of a particle which provides the attachment by the repetitive collision. We define a hydrodynamic interaction as collision if St exceeds St_{cr}. At $St \ll St_{cr} = 1/12$ the notion of an inertiafree flotation is usually used. A particle rebound is possible at $St < St_{cr}$ if $|St - St_{cr}|$ is small.

Thus, as a first approximation,

$$St \geq St_{cr} \qquad (11.83)$$

can be considered as a necessary condition for the attachment by collision. It means that the sequence of repetitive collisions terminates at the collision for which St_r satisfies the condition

$$St_r \sim St_{cr} . \qquad (11.84)$$

The knowledge of the St_r value for the last possible collision enables us to quantify the condition of attachment by repetitive collision. It is sufficient to represent the decreased thickness according the Eqs (11.17) and (11.18) and finally through St_r

$$h_r = \frac{h_o}{3} = \frac{a_p}{3} \sqrt{\frac{3v\eta}{2\gamma\alpha}} = M\sqrt{St_r} \qquad (11.85)$$

where $M = \sqrt{\dfrac{3a_b\eta v\rho}{4\gamma\alpha\rho_p}}$.

Eq. (11.85) describes the minimum thickness of the water interlayer for both h_T and repetitive collision h_{min}^{rep}. Consequently, the ratio of minimum thicknesses given by

$$\frac{h_{min}^{rep}}{h_T} = \sqrt{\frac{St_{cr}}{St}}, \qquad (11.86)$$

where the substitution $St_r = St_{cr}$ is used. For example, if $St = 1$ and $St_{cr} = 1/12$, then h_{min}^{rep} is approximately three times less than h_T ($h_{min}^{rep} = 0.3\mu m$).

We can conclude that the attachment by repetitive collision is possible only when a special roughness or particle shape are favourable for the rupture of a rather thick film.

Eq. (11.75) yields only the value of E_c without any information concerning the possibility of attachment. Now the necessary condition for attachment is given by $St_r \le St_{cr}$ and the particle surface has to meet the condition

$$h_{cr} \ge h_{min}^{rep}. \qquad (11.87)$$

Let us establish a relationship between the Stokes numbers of the initial and last collisions and estimate the collision number. The decrease in the Stokes number, which corresponds to the repetitive collision St_r, can be estimated by substituting the particle velocity v_r (at its repetitive touch with the bubble surface) into the equation for St. v_r can be determined by inserting the time interval T_o between two collisions (cf. Eq. (11.61)) into Eq. (11.60). The second term in Eq. (11.60) can be neglected and we get $\omega_1 \tau_o \cong \pi$. Taking this into account, from Eqs (11.59) and (11.68) we obtain

$$v_r = \tilde{r}v \approx v_o \exp(-\gamma_1 \tau_o) = v_o \exp\left(-\frac{\gamma_1}{\omega_1}\omega_1\tau_o\right) = v_o \exp\left(-\frac{\pi\gamma_1}{\omega_1}\right), \qquad (11.88)$$

which means

$$St_r \cong St \frac{v_o}{v} \exp\left(-\frac{\pi}{\sqrt{12K^*-1}}\right). \qquad (11.89)$$

In the application of this equation to the last jump the approximations $St_r \sim St_{cr} \sim 0.1$ and $\frac{v_o}{v} \approx 0.3$ can be used (in Section 11.2 $\frac{v_o}{v} \approx 0.5$ was used). However, the energy dissipation by the capillary wave caused by a particle impact decreases this ratio $\frac{v_o}{v}$ as one can see from experiments of Stechemesser et al. (1985) and calculations by Rulyov (1988).

An approximate solution of Eq. (11.89) is

$$\overline{St} = 1. \tag{11.90}$$

Since the exact solution Eq. (11.89) depends to some extent on the bubble velocity, which can vary in a narrow range, it is not used. St values exceeding $\overline{St}$ are excluded from further consideration because Eq. (11.62) is not valid. Thus, only one repetitive collision is possible if St<1.

The length of repetitive rebound is determined by Eq. (11.58) when v_0 has to be replaced by v_r and $\omega_1 \tau_0 \cong \pi$. According to Eq. (11.60) v_r is very small at $St_r \approx St_{cr}$. Replacing v_0 by v_r in Eq. (11.58) yields very small terms which can be omitted. The second and third terms in Eq. (11.58) are equal and

$$l_r \approx 2 \frac{E_0}{3} a_b \approx \frac{a_p}{e}. \tag{11.91}$$

The length of the repetitive recoil l_r is smaller than the particle diameter. It measn that the particle and the surface oscillate together including the liquid interlayer. Eq. (11.55) has to be generalised to regard for this effect. These results differ essentially from the results given by Rulyov et al. (1987, 1988).

11.8.3. EFFECT OF PARTICLE SHAPE ON ATTACHMENT BY COLLISION

Rulyov (1988) considered the inertial hydrodynamic interaction of a bubble with a conical particle with its axis perpendicular to the bubble surface. The consideration of BCS (similar to that described in Sections 11.1-11.2) was supplemented by the consideration of ACS and special attention was paid to the energy dissipation.

It was found that the degree of energy dissipation caused by viscous flow in the liquid interlayer approaches 100% at sufficiently small cone angle, which provides high rate of t.p.c. extension. Under these conditions, separation of particle from the bubble surface is prevented. The attachment is provided by first collision of particles with large diameter, i.e. $200 \mu m$.

However, as the dynamic contact angle decreases with increasing cone angle the particle rebounds. At asymmetric cone orientation the increase of angle β between the cone axis and the normal surface supports particle rebound. The particle impact occurs mostly at asymmetric cone orientation. Thus the particle rebounds can predominate even at rather sharp cone angle. Attachment of a particle can occur at a first collision when the angle β is small, and after repetitive collision when the angle β is larger.

11.8.4. SCHULZE'S SUPERPOSITION MODEL

Schulze (1993) notices that "the overall collision probability of single particles should be the algebraic sum of individual collision efficiencies". These are given as the interception effect E_{ic}, the gravitational effect E_g and inertial effect E_{in}. The quality of this approach can be evaluated by comparing it with results from the exact method of grazing trajectory. The differential equation for the calculation of the trajectory of a particle includes four dimensionless numbers $\left(\dfrac{a_p}{a_b}, Re, St, \tilde{g} \right)$. Hence, the collision efficiency is a function of these four parameters.

The representation of a four-dimensional function by a superposition of functions of smaller dimension is not generally justified. However, in extreme cases, at very small values of one of the parameters, the dependence on it can be negligible. For instance, under condition (10.76) the collision efficiency does not depend on St. At $St \geq 0,1$ the dependence on $\dfrac{a_p}{a_b}$ can be neglected. Thus one can conclude that in extreme cases the superposition proposed by Schulze (1993) is valid (for example Sutherland's or Langmuir's equation) and other small terms can be disregarded. But it is not valid for intermediate cases, for instant if the conditions (10.76) and

$$\frac{a_p}{a_b} \ll 1 \tag{11.92}$$

are valid and their values are of the same order. In this case E_c is expressed by Eq. (10.68), where the first factor is caused by the interception, and the second by inertial force. The example shows that the product of function is useful if extreme cases are excluded.

The process of particle rebound is not incorporated in the superposition model by Schulze so that it can only be used for particles which attach at the first collision.

The hydrodynamic field used does not comprise the interesting case of large rising bubbles under the potential flow. E_{ic} and E_{in} are sensitive to the hydrodynamic field and the bubble surface state, which was not yet been taken into account.

To evaluate Schulze's equation as a semi-empirical relationship, let us analyse the experimental method and data. Schulze used flow tube experiments with bubbles of 2 mm in diameter and $Re = 30$. There is a significant difference between experiments by Schulze (1993) and Hewitt et al. (1994) who used single rising bubble with 2 mm diameter, $Re = 600$ and $St = 0.25$. It is unclear if collisions occur in Schulze's experiment at $St \sim 0.07$.

In experiments by Anfruns & Kitchener (1976) no attention was paid to the purification of water. This means that the surfaces of bubbles 1 mm diameter were probably retarded. In SDS solution of concentration above 10^{-3} M there is completely no particle deposition while the contact angle is still large. The pressing hydrodynamic force for the 1 mm bubbles and particles of 40μm diameter in a potential flow is rather strong and can surmount the electrostatic disjoining pressure (cf. Section 10.5.2). On the contrary this force becomes weaker for retarded surfaces and does not suffice to surmount the electrostatic barrier. Thus, the bubble surface was retarded in the experiments by Anfruns & Kitchener (1976). In addition the Stokes number was low and centrifugal forces were absent and particle rebound is unlikely. Thus Schulze used experimental data in the verification of his superposition model, where particle rebound and significant centrifugal force were improbable. However, even not at comparatively high Re-values the Stokes number can reach supercritical values for large particles, and then a particle rebound becomes possible. Thus, particles which cannot be attached by a first collision have to be excluded. Under these two restrictions Schulze's superposition can be considered as being an interpolation.

Additional restrictions allow another test of the validity of Schulze's approximation. It can be applied at potential flow $St > St_{cr}$ if attachment happens during the first impact. Indeed if the particles which cannot attach during the first impact are excluded. This approach is not valid at $St < St_{cr}$, as mentioned above.

Under potential flow E_{ic} can be expressed by Sutherland's equation, E_{in} by Langmuir's equation for a potential flow. Note that Langmuir obtained the equation from numerical calculations of the differential equations for the particle trajectory. In this equation the particle is considered as a point mass, i.e. the particle dimension is absent in Eq. (10.10). This means there is no direct influence of the particle dimension on the trajectory. However, the particle mass, the drag coefficient, and the Stokes number depend on the particle size, only later Langmuir's equation was derived for a finite particle size. This result cannot be used in Schulze's approximation because the influence of $\dfrac{a_p}{a_b}$ is considered in Sutherland's equation.

11.9. INFLUENCE OF DYNAMIC ADSORPTION LAYER ON ATTACHMENT BY COLLISION

DAL influences practically all stages of the elementary flotation act. Buoyancy velocity of bubbles of definite size with retarded and non-retarded surfaces can differ from each another by a factor of about 2 (see Fig. 8.2). According to the theory of quasi-elastic (Section 10.1) and inelastic (Section 10.2) collisions, a smaller film thickness h_o corresponding to the beginning of the impact to the surface is attained when the bubble velocity decreases. The smallest film thickness attained by an impact h_T depends on h_o (i.e., bubble velocity) and, in addition, on the degree of surface retardation. A decrease of the velocity of floating bubbles under the effect of DAL contributes to decrease of h_T.

Retardation of the surface of the film between particle and bubble under the effect of DAL exerts an opposite effect on h_T. Such paired effects of the DAL is possible only if a surfactant fulfils the condition (8.71). Such surfactants can also effect on h_T twice. If the surfactant concentration is low also the surface concentration on the leading bubble surface is low. For the same reason the DAL does not determine the retardation of the surface film but affects h_T only through the bubble buoyancy velocity. The effect of each surfactant on the mobility of the leading bubble surface as a function of concentration and surface activity can be evaluated from the data given in Fig. 8.2.

The effect of DAL on the extension of the three-phase contact is probably more important. The driving force of its extension is the difference between the instantaneous and equilibrium surface energy which depend on the corresponding surface concentration values.

The bubble surface mobility has no effect on surface concentration if Eq. (8.71) holds. However, its strong decrease in the vicinity of the front pole of the bubble is possible under condition (8.72). For each surfactant, a possibly significant surface concentration decrease in the vicinity of the front pole can be estimated from the data given in Fig. 8.2. Thus, a principle possibility exists to formulate a relationship between the equilibrium contact angle $\theta_{R\infty}$ and the size of floating bubbles, the surfactant concentration and its surface activity. At the present stage of the DAL theory such a relationship can be obtained only for weak retardation of the most part of the surface of a large bubble under a potential distribution of the liquid velocity.

It is necessary to develop the theory of DAL for extending liquid interlayers. When the t.p.c. line moves, a transfer of surfactant from the liquid/gas to the solid/liquid interface and vice versa is possible. Thus, there are changes in the interfacial energy and surface tension of the liquid in the region of the moving liquid meniscus which depend on the diffusion rate of surfactant molecules (Schulze 1992). Consequently, the movement of the liquid meniscus can also depend on the kinetics of the surfactant desorption-adsorption. Some additional remarks will be given in Section 12.

452

11.10. ATTACHMENT BY SLIDING AT POTENTIAL FLOW. THE ROLE OF PARTICLE REBOUND AND CRITICAL FILM THICKNESS

11.10.1 CLASSIFICATION OF MULTISTAGE COLLISION / ATTACHMENT PROCESS AT POTENTIAL FLOW AND $St > St_{cr}$

The attachment during the first impact of a particle on a bubble surface is possible for special particle shapes and a high rate of t.p.c. expansion when

$$St > St_{cr}. \tag{11.93}$$

According to Eq. (11.80) the attachment by repetitive impact of a particle with a macroscopically smooth, rough surface is possible for large h_{cr} and high rate of the t. p. c. expansion. Under the realistic condition

$$h_{cr} < h_{cr_{min}}^{coll} \tag{11.94}$$

attachment of rough and smooth particles is possible by sliding, which does not exclude a possible particle rebound. Under condition (11.93) a particle rebound is inevitable with some exclusions discussed in Section 11.8.2. Omitting these special cases we can conclude that under condition (11.94) attachment is possible by sliding after repetitive collision.

Luttrel & Yoon (1992) and Nquen Van & de Kmet (1992) ignore particle rebound before sliding when analysing the sliding process. This is justified only for small particles, while for large particles (when condition (11.93) holds) the attachment by sliding occurs after repetitive collisions.

The approach by Schulze (1992) to quantify the sliding process differs completely from the approach of Luttrel & Yoon (1992). Schulze observed that particle sliding occurs after preceding particle rebound. This is an extremely important observation and initiated a discrimination between different variants of collision/attachment processes. In his review Schulze (1992) considers the attachment by sliding as a two stage process taking into account particle rebound before the sliding stage. The description by Schulze (1992) of the influence of particle recoil on the initial conditions of the sliding process is based on the data given by Dukhin & Schulze (1987). Ye & Miller (1988, 1989) consider attachment by collision too.

11.10.2. INFLUENCE OF RECOIL AND CRITICAL FILM THICKNESS ON ATTACHMENT BY SLIDING

The sequence of the sub-process (Fig 11.8) includes an initial collision, followed by particle rebound, a second collision followed by repetitive rebound and sliding accompanied by drainage of the film between particle and bubble. This process culminates in achievement of a critical thickness and rupture of the film, expansion of the t.p.c. which leads to an equilibrium contact angle. To calculate E_{col} we need to specify the boundary conditions. During sliding the critical film thickness has to be achieved at an angle θ_t,

$$h(\theta_t) = h_{cr},$$
(11.95)

if the rate of the t.p.c. extension is very high. As the expansion of the t.p.c. takes some time the film ruptures at a smaller angle

$$h_r = h(\theta_t - \Delta_{tpc}) = h_{cr}.$$
(11.96)

The t.p.c. extension happens during particle sliding between angles $\theta_t - \Delta_{tpc} < \theta < \theta_t$.

If θ_b is the angle which characterises the boundary between repetitive particle recoil and sliding repetitive particle recoil stops at the maximum distance, given by Eq. (11.91),

$$h(\theta_b) = \frac{a_p}{e}.$$
(11.97)

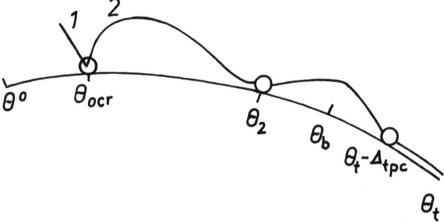

Fig. 11.8.   Illustration of the flotation mechanism by attachment by sliding after the second collision. θ_{0cr}- critical angle of first collision, θ_2- angle at the end of the second rebound and at the beginning of sliding, $\theta_t - \Delta_{tpc}$ angle at which the film ruptures and the t.p.c. extension begins, θ_t- the maximum angle of sliding restricted by centrifugal force influence

The consideration of sliding process enables us to relate the coordinates of the onset of slipping at $\theta_b, h(\theta_b)$ and its end at $\theta_t - \Delta_{tpc}, h_{cr}$,

454

$$\sin\theta_b = \sin(\theta_t - \Delta_{tpc})\left(\frac{h_{cr}}{h(\theta_b)}\right) = \sin(\theta_t - \Delta_{tpc})\left(\frac{eh_{cr}}{a_p}\right)^{1/8}.$$

(11.98)

According to Eqs (11.68) and (11.61), the time between the first collision and the onset of slipping is

$$T + \tau_o = 2K^*\left(\frac{\pi}{\sqrt{2K^*-1}}\right).$$

(11.99)

It enables us to interconnect the angle θ_o and θ_b by substitutions in Eq. (11.73) (T is replaced by $T + \tau_o$ and $\sin\theta_t$ by $\sin\theta_b$, and S is given by Eq. (11.45)),

$$\theta_{ocr} \cong \sin(\theta_t - \Delta_{t.p.c.})\left(\frac{eh_{cr}}{a_p}\right)^{1/8} \exp[-S(T+\tau_o)].$$

(11.100)

11.10.3. THE THEORY OF SHORT RANGE HYDRODYNAMIC INTERACTION (SRHI)

The physical aspects of SRHI were discussed in Section (10.5.2). In the present section the problem is discussed from the mathematical point of view. The first important results were obtained in aerosol science where Natanson (1957) demonstrated the importance of molecular forces for aerosol particle attachment. Natanson did not take into account SRHI. The first rigorous treatment of the problem was given by Derjaguin & Smirnov (1967), who have taken into account both SRHI and molecular forces. A more general approach to the problem of small particle deposition under laminar flow on the surface of a big particle (collector) was elaborated later by Spielman & Goren (1970, 1971). They introduced a local coordinate system for the description of local hydrodynamic field around the small particle and in the film between the particle and the collector and derived an equations for the short range hydrodynamic forces. These results are an important component of the theory of Derjaguin, Dukhin and Rulyov (DDR) on SRHI in flotation (1976, 1977).

When a particle moves along a bubble surface the conditions for the particle/bubble interaction changes and the short range interaction becomes unsteady. However, in local coordinates linked to the particle the interaction can be considered as quasi steady. The system of cylindrical coordinates can be used with its centre on the bubble surface and a z-axis crossing the centre of the moving particle. Such cylindrical coordinates simplify the description of hydrodynamic interaction. The equations for the normal $V_{pz} = \dfrac{dz}{dt}$ and tangential component

$V_{po} = a_b \dfrac{d\theta}{dt}$ of the particle velocity are given by

$$\frac{dz}{dt} = F_n f_1 / b\pi\eta a_p \tag{11.101}$$

and

$$a_b \frac{d\theta}{dt} = u_\theta f_2. \tag{11.102}$$

The equations describe that stage of the particle movement when the liquid interlayer is thin and the distance between the centres of a particle and the bubble is equal to a_b. The rates of interlayer thinning and particle movement are identical and are controlled by the action of the pressing force F_n and the resistance force (product of Stokes' drag coefficient and the dimensionless function $f_1(H)$, $H = \dfrac{z - a_p}{a_p} = \dfrac{h}{a_p}$). The equation of the particle trajectory follows from the equations (11.101), (11.102) after replacement of z by H and exclusion of the time,

$$\frac{dH}{d\theta} = \frac{a_b F_n(\theta, H) f_1(H)}{6\pi a_p^2 \eta u_\theta(\theta, H) f_2(H)}. \tag{11.103}$$

In general, the pressing force is a superposition of several forces (see Eq. (10.54)). A semi-analytical solution of Eq. (11.103) is possible if the radius of action of surface forces is small compared with to the particle size, Brownian motion is not taken into account and a sufficiently high electrolyte concentration is present. This means that only particles of micron size and larger are considered and the thickness of double layer exceeds 1-10 nm. Beyond a distance of 10 nm the particle moves to the surface under the action of the hydrodynamic pressing force (see Section 10.5) and gravity. Gravity will be neglected because in this section a potential hydrodynamic flow is considered. The hydrodynamic pressing force is proportional to the local value of the normal component of hydrodynamic velocity

$$F_n = F_h = 6\pi a_p \eta \cdot u_z(H, \theta) \cdot f_3(H). \tag{11.104}$$

The function $f_3(H)$ yields the dependence on the film thickness. After the substitution into Eq. (11.103) we obtain

$$\frac{dH}{d\theta} = \frac{a_b u_z(\theta, H) f_1(H) f_3(H)}{a_p u_\theta(\theta, H) f_2(H)}. \tag{11.105}$$

The functions f_1, f_2, f_3 are determined by Goren (1970), Goren & O'Neil (1971), Goldman et al. (1967), Spielman & Fitzpatrick (1973). At a distance of the order of the particle size, i.e. at $H \geq 1$, we can assume $f_1 f_2 / f_3 \approx 1$ and the particle trajectory coincides with the liquid streamline.

The theory can be specified for two models of different surface effects on the film drainage:

1. A critical thickness h_{cr} exists. The correlation between h_{cr} and DL thickness is not taken into account and surface forces beyond h_{cr} are neglected. In this model a particle is attached when the boundary condition for film thinning is fulfilled

$$H(\pi/2) = H_{cr}.$$
(11.106)

A particle is attached when $h(\theta) = h_{cr}$ at any θ. However, condition (11.106) separates the grazing trajectory. The result for a potential flow (Dukhin & Rulyov 1977) is

$$E_c = 1.8 \cdot \frac{a_p}{a_b} \left(\frac{h_{cr}}{a_p} \right)^{1/8}$$
(11.107)

2. Notion of h_{cr} is neglected and the attachment is caused by attraction. In the first model (molecular attraction forces are neglected) we cannot consider a particle attachment on the lower surface of the bubble because any particle depart from bubble surface under the action of gravity. When neglecting h_{cr} we consider the particle being almost on the bubble surface is under the action of the attractive force which can exceed gravity. Hence, the model allows us to consider particle attachment on the lower surface of a bubble. The grazing trajectory terminates at the rear stagnant pole of the bubble. Due to cylindrical symmetry we have

$$\left. \frac{dH}{d\theta} \right|_{\theta=\pi} = 0.$$
(11.108)

Comparing the conditions (11.108) and (11.103) we can conclude that the final coordinates of the grazing trajectory are given by

$$\theta_0 = \pi, \; F_n(\pi, H_0) = 0.$$
(11.109)

Equations (10.16) and (10.17) were derived by this method.

11.10.4. INFLUENCE OF PARTICLE AND BUBBLE PROPERTIES ON COLLISION EFFICIENCY

Fig. 11.9 shows differences in the curves characterising the dependence of collision efficiency on particle size which are caused by attachment at the first or second impact and by the surface state of the bubble. The calculation are accomplished for conditions given in experiments by Ralston and his co-workers (Crawford & Ralston 1988, Hewitt et al. 1994) and are discussed in next section.

In these experiments Ralston et al. used bubbles of three different sizes. We choose bubbles of 2 mm in diameter although deviations from spherical shape are possible. In order to use these equations the most important criterion is surface mobility ruther than an exact bubble shape. Surface retardation for such large bubble can be excluded.

The data in Fig. 11.9 correspond to results given in Fig. 3 by Hewitt et al. (1994) where the curves are calculated by means Schulze' superposition model. The solid line is calculated by Hewitt et al. using Schulze's superposition and for a retarded bubble surface of 2 mm diameter. If the bubble surface is free of surfactant Sutherland's Eq. (10.11) yields E_{ic} in Schulze's approximation and Langmuir's equation can be used to estimate the effect by inertia (Section 10.4). E_g is negligible compared to E_{ic} under potential flow. The results of this calculation are plotted as the dotted line. Both, solid and dotted lines correspond to the attachment after a first impact.

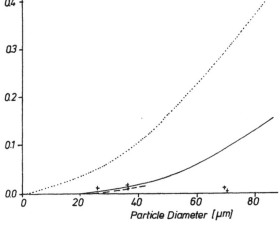

Fig. 11.9.   Collision efficiency calculated under different assumptions; retarded bubble surface and attachment at first collision (——), unretarded bubble surface and attachment at first collision (····), unretarded bubble surface and attachment at second collision (+--+); bubble diameter 2 mm, $\rho'=2.6g/cm^3$, $Re_b = 600$

The results for potential flow and attachment after a second impact are given in the lower part of Fig. 11.9. For sub-critical St values the calculation are accomplished using Eqs (10.141)-

(11.143) and for super-critical Stokes numbers by substituting θ_{ocr} into Eq. (11.74). Thus, according to Eq. (11.100) we get

$$E_{col} = E_1(\theta_o)\cdot\theta_o^2 = E_c E_{at} \tag{11.110}$$

where

$$E_c = e(\theta_o)\sin^2\theta_t \exp[-2s(T+\tau_o)], \tag{11.111}$$

and

$$E_a = \frac{\sin^2(\theta_t - \Delta_{t.p.c.})}{\sin^2\theta_t}\left(\frac{eh_{cr}}{a_p}\right)^{1/4}. \tag{11.112}$$

In Eq. (11.112) the first multiplier characterises the influence of ACS on attachment, and the second factor describes the influence of BCS stage, i.e. the drainage. The function $e(\theta_o)$ takes into account the deviation of the particle trajectory from a straight line before the first impact (cf. Fig. 11.8). $e(\theta_o) < 1$ is the collision efficiency at the first impact and can be calculated independently. Thus, the real collision efficiency E^{real} is lower than that plotted in Fig 11.9 where we omitted $e(\theta_o)$ in Eq. (11.111), and the equation

$$E = \sin^2\theta_t \exp[-2S(T+\tau_o)] \tag{11.113}$$

with

$$S(T+\tau_o) = \left[-1+\sqrt{1+6St\left(1+\frac{\rho}{2\rho_p}\right)\left(1-\frac{9}{2}St\frac{\rho}{\Delta\rho}\right)}\right]\left[\frac{\pi}{\sqrt{12St(1+\rho/2\rho_p)-1}}+1\right] \tag{11.114}$$

was used. The experimental data and equations used in these calculations are summarised in Table 11.1.

There large uncertainty in the calculated results for the super-critical Stokes numbers is due to three reasons. The Basset integral (Basset 1988, Thomas 1992) is not incorporated in the calculation for the trajectory, and the particle tangential velocity according to Eq. (11.76) is used to calculate the centrifugal forces, θ_s and θ_r.

The theories for both sub-critical and super-critical St values, are not valid in the vicinity St_{cr}. Eq. (11.114) can be used only if $\frac{1}{12}\left(1+\frac{\rho}{2\rho'}\right) < S < \frac{2}{9}\frac{\Delta\rho}{\rho}$. In the vicinity of the interval limits its accuracy is low. Fortunately, the St-value corresponding to the experiments by Ralston et al. is within this interval and Eq. (11.110) is valid.

Table 11.1   Data and equations used in the calculations, $\rho_p = 2.6 g/cm^3$, $2a_b = 2000\mu m$, $Re_b = 600$

$a_p[\mu m]$	25	35	equation	$a_p[\mu m]$	75		equation
				y	2/3 S	S	
E_o	0.04	0.05	Sutherland	E_o	0.1	0.1	Sutherland
St	0.02	0.04	11.48	St	0.2	0.2	11.48
				S	0.5	0.5	11.45
β	0.25	0.17	10.72	β	0.3	0.13	11.76
$\sin^2\theta_t$	0.35	0.35	10.71	$\sin^2\theta_t$	0.4	0.2	10.71
E/E_o	0.3	0.3	Dukhin 1982, 1983	$S(T+\tau)$	1.1	1.1	11.114
E	0.012	0.016	$E = (E/E_o)E_o$	E	0.044	0.022	11.113

The collision efficiency for the case of attachment after the first impact (solid and dotted lines in Fig. 11.9) is a monotonous function. The particle jump and the centrifugal force lead to maximum in the curve and to the strong decrease of collision efficiency at supercritical St values (cf. Fig. 11.9).

11.11. INFLUENCE OF DYNAMIC ADSORPTION LAYER ON ATTACHMENT PROCESS BY SLIDING

The DAL affects all microprocesses of flotation. The initial conditions of sliding are determined by the collision process. The DAL effects the collision process and also influences particle sliding.

Particle reflection after a collision is described by Eq. (11.55). Particle movement from the bubble surface after its reflection is retarded by a liquid movement with opposite direction, i.e.

by the normal component of the bubble hydrodynamic field. There is a large difference in the values of this component for a free and a retarded bubble surface. As explained in Section 10.2, the normal component depends linearly on the ratio a_p / a_b for a free bubble surface and is proportional to $(a_p / a_b)^2$ for a strongly retarded surface.

Thus, the retardation of the movement of a reflected particle by the liquid countercurrent is very sensitive to the degree of retardation of the bubble surface, i.e. to the DAL structure in the vicinity of the bubble front pole. The movement is fast at a weak surface retardation and is slow at a strong surface retardation. Thus, at strong surface retardation, the inertial path of the reflected particle can exceed that for the case of weakly retarded surface. The greater the tangential velocity, the shorter is the sliding time.

If the length of the inertial path decreases, the tangential liquid velocity in the vicinity of the particle decreases and leads to a longer sliding time.

There is a direct and an indirect effect of bubble surface retardation on the tangential particle velocity. The direct influence is caused by the dependence of the bubble hydrodynamic fields on the velocity distribution along its surface. The indirect influence is caused by the effect of the inertia path of a reflected particle on its tangential velocity and by the dependence of the path on the bubble surface retardation. The directions of the two effects are opposite. At the transition from a free to a retarded surface, the liquid tangential velocity diminishes at any point and the inertia path grows, which results in an increase in the tangential particle velocity.

Due to opposite directions of the direct and indirect effects of surface retardation, it is difficult to formulate, even a qualitative relationship, for the influence of the DAL on sliding time. However, it is clear that this influence is strong and a development of DAL theory would be useful.

The influence of the DAL on centrifugal forces is clearer. As the surface is mobile, their influence is strong and prevents particles deposition at angles $\theta > \theta_t$ (Section 11.5). The DAL retards the surface movement and decreases centrifugal forces, which can lead to a dominance of the gravity force. In Section 10.15, the bubble size was estimated which provides a domination of centrifugal forces at weak surface retardation (approximately 1 mm). At greater bubble diameter, increase in surfactant concentration and the growth of surface retardation result in a decrease of centrifugal forces, which is favourable for flotation. The mobility of the leading surface of a bubble as a function of surfactant concentration and its surface activity can be estimated by using the data in Fig. 8.2.

Usually the effect of frothers in flotation is explained by their ability to prevent bubble coalescence. Probably frothers can increase bubble surface retardation and thereby decrease the unwanted influence of centrifugal forces.

Attachment by sliding can be controlled by the drainage rate, which goes down with increasing retardation of liquid film surface by DAL. Thus, surface retardation by DAL is favourable because the unwanted action of centrifugal forces can be prevented. On the other hand, it hampers the drainage, which is unfavourable for flotation.

The DAL influence manifests itself in the extension of the t.p.c. As compared with the attachment by collision, this influence can not be crucial because the sliding time exceeds the collision time by orders of magnitude.

11.12. INVESTIGATIONS OF COLLISION AND ATTACHMENT STAGES OF BOTH MICROFLOTATION AND FLOTATION

11.12.1. IMPORTANCE AND DIFFICULTY IN FLOTATION OF A BIG PARTICLE

The upper limit of particle size for flotation is about 200µm (Jain, 1987). It is often pointed in monographs that the particle size does not exceed 100µm. Mechanical and pneumatic flotomachines produce bubbles with diameters of 1-3 mm (Jain, 1987), in agreement with the statement by Schulze (1993) that in flotation the ratio a_p / a_b is always smaller than 0.1. If a bubble is overloaded by captured particles it sinks and aggravates the flotation selectivity. The casual character of particles capture is caused by the possibility that for some bubbles the number of captures particles can exceed by many times the average value. Thus, the a_p / a_b ratio has to be small, which can be arranged by using large bubbles.

This question must be addressed because a enormous discrepancy in literature values exists concerning the upper limits of particle and bubble sizes. Naturally, the overloading of a bubble depends on the particle concentration and density. If both density and concentration are small the ratio a_p / a_b can be decreased.

However, flotation is more economical in concentrated pulps and a presence of bubbles with diameter <0.5 mm is completely unsuitable (Jain, 1987). The optimal flotation rate corresponds to particle size interval of $30 - 40$µm. Apart from this average size the flotation rate decreases very rapidly.

The clarification of reasons for the slow flotation of large particles is very important for the improvement of technology. A priori three factors decrease the flotation rate of bigger particles. The first is connected with the BCS stage, the second with the ACS stage, and the third with the stability of the bubble/particle aggregate. The effect of the first factor is illustrated by the lower curve in Fig 11.9. Beyond a particle size of $40 - 50$µm the collision efficiency drastically decreases due to the combined action of particle rebound and centrifugal force. As it was emphasised in Section 11.7., the larger the particles the faster is the t.p.c. expansion necessary for particle attachment (cf. Eq. (11.78)). This can cause a decrease in flotation rate. The stability of bubble/particle aggregates is controlled by the turbulence, according to Schulze (1993) and Crawford & Ralston (1988). If the particle size exceeds a critical value bubble/particle aggregates become unstable.

11.12.2. SIMILARITY IN MICROFLOTATION AND FLOTATION WITH RESPECT TO ATTACHMENT AND THEIR DIVERGENCE WITH RESPECT TO COLLISION

The separation of flotation and microflotation theories was proposed by Derjaguin & Dukhin (1960) due to a significant divergence in the collision stage, and difficulties in the description for large particles and bubbles. Attention was paid to microflotation because both stages are more simple and the clarification of the attachment mechanism in microflotation can also be used for flotation.

The term attachment is usually associated with bubble/particle interactions through a thin liquid interlayer. The general regularities of hydrodynamics of a liquid interlayer which manifest themselves in so-called lubrication approximation, and liquid film stability, control the attachment process. The regularities are not sensitive to the linear dimension of the interlayer and particle diameter. There is no reason there to be a qualitative distinction of the BCS stage of the attachment process for "small" particles (less than $10 - 30\mu m$). The common BCS mechanism for particles of any size could be established in investigations under microflotation conditions. The experimental investigation of Collins & Jameson (1977) and Anfruns & Kitchener (1977) are in reasonable agreement with the theory of short range interaction which was emphasised in Section 10.5.

The striking similarity in the attachment stage and difference in the preceding stage manifest themselves in equations for collection efficiency at supercritical Stokes numbers, Eqs (11.74) and (11.100c) and for inertiafree flotation under potential flow (Derjaguin et al. 1976), Eq. (11.107).

Both equations are a product of two factors. The first depends on the stage before sliding. For example, Eq. (11.107) is the product of Sutherland's equation and the parameter $\frac{3}{1.8}\left(h_c / a_p\right)^{1/8.15}$ coming from SHRT. Eq. (11.100) has a similar structure.

The large difference in the structure of first factor in the two equations is the manifestation of different mechanisms of the collision stage. It is seen that the second factor in both equations

(11.112) and (11.107) shows a close similarity due to the similar sliding mechanism at short range. As one can see from Eqs (11.112) and (11.107), the dependence of attachment by sliding on h_{cr} is very weak. However, at supercritical St the influence is greater. In addition, the exponent of h_{cr}/a_p is twice higher in Eq. (11.112) which causes a remarkable decrease in attachment efficiency in comparison with microflotation at usually very small h_{cr}/a_p-values. Even for hydrophobic surfaces $h_{cr} \approx 100nm$ that yields $\left(h_{cr}/a_p\right)^{1/4} \approx 0.1$ for $a_p \approx 50\mu m$. Thus, the decrease in the flotation rate can be caused by the low attachment efficiency during sliding of large particles. However, the application of a h_{cr} is not clear for broken particles.

11.12.3. INVESTIGATION OF COLLISION AND ATTACHMENT STAGES FOR BOTH MICROFLOTATION AND FLOTATION

In contrast with microflotation, not all stages of flotation are investigated. For the collision stage of flotation the theory has been elaborated but it is not verified by experiments yet. The decrease of flotation rate for large particle can be explained by this theory (Section 11.1-11.9). However, two alternative explanations exist, which makes complex experimental investigations on collision and attachment stages in microflotation and flotation very important. Recently, systematic studies were performed by Ralston and co-workers. They investigated the flotation rate of particles and bubbles of different size, and with different contact angles provided by different degree of the modification of the surface of broken quartz particles. It is important that experiments with single monodisperse bubbles are accomplished that provide better conditions for an estimation of E_c.

In distinction from experiments by Anfruns & Kitchener (1977) an important novelty is incorporated and increases the exactness of the number of particles captured by a single bubble. A video camera was used to record the bubble electronically so that the number of bubbles could be counted, their velocity and size determined. The total quantity of captured particles over the known number of bubbles yields the number of captured particles per single bubble. This quantity for a bubble diameter of 2 mm and different particle fractions of varying degree of hydrophobicity, given by Hewitt et al. (1994), is shown in Fig 11.10.

The number of captured particles per single bubble can be expressed by the collision efficiency. This quantity, calculated by Hewitt et al. (1994), using Schulze's superposition model (solid line in Fig. 11.9), is also shown in Fig. 11.10. We followed the same procedure using the dashed line of Fig 11.9 and taking into account the unwanted influence of particle rebound (dashed line in Fig. 11.10). The ordinates of dashed line of Fig. 11.10 is calculated from the ordinate of the solid line in Fig. 11.10 as the product with the ratio of ordinates of dashed line over ordinate of solid line in Fig. 11.9. Naturally, all these ordinates correspond to the same abscissa. This procedure is justified by the linear relationship between number of captured particles per bubble and the collision efficiency. Due to the maximum in the particle size distribution, the solid line in Fig. 11.10 has a maximum in contrast to the solid line in Fig. 11.9.

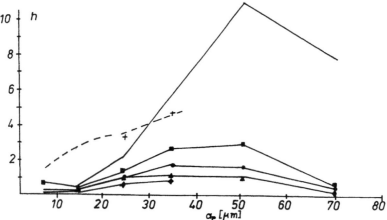

Fig. 11.10.   Number of quartz particles collected by one bubble of 2 mm diameter in water at different degrees of particle hydrophobicity: $\theta = 20°(\blacksquare)$, $\theta = 50°(\bullet)$, $\theta = 65°(\blacktriangle)$, $\theta = 88°(\blacktriangledown)$; the solid line represents the number of collisions per bubble calculated from collision efficiency for attachment during the first collision with a retarded surface (——) and during the second collision with an unretarded surface (+ +)

A conclusion about the state of bubble surface can be derived from information on the sizes and velocities of rising bubbles given by Crawford & Ralston (1988): "Ninety percent of the bubbles had diameters between 10^{-3} m and $1.4 \cdot 10^{-3}$ m, correspondingly to bubble rise $20 \cdot 10^{-2}$ m/s and $30 \cdot 10^{-2}$ m/s". These bubble velocities exceed the velocity given in Table 10.1 by 50-100% which suggests a low retardation of their surface. Similar information are not available from the experiments published by Hewitt et al. (1994). However, a procedure for a deep cleaning of the water is described. The height of the flotation cell used was 33 cm. According to the estimation given in Chapter 8 and experimental data by Loglio et al. (1989) the adsorption layer of large bubbles cannot reach saturation along such short path. Taking this information one can conclude that especially bubbles with a diameter of 2 mm cannot be retarded. Three different fractions of size of broken quartz particles were used in the experiment. The average size for each fraction and the values of St calculated by us from (Eq. (11.48)) and of K* from Eq. (11.50) are also given in Table 1. At a linear dependence of the velocity on the bubble diameter St is independent of a_b. K* depends on a_b because Re_p is a function of the bubble velocity. Thus, the values of St for the first (25μm) and the second (35μm) fractions are smaller than St_{cr} i.e. sub-critical, and for the third fraction (70μm) St is super-critical. Thus, the data for first and second fractions correspond to microflotation modelling and data for third fraction correspond to flotation modelling.

Three theoretical curves are available for the comparison with experimental data. The selection of curves is accomplished by taking into account that the surface of 2 mm bubbles is mobile. The solid line corresponds to a retarded surface and so can be ignored. Whilst the dotted and dashed line corresponds to mobile bubble surface. The attachment after a first impact yields high E_{col}-value (dotted line). There is a large discrepancy between this curve and the experimental data. Thus we can conclude that under the experimental conditions particles attach to a mobile bubble surface during or after the second impact. It means the experimental data have to be compared with dashed line.

These findings are favourable for the theory developed in Chapters 10 and 11. Simultaneously it enables us to conclude that the collision stage, i.e. the particle rebound and the effect of centrifugal force, play a major role in the decrease of measured collection efficiency for particles of supercritical size.

This conclusion can be confirmed by two arguments which follow from the experimental data:

1. The high similarity of the dashed theoretical and experimental curves and the small deviation between them allow to conclude that qualitative differences in flotation of particles of sub-critical and super-critical size is caused by a difference in the collision stage. The experimental data show that at $St < St_{cr}$ the flotation rate grows while at $St > St_{cr}$ it decreases. Thus St_{cr} is the main characteristic parameter for the transition from one flotation regime to the other, determined by the collision stage.

2. The role of SRHI in flotation kinetics can be estimated by a comparison of the curves for different degree of hydrophobicity. The contact angle changes in a wide range leads to a decrease in the flotation rate by a factor of 3. A qualitative change in the hydrophobicity does not influence the flotation rate in the transition region from sub-critical to super-critical particle size. The flotation rate decreases by the same amount for both particle sizes. Hydrophobicity manifests itself in the attachment process. There is no substantial difference in the influence of hydrophobicity for the sub-critical and super-critical particles on the collection efficiency. Thus, qualitative changes in the attachment stage of the different particle sizes are unlikely.

The attachment stage is of major importance at low hydrophobicity when the attachment efficiency is close to zero. Experimental evidence for this phenomenon was shown by Crawford & Ralston (1988) who introduced the important notion of flotability regions. The limit of this region is determined by the attachment stage. It seems possible that this limit can be interpreted by DDR theory of SRHI. According to Eqs (11.107) and (11.100) E_a tends to 0 if $h_{cr} \rightarrow 0$.

The opposite case was investigated by Hewitt et al. (1994). Rather large contact angles were chosen which excluded the effect of the attachment process in the collection efficiency change with particle size.

The experimental data are in qualitative agreement with the DDR theory of SRHI and attachment. In the above discussed experiments a qualitative or semi-qualitative confirmation of the main statements of DDR theory was achieved. If attachment occurs due to a sufficient thinning of the film the absolute value of E_a cannot be more than doubled. Indeed h_{cr} does not vary in a wide range and $h_{cr}^{1/8}$ even less. It is known that $h_{cr} \approx 100nm$ for highly hydrophobic quartz. Still lower h_{cr} values are unreasonable. Instead, a direct influence of molecular attraction forces on the thinning process has to be considered, then values of $h_{cr} = 0.3 \div 1.0nm$ are more reasonable. In that case $h_{cr}^{1/8}$ varies less than 2 times. Changes of h_{cr}-values obtained in experiments Hewitt et al. (1994) are in the range of a factor of 3 which is in good qualitative agreement with the DDR theory. However, the DDR theory underestimates the role of SRHI.

Recently the influence of surface properties on the drainage rate was shown (Hewitt et al. (1993), Fischer et al. (1992)). A quantification of these phenomena by the DDR approach seems to be possible.

There are two more possible reasons of the underestimation of SRHI by the DDR theory. If the ACS stage controls the attachment rate it is useless to use the DDR theory for interpretation of experimental data, as it describes the BCS stage. Moreover, the DDR equations were derived for spherical particles with a smooth surface. The behaviour of broken particles in collision process can be completely different, as it was emphasised in Section 10.8.3. Under a favourable orientation of a broken particle to the surface the attachment can occur already at the first impact. With an unfavourable orientation attachment is impossible. Between these two extreme cases many possibilities of orientation and attachment exist. It is not excluded that a small part of broken particles attach during the first collision, another part during the second impact and more particles even during sliding. There is no theory of attachment of broken particles. Experiments by Hewitt et al. (1994) show that the qualitative predictions of the DDR theory hold also for broken particles, although the quantitative description underestimates the importance of BCS or ACS stages of attachment.

It would be very interesting to use the unique flotation device of Hewitt et al. (1993, 1994) for investigations of spherical monodisperse glass particles of different size. There is hope that experiments with spherical particles can confirm both the theory of collision and attachment. Afterwards a comparison of experimental results for spherical and broken particles could yield valuable, and unique, information concerning the attachment mechanism for broken particles.

Another promising perspective is connected with dynamic adsorption layers of rising bubbles which is discussed in next section.

11.13. INFLUENCE OF DYNAMIC ADSORPTION LAYER ON DETACHMENT

The influence of the DAL on small particle detachment is discussed in Section 10.7. The main qualitative results are also valid for large particle detachment, although there are significant quantitative differences. First, contactless cellectorless flotation is impossible for large particles due to strong detachment forces. Thus, collectors are necessary to provide stability of the bubble-particle aggregates. Second, large contact angles are necessary to provide stability of such particle-bubble aggregates as distinct from the stability of small particle-bubble aggregates. Third, a decrease of detachment forces can be arranged through the control of the DAL. It relates to big bubbles because the lower part of their surface cannot be retarded beyond a narrow rear stagnant cup. A large normal component of the liquid velocity at the mobile part of the lower surface can cause strong detachment forces. This difficulty can be avoided by increasing surfactant concentration which extend the rear stagnant cup. Another reason of frother application in flotation is the influence on bubble surface mobility.

The contact angle within the rear stagnant cup can strongly differ from its equilibrium value because surface concentration exceeds the equilibrium value. In particular, difference between surface concentration within the r.s.c. and the equilibrium value causes a difference in the Stern potential too. Here, the absolute value of the Stern potential exceeds its equilibrium value because the surface concentration of OH^- is higher than at equilibrium. The effect of this difference on the value of the contact angle within r.s.c. can be estimated on the basis of Eq. (10A.6).

Li & Somasundaran (1991, 1992) have established that the absolute value of the negative charge of the water-air interface is sufficiently high already at $10^{-2}-10^{-1}$ M NaCl at sufficiently high pH-value. It leads to the conclusion that OH^--adsorption at the water-air interface satisfies condition (8.72) and its value within the r.s.c. exceeds equilibrium. Consequently, the Stern potential also can exceed the equilibrium. Due to the high sensitivity of the contact angle to the Stern potential, this difference can be important.

Let us assume the absolute value of the Stern potential at the water-air interface Ψ_1 exceeds that of the mineral particle Ψ_2, i.e. $|\Psi_1| > |\Psi_2|$. Thus, within the r.s.c. $|\Psi_1 - \Psi_2|$-values are higher than in equilibrium. Taking into account the fourth degree in Eq. (10A.6), even a small increase in $|\Psi_1 - \Psi_2|$ can cause a big increase of the contact angle within the r.s.c.

The application of cationic surfactants was proposed initially by Derjaguin & Dukhin (1961) in order to decrease the electrostatic barrier which prevents particle from approaching the bubble surface. This barrier is not important in flotation of normal size particles because they overcome the barrier due to gravity. However, the application of cationic surfactants is recommended in flotation of normal size particles, because it increases the stability of the bubble-particle aggregate by the increase the $|\Psi_1 - \Psi_2|$-value.

Thus, the mechanism of the supporting influence of cationic surfactants on microflotation and flotation can be different. In microflotation the electrostatic barrier can be decreased, in flotation the contact angle can be increased. Naturally, both effects manifest themselves simultaneously. Li & Somasundaran(1990, 1992) observed a bubble recharge due to adsorption of multivalent inorganic cations. Thus, their application is recommended in order to increase the contact angle and to stabilise bubble-particle aggregates. Naturally, selective adsorption of multivalent ions at the water-air interface is important. Adsorption of organic or inorganic cations at the r.s.c. can exceed their equilibrium value. This is important for the precalculation of increase of the contact angle caused by cation adsorption.

11.14. SUMMARY

QUALITATIVE CHANGE IN THE COLLISION STAGE OF FLOTATION AT SUPERCRITICAL STOKES NUMBERS

At $St > St_{cr}$ the inertial impact of a particle deforms the bubble surface, creates a thin water layer between the particle and bubble and makes the particle jump back from the bubble surface. This decreases the collision efficiency which otherwise rises with the particles size. For particles of subcritical diameter the collision efficiency increases with particle diameter. The proposed theory of inelastic collision unlike other theories describes the coupling of inertial bubble-particle interaction and water drainage from the liquid interlayer.

The theory makes for a more accurate calculation of the minimum thickness of a liquid film h_{min}, which is reached during a collision. The estimated value of h_{min} of the range of 0.2µm and 1µm exceeds the critical thickness h_{cr} of the spontaneous rupture of a thin water film. This means that attachment by collision is impossible for particles of adequately smooth surfaces.

REPETITIVE COLLISION

We have described the theory of repetitive collision to show that the minimum thickness of the liquid interlayer during a second collision many times exceeds h_{cr}. Thus, the attachment by a second collision is also impossible with particles with surfaces that are too smooth. The derived equation of the particle trajectory between the first and the second collision is restricted to Stokes numbers $St < 1$. Only one repetitive collision is possible under this condition. An additional restriction is given by the difference between St and St_{cr} which must not be too small.

The attachment by a first or repetitive collision is not excluded in the theory for particle having special surface roughness or shape favourable for the rupture of comparatively thick water films.

ATTACHMENT OF PARTICLES WITH SMOOTH SURFACES

The attachment of particles with smooth or slightly rough surfaces is possible only by sliding. Particle rebounds cannot be neglected as they govern the initial conditions of the sliding process which starts after the first or second recoil. The length of the second recoil l_r determines the initial distance to the surface for a sliding process. The joint action of particle rebound and centrifugal forces make a second collision impossible when the first collision takes place too far from the front pole, i.e. at an angle $\theta > \theta_{0cr}$.

QUANTIFICATION OF THE QUALITATIVE DIFFERENCE BETWEEN THE COLLISION EFFICIENCY AT SUPERCRITICAL STOKES NUMBERS AND THE EXPERIMENTS BY RALSTON AND CO-WORKERS

The equation for θ_{0cr} and the collision efficiency is obtained by taking into account the joint negative influence of particle rebound and centrifugal forces on particle-bubble collision. The experiments of Hewitt et al. (1994) show the flotation rate's complete dependence on the diameter of highly hydrophobic particles and large contact angles. This and the weak sensitivity to hydrophobicity and to the magnitude of contact angle means that the extreme in the dependence of flotation rate on particle diameter is caused by the collision stage. At subcritical Stokes numbers the measured capture efficiency increases with particle diameter while for supercritical St the opposite tendency is observed in agreement with the theory.

INFLUENCE OF THE DAL ON ATTACHMENT BY COLLISION OR SLIDING

The DAL influences practically all sub-processes which manifest themselves in attachemnt by collision and sliding. Surface retardation by a DAL affects the bubble velocity and consequently the bubble-particle inertial hydrodynamic interaction. It also affects the drainage and thereby the minimum thickness of the liquid interlayer achieved during first or second collision and sliding. Surface retardation influences also on the hydrodynamic field around the bubble and on the centrifugal forces and length of the particle recoil. As the result the particle trajectory between the first and second collision as well as the collision efficiency are very sensitive to the DAL structure. There is big difference in the DAL influence on the attachment between conditions (8.71) and (8.72). There is also a significant effect on the rate of extension of the t.p.c. The sub-processes contribute to the capture efficiency in opposite directions which makes even a qualitative prediction of the DAL effect very complicated.

INFLUENCE OF DYNAMIC ADSORPTION LAYER ON DETACHMENT

In distinction from microflotation, contactless collectorless flotation is impossible for large particles due to strong detachment forces. Thus, collectors and large contact angles are necessary to provide stability of such particle-bubble aggregates as distinct from the stability of small particle-bubble aggregates. A decrease of detachment forces can be arranged through the control of the DAL. It relates to big bubbles because the lower part of their surface cannot be always retarded beyond a narrow rear stagnant cup. A large normal component of the liquid velocity at the mobile part of the lower surface can cause strong detachment forces. This difficulty can be avoided by increasing surfactant concentration which extend the rear stagnant cup.

The contact angle within the rear stagnant cup can strongly differ from its equilibrium value because surface concentration exceeds the equilibrium value. In particular, difference between surface concentration within the r.s.c. and the equilibrium value causes a difference in the Stern potential too. Here, the absolute value of the Stern potential can exceed its equilibrium value because the surface concentration of OH^- is higher than at equilibrium. This holds true as long as the presence of ionic surfactants does not complicate the situation.

This phenomenon becomes even more important after the results published by Li & Somasundaran (1991, 1992). They have established that the absolute value of the negative charge of the water-air interface is sufficiently already high at 10^{-2} - 10^{-1} M NaCl at sufficiently high pH-value. From a comparison with the DAL theory we can draw the conclusion that OH^--adsorption within the r.s.c. exceeds equilibrium. Consequently, the Stern potential also can exceed the equilibrium. Due to the high sensitivity of the contact angle to the Stern potential, this difference can be important.

Thus, it cannot be excluded that OH^- ions can serve as ionic collectors if all conditions are satisfied (sufficiently high pH, not too high electrolyte concentration and definite electrical surface properties of the particles). A quantitative evaluation of this phenomenon appears to be impossible because of lack of a DAL theory for large Reynolds numbers. At adsorption of cationic surfactants the absolute value of the potential Ψ_1 of a water-air interface can be smaller than that of the particle Ψ_2, i.e. $|\Psi_1| < |\Psi_2|$. Within the r.s.c. a significant decrease in Ψ_1 is possible and simultaneously $|\Psi_1 - \Psi_2|$ increases. Taking into account the fourth degree in Eq. (10D.6), even a small increase in $|\Psi_1 - \Psi_2|$ can cause a big increase of the contact angle within the r.s.c.

Thus, the mechanism of the supporting influence of cationic surfactants on microflotation and flotation can be different. In microflotation the electrostatic barrier can be decreased, in flotation the contact angle can be increased. Naturally, both effects manifest themselves simultaneously. Li & Somasundaran (1990, 1992) observed a bubble recharge due to adsorption of multivalent inorganic cations. Thus, their application is recommended in order to increase the contact angle and to stabilise bubble-particle aggregates. Naturally, selective adsorption of multivalent ions at the water-air interface is important. But even in the absence of adsorption selectivity under equilibrium conditions a deviation from equilibrium can happen due to the increase of adsorption within the r.s.c. This is important for the precalculation of increase of the contact angle caused by cation adsorption.

11.15. REFERENCES

Anfruns, J.P. and Kitchener, J.A., Trans. Inst Mining and Mat., 86(1977)C9

Basset, A.B., A Treatise on Hydrodynamics, (Cambrigde, Beighton Bell), Cambrigde 2(1988)

Bergelt, H., Stechemesser, H. and Weber, K., Intern. J. Miner. Process., 34(1992)321

Brauer, H. Grundlagen der Ein- und Mehrphasenströmungen. Aarau: Verl. Sauerlander, 1971

Collins, C.L. and Jameson, G.L., Chem. Eng. Sci., 32(1977)239

Crawford, R. and Ralston J., Int. J. Miner. Process., 23(1988)1

Derjaguin, B.V. and Kusakov, N.N. Acta Physicochim USSR, 10(1939)25

Derjaguin, B.V. and Smirnov, L.P., in book Issledovania voblasti povorkhnortnik sil M. Nauka, (1967)188

Derjaguin, B.V., Dukhin, S.S. and Rulyov, N.N., Kolloidn. Zh., 38(1976)251

Dimitrov, D.S. and Ivanov, I.B., J. Colloid Interface Sci., 64(1978)971

Dukhin, S.S., Rulyov, N.N. and Dimitrov, D.S., Coagulation and Dynamics of Thin Films, Naukova Dumka, Kiev, 1986

Dukhin, S.S. and Schulze J., Kolloidn. Zh., 49(1987)644

Evans, L., Ind. Eng. Chem., 46(1954)2420

Fisher L.R., Hewitt D., Mitchel E.E., Ralston J., Wolfe E.., Adv Colloid Interface Sci., 39(1992)397

Fuchs, N.A., Uspekhi Mekhaniki Aerosoley, Izd-vo AN USSR, Moscow (1961), p.158

Goldman, A.L., Cox, R.G. and Brenner, H., Chem. Eng. Sci., 22(1967)637

Goren, S.L., J. Fluid Mech., 41(1970)619

Goren, S.L. and O'Neill, M.E., Chem. Eng. Sci., 26(1971)325

Hewitt, D., Fornasiaro, D., Ralston, J., and Fisher,L., J. Chem. Soc. Farady Trans., 89(1993)817

Hewitt, D., Fornasiaro, D. and Ralston, J., Mining Eng. Minerals Engineering, 7(1994)657

Hopf, W. and Geidel, Th., Colloid. Polym. Sci., 265(1987)1075

Hopf, W. and Stechemesser, H., Colloid & Surfacec, 33(1988)25

Hornsby, D. and Leya, J., in Surface and Colloid Science. (Ed. E. Matijevic), N.Y., Wiley Interscience, 12 (1982)217

Ivanov, I.B., Dimitrov, D.S. and Radoev, B.P., Kolloidn. Zh., 41(1979)36

Jain, S.K., Ore processing, A. A. Balkema, Rotterdam, 1987

Luttrel, G.H. and Yoon, R.H., J.Colloid Interface Sci., 154(1992)129

Natanson, G.L., Dokl. Acad. Nauk SSSR, 112(1957)110

Nquen Van, A. and de Kmet, S., Intern. J. of Mineral Processing, 35(1992)205

Reynolds, O., Phil. Trans. Rog. Soc. London, 177(1986)157

Rulyov, N.N., Kolloidn Zh., 50(1988)1151

Rulyov, N.N. and Dukhin, S.S., Kolloidn. Zh., 38(1986)302

Rulyov, N.N., Dukhin, S.S. and Chaplygin, A.G., Kolloidn. Zh., 49(1987)939

Rulyov, N.N. and Chaplygin, A.G., Kolloidn. Zh., 50(1988)1144

Schulze, H.J., Int. J. Miner. Process, 4(1977)241

Schulze, H.J. and Gottschalk, G., Kolloidn Zh., 43(1981)934

Schulze, H.J. and Dukhin, S.S., Kolloidn Zh., 44(1982)1011

Schulze, H.J. and Birzer, O., Colloids Surf., 24(1987)607

Schulze, H.J., Radoev, B., Geidel, Th., Stechemesser, H. and Töpfer, E., Int. J. Miner. Process, 27(1989)263

Schulze, H.,J. in Flothing in Flotation, (J.S. Laskowski, Ed.), Gordon and Breach, Glasgow, 1989, Chapter 3.

Schulze, H.J., Adv. Colloid Interface Sci., 40(1992)283

Schulze, H.J. in B. Dobias (Ed.), Coagulation and Floculation, Marcel Decker, New York, 1993

Scheludko, A, Tchaljovska, S. and Fabrikand, A.M., Faraday Spec. Discuss. 1, A(1970)112

Li, Ch. and Somasundaran, P., J. Colloid Interface Sci., 146(1991)215

Li, Ch. and Somasundaran, P., J. Colloid Interface Sci., 148(1992)587

Spedden, H.R. and Hannan, W.S., AIME Technical Publication 2534(1948)37

Spielman, L.A. and Goren, S.L., Environ. Sci. and Technol. 4(1970)135, 5(1971)85

Spielman, L.A. and Fitzpatrick, L.A., J. Colloid Interface Sci., 42(1973)607

Stechemesser, H.J., Geidel, T. and Weber, K., Colloid Polym.Sci., 258(1980)109

Stechemesser, H.J., Geidel, T. and Weber, K., Colloid Polym.Sci., 258(1980)1206

Stechemesser, H.J., Freiberger Forschungshefte, A790 Aufbereitungtechnick, (1989)

Stechemesser, H.J., Schulze, H.J. and Radoev, B., Thesis of International Conference Surface Forces, Moscow, (1985)64

Thomas, P.J., Phsy. Fluids, 4(1992)2090

Tschaliovska, S., Alexandrova, L.B., Godishnik na Sofisky Universitet, 72(1977/78)45

Tschaliovska, S. Thesis,Univ.Sofia,Fac.Chemie, 1988

Van Nguen, A., Int. J. Miner. Process., 37(1993)1

Whelan, P.F. and Brown, D.J., Trans. Inst. Mining and Met., 65(1956)181

Ye, Y. and Miller, J.D., Coal. Prep., 5(1988)147

Ye, Y. and Miller, J.D., Int. J. Miner. Process., 25(1989)199

Yoon, R.H. and Luttrel, G.H., Mineral Processing and Extructive Metallurgy Review, 1989, Vol 5, 101

CHAPTER 12

12. NON-EQUILIBRIUM SURFACE FORCES CAUSED BY DYNAMIC ADSORPTION LAYERS AND THEIR RELEVANCE IN FILM STABILITY AND FLOTATION

DAL generates surface forces which can be naturally named non-equilibrium forces since they arise due to a deviation of the adsorption layer from equilibrium. The effect of non-equilibrium surface forces on the dynamics of these layers is substantially different to that of equilibrium ones. In many cases, the radius of their actions is much greater than the radius of action of equilibrium surface forces since they are localised within the diffusion boundary layer. Approaching a surface, particles pass first of all the diffusion layer, so that in many cases the possibility of coagulation is determined by the action of non-equilibrium surface forces. In other situations it is connected with the action of equilibrium surface forces while non-equilibrium forces influence the rate of the process.

12.1. THE EFFECT OF THE DYNAMIC ADSORPTION LAYER ON COAGULATION

Both the surface concentration of adsorbed material and the surface tension are functions of the position on the dynamic surface. Each fluid surface element experiences a force directed along the gradient of increasing surface tension. The tangential liquid motions caused by these surface tension gradients can either hinder or accelerate the thinning of films (Alan & Mason 1962; Charles & Mason 1960; Groothius & Zuiderweg 1960). Among these effects, the Marangoni-Gibbs effect is the best known and plays an important role in stabilisation of foams and emulsions.

As an example, suppose a soap film is stretched and thinned. The increase in surface area causes a drop in surface concentration of adsorbed surfactant and a consequent rise in surface tension, an effect which tends to resist the process of film stretching.

The kinetics of coalescence of liquid emulsion droplets (phase 1) suspended in a second immiscible liquid (phase 2) is influenced by the presence of a third component at the interface, soluble in both phases and able to lower the interfacial tension. It is found experimentally, that if the third component is removed from the interface by diffusion into phase 2, the coalescence rate increases, while if it diffuses into phase 1, the rate decreases. This observation may be explained by noting that diffusion into the outer phase causes an increase in surfactant concentration in the gap between two approaching droplets, while diffusion into the inner phase has the opposite result. The gap region in the former case, enriched with surface active material, produces a lower local interfacial tension, the surface spreads away from the gap region, which then thins and increases the rate of droplet collision and coalescence. Enrichment of surface active material between two approaching droplets increases the film thickness due to surface motion and coalescence is correspondingly inhibited.

These effects have been thoroughly investigated by MacKay and Mason (1963), who studied the kinetics of layer thinning between a flat surface and an approaching droplet. An interferometric method was used and the results compared with theoretical predictions based upon laminar flow theory for the bulk fluid in the gap and stability theory for the fluid interface. The theory was simplified by setting the gap equivalent to that between two parallel flat plates.

As a first approximation, the experimental results support the theory in so far as the viscous resistance of the thinning gap is concerned. The effect of a third soluble component on the kinetics of thinning was also at least qualitatively in accordance with expectation for a diffusion model, but quantitatively it was not possible to overcome the mathematical difficulties associated with surface motion caused by surface tension gradients. The fact that such a gradient is present was confirmed by microscopic observation of dust particles at the droplet interface.

Thiessen (1963) presents results of a similar nature. The effect of surfactant diffusion on droplet coalescence confirms the results of MacKay & Mason (1963). In addition, the influence of surface inactive materials such as inorganic salts was investigated. As expected, the sign of the effect is reversed when a surface inactive substance is substituted for a surface active one. Thiessen (1963) points out the relevance of studies of this type to the extraction of salts from aqueous solutions using organic solvents. The isolation of metals by this technique is currently a popular problem.

With regard to foam stability, the attention of investigators has been drawn (Rebinder & Vestrem 1930, Venstrem & Rebinder 1931, Allan et al. 1964) to the process of bubbles approaching a gas-liquid interface, the thinning and destruction of the interphase film, its kinetics, and the influence of surface active substances upon these processes. Predictions for the thinning rate, based on viscous flow theory, have been confirmed experimentally. The theory takes into account surface tension gradients generated by surface active materials in the system which act to reduce the thinning velocity. Klassen's results (1949) regarding the thinning kinetics between two converging bubbles are in good agreement with the idea of surface tension gradients coupled with viscous forces.

As the inertia forces for tiny particles are small compared to viscous forces, Sutherland & Work (1960) have hypothesised that attraction forces exist between the particle and the bubble causing their mutual adhesion. These forces are sufficiently long-range that the thickness of the boundary layers is exceeded.

This theory has been employed to explain the results obtained by Ewans & Ewers (1953). In their experiments a small section of the surface of an inclined solid was coated with a thick layer of an organic, soluble surfactant. The surface was flushed with water and the thickness of the draining water film proved to be uniform over the inclined surface except for the section containing surfactant, where the draining film was much thinner. This occurs in the transition region between thick and thin films and the air/water interface must exhibit a local curvature. The investigators concluded that the resultant capillary pressure was compensated by long-range forces originating in the region containing surfactant, of longer range that the film thickness equal to tens of microns. While Sutherland & Work (1968) did not explain the nature of these long-range forces, they pointed out the significance of their findings for the theory and practice of flotation. The long-range effect which causes the thinning, is associated with the desorption of the surfactant from the solid surface and its diffusive transport to the air/water interface. The local lowering of surface tension generates a Marangoni effect with an outward flux of the film directed away from the area containing surfactant towards the thicker regions of the water film, thus enhancing film thinning.

The union of long-range actions arising under conditions of this experiment and the quantitative theory of the phenomenon were predicted for the first time by Dukhin (1960, 1961, 1963) whose main results are discussed in the following sections.

12.2. THE INFLUENCE OF THE IONIC ADSORPTION LAYER UPON COAGULATION PROCESSES

Any portion of a dynamic ionic adsorption layer leads to an electrical double layer out of electroneutrality. The adsorbed layer acquires the charge of the fast diffusing ion, while the diffusion layer takes the charge of the slow diffusing ion. It is possible to describe qualitatively the adsorption layer interactions and their kinetics without rigorous mathematical analysis. The initial adsorption of surface active ions is followed by the adsorption of the counter ions which reside in the diffuse double layer. Macroscopically equivalent numbers of oppositely charged ions are involved to preserve overall electric neutrality, each ion is transported by diffusion.

Counter-ions diffuse rapidly compared with the surface active species, but the build-up of the electric fields at the interface modifies the transport in such a way that it accelerates the slowly diffusing species and decelerate the rapidly diffusing ones so that finally the fluxes of each are equal. Thus, the condition of approximate equality of the currents of adsorbing ions and counter-ions is provided by the appearance of an excessive surface charge the sign of which coincides with that of the fast-diffusing ions.

Finally, the effect of the ionic character of the surfactant upon the surface forces. Let us consider first of all the diffusion - electric analog of the Marangoni-Gibbs effect. Since stretching of the film results in adsorption, both film surfaces are charged simultaneously and are pushed away from each other. This makes film thinning difficult. This repulsion of a non-equilibrium double layers takes place at distances which are many times greater than the double layer thickness.

In the experiment of Ewans & Ewers (1953), ionic surfactants were used, so that the mechanism of film thinning is more complicated. The solid substrate from which the ionic surfactant is desorbed acquires a charge opposite to that generated by the adsorption of the surfactant at the air/water interface. The resulting electrostatic attraction contributes to film thinning (Dukhin 1960). This effect is accompanied by the Marangoni hydrodynamic thinning. The combined result is understandable only after thorough mathematical analysis of the role played by each mechanism. Only in the case of high concentrations and high surface activity can the electrostatic contribution to the thinning kinetics be neglected. Thiessen (1963), in discussing the kinetic influence of inorganic electrolyte, needed to invoke the arguments of Dukhin (1963) to explain her results.

12.3. THE LIQUID INTERLAYER STABILISATION BY DYNAMIC ADSORPTION LAYERS IN ELEMENTARY FLOTATION ACT

Rather high electrolyte concentrations characteristic of natural and waste waters, substantially weaken non-equilibrium surface forces of the diffusion - electric nature but scarcely affect non-equilibrium surface forces caused by the dynamic adsorption layer of nonionic surfactant (Dukhin 1981). Therefore, we consider such forces are important and their mechanism deserves special attention. The flotation of tiny particles is, nevertheless, possible if the distance between the bubble and particle surfaces becomes smaller than a critical value h_{cr}. The film of thickness h_{cr} is thinned, becomes unstable, and collapses if long-range attractive forces exist between the particle and the bubble, drawing them together.

The adsorption change along a bubble surface leads to a change of adsorption in the film (water interlayer between the bubble and the approaching particle) as well. The adsorption in the centre of the liquid interlayer is smaller than in its periphery so that the central surface tension is greater than the peripheral one. The increase in surface tension towards the centre of the film should be followed by a surface flow in the same direction involving the entire film liquid. The liquid flow in the gap between particle and bubble prevents film thinning and flotation. It is reasonable to call this effect liquid interlayer stabilisation by a dynamic adsorption layer in the elementary flotation act. In the elementary flotation act there is an effect which is analogous to the Marangoni-Gibbs effect. This effect may be used to explain how a dynamic adsorption layer can contribute to the stabilisation of a film at the interface via the Marangoni-Gibbs effect. This effect is zero if the surface concentration on the bubble remains uniform.

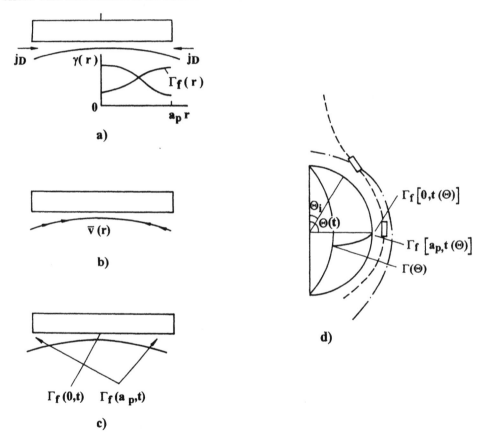

Fig. 12.1.   Liquid interlayer stabilisation by a dynamic adsorption layer; trajectory of a flat particle along the bubble surface (a), adsorption distribution Γ_f (b), velocity $\bar{v}(r)$ in the interphase film (c), j_D is the diffusion flux (d)

The quantitative analysis of this effect requires the calculation of adsorption changes between the centre of the film $\Gamma_f(0,t)$ and its periphery $\Gamma_f(a_p,t)$ (cf. Fig. 12.1). Calculations have been

478

performed for disk-like particles, because the film can be regarded as plane-parallel, and the diffusion process inside the film has an axisymmetry. Here, θ_1 is the angle which characterises the particle entering the bubble diffusion layer at the moment t=0, $\Gamma(\theta)$ is the adsorption distribution along the bubble surface.

Thus, the equation of the non-steady diffusion reads

$$\frac{\partial \Gamma}{\partial t} = D_s \left(\frac{\partial^2 \Gamma_f}{\partial r^2} + \frac{1}{r} \frac{\partial \Gamma_f}{\partial r} \right), \tag{12.1}$$

in which D_s is the coefficient of the surface diffusion of the surfactant molecules. The initial and boundary conditions are, respectively

$$\frac{\partial \Gamma_f}{\partial r}(0,t) = 0, \quad \Gamma_f(a,t) = \Gamma(t), \quad \Gamma_f(a,0) = \Gamma(\theta_1). \tag{12.2}$$

The solution of equation (12.1) can be presented with the help of Duamel integral

$$\delta \Gamma_f(r,t) = \sum_{n=1}^{\infty} \frac{2 J_0 (v_n r / a)}{v_n J_1 (v_n)} \chi_n(t), \tag{12.3}$$

where

$$\chi_n(t) = \frac{v_n^2}{\tau_D} \int_0^t \left(\Gamma(\tau) - \Gamma(\theta_1) \right) e^{-v_n^2 (\tau - t)/\tau_D} d\tau + \Gamma(t) - \Gamma(\theta_1). \tag{12.4}$$

J_0 and J_1 are the Bessel functions of the first type, v_n are the roots of J_0, $\tau_D = a^2 / D_s$ is the time relevant to the diffusion in the film. The stabilising effect of the non-equilibrium adsorption layer can be neutralised by pressing forces due to the normal component of the liquid velocity. The liquid moves towards the leading bubble surface. In doing this the liquid flows around the particle, as though it is pressed to the bubble. The force which forces a disk-like particle with thickness b to the bubble surface at Re»1 is given by

$$F_1 = 48 \eta \frac{abv}{a_b} \cos\theta \tag{12.5}$$

using the equation of hydrodynamic force for disks (Happel & Brenner, 1976).

The flow forcing a particle to the bubble surface also causes excess pressure in the liquid interlayer $\delta p(r)$ which should compensate for F_1

$$F_2 = \int_0^a 2\pi r \delta p(r) \, dr. \tag{12.6}$$

The minimum value of the film thickness h_{lim}. At $h > h_{lim}$ the film gets thinner due to the pressing force. At $h = h_{lim}$ the film thickness is stabilised because the out-flow through the gap caused by the pressure drop is balanced by the in-flow due to the surface-tension gradient. The minimum film thickness for the case of a plane-parallel gap between the disk-like particles and the almost flat ones of the greater surface section bubble can be calculated and is determined by the stability condition for film thickness. It corresponds to the equation of zero liquid flow through any cylindrical section of a film

$$2\pi r \int_{0}^{h_{lim}} v(r,z)\,dz = 0,\tag{12.7}$$

in which the radial tangential velocity distribution $v(r,z)$ is determined by the co-action of the pressure gradients and the surface tension. The influence of $\partial p / \partial r$ on $v(r,z)$ is described in the Navier-Stokes equation where the influence of $\partial \gamma / \partial r$ is considered while it is solved within the boundary condition (12.9), because $\partial \gamma / \partial r$ must be balanced by the liquid viscous tension at the surface-air interface. A simplification of the Navier-Stokes equations due to the use of approximations for the thin films hydrodynamics is favourable for the determination of the function $v(r,z)$,

$$\eta \frac{\partial^2 v}{\partial z^2} = \frac{\partial p(r)}{\partial r},\tag{12.8}$$

$$v_r\big|_{z=h} = 0; \quad \frac{\partial v_r}{\partial z}\bigg|_{z=0} = \eta^{-1}\frac{\partial \gamma}{\partial r}.\tag{12.9}$$

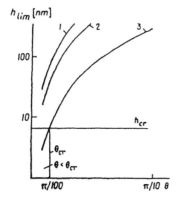

Fig. 12.2.   Minimum thickness of the liquid interlayer h_{lim} for different bubble surface sections characterised by the angle θ, The region where the impossibility of deposition has not been proved, $\theta < \theta_{cr}$.

Here $z = 0$ and $z = h$ correspond to the surfaces of the liquid interlayer between the bubble and the solid particle. Two unknown constants which appear in the solution of Eq. (12.8) are determined by the boundary condition (12.9). Substitution of the expression for $v(r,z)$ in Eq. (12.7) and integration gives a relation between pressure and adsorption distribution within the interphase film,

$$\delta p(r) = \frac{3}{2h_{lim}} \frac{\partial \gamma}{\partial r} d\Gamma(r). \tag{12.10}$$

This equation can lead to a formula for h_{lim}, because the adsorption distribution can be expressed by Eq. (12.3), and the pressure distribution is connected through (12.6) with the pressure force,

$$-F_1 = F_2 = \frac{3}{2h_{lim}} \int_0^a 2\pi r \, \delta\Gamma(r) \, dr. \tag{12.11}$$

The left-hand side of this equation can be expressed by Eq. (12.6) and the well-known formula for the bubble velocity

$$v = \frac{1}{9} ga^2 / v. \tag{12.12}$$

The right-hand side can be expressed by the adsorption distribution given by (12.3) and integration leads to

$$\int_0^a J_0\left(\frac{v_n r}{a}\right) r \, dr = \frac{a}{v_n} J_1(v_n). \tag{12.13}$$

The obtained relations allow to express h_{lim}

$$h_{lim}(\theta) = \frac{\chi_b}{\eta} \frac{3\pi a_b}{16} \frac{a}{b} \sum_{n=1}^{\infty} \frac{\chi_n(t)}{v_n}, \tag{12.14}$$

where χ_b is given by Eq. (8.35).

Now, we use the dependence $\theta(t)$ which characterises the liquid interlayer and the particle transport along the bubble surface

$$tg\frac{\theta}{2} = tg\frac{\theta_1}{2} e^{t/\tau_b}, \tag{12.15}$$

where $\tau_b = a_b / v_o$ is the characteristic time of the bubble surface motion, $\theta_1 = \theta|_{t=0}$.

This dependence can be obtained by integrating the equation of motion of a bubble surface section

$$a_b \frac{d\theta}{dt} = v_o \sin\theta. \tag{12.16}$$

The angular dependence of the adsorption on the upper half of the bubble surface can be approximated by

$$\Gamma(\theta) = \Gamma_o - \Delta\Gamma \cos\theta, \tag{12.17}$$

in which

$$\Delta\Gamma = c_o \left(\frac{\Gamma_o}{c_o}\right) \frac{\sqrt{Pe}}{a_b}. \tag{12.18}$$

Substitution of Eq. (12.17) into (12.4) and according to Eq. (8.69) integration gives

$$\chi_n(t) = 2\Delta\Gamma tg^2 \frac{\theta_1}{2} \left(e^{-2t/\tau_b} - e^{-v_n^2 t/\tau_D}\right) \frac{1}{1 + v_n^2 \tau_b / 2\tau_D}. \tag{12.19}$$

By simple algebraic transformations using Eqs (12.14), (12.15), (12.18), and (12.19) a more convenient equation for h_{lim} is obtained

$$h_{lim} = \frac{\chi_b}{\eta} \frac{9\pi a_b}{48} \frac{a}{b} tg^2 \frac{\theta}{2} \sum_{n=1}^{\infty} \frac{1 - \left[\frac{tg(\theta_1/2)}{tg(\theta/2)}\right]^{2\left(1 + v_n^2 \tau_b / 2\tau_D\right)}}{v_n^2 \left(1 + v_n^2 \tau_b / 2\tau_D\right)}. \tag{12.20}$$

This equation considerably simplifies at $\theta > 2\theta_1$, which allows to cancel the term depending on θ. Dependencies of $h_{lim}(\theta)$ at $\theta > 2\theta_1$ with

$$\frac{\chi_b}{\eta} = 0.3, \quad \frac{a}{b} = 3, \quad a_b = 0.1\,cm, \quad D_s = 10^{-5}\,\frac{cm^2}{s}, \quad \tau_b = 3\cdot10^{-3}\,s, \quad v_o = \frac{3}{2}v = 30\,cm/s, \tag{12.21}$$

are calculated and shown in Fig. 12.2 for the values τ_b/τ_D=3; 0.3; 0.03 which correspond to particle radii of 1 µm, 3.3 µm, 10 µm, respectively. The particle approaches the bubble at a distance less than h_{cr} (which is necessary for flotation) only when it is close to the front pole of the bubble.

The angle region determined by the condition

$$h_{lim}(\theta_{cr}) = h_{cr} \tag{12.22}$$

is so narrow (cf. Fig. 12.2) that the flotation efficiency is insubstantial. The formula for θ_{cr} can be easily obtained from Eq. (12.20) if we put $h_{lim} = h_{cr}$,

$$\theta_{cr} = \sqrt{\frac{\dfrac{\eta b}{\chi_b a} \dfrac{64 h_{cr}}{3\pi a_b}}{\displaystyle\sum_{n=1}^{\infty} \dfrac{1}{v_n^2 \left(1 + v_n^2 \tau_b / 2\tau_D\right)}}}. \tag{12.23}$$

The time of existence of the liquid interlayer (the time of the bubble motion with the adjacent particle surface section) is of the order of τ_b. The time of equalisation of the concentration in the liquid interlayer by diffusion is of the order of τ_D. If $\tau_D/\tau_b \gg 1$ there is no time for diffusion to restore a concentration gradient. So, at large values of τ_D/τ_b its influence on stabilisation is significant (cf. Eq. (12.23)).

From Eq. (12.16) it can be seen that at $\theta \to 0$ the rate of surface motion and accordingly the particle with the liquid interlayer is reduced. Therefore, the closer the particle to the upper pole of the bubble, the less pronounced is the concentration gradient in the interphase film. Consequently, h_{lim} is reduced with θ, given by Eq. (12.20), and the particle deposition is possible near the front pole.

During the transition from a plate-like particle to a spherical one a fast weakening of the liquid interlayer stabilisation effect is expected. With the same excessive pressure in the gap, caused by the pressing force, the film thinning would be more intensive, because the thickness of the liquid interlayer (in case of a sphere) is rapidly increased, directed from its centre. This supports the outflow of liquid.

12.4. ESTIMATION OF EFFECTIVENESS OF PARTICLES CAPTURE CONSIDERING LIQUID INTERLAYER STABILISATION BY DYNAMIC ADSORPTION LAYERS

The equation for the streamlines of a potential flow near a bubble reads

$$b^2 = \sin^2 \theta \, r^2 \left(1 - a_b^3 / r^3\right). \tag{12.24}$$

This formula is simplified for $r \approx a_b$,

$$b^2 = \sin^2 \theta (H + 1) 3 a_b a, \tag{12.25}$$

in which $H = h / a$. h is the shortest distance between the centre of the particle and the bubble. At small distances the particle trajectory deviates from the streamlines because of hydrodynamic interaction which requires a more precise definition (Derjaguin et al. 1976),

$$\frac{1}{H+1}F_p(H)\frac{dH}{d\theta}+2\frac{\cos\theta}{\sin\theta}=0. \tag{12.26}$$

In the absence of the already mentioned phenomenon of liquid interlayer stabilisation the extreme trajectory is determined by the fact that a particle approaches the bubble at a distance of H_{cr} at $\theta=\pi/2$. Integration results in

$$\int_{H_{cr}}^{H}\frac{F_p(H)dH}{H+1}=\ln\frac{1}{\sin^2\theta}. \tag{12.27}$$

Splitting integration in two intervals $(H_{cr}, 1)$ and $(1, H)$, and presuming that $F_p(H)=1$ at $H>1$ we can apply the method used by Derjaguin & Smirnov (1967). At $H>1$ we get

$$\int_{H}^{1}\frac{F_p(H)dH}{H+1}=\ln\frac{2}{\sin^2\theta(H+1)}. \tag{12.28}$$

Since the hydrodynamic interaction in the region $H>1$ is negligible, Eq. (12.28) is identical to the critical streamline given by Eq. (12.25) at $b=b_{cr}$.

Combining Eqs (12.27), (12.28), and (12.24) we get

$$E_p=6\frac{a}{a_b}e^{-I_p}, \tag{12.29}$$

in which

$$I_p=\int_{H_{cr}}^{1}\frac{F_p(H)dH}{1+H}. \tag{12.30}$$

The influence of a stabilising phenomenon restricts the extreme trajectory to $\theta_{cr}<\pi/2$. Considering this boundary condition we get

$$\int_{H_{cr}}^{H}\frac{F_p(H)dH}{1+H}=\ln\frac{\sin^2\theta_{cr}\cdot 2}{\sin^2\theta} \tag{12.31}$$

and, instead of Eq. (12.29),

$$E_p^*=6\frac{a}{a_b}e^{-I_p}\sin^2\theta_{cr}. \tag{12.31}$$

Thus, the decrease of the capture effectiveness caused by the interlayer stabilisation is expressed by

$$\frac{E_p^*}{E_p} = \sin^2 \theta_{cr} \approx \theta_{cr}^2 \qquad (12.32)$$

in which θ_{cr} is determined by Eq. (12.23).

The theory of Dukhin (1981) was generalised by Listovnichiy & Dukhin (1986) where the effect of stabilisation is considered under arbitrary hydrodynamic flow conditions around a bubble and the effect of convective transfer of surfactant into the adsorption layer was taken into account. Numerical estimations of θ_{cr}^2 were carried out and it was shown that the effect of the liquid interlayer stabilisation by a DAL decreases the flotation efficiency over a wide range of system parameters by more than an order of magnitude. Numerical estimations also point to the fact that the effects under consideration have a much smaller influence on flotation of spherical particles than of disk-shaped particles.

12.5. Non-equilibrium Surface Forces of Diffusion-Electrical Nature in Flotation

The electrical double layer varies along the bubble surface as a function of changes in the surface charge and the electrolyte concentration. For each surface location the surface structure is described by the theory of an equilibrium DL. Thus, one may consider the elementary flotation in terms of Derjaguin's heterocoagulation theory. However, qualitatively new effects are caused by the deformation of the DL. It is known that diffusion in an electrolyte solution is usually accompanied by the appearance of an electric field which is called the electro-diffusion potential. The vector lines of the concentration gradient grad C coincide with the vector lines of the electric field E_D which is expressed in terms of the local values of the concentration gradient for binary symmetric electrolyte (Derjaguin et al., 1960; Dukhin & Derjaguin, 1976),

$$E_D = \left(\frac{D^+ - D^-}{z^+D^+ + z^-D^-}\right)\left(\frac{RT}{F}\right)\left(\frac{\text{grad}C}{C}\right). \qquad (12.33)$$

As a result, an electric field must arise within the limits of the diffusion boundary layer which is associated with the dynamic adsorption layer of ionic surfactants. The electric field is caused by the deformation of the DL, i.e. by its slight deviation from electroneutrality.

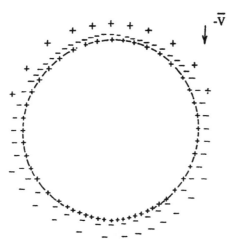

Fig. 12.3.   The distribution of charge signs in the primary and secondary double layers in the presence of

cation-active substances of rather low surface activity

The vector lines E_D, which originate (or terminate) at the outer boundary of the quasi equilibrium DL are oriented approximately normal to the surface and terminate (or originate) at some point of the diffusion layer. It is known that the sources of the lines of the electric field intensity are the positive charges and the sinks are the negative charges. Hence, it follows that the charges of opposite sign, which approximately counterbalance each other, are located within the limits of the quasi-equilibrium DL and in the diffusion layer (Fig. 12.3). These two layers of charges represent a single system in a sense that a variation in the distribution of charges in one layer is necessarily accompanied by redistribution of charges in the other layer. This system of charges which arises from the deformation of the equilibrium DL is designed as a secondary double layer (Dukhin 1960, 1961). Since the thickness of the secondary DL exceeds that of the equilibrium DL by several orders of magnitude (with the exception of low electrolyte concentrations), the deposition of particles on a bubble surface can be controlled under certain conditions by the passage of particles through the diffusion layer (Dukhin 1963). The consideration of this problem is simplified if the particle size is smaller than δ_D. When passing through the diffusion layer such a particle is subjected to uniform electric and diffusion fields and, consequently undergoes electrophoresis and diffusiophoresis (Derjaguin et al. 1961).

Derjaguin & Samygin (1962) investigated the gravity deposition of galenite particles of different size on the mobile surface of a fixed bubble from a vertical flow of liquid. They found that the particles smaller than some critical size a_{pcr} do not reach the bubble surface. Using the measured values of the potential it was obtained that for $a_p < a_{pcr}$ the joint effect of electro- and diffusiophoresis determines the repulsion force between a particle and a bubble, which exceeds the gravity force. As the electrolyte concentration increases, this repulsion force becomes weaker, and a decrease in a_{pcr} is an experimental manifestation of this phenomenon.

Because of its large velocity, a freely rising bubble has a diffusion layer much thinner that in the described experiments. This effect can manifest itself only if the particles are small enough so that their thermal motion becomes significant. Thus, electro- and diffusiophoresis should be taken into account in describing the Brownian diffusion of sub-micron particles towards the bubble's mobile surface under the conditions of a sufficiently low electrolyte concentration. The influence of diffusiophoretic transport to the surface of a rising bubble through its diffusion layer is theoretically proved by Zholkovsky et al. (1983).

12.6. SUMMARY

When drops or bubbles approach each other their diffusion layers overlap. This leads to local changes in the surface concentration and surface tension which causes liquid to flow into or out of the thick liquid interlayer. The dynamic adsorption layer and its diffusion layer deviate from electroneutrality and contain charges of opposite sign. The charged dynamic adsorption layer and the oppositely charges diffusion layer as a whole are electroneutral. This ensemble can be called the secondary electrical double layer.

The overlap of the secondary double layer of approaching drops or bubbles causes electrostatic interaction before their diffuse layers overlap.

If a solid particle crosses the diffusion layer of a bubble or drop it also includes the long range interaction caused by a local disturbance of the adsorption layer. This leads to Marangoni effects and influences the film drainage between particle and bubble or drop. Local desorption of surfactant from one surface and its adsorption on the other also causes interaction.

These non-equilibrium forces of diffusion-electrical nature can be suppressed by an increased electrolyte concentration and can be of importance in water/oil emulsions.

In case of nonionics non-equilibrium surface forces are suppressed together with surface retardation effects by the surfactant adsorption which hampers Marangoni effects.

12.7. REFERENCES

Allan, K.S., Charles, G.E. and Mason, S.G., J. Colloid Sci., 16(1964)150

Allan, K.S. and Mason, S.G., J. Colloid. Sci., 17(1962)383

Charles, G.E. and Mason, S.G., J. Colloid. Sci., 15(1960)236

Derjaguin, B.V., Dukhin, S.S. and Korotkova, A.A., Kolloidn. Zh., 23(1961)409

Derjaguin, B.V., Dukhin, S.S. and Lisichenko, V.A., Zh. Fiz. Khim., 34(1960)524

Derjaguin, B.V., Dukhin, S.S. and Rulev, N.N., Kolloidn. Zh., 38(1976)251

Derjaguin, B.V. and Samigin, V.D., Collection of Papers of Ginzvetmet, No. 9, Metallurgizdat, (1962)840

Derjaguin, B.V. and Smirnov, L.P., "Sb. Issledovaniya v oblasti poverkhnostnikh sil", Nauka, Moscow, (1967)188

Dukhin, S.S., Kolloidn. Zh., 23(1961)409

Dukhin, S.S., "Sb. Issledovanija v oblasti poverkhnostnikh sil", Moscow, (1961)38

Dukhin, S.S., Dokl. AN SSSR, 130(1960)1298

Dukhin, S.S. and Derjaguin, B.V., Electrophoresis, Nauka, 1976, Moscow

Dukhin, S.S., Zh. Fiz. Khim., 34(1960)1053

Dukhin, S.S., in Research in Surface Forces, Vol. 1, Derjaguin B.V. (Ed.) Consultants Bureau, N.Y., (1963)27

Dukhin, S.S., in The Modern Theory of Capillarity, Eds Goodrich F. and Rusanov A., Akademie Verlag, Berlin, 1981

Ewans, L.F. and Ewers, W.E., "Recent Developments in Mineral Dressing" (London), Ind. Eng. Chem. Institution of Mining and Metallurgy, 46(1953)2420

Groothius, H. and Zuideriweg, F.S., Chem. Eng. Sci., 12(1960)289

Happel, J. and Brenner, G., "Gidrodinamika pri malikh chislakh Reynoldsa", MIR, Moscow, (1976)174

Klassen, V.I., Voprosy teorii aeratsii i flotatsii. Goskhimizdat, M., 1949

Listovnichiy, A.V. and Dukhin, S.S., Kolloidn. Zh., 48(1986), 1184

MacKay, D.C. and Mason, S.G., J. Colloid. Sci., 18(1963)674

Rehbinder, P.A. and Venstrem, E.K., Kolloidn. Zh., 53(1930)145

Sutherland, K.C. and Work, E.V., "Printsipy flotatsii". Metallurgizdat, M., 1968 (in Russian)

Thiessen, D., Z. Phys. Chem., Lzg., 223(1963)218

Venstrem, E.K. and Rehbinder, P.A., Zh. Fiz. Khim., 2(1931)754

Zholkovskij, E.K. Listovnichij, A.V. and Dukhin, S.S., Kolloidn. Zh. 47(1985)517

APPENDIX 2A: GENERAL PRINCIPLES OF THE DEGREES OF FREEDOM OF INTERFACES

A most often observed fact of colloid and surface chemistry is that work must be done in order to create a new surface. This law is a basic principle not only valid for liquid interfaces, as shown in Chapter 1, but also for solid bodies; work is necessary for grinding and crushing for example. Surface thermodynamics starts from the fundamental principles of the general thermodynamics and includes equilibrium and non-equilibrium states.

For the subject of this book, Langmuir's (1933) extension of the phase rule for adsorption under equilibrium and non-equilibrium conditions is placed at the beginning of the treatment of surface thermodynamics. In the study of heterogeneous equilibrium by Gibbs' methods, the term phase is used for a homogeneous part of a system without regarding for quantity or form.

Langmuir introduced the concept of fields to distinguish homogeneous parts of a surface on which adsorption can occur and which differ in structure or composition regardless of the size or shape of the area involved. A simple example of a two field system is a surface of two types of crystal faces with their surface lattices giving them different properties.

A single field is the normal situation. Therefore, a surface phase can be defined as a homogeneous part of a system which extends over a surface field separated from other parts by a boundary. It is essential in the concept of phases that the properties of each phase have a definite number of parameters or degrees of freedom.

Langmuir's surface phase rule reads: "All the intrinsic properties of a surface phase posseses $C+E+1$ degrees of freedom even under non-equilibrium conditions". Here C represents the number of components in the system in the same sense as in the phase rule. The symbol E is used to denote the degree of freedom corresponding to the application of an external electric force to the surface field. The phase rule for the equilibrium between R_V volume phases and R_S surface phases in S fields is, according to Langmuir (1933),

$$F_N = (R_V + R_S)(C+1). \tag{2A.1}$$

Equilibrium conditions refer to the following properties:

1. temperature equilibrium; all $R_V + R_S$ phases must have the same temperature, which gives $R_V + R_S - 1$ conditions;
2. pressure equilibrium among the volume phases; $R_V - 1$ conditions;
3. spreading force(s) equilibrium between the surface phases in each of the fields;
4. concentration equilibrium.

As shown later, Gibbs' thermodynamics require that the chemical potential of each component is the same in all phases. This all leads to $(R_S + R_V - 1)C$ conditions. With K fields and the fulfilled statements 1 to 4 we have a total of $(R_V + R_S + 1)(C+2) - K$ variables defined in the system.

Subtracting these defined variables from the total degrees of freedom F_N leads to F_o, the degrees of freedom for the whole system under equilibrium conditions,

$$F_o = C + S - R_V - R_s + 2.$$
(2A.2)

F_o is reduced to the ordinary phase rule if we put $S = R_S = 0$. Langmuir pointed out that non-equilibrium states are of two kinds: steady states, in which the intrinsic properties of all phases and the relative amounts of the phases do not vary with time, and transient states in which at least one of these variables change with time.

According to the definitions 1 to 4 for an equilibrium state we obtain for the non-equilibrium state

$$F = F_o + \Delta F$$
(2A.3)

where ΔF characterises all transient phenomena:

1. partial thermal equilibrium;

2. pressure equilibrium;

3. partial spreading force equilibrium;

4. partial concentration equilibrium.

A variable electric field of the same source acting at all surface phases increases F_o by one.

Non-equilibrium adsorption layers of surfactants are mainly characterised by points 3 and 4 of Langmuir's definition: the partial spreading due to the action of an external hydrodynamic shear field and the partial concentration equilibrium due to the time-dependent adsorption of surfactants at a freshly-formed surfaces or due to expansion or compression of adsorption layers.

APPENDIX 2B: DISCUSSION OF FURTHER ADSORPTION ISOTHERMS

2B.1.   HÜCKEL-CASSEL ISOTHERM

According to the non-ideality of an adsorption layer the self area of a molecule β was introduced by Volmer (1925)

$$\pi (A - \beta) = RT. \tag{2B.1}$$

The derivation $\dfrac{d\gamma}{dc}$ yields

$$\frac{d\gamma}{dc} = \frac{a}{b+c}, \tag{2B.2}$$

where a and b are constants. Together with the fundamental equation of Gibbs (2.33) we obtain an isotherm which has exactly the same structure as the Langmuir isotherm (1918) if we set $\dfrac{a}{RT} = \Gamma_o$ and $\dfrac{1}{b} = k$,

$$\frac{\Gamma}{\Gamma_o} = \Theta = \frac{kc}{1+kc}. \tag{2B.3}$$

In analogy to a three dimensional real gas the van der Waals analogous state of the interface can be described by

$$\left(\pi + a / A^2\right)\left(A - \beta\right) = RT. \tag{2B.4}$$

By writing the Gibbs adsorption isotherm in the following form

$$\Gamma = \frac{c}{RT} \frac{d\pi}{dA} \frac{dA}{dc}, \tag{2B.5}$$

and by explaining the two-dimensional van der Waals equation (2B.4) in the form

$$\frac{d\pi}{dA} = -\frac{RT}{(A - \beta)} + \frac{2\alpha}{A^3} \tag{2B.6}$$

with $\Gamma = \dfrac{1}{A}$, $\Gamma_o = \dfrac{1}{\beta}$ and k' as integration constant, we obtain the equation of Hückel (1932) and Cassel (1944),

$$c = k' \frac{\Theta}{1-\Theta} e^{\Theta/(1-\Theta)} e^{-2\alpha\Gamma/RT}. \tag{2B.7}$$

Examples for the application of this equation to dynamic processes at interfaces are given by de Boer (1953) and Kretzschmar (1975, 1976).

With the help of the three-dimensional van der Waals-plot, the critical pressure, critical area and critical temperature can be defined by the setting $\dfrac{d\pi}{dA} = 0$ and $\dfrac{d^2\pi}{dA^2} = 0$,

$$\pi_c = \frac{\alpha}{27\,\beta^2}, \, F_c = 3\beta, \, T_c = \frac{8\alpha}{R\beta}. \tag{2B.8}$$

The relation of Eq. (2B.7) to the Langmuir adsorption isotherm can be shown as follows. With $\Theta = \Gamma/\Gamma_o$, $A = 1/\Gamma$ and $\beta = 1/\Gamma_o$ we obtain

$$c = \frac{k'}{F-\beta} e^{\beta/F-\beta} \, e^{-2\alpha\Gamma/RT}. \tag{2B.9}$$

For large values of A we find

$$c \cong \frac{k'}{A-\beta}\left(1+\frac{\beta}{A-\beta}\right)\left(1-\frac{2\alpha\Gamma}{RT}\right), \tag{2B.10}$$

which is equivalent to

$$c = \frac{k'}{A\left[2\beta - 2\alpha/RT - (\beta - 2\alpha/RT)^2/A\right]}. \tag{2B.11}$$

By the A»100Å^2 / molecule we get

$$c = \frac{k'}{A - (2\beta - 2\alpha/RT)}. \tag{2B.12}$$

Eq. (2B.12) equals the structure of a Langmuir adsorption isotherm.

To sum up we can say that Langmuir's adsorption isotherm derived in 1918 semi-empirically on the basis of adsorption experiments of gases at solid surfaces was physically founded by Volmer (1925) on kinetics models. De Boer (1953) then gave a general interpretation of the Langmuir isotherm, which is identical in a limit range to a van der Waals analogous equation of state. The Langmuir and von Szyszkowski adsorption isotherms are consistent with the Gibbs adsorption equation (Davies & Rideal 1961). When the self area of molecules occupied at a liquid interface and the interaction forces between adsorbed molecules are taken into account they produce an isotherm first published by Frumkin (1925) and now in common use.

2B.2. THE VOLMER ADSORPTION ISOTHERM

Another adsorption isotherm was derived by Volmer (1925). It assumes a self area of the adsorbed molecules

$$\frac{\Theta}{1-\Theta}\exp\left(\frac{\Theta}{1-\Theta}\right) = bc \qquad (2B.13)$$

or

$$c = \frac{1}{b}\frac{\Theta}{1-\Theta}\exp\left(\frac{\Theta}{1-\Theta}\right). \qquad (2B.14)$$

At large area per adsorbed species this equation transfers into the Langmuir isotherm.

2B.3. THE BUTLER-VOLMER-EQUATION

Bockris & Reddy (1970) describes the Butler-Volmer-equation as the "central equation of electrode kinetics". In equilibrium the adsorption and desorption fluxes of charges at the interface are equal. There are common principles for the kinetics of charge exchange at the polarisable mercury/water interface and the adsorption kinetics of charged surfactants at the liquid/fluid interface. Theoretical considerations about the electrostatic retardation for the adsorption kinetics of ions were first introduced by Dukhin et al. (1973).

In the derivation of the Butler-Volmer equation (1924) the same kinetic equations for the adsorption and desorption fluxes are used as for the derivation of adsorption isotherms by Langmuir (1916, 1917, 1918), Milner (1907), Ward & Tordai (1946), Baret et al. (1968, 1969). The advantage of the polarisable mercury electrode lies in the fact that the electric potential difference across the interface are directly adjustable, whereas the electric potential at the water/fluid interface can only obtained via adsorption isotherms or direct spreading experiments. In both cases there are complications which are discussed in more detail elsewhere. Let us return to the Butler equation. After Bockris & Reddy (1973) the interface constitutes the essential part of an electrochemical system. It is the place where the charge is pumped in and out of the system. The system is therefore mainly described by the charge-transfer reactions occuring at the interface. Charge-transfer reactions of this type has been expressed by the Butler-Volmer relationship for electrode processes.

$$i = i_0\left[e^{[1-\beta]F\eta/RT} - e^{\beta F\eta/RT}\right] \qquad (2B.15)$$

with

$$i_o = F\bar{k}c_A^o e^{-\beta F\Delta\Phi/RT} = F\bar{k}c_D^o e^{(1-\beta)\Delta\Phi_e/RT}. \qquad (2B.16)$$

β is called the symmetry factor, defined by the ratio of the distances across the DL up to the summit over the total DL thickness. The electric field at the interface is a vector. $\Delta\Phi_e$ represents a characteristic equilibrium potential difference across the interface and is characteristic for the reaction,

$$\bar{k} = \frac{kT}{e} e^{-\Delta\bar{G}^{0*}/RT}. \qquad (2B.17)$$

This situation is demonstrated in Fig. 2B.1.

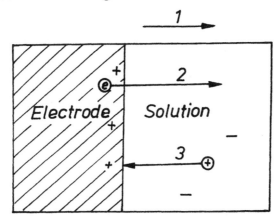

Fig. 2B.1.   Electric field across the interface hinders further charge transfer, after Bockris (1970)

$\beta\Delta\Phi F$ is the magnitude at which the energy barrier for the ion-electrode transfer is lowered, and consequently, $(1-\beta)\Delta\Phi F$ is the action for the raise of the metal-solution reaction. In conclusion we can say that in the presence of an electric field, the total free energy of activation for the electrode reaction is equal to the chemical free energy of activation.

APPENDIX 2C.   NON-EQUILIBRIUM SURFACE THERMODYNAMICS

The equilibrium of the surface, the state without any transport processes, is characterised in the framework of irreversible thermodynamics by the complete absence of fluxes and forces acting at the interface

$$J_k^{eq} = 0 \text{ and } X_k^{eq}. \qquad (2C.1)$$

494

Restricted to Defay et al. (1977) Gibbs' surface law for a non-equilibrium state is developed
by the basic entropy equation for non-equilibrium processes

$$dS = d_e S + d_i S,$$

(2C.2)

where $d_e S$ represents the entropy flow and $d_i S$ the entropy production, respectively.

The last term is characteristic of the thermodynamics of irreversible processes. Its magnitude
becomes positive if the system's processes are irreversible. Typical irreversible processes are
the adsorption or desorption of surfactants at liquid interfaces. The derivative of the second
term of Eq. (2C.2), as local entropy production is

$$\frac{d_i S}{dt} = \int \sigma(s)dV = \int \sum_k J_k X_k dV \geq 0$$

(2C.3)

with $\sigma(s) = \sum_k J_k X_k$. Here J_k are the generalised fluxes and X_k the generalised
thermodynamic forces, which may be either gradients or chemical affinities.

In systems, like adsorption layers, in thermal and mechanical equilibrium the global entropy
production is

$$T^{-1}\sum_{\gamma\alpha} A_\gamma^\alpha \frac{d\xi_\gamma^\alpha}{dt} + T^{-1}\sum_p A_p \frac{dF}{dt} \geq 0.$$

(2C.4)

A_γ^α represents the affinity of adsorption of component γ from the phase α to the surface, A_p is
the affinity of chemical reactions p, ξ_γ^α, ξ_p are the coordinates of adsorption and reaction,
respectively. The significance of other terms was mention above.

Defay et al. (1977) explained that in a non-equilibrium state the surface energy is determined
differently from in an equilibrium. Non-equilibrium adsorption layers can only be described
exactly by accounting for the interactions between the molecules in the surface and bulk
phases.

The surface energy depends not only on the composition of the surface layer, but also on the
compositions of the bulk phases. Bulk phases can be declared autonomous, while surface
phases are non-autonomous. This distinction is the origin for dynamic surface tension, e.g. for
liquid two component systems, as intensively studied in the classical monograph by Defay et
al. (1966) and demonstrated in Fig. 2C.1.

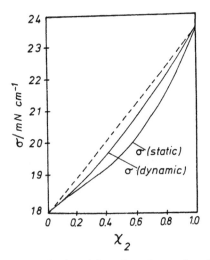

Fig. 2C.1.   Static and dynamic surface tension of mixtures of diethyl ether and acetone vs. mol fraction of acetone

Defay et al. (1966) defined

$$\varepsilon_\gamma^\alpha = \delta f^\Omega / \delta C_\gamma^\alpha \qquad (2C.5)$$

with

$$f^\Omega = f^\Omega\left(T, C_{1.......\gamma}^{1.......\alpha}, \Gamma_{1.......\gamma}\right) \qquad (2C.6)$$

and with the extension for surfaces

$$s^\Omega = s^\Omega\left(u^\Omega, \Gamma_{1.....\gamma}, C_{1.......\gamma}^{1.......\alpha}\right). \qquad (2C.7)$$

The fundamental Gibbs equation for surface tension can be written for the non-equilibrium state as

$$d\gamma = -s^\Omega dT - \sum_\gamma \Gamma_\gamma d\mu_\gamma^\Omega + \sum_{\gamma\alpha} \varepsilon_\gamma^\alpha dC_\gamma^\alpha. \qquad (2C.8)$$

with

$$\mu_\gamma^\Omega = \delta f^\Omega / \delta \Gamma_\gamma. \qquad (2C.9)$$

When the system is in equilibrium the Gibbs' fundamental equation (2.33) is obtained.

Further symbol used in this appendix are:

C_γ^α concentration of component γ in the adjacent bulk phase

d_e entropy flow

d_i entropy production

f^Ω unit area

s^Ω, u^Ω entropy and internal energy per unit area on the surface

u internal energy density

v molecule volume

X_k corresponding generalised thermodynamic force (gradient or chemical affinity)

Γ_γ the Gibbs adsorption of component γ defined by $\mu_\gamma^\Omega = \delta f^\Omega / \delta \Gamma_\gamma$

$\varepsilon_\gamma^\alpha$ cross chemical potential

$\varepsilon_\gamma^\alpha = 0$ equilibrium condition of the system

δ^Ω / dt total time derivative at the surface

μ_γ chemical potential in the bulk phase

APPENDIX 2D: THERMODYNAMICS OF THIN LIQUID FILMS

The coalescence of disperse systems, such as foams and emulsions, and the contact of air bubbles with solid particles, e.g. in the process of flotation, takes place in two steps. The first is characterised by a flocculation of the system, the formation of thin liquid films with an equilibrium thickness. In the second step the film becomes thin enough for the interparticular attractions to overcome the film state so the two separated interfaces form a new interface. The situation where a small bubble attaches a liquid interface is shown in Fig. 2D.1.

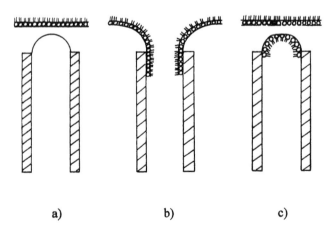

a)   b)   c)

Fig. 2D.1.   Bubble attached to a liquid interface, a) common film, b) coalescence, c) black film; according to Richter et al. (1987)

Thin films are the subject of world-wide scientific interest. The foundation of the related theory and basic experiments have been published most of all by scientific schools from Holland, Russia, Ukraine and Bulgaria. Research literature is enormous. The textbooks by Scheludko (1966) and Sonntag & Strenge (1970) contain a brief description of this topic. "Thin Liquid Films" edited by Ivanov (1988) goes into greater detail and in the modern description of this area by Hunter (1992), a book devoted to dynamic properties of interfaces. The formation of thin liquid films is a time-dependent process, because the thinning of liquid films up to its equilibrium state can only take place when there is a drainage of liquid from the film into the border region. The thinning rate of liquid films decreases by the shear tension at the two film boundaries; their surface rheological properties make them rigid.

At this point we want to describe briefly the thermodynamic precondition for the formation of thin liquid films and some of their physical properties which differ from the related liquid bulk phase. Further we report on the present state of knowledge on the rate of thinning of liquid films dependent on the related surface rheological properties.

The situation of an interlayer between two liquid phases is given by Derjaguin (1993) (Fig. 2D.2). This makes obvious what happens when the peculiarities of thin liquid films are neglected.

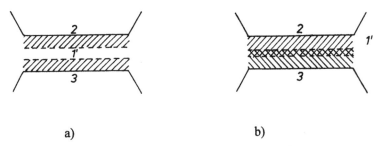

a)   b)

Fig. 2D.2.   Sketch on the features of thin liquid films, a) DLVO without structural component, b) DLVO with structural component, after Derjaguin (1993)

The topics of the early scientific work of Derjaguin and his collaborators was the evaluation of the term "disjoining pressure" as basic property of a thin liquid film. Derjaguin & Obuchov (1936) and Derjaguin & Kussakov (1939) have detected the growth of repulsive forces in such films as the film becomes thinner. The classic thermodynamics of Gibbs was extended by the thermodynamic formulation of the disjoining pressure concept.

Staring from the Gouy-Chapman theory the observed repulsion force in thin liquid films must be of electrostatic nature caused by the overlap of the corresponding diffuse electrical double

498

layers. The basic idea consists of a splitting of the disjoining pressure π in a "repulsive" electrostatic term and an "attractive" van der Waals term,

$$\pi = \pi_{el} + \pi_{vdW}.\qquad(2D.1)$$

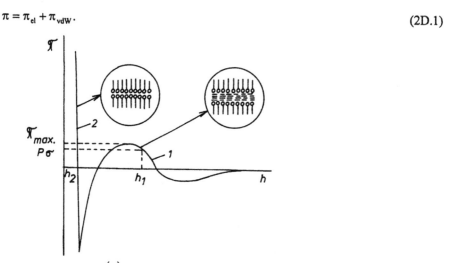

Fig. 2D.3.   Typical $\pi(h)$ isotherm formed of the solution of an ionic surfactant, h_1 - stable film, 1 - common black film, 2 - Newton black film

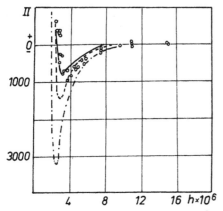

Fig. 2D.4. $\pi(h)$ isotherm measured for a $5 \cdot 10^{-4}\%$ Saponin solution in 0.01M KCl, dashed lines correspond to the DLVO theory with different Hamaker constants, according to Scheludko et al. (1969)

As an example the resulting $\pi(h)$ isotherm, using realistic values for both components, is shown in Fig. 2D.3. It is characterised by two pronounced minima. Fig. 2D.4. shows an experimental example for an aqueous Saponin solution (Scheludko et al. 1969).

The classical quantitative calculations of interaction forces in thin liquid films were made during the second world war by Derjaguin & Landau (1941) and in a monograph by Verwey & Overbeek (1948). This theory for thin liquid films is therefore known as the Derjaguin, Landau, Verwey and Overbeek theory (DLVO).

The disjoining pressure π is fully expressed by the change of the chemical potential $\Delta\mu(h)$ of a substance in a thin film with respect to the chemical potential of the same substance in an infinitely extended phase,

$$\Delta\mu(h) = \pi v, \tag{2D.2}$$

where v is the molar volume. The result that the film tension of a thin liquid film is not twice the surface tension of the bulk phase is the reason for the contact angle between the flat part of the thin liquid film and the adjacent bulk phase, as shown schematically in Fig. 2D.5.

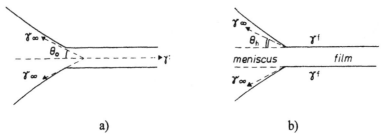

a)   b)

Fig. 2D.5.   Contact angle between the flat part and the Plateau border of a thin liquid film, a) definition of Θ_o, b) definition of Θ_h

The thermodynamic analysis of the liquid film state on the basis of Equ. (2D.2) leads to the relationship for the symmetric film tension and the disjoining pressure isotherm

$$\gamma^f = 2\gamma^\infty - \int_\infty^h \pi dh + \pi h = 2\gamma^\infty + \Delta f(h) + \pi h. \tag{2D.3}$$

Beside direct disjoining pressure measurements an interferometric determination of this contact angle provides another way of examining thin liquid films. The Young equation applied here reads

$$2\gamma^\infty \cos\Theta_h = \gamma^f. \tag{2D.4}$$

The combination of Eqs (2D.3) and (2D.4) gives

$$2\gamma^\infty \cos\Theta_o = 2\gamma^\infty - \int_\infty^h \pi dh + \Delta f(h) + \pi h \tag{2D.5}$$

500

or

$$\bar{\Delta} = 2\gamma^{\infty}(1 - \cos\Theta_{o}).$$ (2D.6)

The critical thickness for the rupture of such thin film was formulated by de Feijter (1988) using the Mandelstam's concept,

$$\frac{d\pi}{dh} \geq \frac{dP_{\gamma}}{dh}.$$ (2D.7)

The concept for the rupture of a free liquid film or a thin liquid film on a solid substrate is used in some applications, for example in the flotation process. Scheludko et al. (1968) have published first contact angle measurements for liquid films.

As the main components the disjoining pressure contains an electrostatic part and the van der Waals attraction forces. The electrostatic part is the result of the overlap of the diffuse electrical double layers in a thin liquid film as shown in Fig. 2D.6. including the coordinates in the thin film.

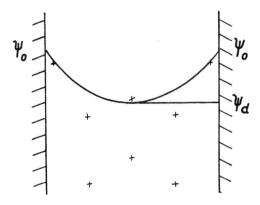

Fig. 2D.6.   Situation of overlapping diffuse electric double layers

The complex problem of calculation of the potential profile in the case of overlapping diffuse double layers was solved on the basis of these coordinates and a relationship of the electric charge at the interface was obtained. Solutions to the Gouy-Chapman equation are restricted to some boundary conditions analysed by Grimson et al. (1988) for the case of overlapping diffuse double layers. These boundary conditions focus on the electrostatic disjoining pressure with charge regulation. It means that by thinning of the liquid film the surface charge σ or the surface potential Ψ_{o} are kept constant. Attention is also paid to discrete charge effects and structure components due to electrolytes.

Instead of using the middle potential $\Psi(0)$ and the potential $\Psi(1)$, it is useful to perform the calculations with the dimensionless potentials u_{o} and u_{1} and a dimensionless surface charge α. A solution to the problem of pressure due to overlapping electrical double layers in a thin liquid film was first given by Langmuir in form of an approximation

$$\pi(z) - \pi(\infty) = P = \frac{2n_o}{\beta}(\cosh u_o - 1). \tag{2D.8}$$

Later the equation of Verwey & Overbeek (1948) was derived

$$a^2 = 2(\cosh u_1 - \cosh u_o). \tag{2D.9}$$

Fig. 2D.3. demonstrates that an equilibrium film, the so-called common black film can reach a critical thickness at which it ruptures due to surface disturbances. Vrij (1966) studied surface fluctuations theoretically on the basis of Mandelstam's theory and computer simulations. Newton black film rupture was studied experimentally and theoretically by Exerowa et al. (1982) and Exerowa & Kachiev (1986). They assume the existence of vacancies in the film. The mobility of these vacancies is the mechanism which controls the film stability (Fig. 2D.7),

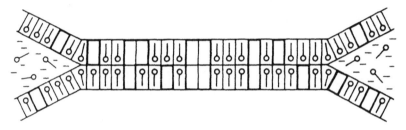

Fig. 2D.7.   Schematic of a Newton black film, empty boxes are vacancies, after Exerowa et al. (1982)

Scheludko's theory of thin film formation and stability are a significant contribution to the modern flotation theory (Fijnaut & Joosten 1978). Chapters 10 and 11 present a detailed description of this theory especially with respect to the influence of dynamic adsorption layers.

APPENDIX 3A: SOUND PROPAGATION IN LIQUID/FLUID DISPERSE SYSTEMS AND CHEMICAL REACTION

Let us start from the principle of de Donder (1927) for the definition of the affinity $\overline{A}_i$ as the sum of the atomic chemical potentials $\overline{\mu}_k$ of the component k and $\overline{\Delta G}_i$ as the change of the Gibbs free energy of the given reaction. For proceeding reactions $\overline{\Delta G}_i$ is necessarily negative.

$$\overline{A}_i = \sum \overline{A}_{ik}\,\overline{\mu}_k \tag{3A.1}$$

The entropy production S for irreversible processes is defined as positive. If we consider one chemical reaction , the flux j and the affinity $\overline{A}$ have opposite signs, so that we get

$$S = -\frac{1}{T}\sum_i j_i \overline{A}_i. \tag{3A.2}$$

In contrast to a chemical reaction, dynamic surface processes are mainly characterised as "incomplete". This means that one or more parameters, necessary for the equilibrium state, are required to describe the distance of the instantaneous state of the entire system from equilibrium. A chemical reaction can be described by a degree of advancement ε. The same procedure is possible for surface chemical reaction including mass transfer and processes of formation and dissolution of associates. ε varies between +1 and -1 and depends on the equilibrium state and affinity $\overline{A}$. The differential quotient of affinity to the degree of advancement was defined by Yao (1981) as the ordering coefficient

$$\left(\partial \overline{A} / \partial \varepsilon \right)_{p,T}.$$ (3A.3)

From the definition of a negative ordering coefficient the specific heat of constant pressure becomes

$$\Delta C_p = C_{pA} - C_{p\xi} = T\left(\frac{\partial S}{\partial T}\right)_{p,A} - T\left(\frac{\partial S}{\partial T}\right)_{p,\xi} = T\left(\frac{\partial S}{\partial \varepsilon}\right)_{p,T}\left(\frac{\partial \varepsilon}{\partial T}\right)_{p,A}.$$ (3A.4)

Starting from the basic equation

$$\left(\frac{\partial G}{\partial \varepsilon_k}\right)_{T,p,\varepsilon_{l,i\neq k}} = -A_k,$$ (3A.5)

two reactions will be distinguished, and a strong analogy can be found for surface chemical reactions. At first the chemical potential is not limited by partial Gibbs free energy at constant temperature and pressure and we have

$$-A_k = \left(\frac{\partial G}{\partial \varepsilon_k}\right)_{T,p} = \left(\frac{\partial F}{\partial \varepsilon_k}\right)_{T,V} = \left(\frac{\partial U}{\partial \varepsilon_k}\right)_{S,V} = \left(\frac{\partial H}{\partial \varepsilon_k}\right)_{S,p}.$$ (3A.6)

At second, for the limitation to one chemical reaction not far from equilibrium (important for surface chemistry) we get

$$A = A_o - A_{eq} = \Delta A = \left(\frac{\partial A}{\partial \varepsilon}\right)\Delta \varepsilon$$ (3A.7)

with

$$j_i = \rho \frac{d\varepsilon_i}{dt} = -\sum_k \frac{L_{ik} A_k}{T} \tag{3A.8}$$

From here it follows

$$\frac{\partial \varepsilon}{\partial t} = \frac{j}{\rho} = -\frac{L}{\rho T} \left(\frac{\partial A}{\partial \varepsilon} \right)_{T,p} \Delta\varepsilon. \tag{3A.9}$$

By definition $\dfrac{L}{\rho T} = B$ and after linearisation $\dfrac{\partial \varepsilon}{\partial t} = \dfrac{\Delta\varepsilon}{\tau_{T,p}}$ it follows

$$\frac{\partial \varepsilon}{\partial t} = -B \left(\frac{\partial A}{\partial \varepsilon} \right)_{T,p} \Delta\varepsilon \tag{3A.10}$$

and

$$\left(\frac{\partial A}{\partial \varepsilon} \right)_{T,p} = -\frac{1}{B\tau_{T,p}}. \tag{3A.11}$$

Here τ is the relaxation time corresponding to a decay of the degree of advancement down to $(1/e)$th of its original value.

Summarising we can conclude that for a simple reaction four relaxation times can be defined,

$$\tau_{T,p}, \tau_{T,V}, \tau_{S,p}, \tau_{S,T}. \tag{3A.12}$$

On the basis of irreversible thermodynamics it can be shown that the sound propagation in ideal fluids, obeying the disturbance relation

$$\frac{I}{I_o} = \exp\left[i(\kappa \cdot x - \omega t) \right], \tag{3A.13}$$

cannot account for sound absorption (I_o is the sound density at zero time). The sound propagation in fluids with chemical irreversibility is one of the best demonstrations of the relaxation processes. The theory of irreversibility can be applied correctly to dynamic surface processes.

The present state of surface rheology and surface light-scattering research into relaxation times at liquid interface will be dealt with later. Famous scientists like Einstein (1920), Liebermann (1949), Frenkel & Obraztsor (1940) and Yao (1981) developed mathematical algorithms to transfer the theory of sound propagation in fluids with chemical irreversibility to liquid interfaces covered with adsorption layers.

Surface chemistry can be restricted simply to bulk chemistry if we assume that a "reaction" is characterised by a time-dependent change of components A and B. In terms of surfaces science A can be the monomer and B the associated state of an adsorbed species. In other words, sound propagation in bulk phases and hydrodynamic stresses act in the same way. Considering only one chemical reaction the plot $\ln \varepsilon - t$ with V and ε as independent variables obtains the form

$$
\begin{bmatrix} dp \\ dA \end{bmatrix} = \begin{bmatrix} \left(\dfrac{\partial p}{\partial V}\right)_{S,\varepsilon} & \left(\dfrac{\partial p}{\partial \varepsilon}\right)_{S,V} \\ \left(\dfrac{\partial A}{\partial V}\right)_{S,\varepsilon} & \left(\dfrac{\partial A}{\partial \varepsilon}\right)_{S,V} \end{bmatrix} \begin{bmatrix} dV \\ d\varepsilon \end{bmatrix}.
\tag{3A.14}
$$

Under linearised conditions the second row of the matrix can be neglected. From

$$
\left(\frac{\partial A}{\partial \varepsilon}\right)_{\alpha} = -\frac{1}{B\tau_{\alpha}},
\tag{3A.15}
$$

(α and β are fixed but unspecified state variables) we obtain

$$
\frac{\partial \varepsilon}{\partial t} = -BA.
\tag{3A.16}
$$

By substitution of Eq. (3A.16) into (3A.11) it follows

$$
-i\omega\varepsilon_{o} = -BA_{o},
\tag{3A.17}
$$

and

$$
\frac{A_{o}}{\varepsilon_{o}} = \left(\frac{\partial A}{\partial \varepsilon}\right)_{\alpha,B} = \frac{i\omega}{B}.
\tag{3A.18}
$$

From the first row of the matrix we get

$$
\left(\frac{\partial p}{\partial V}\right)_{S,\beta} = \left(\frac{\partial p}{\partial V}\right)_{S,\varepsilon} + \frac{\left[\left(\dfrac{\partial p}{\partial V}\right)_{S,A} - \left(\dfrac{\partial p}{\partial V}\right)_{S,V}\right]}{1 + i\omega\tau_{S,V}},
\tag{3A.19}
$$

where β refers to a particular stage of development in dependence of ω

$$\chi(\omega) = \left(\frac{\partial p}{\partial V}\right)_{S,\varepsilon} + \frac{\left[\left(\frac{\partial p}{\partial V}\right)_{S,A} - \left(\frac{\partial p}{\partial V}\right)_{S,\varepsilon}\right]}{1 + i\omega\tau_{S,V}}. \tag{3A.20}$$

At zero frequency at equilibrium Eq. (3A.20) reduces to

$$\chi(0) = \left(\frac{\partial p}{\partial V}\right)_{S,A} = -\frac{1}{V_{S,A}} < 0 \tag{3A.21}$$

and at infinite frequency when the process is frozen

$$\chi(\infty) = \left(\frac{\partial p}{\partial V}\right)_{S,\varepsilon} = -\frac{1}{V_{S,\varepsilon}} < 0. \tag{3A.22}$$

χ is the differential quotient of pressure to volume.

We can treat the sound propagation in a simple fluid as viscous-irreversible instead of chemically irreversible. The entropy production Δs_i for the chemical irreversibility is

$$\frac{\partial \Delta s_i}{\partial t} = \frac{1}{\rho BT}\left(\frac{\partial \varepsilon}{\partial V}\right)_{T,p}^2 (\text{div } V)^2 \tag{3A.23}$$

and for a viscous irreversibility

$$\frac{\partial \Delta s_i}{\partial t} = \frac{\eta V}{T}(\text{div } V)^2. \tag{3A.24}$$

Comparing (3A.23) and (3A.24) we obtain

$$\eta v = \frac{1}{\rho B}\left(\frac{\partial \varepsilon}{\partial v}\right)_{T,p}^2, \tag{3A.25}$$

where v is the velocity, and ηv the bulk viscosity coefficient.

It can be shown that with a single relaxation time the kinetic equation has an exponential decay. By solution of the linearised version of Eq. (3A.10) with $\varepsilon_0 - \varepsilon = \Delta\varepsilon$ it follows

$$\varepsilon(t) = \exp\left(\frac{-t}{\tau}\right) + \varepsilon_0 \tag{3A.26}$$

The quantity of $d\ln\varepsilon / dt$ represents an activation energy barrier against the flow. A spectrum of relaxation times increases the order of the differential equation to n, where n is the number of relaxation times. The equation of kinetics is no longer exponential and all interpretations become difficult, for example

$$\alpha_1 s + \alpha_2 \dot{s} = \alpha_3 \varepsilon + \alpha_4 \dot{\varepsilon} \qquad (3A.27)$$

Here s is the stress and ξ is the strain, respectively. If $\dot{s} = 0, \dot{\varepsilon} = 0$, we have an elastic element or a spring, if $\dot{s} = 0, \varepsilon = 0$, we have a viscous element or a dashpot. If $\varepsilon = 0$, we have an elastic element and a viscous element in series, which is called visco-elastic element.

This derivation on the basis of irreversible thermodynamics is in agreement with the classical treatment we mentioned above.

APPENDIX 3B: DYNAMIC CONTACT ANGLES

Wetting phenomena are mainly characterised by the contact angle at the three phase contact line solid/liquid/air. In equilibrium, the contact angle Θ is well described by the Young equation (1.9). There are many problems in the way of obtaining correct experimental data and calculating the surface tension of the solid substrate by using Eq. (1.9) as controversy in the literature shows. Neumann (1972) has given an introduction to contact angles on imperfect solids.

Some important industrial processes are controlled by the contact angle, most of all by dynamic contact angles, for example the flotation or coating process. These processes are not in thermodynamic equilibrium, represented by Eq. (2.8). At high rates of sliding of the three phase contact line, we observe a dynamic contact angle. For example, the contact angle increases when a liquid drop spreads over a solid surface. Similar observations exist for the dipping of a plate into a liquid. Depending on the experimental conditions we observe advancing or receding contact angles. The dynamics or kinetics of wetting is a field of intensive research. Beside the use of empirical equations, two basic propositions for the theoretical calculations have been developed. One of them is based on the molecular kinetics in the three phase contact zone and the other one deals with the hydrodynamics of the streaming in the meniscus.

According to Blake & Haynes (1969), wetting is a stress modified molecular rate process described by Eyring's classical theory. They assume a displacement of molecules within the three-phase zone. Activation energies for viscous flows play a determining role in the theories of Cherry & Holmes (1969), Blake (1973), and Hoffmann (1983). Blake (1988) takes into consideration the capillary number

$$Ca = (\eta v / \gamma).$$
(3B.1)

Without any molecular kinetics assumption Joos et al. (1990) found a linear relationship between the cosine of the dynamic angle and a web speed. This relationship can be extrapolated to zero speed giving the static contact angle; extrapolation to high speeds yields contact angle of 180 degrees. This simple equation can be written, after Joos et al. (1990), as

$$\cos\Theta = \cos\Theta_0 - Kv^{1/2}$$
(3B.2)

or

$$\cos\Theta = \cos\Theta_0 - \gamma \left(\frac{\eta v}{\gamma_1}\right)^{1/2},$$
(3B.3)

where K is the rate constant, η is the viscosity and v is the speed of the three phase contact line.

Related experiments are based on coating of liquid at a moving web. This procedure refers to applied technology in coating processes, e.g. film coating and coating of plastic sheets. The coater has a fixed position and the substrate to be coated with a thin layer is moving along the coater. In the contact zone of the liquid with the substrate a three phase contact line is formed having a dynamic contact angle which is controlled not only by the Young relation but also by the hydrodynamic shear stresses. The dynamic contact angle is therefore influenced by surface energetic properties of the contact zone as well as by the hydrodynamic forces acting at higher speed (several meters per second).

It is important to remember that many dynamic wetting processes act in the presence of wetting agents as a special type of surfactants. In contrast to very low displacements of the contact line where no shear stresses exist and the adsorption equilibrium is established, at higher speeds the modelling of the process (overlap of surface energetic and hydrodynamic forces) becomes very difficult and a lot of boundary conditions must be simplified.

Voinov (1976) introduced a practicable definition of the dynamic contact angle. The situation is illustrated in Fig. 3B.1. The tangent is placed such that it contacts the meniscus at a distance H from the solid; this distance H is defined as the lowest high where the meniscus just becomes visible.

508

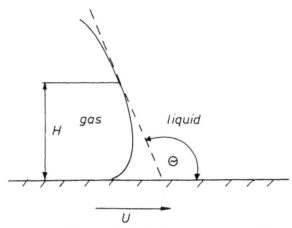

Fig. 3B.1.   Dynamic contact angle after the definition by Voinov (1977)

The problem of modelling the dynamic contact angle Θ_d by using various boundary conditions, surface shapes, geometries of streaming and stream lines, has been treated by some distinguished scientists: Moffat (1964), Fritz (1965), Ludviksson & Lightfoot (1968), Yin (1969), Hansen & Toong (1971), Huh & Scriven (1971), Voinov (1976), Huh & Mason (1977), Petrov & Radoev (1981), Giordano & Slattery (1983). Kretzschmar (1983) has recently given a semi-quantitative calculation of the influence of the adsorption kinetics of wetting agents on a dynamic contact angle using numeric results by Miller (1990) and Grader (1985).

APPENDIX 3C: MARANGONI-INSTABILITIES AND DISSIPATIVE STRUCTURES

While the Bénard (1901) effect is stimulated mainly by gradients of liquid densities, (for example by heating a vessel with liquid from the bottom), the Marangoni effect is the result of a gradient in interfacial tension. Such gradient can be stimulatet by an unequal distribution of surfactant or by differences in temperature at the liquid interface. As shown in Section 3.3. such gradient can only exist by the action of opposite forces, like hydrodynamic or aerodynamic shear stresses at an interface covered by an adsorption layer. This condition leads to a deformation of the liquid interface. A surface tension gradient can also be the source of structure formation by coupling the interfacial flow with the flow of the adjacent liquid.

From the thermodynamical point of view the formation of dissipative structures is entropy driven as intensively explained by Prigogine & Glansdorf (1971). The criteria for surface instabilities due to mass transfer across a liquid interface were evaluated by Sternling & Scriven (1959). The typical Marangoni instability starts on surfactant concentration or temperature differences between two phases. Surface tension differences along the surface are amplified beginning from initial disturbances as fluctuations, if no other rough disturbances exist.

A classical form of dissipative structures are the so-called first order or higher order roll-cells. Fig. 3C.1 shows the stream lines of a first order roll-cell stimulated by mass transfer (Linde 1978). A magnified section of this picture is given in Fig. 1.19. First "Schlieren"-optical observations of hydrodynamically caused structure formation during mass transfer across a liquid/liquid interface were published by Linde (1959) and Linde & Kretzschmar (1962).

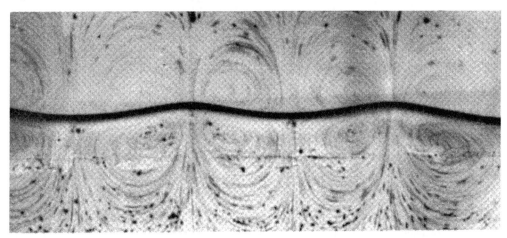

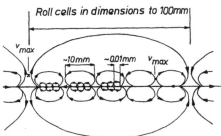

Fig. 3C.1.   Stream lines of first order roll-cell during transfer of SDS across the iso-amylol/water interface, according to Linde (1978)

Fig 1.19 describes the same structure as alignment of small particles along the trajectories of a first order roll-cell in a vertically oriented capillary gap. The curvature of the liquid meniscus controls the capillary pressure differences.

A schematic of an instability caused by temperature gradients between the two phases resulting in surface tension gradient along the interface is shown in   Fig. 3C.2. A small initial disturbance like interfacial convection from R to S will be amplified by additional mass transport at higher temperature, because this leads to $\gamma_R < \gamma_S$ on condition $d\gamma / dT < 0$.

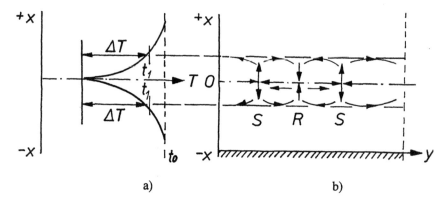

Fig. 3C.2.   Schematic of instability caused by the temperature gradient of a sink, a) temperature gradients, b) sceme of initial interfacial convection

Roll-cells of any order can be instationary due to relaxation oscillation. That means the roll-cells are either instationary amplified in their amplitudes or break down. A new amplification and the subsequent breakdown are more and more pronounced. At highly intensive circulations it leads to turbulence-like chaotic behavoir, well-known as interfacial convection. At low intensity it leads to the development of regulary drifting convection units, which behave like the well-known autowaves of Belousov-Zhabotinski reaction (BZR). Concentric rings and even spirals have been observed formed by travelling autowaves (Fig. 3C.3a, b, left side). Fig. 3C3a, b (right side) shows a comparable autowave of a BZR.

If we change the direction of mass or heat transfer in liquid/gas systems by keeping the ondition of $\frac{\Delta\gamma}{\Delta c} < 0$ and $\frac{\Delta\gamma}{\Delta T} < 0$, an oscillatory regime is espected from theory and was found experimentally by Linde and co-worker.

Recently it could be show that such oscillations create traines of soliton-like non-linear waves. The non-linear interaction between them or with a wall shows different effects: at acute angle interaction a negative phase shift (Fig. 3C.4a) is observed. After the head-on collision (Fig. 3C.4b) the pattern of the acute angle crossing is preserved whilst the other crossing changes to the obtuse angle crossing with a resonant interaction and the third wave (Mach-stem) due to the positive phase shift (Fig 3C.5).

The incident wave interacts with the reflected wave forming the third wave which travels rectangular to the wall.

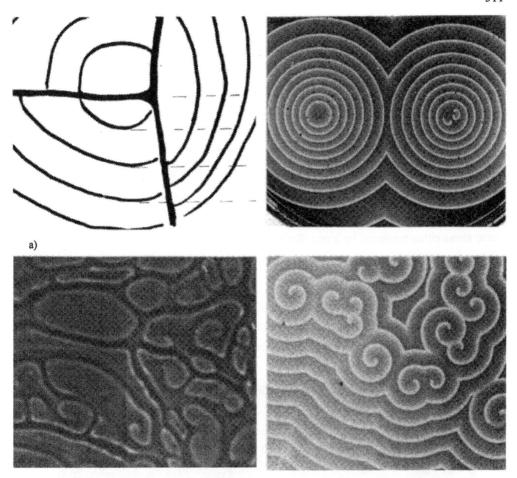

a)

Fig. 3C.3.   Example of autowaves of first-order roll-cells of Marangoni instability (left pictures) and of BZR (right pictures), a) concentric rings, b) spirals, according to Linde and co-workers

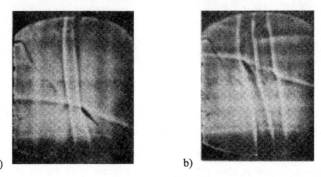

a)   b)

Fig.3C.4.   Interaction of three Marangoni waves; a) collision of two of them head-on with negative phase shift, b) acute angle crossing after collision; according to Linde and co-workers

512

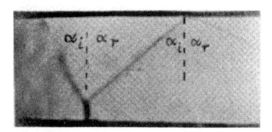

Fig.3C.5.   Mach-reflection of Marangoni waves, according to Linde and co-workers

Another complicated interfacial dynamic instability is observed when the surfactant covers the liquid surface in a frame, for example on a Langmuir trough. If the surface is subjected to a shear stress either produced by a one-dimensional flow in the liquid bulk or by an air flow the surface instability observed looks like a hair-needle-like flow (Linde & Shuleva 1970, Linde & Friese 1971, Schwartz et al. 1985).

Wassmuth et al. (1990) modelled this type of surface instabilities, denoted as Linde instability. The resulting hair needle-like flow pattern in the surface is shown in Fig. 3C.6. The condition of two dimensional continuity of flow is fulfilled in the surface.

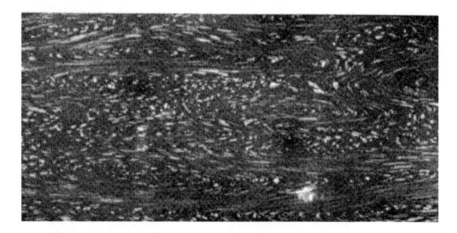

Fig. 3C.6.   Hair-needle-like flow pattern in a surface itself, according to Linde and co-workers

The field of Marangoni instabilities shows a large variety of dissipative structures, including the principle of stationary structures, hierarchical structures with limited self-similarity, relaxation oscillations and regular behavior of travelling autowaves with chaotic turbulence-like behaviour. There is also the oscillatory regime with trains of waves with soliton-like behaviour of each wave. Anormal as well as normal dispersion of these waves have recently been proofed. This field is of importance for a better understanding of the molecular behaviour and kinetics of interfacial dynamics (Linde & Schwartz 1989, Linde & Engel 1991, Linde & Zirkel 1991, Linde et al. 1993, Weidman et al. 1992, Velarde & Normand 1980, Velarde & Chu 1989)

APPENDIX 3D: LATERAL TRANSPORT PHENOMENA

Lateral transport phenomena can be observed for example as surface self-diffusion (Vollhardt et al. 1980), from concentration gradients in surface films as the result of compression or expansion of soluble or insoluble monolayers (Dimitrov et al. 1978), from the effect of aggregation or domain formation in such monolayers (Lucassen-Reynders 1987), or from domain movements in monolayers induced by an electric field (Heckl et al. 1988).

Surface self-diffusion is the two-dimensional analogue of the Brownian motion of molecules in a liquid bulk. Measurements of self-diffusion have to be performed in complete absence of any Marangoni flow caused by surface tension differences. Such experimental conditions are best established in an insoluble monolayer where one part consists of unlabelled and the other of radio-tracer labelled molecules. The movement of molecules within the surface monolayer can be now observed by using a Geiger-Müller counter. There are possible effects of liquid convective flow in the sublayer which was discussed for example by Vollhardt et al. (1980a). With e special designed apparatus Vollhardt et al. (1980b) studied the self-diffusion of different palmitic and stearic acid and stearyl alcohol and obtained self-diffusion coefficients between $1 \div 4 \cdot 10^{-5}\,\mathrm{cm}^2 / \mathrm{s}$.

One of the recent techniques most commonly used to measure diffusion in 2-dimensions is fluorescence recovery after photobleaching (FRAP). This method uses a laser beam focused through a fluorescence microscope to follow the diffusion of fluorescent molecules in a plane perpendicular to the laser beam. The fluorescence intensity from a laser spot of known diameter, typically a few microns, is measured. The laser intensity is then increased by approximately 1000 times. This irreversibly photobleaches any fluorophore in the spot. The intensity is then decreased again and the recovery in fluorescence intensity measured as unbleached molecules diffuse into the spot. The time function of fluorescence intensity is then analysed to give surface self-diffusion coefficient (Clark et al. 1990a, b, Wilde & Clark 1993, Ladha et al. 1994).

Dynamics in insoluble monolayers were studied for example by Dimitrov et al. (1978). They observed a Marangoni effect during a continuous compression and described it by a mechanism developed for particle movements. This mechanism is based on an elasticity coefficient and the Bressler-Talmud coefficient of the monolayer and the subphase. The work of Dimitrov et al. (1978) is probably the first experimental observation of an unisotropic film pressure in an insoluble monolayer, which lasted over some seconds.

Irregularities in dynamic surface tensions of adsorption layers of soluble surfactants were discussed by Lucassen-Reynders (1987) in terms of aggregation phenomena of adsorbed molecules. She gave a theoretical model for the frequency spectrum of surface dilational properties.

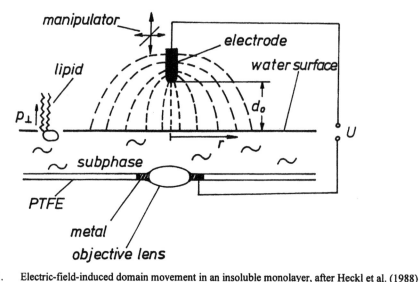

Fig. 3D.1.   Electric-field-induced domain movement in an insoluble monolayer, after Heckl et al. (1988)

The effect of an electric field on the movement of domains in lipid monolayers was studied by Heckl et al. (1988) (cf. Fig. 3D.1). The mobility of domains in an inhomogeneous electric field depends on the surface viscosity and leads to special cluster formation.

APPENDIX 4A: NUMERICAL SOLUTION OF THE INTEGRAL EQUATION OF WARD AND TORDAI

To solve Eq. (4.1) numerically, an approximation of the integral is needed. The simplest integration formula is the trapezium law, which devides the integration interval into n equidistant subintervals and approximates the integral in each of the subintervals by a trapezium. If we define the n intervals by

$$t_1 = 0, t_i = t_{i-1} + \Delta t \; ; i = 2, ..., n+1, \tag{4A.1}$$

the integral is approximated by

$$\int_0^{\sqrt{t}} c(0, t-\tau)d\sqrt{\tau} =$$

$$\sum_{i=1}^{n} \int_{\sqrt{t_i}}^{\sqrt{t_{i+1}}} c(0, t-\tau)d\sqrt{\tau} \approx \frac{1}{2}\sum_{i=1}^{n}[c(0, t-t_{i+1}) + c(t-t_i)]\left[\sqrt{t_{i+1}} - \sqrt{t_i}\right]. \qquad (4A.2)$$

Inserting the approximation (4A.2) into Eq. (4.1) and rearranging lead to the following relation,

$$\Gamma(t_{n+1}) \approx a_1 c(0, t_{n+1}) + a_2 (c(0, t_1), ..., c(0, t_n)). \qquad (4A.3)$$

Starting from the initial value of $c(0,0)$ given by the initial condition Eq. (4.17b), the values of $c(0, t_i)$ can be calculated successively when an adsorption isotherm has been assumed as a relation between $\Gamma(t)$ and $c(0, t)$. As one of the most frequently used isotherms, a Langmuir isotherm is used here,

$$\Gamma(t_{n+1}) = \Gamma_\infty c(0, t_{n+1}) / (a_L + c(0, t_{n+1})). \qquad (4A.4)$$

Replacing $\Gamma(t_{n+1})$ by the left hand side of Eq. (4A.4) a quadratic equation in $c(0, t_{n+1})$ results for each subsequent time t_{n+1},

$$c^2(0, t_{n+1}) + (a_L + \frac{a_2}{a_1} - \frac{\Gamma_\infty}{a_1})c(0, t_{n+1}) + \frac{a_L a_2}{a_1} = 0. \qquad (4A.5)$$

Only one of the two roots give a physically sensible value of $c(0, t_{n+1})$. The corresponding value of $\Gamma(t_{n+1})$ results from Eq. (4A.4). Any forms of adsorption isotherms can be used instead of the chosen Langmuir equation and the given algorithm will work.

APPENDIX 4B: NUMERICAL SOLUTION OF THE FUNCTION EXP(X²)ERFC(X)

The error function erfc(x) is defined in the following way,

$$erfc(x) = 1 - erf(x) = \frac{2}{\sqrt{\pi}}\int_x^\infty \exp(-\xi^2)d\xi. \qquad (4B.1)$$

Values of this function are difficult to determine and available usually from table. Still more complicated is the calculation of values of the product $\exp(x^2)\cdot erfc(x)$, a standard function in diffusion and heat transfer problems. In chapter 4 approximate solutions were given for small and large x-values (cf. Eqs (4.81) and (4.82)). Especially for the intermediate interval, $0.1 < x < 3$, the approximate equation (4.81) and (4.82) are not valid. One possible way to calculate the function is the expansion of $\exp(-\xi^2)$ in a series

$$\exp(-\xi^2) = \sum_{i=0}^{\infty} (-1)^i \frac{\xi^i}{i!}. \tag{4B.2}$$

Inserting into Eq. (4B.1) and integration leads to

$$\text{erfc}(x) = 1 - \frac{2}{\sqrt{\pi}} \sum_{i=0}^{\infty} \frac{(-1)^i}{i!} \frac{x^{2i+1}}{2i+1}. \tag{4B.3}$$

Due to the theorem of Leibnitz, the maximal error, after breaking off the series at $i = n$, can be estimated by,

$$R_n \le \left| \frac{x^{2n+3}}{(n+1)!(2n+3)} \right|. \tag{4B.4}$$

Finally, $\exp(x^2)\text{erfc}(x)$ can be calculated from Eq. (4B.3) with the factor $\exp(x^2)$, using all members of the series up to n with $R_n < 10^{-m}$. If $m = 4$, and $x = 5$ already 76 items of the series have to be calculated. The accuracy of calculating an item at higher powers become very low and this kind of solution become very inexact.

Much more precise approximation forms were given by Cody, who used Chebyshev approximations. Three rational approximations for the function $\exp(x^2)\text{erfc}(x)$ with maximal relative errors ranging down to 10^{-19} were derived. These forms and the corresponding intervals are,

$$\exp(x^2)\text{erfc}(x) \approx \exp(x^2)\left(1 - x \frac{\sum_{i=0}^{4} p_i x^{2i}}{\sum_{i=0}^{4} q_i x^{2i}}\right), \qquad |x| < 0.5, \tag{4B.5}$$

$$\exp(x^2)\text{erfc}(x) \approx \frac{\sum_{i=0}^{7} p_i x^i}{\sum_{i=0}^{7} q_i x^i}, \qquad 0.46875 \le x \le 4, \tag{4B.6}$$

$$\exp(x^2)\text{erfc}(x) \approx \frac{1}{x}\left(\frac{1}{\sqrt{\pi}} + \frac{1}{x^2} \frac{\sum_{i=0}^{3} p_i x^{-2i}}{\sum_{i=0}^{3} q_i x^{-2i}}\right), \qquad x \ge 4. \tag{4B.7}$$

The coefficients of the polynomials p_i and q_i are listed in the following table. The relation $\text{erfc}(-x) = 2 - \text{erfc}(x)$ can be used to evaluate the function for negative arguments.

Table 4B.1   Polynomial coefficients for the Chebyshev approximations for |x| < 0.5

i	p_i	q_i
0	242.667955230532	215.058875869861
1	21.9792616182942	91.1649054045149
2	6.99638348861914	15.0827976304078
3	-0.0356098437018154	1.00000000000000

Table 4B.1   Polynomial coefficients for the Chebyshev approximations for $0.46875 \le x \le 4$

i	p_i	q_i
0	300.459261020162	300.459260956983
1	451.918953711873	790.950925327898
2	339.320816734344	931.35409485061
3	152.98928504694	639.980264465631
4	43.1622272220567	277.585444743988
5	7.21175825088309	77.0001529352295
6	0.564195517478974	12.7827273196294
7	1.36864857382717E-007	1.00000000000000

Table 4B.1   Polynomial coefficients for the Chebyshev approximations for $x \ge 4$

i	p_i	q_i
0	-0.00299610707703542	0.0106209230528468
1	-0.0494730910623251	0.19130892610783
2	-0.226956593539687	1.0516751076793
3	-0.278661308609648	1.98733201817135
4	-0.0223192459734185	1.000000000000000

518

APPENDIX 4C: FINITE DIFFERENCE SCHEME TO SOLVE THE INITIAL AND BOUNDARY CONDITION PROBLEM OF A DIFFUSION CONTROLLED ADSORPTION MODEL

The finite difference scheme is one of the direct methods to solve the initial and boundary value problem of a diffusion controlled adsorption model. To minimize the numerical efforts in solving the present complex problem with several independent physical parameters, it is efficient to use dimensionless variables. The following transforms are used for the present transport problem, which was defined in chapter4,

$$\Theta = \frac{Dt}{K^2}, \ X = \frac{x}{K}, \ K = \frac{\Gamma_0}{c_0}, \ C = \frac{c}{c_0}, \ \bar{\Gamma} = \frac{\Gamma}{\Gamma_0}. \tag{4C.1}$$

The transport equation and all further boundary and initial conditions then read:

$$\frac{\partial C}{\partial \Theta} = \frac{\partial^2 C}{\partial X^2}, \ X > 0, \Theta > 0, \tag{4C.2}$$

$$\frac{\partial \bar{\Gamma}}{\partial C} \frac{\partial C}{\partial \Theta} = \frac{\partial C}{\partial X}, \ X = 0, \Theta > 0, \tag{4C.3}$$

$$\lim_{X \to \infty} C = 1, \Theta > 0 \quad , \tag{4C.4}$$

$$C = 1, X > 0, \Theta = 0. \tag{4C.5}$$

The general idea of the finite difference method is to replace differential quotients by a difference quotient. For example, the derivative of c with respect to time t is approximated by:

$$\frac{\partial c}{\partial t} \approx \frac{\Delta c}{\Delta t} = \frac{c(x,t_2) - c(x,t_1)}{t_2 - t_1}. \tag{4C.6}$$

In the same way, the first and second derivatives of a function are approximated. The intervals of all coordinates, in the present case for x and t, are divided into equidistant subintervals, and for each of the intervals a difference equation is obtained. At the boundaries of the intervals, the necessary relations are given by the corresponding boundary conditions.

Step by step, the change of the function with time can be calculated knowing the initial values of the function.

Of course, if a differential quotient is replaced by a difference quotient with respect to one of the coordinates, the remaining coordinates have to be kept constant. There are for example three ways to define the difference quotient of $\dfrac{\partial c}{\partial x}$ (cf. for example Kiesewetter & Maess, 1974),

$$\frac{\partial c}{\partial x} \approx \frac{c(x_2,t_1)-c(x_1,t_1)}{x_2-x_1}, \tag{4C.7}$$

$$\frac{\partial c}{\partial x} \approx \frac{c(x_2,t_2)-c(x_1,t_2)}{x_2-x_1}, \tag{4C.8}$$

$$\frac{\partial c}{\partial x} \approx \alpha\frac{c(x_2,t_1)-c(x_1,t_1)}{x_2-x_1}+(1-\alpha)\frac{c(x_2,t_2)-c(x_1,t_2)}{x_2-x_1};0<\alpha<1. \tag{4C.9}$$

The simplest version is the quotient (4C.7) leading to explicit equations for each time interval, while the other two lead to implicit equations. If we use the scheme (4C.9) with $\alpha = 0.5$, an arithmetic average of the two concentration gradients at the time level t_1 and t_2, the following linear equation system results from Eq. (4C.2):

$$C_{i-1,j+1}-2(1+\frac{\Delta X^2}{\Delta\Theta})C_{i,j+1}+C_{i+1,j+1}=-C_{i-1,j}+2(1-\frac{\Delta X^2}{\Delta\Theta})C_{i,j}-C_{i+1,j};$$
$$i=2,3,...,N; j=2,3,... \tag{4C.10}$$

The indices are defined by the following relation:

$$C_{i,j}=C(X_i,\Theta_j)=C((i-1)\Delta X,(j-1)\Delta\Theta). \tag{4C.11}$$

At time $\Theta = 0$, i.e. for $j=1$, the initial conditions (4C.5) yields

$$C_{1,1}=0, C_{i,1}=2,3,...,N. \tag{4C.12}$$

For each time step j, the value of $C_{1,j+1}$ is given by the respective difference scheme, derived from the boundary condition (4C.3):

$$\frac{1}{2\Delta\Theta}(\frac{\partial\overline{\Gamma}}{\partial C}_{(1,j+1)}+\frac{\partial\overline{\Gamma}}{\partial C}_{(1,j)})(C_{1,j+1}-C_{1,j})=\frac{1}{2\Delta X}(C_{2,j+1}+C_{2,j}-C_{1,j+1}-C_{1,j}). \tag{4C.13}$$

Here again, an average of the difference schemes of the two subsequent time levels is used. Rearrangement of Eq. (4C.13) results in

$$
C_{1,j+1} = \frac{\left(C_{1,j} + \frac{\Delta\Theta}{\Delta X}(C_{2,j+1} + C_{2,j} - C_{1,j})/(\frac{\partial\overline{\Gamma}}{\partial C_{(1,j+1)}} + \frac{\partial\overline{\Gamma}}{\partial C_{(1,j)}})\right)}{\left(1 + \frac{\Delta\Theta}{\Delta X}/(\frac{\partial\overline{\Gamma}}{\partial C_{(1,j+1)}} + \frac{\partial\overline{\Gamma}}{\partial C_{(1,j)}})\right)}, \tag{4C.14}
$$

an implicit relation as it still contains the unknown values at the time step (j+1). To solve such implicit linear equation systems, usually a simple iteration can be made by assuming as start values of the unknown values of the time step (j+1) the values at the previous time step j. For the present type of problems, no more than three iterations are necessary until the values of the new time step (j+1) become stable.

The dimensionless form of the boundary condition (4.37), considering the change of the interfacial area with time, reads,

$$
\frac{d\overline{\Gamma}}{dC}\frac{dC}{d\Theta} + \frac{\overline{\Gamma}}{A}\frac{dA}{d\Theta} = D\frac{dC}{dX}; X = 0; \Theta > 0. \tag{4C.15}
$$

A good approximation of the second term at the right hand side of Eq. (4C.15) is,

$$
\frac{\overline{\Gamma}}{A}\frac{dA}{d\Theta} \approx \frac{\overline{\Gamma}_j + \overline{\Gamma}_{j+1}}{A_j + A_{j+1}}\frac{A_{j+1} - A_j}{\Delta\Theta}. \tag{4C.16}
$$

Instead of Eq. (4C.14), the following equation results to calculate the boundary value $C_{1,j+1}$, the value at x=0 and the time step (j+1),

$$
C_{1,j+1} = \frac{\left(C_{1,j} - 2\frac{\overline{\Gamma}_j + \overline{\Gamma}_{j+1}}{A_j + A_{j+1}}\frac{A_{j+1} - A_j}{(\frac{\partial\overline{\Gamma}}{\partial C_{(1,j+1)}} + \frac{\partial\overline{\Gamma}}{\partial C_{(1,j)}})} + \frac{\Delta\Theta}{\Delta X}(C_{2,j+1} + C_{2,j} - C_{1,j})/(\frac{\partial\overline{\Gamma}}{\partial C_{(1,j+1)}} + \frac{\partial\overline{\Gamma}}{\partial C_{(1,j)}})\right)}{\left(1 + \frac{\Delta\Theta}{\Delta X}/(\frac{\partial\overline{\Gamma}}{\partial C_{(1,j+1)}} + \frac{\partial\overline{\Gamma}}{\partial C_{(1,j)}})\right)}
$$

$$
\tag{4C.17}
$$

APPENDIX 4D: F INITE DIFFERENCE SCHEME TO SOLVE THE INITIAL AND BOUNDARY CONDITION PROBLEM OF A DIFFUSION CONTROLLED ADSORPTION MODEL FOR A TWO COMPONENT SURFACTANT SYSTEM

The solution of the adsorption kinetics problem of a surfactant mixture can be performed numerically with the same difference scheme. There are some details which have to be allowed for when calculating the adsorption of the individual components.

For each component a system of linear equations results, analogous to Eq. (4C.10). The definitions of the concentration values as a function of space and time for a two component system read,

$$C1_{i,j} = C1(X_i, \Theta_j) = C1((i-1)\Delta X, (j-1)\Delta\Theta), \tag{4D.1}$$

$$C2_{i,j} = C2(X_i, \Theta_j) = C2((i-1)\Delta X, (j-1)\Delta\Theta). \tag{4D.2}$$

The boundary conditions for each component, equivalent to Eq. (4C.13) contains the derivatives $\dfrac{d\bar{\Gamma}1}{dC1_{(i,j+1)}}$ and $\dfrac{d\bar{\Gamma}2}{dC2_{(i,j+1)}}$, which depend on each other via the generalised adsorption isotherm. Thus, to calculate the evolution of C1, Γ1 the actual values of C2, Γ2 are required, and vice versa. As only slow changes in these functions are expected, the values of the previous time layer can be used as a first approximation and then, an iteration of all values performed, using the most recent values of the corresponding functions C1, Γ1 or C2, Γ2. This internal iteration for each time step requires usually only two to three cycles and no chages of the results appear within the necessary accuracy. The results presented in Fig. 4.7 are obtained from such calculations, using a generalised Langmuir isotherm.

APPENDIX 4E: A PPLICATION OF THE LAPLACE TRANSFORM TO SOLVE THE DIFFUSION-CONTROLLED ADSORPTION KINETICS MODEL

The advantage of the Laplace transform calculus for solving ordinary and partial differential equations consists in its property to reduce the degree of complexity of the original problem by transforming functions into another functional space. The transformation is performed via the relation

$$\mathsf{L}(f) = \int_0^\infty \exp(-st)f(t)\, dt = F(s), \tag{4E.1}$$

where f(t) is the original and F(s) the image function (see for example Bronstein & Semendjajew 1960). For solving differential equations, the following rules are very useful:

transformation of the derivative of the function f(t) -

$$L(\frac{df}{dt}) = s\,F(s) - f(0)$$

(4E.2)

transformation of a second derivative of the function f(t) -

$$L(\frac{d^2f}{dt^2}) = s^2\,F(s) - s\,f(0) - \frac{df}{dt}(0)$$

(4E.3)

transformation of the convolution integral of the functions f_1 and f_2 -

$$L\left(\int_0^t f_1(t-\tau)\,f_2(\tau)\,d\tau\right) = \frac{1}{s} F_1(s)\,F_2(s).$$

(4E.4)

The application of the calculus, with respect to the time coordinate t, to the transport equation (4.9) and the boundary conditions (4.11) and (4.15), making use of the initial conditions (4.17a) and (4.17b) leads to the following ordinary differential equation

$$D\frac{d^2C}{dx^2} = sC - c_o,\ x > 0,$$

(4E.5)

with the boundary conditions

$$D\frac{dC}{dx} = s\upsilon - \Gamma_d,\ x = 0$$

(4E.6)

and

$$\lim_{x \to \infty} C(x,s) = \frac{c_o}{s}.$$

(4E.7)

The functions $C(x,s)$ and $\upsilon(s)$ are the image functions of $c(x,t)$ and $\Gamma(t)$.

The general solution of the inhomogeneous differential equation (4E.5) is given by the general solution of corresponding homogeneous one and a special, non-trivial solution of the inhomogeneous equation. Such special solution is

$$C(x,s) = \frac{c_o}{s}.$$

(4E.8)

The general solution of the linear second order differential equation is

$$C(x,s) = \exp(\lambda x). \tag{4.E.9}$$

Inserting the derivatives of $C(x,s)$ into Eq. (4E.5) leads to an algebraic equation in λ and finally to the solution of the differential equation

$$C(x,s) = \frac{c_o}{s} + A_1 \exp(\sqrt{\frac{s}{D}}x) + A_2 \exp(-\sqrt{\frac{s}{D}}x). \tag{4E.10}$$

The coefficients A_1 and A_2 have to be determined from the boundary conditions. Condition (4E.7) immediately results in $A_1 = 0$, as

$$\lim_{x \to \infty} \left(C(x,s) - \frac{c_o}{s}\right) = 0. \tag{4E.11}$$

The value of A_2 results from the second boundary condition Eq. (4E.6)

$$A_2 = -\sqrt{\frac{s}{D}}\upsilon + \frac{\Gamma_d}{\sqrt{sD}}. \tag{4E.12}$$

The final solution now reads

$$C(x,s) = \frac{c_o}{s} - \left(\sqrt{\frac{s}{D}}\upsilon + \frac{\Gamma_d}{\sqrt{sD}}\right)\exp\left(-\sqrt{\frac{s}{D}}x\right). \tag{4E.13}$$

To obtain the solution in the functional space of time t, the solution must be re-tranformed, using the inverse Laplace transform (cf. for example Oberhettinger & Bordii, 1973). At $x = 0$ this inverse transform finally leads to

$$c(0,t) = c_o - \frac{2}{\sqrt{D\pi}}\frac{d}{dt}\int_0^{\sqrt{t}} \Gamma(t-\tau)d\sqrt{\tau} + \frac{\Gamma_d}{\sqrt{\pi Dt}}. \tag{4E.14}$$

The classical Ward & Tordai equation results when Eq. (4E.13) is rearranged to

$$\upsilon(s) = \left(\left(\frac{c_o}{s} - C(x,s)\right)\exp\left(\sqrt{\frac{s}{D}}x\right) + \frac{\Gamma_d}{\sqrt{sD}}\right)\sqrt{\frac{D}{s}}, \tag{4E.15}$$

which finally leads to

$$\Gamma(t) = 2\sqrt{\frac{Dt}{\pi}}\left(c_o\sqrt{t} - \int_0^{\sqrt{t}} c(0,t-\tau)d\sqrt{\tau} + \Gamma_d\right). \tag{4E.16}$$

Both, Eqs (4E.14) and (4E.16) are equivalent and describe diffusion controlled adsorption. As mentioned before, its derivation is independent of any adsorption isotherm and, therefore, these equations present a general solution to the problem. Only after the selection of an equilibrium adsorption isotherm or a non-equilibrium relation between $\Gamma(t)$ and $c(0,t)$, a specific model results.

APPENDIX 4F: POLYNOMIAL PARAMETERS OF THE COLLOCATION SOLUTION EQ. (4.25) AFTER ZILLER & MILLER (1986)

c_o/a_L	ξ_1	ξ_2	ξ_3
0	1.138343	0.181811	-0.320154
0.1	1.240679	-0.320797	0.080118
0.2	1.291053	-0.522419	0.231366
0.3	1.319633	-0.628327	0.308694
0.4	1.337294	-0.688920	0.351626
0.5	1.348353	-0.723634	0.375281
0.6	1.355013	-0.742049	0.387036
0.7	1.358563	-0.749568	0.391005
0.8	1.359828	-0.749516	0.389688
0.9	1.359368	-0.744064	0.384696
1.0	1.357582	-0.734697	0.377115
1.5	1.336915	-0.657505	0.320590
2.0	1.307549	-0.564116	0.256567
2.5	1.276178	-0.472994	0.196816
3.0	1.245410	-0.389817	0.144407
3.5	1.216315	-0.316119	0.099804
4.0	1.189300	-0.251908	0.062608
4.5	1.164460	-0.196610	0.032150
5.0	1.141751	-0.149457	0.007706
6.0	1.102232	-0.076390	-0.002584
7.0	1.069598	-0.026782	-0.042816
7.5	1.055498	-0.009194	-0.046304
8.0	1.020524	0.004298	-0.046996
9.0	1.020524	0.020802	-0.041326
10.0	1.002232	0.025864	-0.028096

APPENDIX 5A: CORRECTION FACTORS AFTER WILKINSON (1972) IN THE FORM r_{cap}/a AS A FUNCTION OF $r \cdot V^{-1/3}$

$r.V^{-1/3}$	0	1	2	3	4	5	6	7	8	9
0.10	0.0535	0.0543	0.0550	0.0558	0.0566	0.0574	0.0582	0.0590	0.0598	0.0606
0.11	0.0614	0.0622	0.0630	0.0638	0.0646	0.0654	0.0662	0.0671	0.0679	0.0687
0.12	0.0695	0.0703	0.0712	0.0720	0.0728	0.0737	0.0745	0.0754	0.0762	0.0770
0.13	0.0779	0.0787	0.0796	0.0804	0.0813	0.0822	0.0830	0.0839	0.0847	0.0856
0.14	0.0865	0.0873	0.0882	0.0891	0.0900	0.0908	0.0917	0.0926	0.0935	0.0944
0.15	0.0953	0.0962	0.0971	0.0979	0.0988	0.0997	0.1006	0.1015	0.1025	0.1034
0.16	0.1043	0.1052	0.1061	0.1070	0.1079	0.1089	0.1098	0.1107	0.1116	0.1126
0.17	0.1135	0.1144	0.1154	0.1163	0.1172	0.1182	0.1191	0.1201	0.1210	0.1219
0.18	0.1229	0.1238	0.1248	0.1258	0.1267	0.1277	0.1286	0.1296	0.1306	0.1315
0.19	0.1325	0.1335	0.1344	0.1354	0.1364	0.1374	0.1384	0.1393	0.1403	0.1413
0.20	0.1423	0.1433	0.1443	0.1453	0.1463	0.1473	0.1483	0.1493	0.1503	0.1513
0.21	0.1523	0.1533	0.1543	0.1553	0.1563	0.1573	0.1583	0.1594	0.1604	0.1614
0.22	0.1624	0.1635	0.1645	0.1655	0.1665	0.1676	0.1686	0.1696	0.1707	0.1717
0.23	0.1728	0.1738	0.1748	0.1759	0.1769	0.1780	0.1790	0.1801	0.1812	0.1822
0.24	0.1833	0.1843	0.1854	0.1865	0.1875	0.1886	0.1897	0.1907	0.1918	0.1929
0.25	0.1939	0.1950	0.1961	0.1972	0.1983	0.1993	0.2004	0.2015	0.2026	0.2037
0.26	0.2048	0.2059	0.2070	0.2081	0.2092	0.2103	0.2114	0.2125	0.2136	0.2147
0.27	0.2158	0.2169	0.2180	0.2191	0.2202	0.2213	0.2224	0.2236	0.2247	0.2258
0.28	0.2269	0.2281	0.2292	0.2303	0.2314	0.2326	0.2337	0.2348	0.2360	0.2371
0.29	0.2382	0.2394	0.2405	0.2417	0.2428	0.2439	0.2451	0.2462	0.2474	0.2485
0.30	0.2497	0.2508	0.2520	0.2532	0.2543	0.2555	0.2566	0.2578	0.2590	0.2601

r.V$^{-1/3}$	0	1	2	3	4	5	6	7	8	9
0.31	0.2613	0.2625	0.2636	0.2648	0.2660	0.2671	0.2683	0.2695	0.2707	0.2719
0.32	0.2730	0.2742	0.2754	0.2766	0.2778	0.2790	0.2802	0.2813	0.2825	0.2837
0.33	0.2849	0.2861	0.2873	0.2885	0.2897	0.2909	0.2921	0.2933	0.2945	0.2957
0.34	0.2969	0.2982	0.2994	0.3006	0.3018	0.3030	0.3042	0.3054	0.3067	0.3079
0.35	0.3091	0.3103	0.3115	0.3128	0.3140	0.3152	0.3165	0.3177	0.3189	0.3202
0.36	0.3214	0.3226	0.3239	0.3251	0.3263	0.3276	0.3288	0.3301	0.3313	0.3326
0.37	0.3338	0.3351	0.3363	0.3376	0.3388	0.3401	0.3413	0.3426	0.3438	0.3451
0.38	0.3464	0.3476	0.3489	0.3502	0.3514	0.3527	0.3540	0.3552	0.3565	0.3578
0.39	0.3590	0.3603	0.3616	0.3629	0.3641	0.3654	0.3667	0.3680	0.3693	0.3706
0.40	0.3718	0.3731	0.3744	0.3757	0.3770	0.3783	0.3796	0.3809	0.3822	0.3835
0.41	0.3848	0.3861	0.3874	0.3887	0.3900	0.3913	0.3926	0.3939	0.3952	0.3965
0.42	0.3978	0.3991	0.4004	0.4017	0.4031	0.4044	0.4057	0.4070	0.4083	0.4096
0.43	0.4110	0.4123	0.4136	0.4149	0.4163	0.4176	0.4189	0.4202	0.4216	0.4229
0.44	0.4242	0.4256	0.4269	0.4282	0.4296	0.4309	0.4323	0.4336	0.4350	0.4363
0.45	0.4376	0.4390	0.4403	0.4417	0.4430	0.4444	0.4457	0.4471	0.4484	0.4498
0.46	0.4512	0.4525	0.4539	0.4552	0.4566	0.4580	0.4593	0.4607	0.4620	0.4634
0.47	0.4648	0.4662	0.4675	0.4689	0.4703	0.4716	0.4730	0.4744	0.4758	0.4771
0.48	0.4785	0.4799	0.4813	0.4827	0.4841	0.4854	0.4868	0.4882	0.4896	0.4910
0.49	0.4924	0.4938	0.4952	0.4966	0.4980	0.4994	0.5007	0.5021	0.5035	0.5049
0.50	0.5064	0.5078	0.5092	0.5106	0.5120	0.5134	0.5148	0.5162	0.5176	0.5190
0.51	0.5204	0.5218	0.5233	0.5247	0.5261	0.5275	0.5289	0.5304	0.5318	0.5332
0.52	0.5346	0.5361	0.5375	0.5389	0.5403	0.5418	0.5432	0.5446	0.5461	0.5475
0.53	0.5489	0.5504	0.5518	0.5533	0.5547	0.5561	0.5576	0.5590	0.5605	0.5619
0.54	0.5634	0.5648	0.5663	0.5677	0.5692	0.5706	0.5721	0.5735	0.5750	0.764
0.55	0.5779	0.5794	0.5808	0.5823	0.5837	0.5852	0.5867	0.5881	0.5896	0.5911

r.V$^{-1/3}$	0	1	2	3	4	5	6	7	8	9
0.56	0.5926	0.5940	0.5955	0.5970	0.5984	0.5999	0.6014	0.6029	0.6044	0.6058
0.57	0.6073	0.6088	0.6103	0.6118	0.6133	0.6147	0.6162	0.6177	0.6192	0.6207
0.58	0.6222	0.6237	0.6252	0.6267	0.6282	0.6297	0.6312	0.6327	0.6342	0.6357
0.59	0.6372	0.6387	0.6402	0.6417	0.6432	0.6447	0.6462	0.6478	0.6493	0.6508
0.60	0.6523	0.6538	0.6553	0.6569	0.6584	0.6599	0.6614	0.6630	0.6645	0.6660
0.61	0.6675	0.6691	0.6706	0.6721	0.6737	0.6752	0.6767	0.6783	0.6798	0.6814
0.62	0.6829	0.6844	0.6860	0.6875	0.6891	0.6906	0.6922	0.6937	0.6953	0.6968
0.63	0.6984	0.6999	0.7015	0.7030	0.7046	0.7062	0.7077	0.7093	0.7108	0.7124
0.64	0.7140	0.7155	0.7171	0.7187	0.7202	0.7218	0.7234	0.7250	0.7265	0.7281
0.65	0.7297	0.7313	0.7328	0.7344	0.7360	0.7376	0.7392	0.7408	0.7424	0.7439
0.66	0.7455	0.7471	0.7487	0.7503	0.7519	0.7535	0.7551	0.7567	0.7583	0.7599
0.67	0.7615	0.7631	0.7647	0.7663	0.7679	0.7695	0.7712	0.7728	0.7744	0.7760
0.68	0.7776	0.7792	0.7809	0.7825	0.7841	0.7857	0.7873	0.7890	0.7906	0.7922
0.69	0.7939	0.7955	0.7971	0.7988	0.8004	0.8020	0.8037	0.8053	0.8069	0.8086
0.70	0.8102	0.8119	0.8135	0.8152	0.8168	0.8185	0.8201	0.8218	0.8234	0.8251
0.71	0.8268	0.8284	0.8301	0.8317	0.8334	0.8351	0.8367	0.8384	0.8401	0.8417
0.72	0.8434	0.8451	0.8468	0.8484	0.8501	0.8518	0.8535	0.8552	0.8568	0.8585
0.73	0.8602	0.8619	0.8636	0.8653	0.8670	0.8687	0.8704	0.8721	0.8738	0.8755
0.74	0.8772	0.8789	0.8806	0.8823	0.8840	0.8857	0.8874	0.8891	0.8908	0.8925
0.75	0.8943	0.8960	0.8977	0.8994	0.9011	0.9029	0.9046	0.9063	0.9080	0.9098
0.76	0.9115	0.9132	0.9150	0.9167	0.9184	0.9202	0.9219	0.9237	0.9254	0.9272
0.77	0.9289	0.9307	0.9324	0.9342	0.9359	0.9377	0.9394	0.9412	0.9430	0.9447
0.78	0.9465	0.9482	0.9500	0.9518	0.9536	0.9553	0.9571	0.9589	0.9607	0.9624
0.79	0.9642	0.9660	0.9678	0.9696	0.9714	0.9731	0.9749	0.9767	0.9785	0.9803
0.80	0.9821	0.9839	0.9857	0.9875	0.9893	0.9911	0.9929	0.9948	0.9966	0.9984

r.V$^{-1/3}$	0	1	2	3	4	5	6	7	8	9
0.81	1.0002	1.0020	1.0038	1.0057	1.0075	1.0093	1.0111	1.0130	1.0148	1.0166
0.82	1.0184	1.0203	1.0221	1.0240	1.0258	1.0276	1.0295	1.0313	1.0332	1.0350
0.83	1.0369	1.0387	1.0406	1.0424	1.0443	1.0462	1.0480	1.0499	1.0518	1.0536
0.84	1.0555	1.0574	1.0592	1.0611	1.0630	1.0649	1.0668	1.0686	1.0705	1.0724
0.85	1.0743	1.0762	1.0781	1.0800	1.0819	1.0838	1.0857	1.0876	1.0895	1.0914
0.86	1.0933	1.0952	1.0971	1.0991	1.1010	1.1029	1.1048	1.1067	1.1087	1.1106
0.87	1.1125	1.1145	1.1164	1.1183	1.1203	1.1222	1.1241	1.1261	1.1280	1.1300
0.88	1.1319	1.1339	1.1358	1.1378	1.1398	1.1417	1.1437	1.1456	1.1476	1.1496
0.89	1.1516	1.1535	1.1555	1.1575	1.1595	1.1615	1.1634	1.1654	1.1674	1.1694
0.90	1.1714	1.1734	1.1754	1.1774	1.1794	1.1814	1.1834	1.1854	1.1874	1.1894
0.91	1.1915	1.1935	1.1955	1.1975	1.1995	1.2016	1.2036	1.2056	1.2077	1.2097
0.92	1.2117	1.2138	1.2158	1.2179	1.2199	1.2220	1.2240	1.2261	1.2281	1.2302
0.93	1.2323	1.2343	1.2364	1.2385	1.2405	1.2426	1.2447	1.2468	1.2489	1.2509
0.94	1.2530	1.2551	1.2572	1.2593	1.2614	1.2635	1.2656	1.2677	1.2698	1.2719
0.95	1.2740	1.2761	1.2783	1.2804	1.2825	1.2846	1.2868	1.2889	1.2910	1.2932
0.96	1.2953	1.2974	1.2996	1.3017	1.3039	1.3060	1.3082	1.3103	1.3125	1.3146
0.97	1.3168	1.3190	1.3211	1.3233	1.3255	1.3277	1.3298	1.3320	1.3342	1.3364
0.98	1.3386	1.3408	1.3430	1.3452	1.3474	1.3496	1.3518	1.3540	1.3562	1.3584
0.99	1.3606	1.3629	1.3651	1.3673	1.3695	1.3718	1.3740	1.3762	1.3785	1.3807
1.00	1.3830	1.3852	1.3875	1.3897	1.3920	1.3942	1.3965	1.3988	1.4010	1.403
1.01	1.406	1.408	1.410	1.413	1.415	1.417	1.4200	1.422	1.424	1.427
1.02	1.429	1.431	1.433	1.436	1.438	1.440	1.443	1.445	1.447	1.450
1.03	1.452	1.454	1.457	1.459	1.461	1.464	1.466	1.469	1.471	1.473
1.04	1.476	1.478	1.480	1.483	1.485	1.487	1.490	1.492	1.495	1.497

APPENDIX 5B: DENSITY AND VISCOSITY OF SELECTED LIQUIDS

Liquid	Temperature [°C]	Density [g/cm³]	Viscosity [mm²/s]	Literature
Water	20	0.998	1.0	Handbook 1983
	25	0.997	0.89	Handbook 1983
Glycerine	15	1.264*	245*	Miller et al. 1994
Hexane	20	0.660	0.326	Handbook 1983
Heptane	20	0.684	0.409	Handbook 1983
Oktane	20	0.703	0.542	Handbook 1983
Nonane	20	0.718	0.711	Handbook 1983
Decane	20	0.730	0.92	Handbook 1983
Dodecane	20	0.7511	1.35	Handbook 1983
Chloroform	15	1.498		Handbook 1983
Diethanolamine	30	1.097	303	Miller et al. 1994
Castor oil	20		986	Miller et al. 1994
Ethanol	20	0.789		Handbook 1983
n-Butanol	20	0.810	2.948	Handbook 1983
n-Hexanol	20	0.814		Handbook 1983
Octanol	20	0.827		Handbook 1983
Decanol	20	0.830		Handbook 1983
Benzene	20	0.877	0.652	Handbook 1983
Toluene	20	0.867	0.59	Handbook 1983
Dioxane	20	1.034		Handbook 1983

* not free of water

APPENDIX 5C: **SURFACE TENSION OF SELECTED LIQUIDS AND THEIR INTERFACIAL TENSIONS TO WATER**

Liquid	Temperature [°C]	Surface Tension [mN/m]	Interfacial Tension [mN/m]	Reference
Water	20	72.4	-	Handbook 1983
Glycerine	20	63.4	-	Handbook 1983
Hexane	20	18.4	51.1	Handbook 1983
Heptane	20	19.7	51.0	Handbook 1983
Oktane	20	21.6	50.8	Handbook 1983
Nonane	20	22.7	51.5	Miller et al. 1994
Decane	20	23.8	51.8	Handbook 1983
Dodecane	20	24.9	52.1	Handbook 1983
Chloroform	20	27.2	36.1	Miller et al. 1994
Castor oil	30	36.4	-	Miller et al. 1994
Ethanol	20	22.0	-	Handbook 1983
Hexanol	20	25.8	6.8	Handbook 1983
Octanol	20	27.5	8.5	Handbook 1983
Benzene	20	28.9	35.0	Handbook 1983
Toluene	20	28.5	36.1	Handbook 1983
Dioxane	20	35.4	-	Handbook 1983

APPENDIX 5D: **ISOTHERM PARAMETERS OF SELECTED SURFACTANTS**

Substance	a_F [mol/cm³]	Γ_∞ [mol/cm²]	a' [mN/m]	Reference
octyl sulphate	$3.5 \cdot 10^{-5}$	$1 \cdot 10^{-9}$	0	Wüstneck et al. 1992
decyl sulphate	$1.3 \cdot 10^{-5}$	$1 \cdot 10^{-9}$	6.5	Wüstneck et al. 1992
dodecyl sulphate	$2.1 \cdot 10^{-6}$	$6.8 \cdot 10^{-10}$	11.8	Wüstneck et al. 1992
tetradecyl sulphate	$6.0 \cdot 10^{-7}$	$7.5 \cdot 10^{-10}$	17.3	Wüstneck et al. 1992
dodecyldimethyl ammonio acetic acid bromide	$3.23 \cdot 10^{-8}$	$3.77 \cdot 10^{-10}$	0	Wüstneck et al. 1993
tetradecyldimethyl ammonio acetic acid bromide	$4.76 \cdot 10^{-9}$	$3.96 \cdot 10^{-10}$	3.6	Wüstneck et al. 1993
hexadecyldimethyl ammonio acetic acid bromide	$1.16 \cdot 10^{-9}$	$3.87 \cdot 10^{-10}$	9.0	Wüstneck et al. 1993
hexyl dimethyl phosphine oxide	$3.19 \cdot 10^{-6}$	$3.23 \cdot 10^{-10}$	0	Lunkenheimer et al. 1987b
octyl dimethyl phosphine oxide	$3.73 \cdot 10^{-7}$	$3.66 \cdot 10^{-10}$	0	Lunkenheimer et al. 1987b
decyl dimethyl phosphine oxide	$4.96 \cdot 10^{-8}$	$3.85 \cdot 10^{-10}$	0	Lunkenheimer et al. 1987b

Substance	a_F	Γ_∞	a'	Reference
	[mol/cm³]	[mol/cm²]	[mN/m]	
dodecyl dimethyl phosphine oxide	$4.98 \cdot 10^{-9}$	$4.10 \cdot 10^{-10}$	0.5	Lunkenheimer et al. 1987b
octyl diethyl phosphine oxide	$9.31 \cdot 10^{-8}$	$3.08 \cdot 10^{-10}$	0	Lunkenheimer et al. 1987b
decyl diethyl phosphine oxide	$1.04 \cdot 10^{-8}$	$3.30 \cdot 10^{-10}$	0	Lunkenheimer et al. 1987b
dodecyl diethyl phosphine oxide	$1.34 \cdot 10^{-9}$	$3.60 \cdot 10^{-10}$	0	Lunkenheimer et al. 1987b
1,2-dodecandiol	$8.72 \cdot 10^{-8}$	$5.6 \cdot 10^{-10}$	29.8	Lunkenheimer & Hirte 1992
decanoic acid	$9.44 \cdot 10^{-8}$	$6.1 \cdot 10^{-10}$	21.8	Lunkenheimer & Hirte 1992
Triton X-45	$2.1 \cdot 10^{-8}$	$4.7 \cdot 10^{-10}$	-	Fainerman et al. 1994c
Triton X-100	$1.3 \cdot 10^{-8}$	$3.3 \cdot 10^{-10}$	-	Fainerman et al. 1994c
Triton X-114	$1.2 \cdot 10^{-8}$	$3.1 \cdot 10^{-10}$	-	Fainerman et al. 1994c
Triton X-165	$8.3 \cdot 10^{-9}$	$2.8 \cdot 10^{-10}$	-	Fainerman et al. 1994c
Triton X-305	$2.4 \cdot 10^{-9}$	$1.8 \cdot 10^{-10}$	-	Fainerman et al. 1994c
Triton X-405	$1.4 \cdot 10^{-9}$	$1.4 \cdot 10^{-10}$	-	Fainerman et al. 1994c

The definition of the parameters of the Frumkin isotherm, which coincide with the Langmuir isotherm at a' = 0, is given in Chapter 2 by Eqs (2.43) and (2.44).

APPENDIX 5E: **MUTUAL SOLUBILITY OF ORGANIC SOLVENTS AND WATER**

Liquid	Temperature [°C]	Solubility in water [mol-%]	Water in org. solvent [mol-%]	Reference
Hexane	20	$3 \cdot 10^{-4}$	0.043	Sörensen & Arlt 1979
Heptane	20	$5 \cdot 10^{-5}$	0.070	Sörensen & Arlt 1979
Oktane	20	$1 \cdot 10^{-5}$	0.060	Sörensen & Arlt 1979
Decane	20	$2 \cdot 10^{-7}$	0.057	Sörensen & Arlt 1979
Dodecane	20	$4 \cdot 10^{-8}$	0.061	Sörensen & Arlt 1979
Chloroform	15	0.15	0.517	Sörensen & Arlt 1979
Diethanolamin	20	∞	∞	Sörensen & Arlt 1979
Hexanol	20	0.1	29.0	Sörensen & Arlt 1979
Octanol	20	0.06	19.4	Sörensen & Arlt 1979
Benzene	20	0.04	0.25	Sörensen & Arlt 1979
Toluene	20	0.01	0.24	Sörensen & Arlt 1979
Dioxane	20	∞	∞	Timmerman 1950

APPENDIX 5F: **NUMERICAL ALGORITHM TO SOLVE THE GAUSS-LAPLACE EQUATION**

There are several numerical algorithms to solve the Gauss-Laplace equation in order to fit it to experimental drop shape coordinates. Cheng (1990) developed a routine on the basis of a Runge-Kutta method to integrate the set of differential equations (5.23) - (5.25), using finite arc length steps. This technique is very fast and yields very accurate numeric results. However, Lohnstein (1906) described already an algorithm to solve the Gauss-Laplace equation which at that time was used for manual calculations. Compared to results obtained by modern fast computers, the calculations of Lohnstein were surprisingly accurate. The procedure proposed by Lohnstein (1906) starts from the following form of the Gauss-Laplace equation

$$\frac{y''}{\left(1+y'^2\right)^{3/2}} + \frac{y'}{x\sqrt{1+y'^2}} = \frac{2}{a^2}(R-y), \tag{5E.1}$$

where R is the radius of curvature in the apex of the pendent drop. For a sessile drop only the sign in the expression on the right hand side is changed,

$$\frac{y''}{\left(1+y'^2\right)^{3/2}} + \frac{y'}{x\sqrt{1+y'^2}} = \frac{2}{a^2}(R+y). \tag{5E.2}$$

The second order ordinary differential equation (5E.1) can be rewritten in form of a system of first order differential equations (Lohnstein 1906),

$$\frac{d(xu)}{dx} = \frac{2x}{a^2}(R-y), \tag{5E.3}$$

$$\frac{dy}{dx} = \frac{u}{\sqrt{1-u^2}}, \tag{5E.4}$$

where the function u is defined by

$$u = \frac{y'}{\sqrt{1+y'^2}}. \tag{5E.5}$$

At the point of maximum diameter the function u tends to 1 and the system becomes unstable. Except for the pole region, another differential equation system is stable over the whole range of x and y,

$$\frac{dv}{dy} = \frac{\sqrt{1-v^2}}{x} - \frac{2}{a^2}(R-y), \tag{5E.6}$$

$$\frac{dx}{dy} = \frac{v}{\sqrt{1-v^2}}, \tag{5E.7}$$

where the function v is defined by

$$v = \sqrt{1-u^2}. \tag{5E.8}$$

Both differential equation systems can be solved easily by using finite differences instead of the differential quotients:

$$\frac{dy}{dx} \approx \frac{\Delta y}{\Delta x}, \frac{du}{dx} \approx \frac{\Delta u}{\Delta x}, \frac{dx}{dy} \approx \frac{\Delta x}{\Delta y}, \frac{dv}{dy} \approx \frac{\Delta v}{\Delta y}. \qquad (5E.9)$$

There are different formulas to express the difference quotient and each of them works, but has different accuracy. For example, using a simple backwards difference we obtain

$$\frac{dy}{dx} \approx \frac{y_{i+1} - y_i}{x_{i+1} - x_i}. \qquad (5E.10)$$

The integration of Eq. (5E.1) can now be performed by starting with the set of equations (5E.3), (5E.4) until a certain point has been reached (for example u=0.5). Then, further calculations are done on the basis of Eqs (5E.6) and (5E.7).

APPENDIX 5G: DYNAMIC SURFACE TENSIONS IN THE SUB-MILLISECOND RANGE

Using the experimental set-up MPT1 from LAUDA, it is possible to reach adsorption times of the order of milliseconds (Fainerman 1992, Fainerman et al. 1993a, b, 1994a, b, c, d). This is made possible by a special design of the measuring cell which has a gas volume much larger than the volume of a single bubble. In that case, the pressure inside the cell does not change when a bubble detaches. From a pressure-gas flow rate dependence P(L) a characteristic point results which allows to calculate the so-called dead time of the system. In recent experiments the main parameters controlling the dead time were investigated (Fainerman & Miller 1994b): capillary radius, length of the capillary, size of the bubbles.

In the standard version of the MPT1 measuring cell the dead time is of the order of 70-80 ms. To decrease τ_d down to 10 ms it is necessary the decrease the length of the capillary l and the volume of detaching bubbles. The bubble volume can be controlled by the distance between the capillary tip and the electrode located opposite to it (cf. Fig. 5.12). For different lengths of the capillary reproducible and accurate bubble formation was possible under the conditions summarised in Table 5F.1.

Another very important problem in maximum bubble pressure measurements at very short adsorption times is the initial amount of adsorbed material at the surface of a newly formed bubble. The larger this amount the higher is the initial surface pressure $\Pi = \Pi_o = \gamma_o - \gamma(t = 0)$. The time interval during which the bubble grows continuously from its hemispherical to the final size is τ_d. The relative expansion θ of the bubble surface area A is given by

$$\theta = \frac{d \ln A}{dt} = \xi / t. \qquad (5G.1)$$

Table 5F.1   Characteristic parameter in the critical point determined for capillaries with three different lengths and constant inner tip diameter of $2r_{cap} = 0.15mm$, Fainerman & Miller (1994b)

length l [mm]	dead time τ_d [ms]	bubble volume [mm³]	critical flow rate L_c [mm³/s]
15	80-100	4-7	55-60
10	30-40	2-3	75-80
7	8-10	0.8-1.2	100-105

The coefficient ξ takes values between $\xi = 2/3$ (ideal sphere) and $\xi = 1$ (conditions in the MPT1 (Fainerman et al. 1994a). In the moment of bubble detachment this coefficient can reach values $\xi > 1$. Hence, the effective life time of the bubble during this second stage of its growth (effective dead time) results to

$$\tau_{d,eff} = \frac{\tau_d}{2\xi + 1} \leq \frac{\tau_d}{3}. \tag{5G.2}$$

To calculate the initial value of the surface pressure $\Pi = \Pi_o$ after a time $\tau_{d,eff}$ we use a diffusion controlled adsorption model and $\xi = 1$. In the range of short adsorption time the surface pressure is given by the approximate solution (Fainerman et al. 1994d),

$$\Pi = 2RTc_o\sqrt{\frac{Dt_{eff}}{\pi}}. \tag{5G.3}$$

The evaluation for a common surfactant $(D = 10^{-6} cm^2/s, c_o = 10^{-6} mol/cm^3)$ using Eq. (5G.3) and a dead time $\tau_d = 100ms$ yields $\Pi_o = 4mN/m$, while for $\tau_d = 10ms$ the initial surface pressure reduces to $\Pi_o = 1mN/m$. In case of non-diffusional adsorption kinetics this value becomes even smaller. For example, assuming a pure kinetic-controlled adsorption mechanism, Π is proportional to t, and Π_o reduces by one order of magnitude. Thus, to reach extremely short adsorption times with maximum bubble pressure measurements very short dead times, i.e. high bubble formation frequencies, must be arranged. If higher concentrations are to be measured a quantitative interpretation is impossible and requires accurate consideration of the initial coverage of the bubble surface at time t=0.

APPENDIX 6A: **APPLICATION OF SYSTEM THEORY FOR THE DETERMINATION OF INTERFACIAL TENSION RESPONSE FUNCTIONS TO SMALL INTERFACIAL AREA DISTURBANCES**

In the linear case, non-equilibrium properties of adsorption layers at fluid interfaces can be quantitatively described by the interfacial thermodynamic modulus (Defay, Prigogine & Sanfeld 1977),

$$E_0 = A(\frac{\partial^2 G}{\partial A^2})_{T,p,n}.$$

(6A.1)

According to Loglio et al (1979) the modulus can be expressed in the following form,

$$E_c(i\omega) = \mathbf{F}(\Delta\gamma(t)) / \mathbf{F}(\Delta \ln A(t)),$$

(6A.2)

with $\mathbf{F}$ being the Laplace-Fourier oparator defined by,

$$\mathbf{F}(f(t)) = \int_0^\infty f(t)\exp(-i\omega t)dt.$$

(6A.3)

The function $F(i\omega)=\mathbf{F}\{f(t)\}$ is the image function of $f(t)$. After rearrangement of Eq. (6A.2), using the laws of the Laplace-Fourier transformation (cf. Appendix 4E), the measurable surface tension response $\Delta\gamma(t)$ of the system to the area disturbance $\Delta \ln A(t)$ is obtained,

$$\Delta\gamma(t) = \mathbf{F}^{-1}\left(E_c(i\omega)\mathbf{F}(\Delta \ln A(t))\right).$$

(6A.4)

The inverse of the oparator $\mathbf{F}$ is denoted by $\mathbf{F}^{-1}$. Alternatively, by using the convolution theorem the following relation is obtained (Oberhettinger & Bardii 1973),

$$\Delta\gamma(t) = \int_0^t \mathbf{F}^{-1}(\varepsilon(i\omega), \tau)\, \Delta \ln A(t-\tau)\, d\tau.$$

(6A.5)

APPENDIX 6B: **INTERFACIAL TENSION RESPONSE FUNCTIONS $\Delta\gamma(T)$ TO HARMONIC AND SEVERAL TYPES OF TRANSIENT AREA DISTURBANCES**

To calculate the interfacial tension response to any small area changes, the Fourier transform of the time function $\Delta \ln A(t)$ has to be determined. For a periodic area change

$$A(t) = A_0 + \Delta A \cos(\omega_a t)$$

(6B.1)

the corresponding **F**-transform is

$$F(\Delta \ln A(t)) = -\frac{\Delta A}{A_o} \omega_a \frac{\omega_a}{(i\omega)^2 + \omega_a^2} . \tag{6B.2}$$

For a step type change of A (cf. Fig. 6.1, (a))

$$A(t) = \begin{cases} A_o & \text{at } t < 0 \\ A_o + \Delta A & \text{at } t > 0 \end{cases} \tag{6B.3}$$

the **F**-transformation results in

$$F(\Delta \ln A(t)) = \frac{\Delta A}{A_o} \frac{1}{i\omega} . \tag{6B.4}$$

Since a pulse is hard to produce under experimental conditions, it is easier to perform a ramp type area change, a linear area change in a limited time interval $0 < t < t^*$ (cf. Fig. 6.1, (b)). Thus we get

$$F(\Delta \ln A(t)) = \frac{\Omega}{(i\omega)^2} (1 - \exp(-i\omega t^*)) \tag{6B.5}$$

with

$$\Omega = \frac{1}{t^*} \ln(1 - \frac{\Delta A}{A_o}). \tag{6B.6}$$

The form of the function $\Delta \ln A(t)$ results from the condition $\Delta \ln A(t) \sim t$ in the time range $0 < t < t^*$. After simple integration Eq. (6B.6) is obtained.

More convenient for relaxation studies are the square pulse (cf. Fig. 6.1, (c)) or its practical equivalence the trapezoidal area change (cf. Fig. 6.1, (d)). The **F**-transforms of the respective area changes are as follows (cf. Miller et al. 1991):

square pulse -

$$F(\Delta \ln A(t)) = \frac{\Delta A}{A_o} \frac{1}{i\omega} (1 - \exp(-i\omega t_2)) \tag{6B.7}$$

trapezoidal area change -

$$F\{\Delta \ln A(t)\} = \frac{\Omega}{(i\omega)^2} (1 - \exp[-i\omega t_1] - \exp[-i\omega(t_1 + t_2)] - \exp[-i\omega(2t_1 + t_2)]) \tag{6B.8}$$

with

$$\Omega = \frac{1}{t_1} \ln\left(1 - \frac{\Delta A}{A_o}\right).$$

(6B.9)

The interfacial tension response of a system now results from the appropriate combination of an exchange function $\varepsilon(i\omega)$ with the actual area change function $F(\Delta \ln A(t))$ according to Eq. (6A.5).

For a diffusional exchange of matter the finction $\varepsilon(i\omega)$, derived in chapter 6, reads

$$E_c(i\omega) = E_o \frac{\sqrt{i\omega}}{\sqrt{i\omega} + \sqrt{2\omega_o}}.$$

(6B.10)

The following interfacial tension responses are obtained:

periodic area change -

$$\Delta\gamma(t) = E_o \omega_a^2 \frac{\Delta A}{A_o} \int_0^t \exp(2\omega_o\tau)\mathrm{erfc}\left(\sqrt{2\omega_o\tau}\right)\cos(\omega_a(t-\tau))d\tau,$$

(6B.11)

step function -

$$\Delta\gamma(t) = E_o \frac{\Delta A}{A_o} \exp(2\omega_o\tau)\mathrm{erfc}(\sqrt{2\omega_o\tau}),$$

(6B.12)

ramp function -

$$\Delta\gamma_1(t) = E_o \frac{\Omega}{2\omega_o}\left(\exp(2\omega_o\tau)\mathrm{erfc}(\sqrt{2\omega_o\tau}) - 1\right) + \frac{2E_o\sqrt{t}}{\sqrt{2\pi\omega_o}} \qquad \text{at } 0<t<t^*,$$

(6B.13)

$$\Delta\gamma_2(t) = \Delta\gamma_1(t) - \Delta\gamma_1(t-t^*) \qquad\qquad\qquad\qquad \text{at } t>t^*,$$

(6B.14)

square pulse -

$$\Delta\gamma_1(t) = E_o \frac{\Delta A}{A_o} \exp(2\omega_o t)\mathrm{erfc}(\sqrt{2\omega_o t}) \qquad\qquad \text{at } 0<t<t_2,$$

(6B.15)

$$\Delta\gamma_2(t) = \Delta\gamma_1(t) - \Delta\gamma_1(t-t_2) \qquad\qquad\qquad\qquad \text{at } t>t_2,$$

(6B.16)

The result for the trapezoidal area change is given in Chapter 6.

APPENDIX 6C: INTERFACIAL PRESSURE RESPONSE TO AREA DISTURBANCES IN PRESENCE OF INSOLUBLE MONOLAYERS

In a very recent development the pendent drop technique was used to study the static and dynamic behaviour of insoluble monolayers. Kwok et al. (1994a) were the first who performed measurements of surface pressure - area isotherms of octadecanol monolayer by using a pendent water drop and ADSA (cf. Section 5.4) as a film balance. Consistent results with classical Langmuir-Wilhelmy film balance measurements reveal ADSA as a powerful tool also for monolayers.

The general idea is to spread very small amounts of a spreading solvent, containing the definite amount of insoluble surfactant, onto the surface of a pendent drop, for example a water drop. Then, by changing the volume of the pendent drop the area of the drop surface can be compressed or expanded. ADSA provides surface tension, surface area and drop volume simultaneously at any time an image of the drop is acquired. Thus, exact surface pressures and surface areas are available continuously, and isotherms can be measured.

In a paper by Li et al. (1994a), the dynamic behaviour of surfactant monolayers of two phospholipids, DMPE and DPPC, were studied by using this technique. While the octadecanol isotherm covers only an area interval between 30Å²/molecule and 18Å²/molecule, the complete isotherm of DMPE and DPPC develops over an area change of 60 Å²/molecule. The distinct phase transitions, from the liquid expanded (LE) - liquid condensed (LC) coexistence range, to the liquid condensed phase and to the final collapse were obtained and compared with classic Langmuir balance measurements. Extremely high collapse pressures as observed with octadecanol monolayers (Kwok et al. 1994b) were obtained for phospholipid monolayers too. Depending on the compression rate, collapse pressures up to almost 70mN/m were observed (Li et al. 1994a, Kwok et al. 1994b).

The pendent drop technique has also been extended to studies of insoluble monolayers of phospholipids at the water/n-dodecane interface (Li et al. 1994b). In these experiments first a monolayer is produced on a water drop surface as described above. Then, the water drop is gently immersed into the second liquid, for example n-dodecane. Then the change of the drop size enables one to compress and expand the interfacial film. Again, the isotherm obtained with ADSA shows the same type of dynamic behaviour as measured with the classic Langmuir-Blodgett trough technique (Thoma & Möhwald 1994).

In the pendent drop experiments, the surrounding temperature is easily controlled by using a small quartz cuvette. The cuvette can be sealed during the measurements to avoid evaporation effects. This makes the pendent drop technique particularly suited for measurements of very wide temperature ranges in air/liquid or liquid/liquid systems and over long time intervals. The conventional LB balance, as an open system, has objective difficulties to realise measurements at temperatures above 50 °C. In addition, the small drop dimensions ensure good temperature homogeneity of the system. The latter has sometimes been claimed to be responsible for a limited isotherm slope within the phase coexistence range. Moreover, changes of the drop surface area are much more isotrope than the area changes on a Langmuir trough. Surface pressure gradients along the monolayer may be the reason for the significant differences in collapse pressures obtained from the two techniques.

The main limitation of ADSA as a film balance is caused by the accuracy of the spread amount. To spread a big amount of surfactants will either make the drop fall down or make the isotherm enter the phase transition region too early. Considering the restricted surface area, a small amount of surfactant has to be spread which may cause relatively large errors in the calculation of the absolute molecular area of the surfactant onto the drop surface. This will lead to a shift of the isotherm along the x-axis and its stretching. So it is necessary to take particular care in delivering the substance onto the pendent drop.

APPENDIX 7A: THE APPROXIMATE INTEGRATION OF THE DIFFERENTIAL EQUATION OF ADSORPTION OF MULTIVALENT IONS

The concentration $c(\kappa^{-1}, t)$ in Eq. (7.40) has to be expressed in terms of $\overline{\Gamma}(t)$. The general solution for molecular adsorption kinetics was derived by Ward & Tordai (1946) in form of Eq. (4.1). This equation can be used for of ionic adsorption too (Miller et al. 1994). It represents the solution of the non-steady diffusion problem, given by the differential equation (4.9) with the boundary conditions (4.11) and (4.15), and describes the concentration distribution outside the DL. There has to be some explanation of the boundary condition (4.11), which reads in the present case,

$$\frac{d\Gamma(t)}{dt} = D_{eff} \left. \frac{\partial c(x,t)}{\partial x} \right|_{x=0}.$$

(7A.1)

The flux from the bulk to the surface leads to changes in the adsorption $\Gamma(t)$ and the content of adsorbing ions in the diffuse layer changes too. Thus, a more exact boundary condition should consider this change of adsorbing ions in the diffuse layer. Under condition (7.11) adsorbing ions are only the coions and their content in the diffuse layer is negligible.

The striking peculiarity in the boundary condition (7A.1) for ion adsorption is the substitution of the surfactant diffusion coefficient D by an effective one D_{eff}, defined in Eq. (7.18). This is caused by the fact that the diffusional migration of ions in the electric field supplements to the ion transport inside the diffuse layer.

The substitution of Eq. (4.1), or better of the equivalent equation (4.18), into Eq. (7.40) yields,

$$\frac{d\overline{\Gamma}}{d\theta} + \frac{\exp(-z(\overline{\Psi}_{St} - \overline{\Psi}_{Sto}))\overline{\Gamma}(\theta)K(\psi_{Sto})}{K(\psi_{St})} = 1 - \frac{2}{\sqrt{\pi D_{eff}}}\frac{d}{d\theta}\int_0^{\sqrt{\theta}}\overline{\Gamma}(\theta - \tau)d\sqrt{\tau} \tag{7A.2}$$

From Eq. (7.34) follows,

$$\exp\left(\frac{z^+(\overline{\Psi}_{Sto} - \overline{\Psi}_{St})}{2}\right) = \frac{\Gamma}{\Gamma_o} = \overline{\Gamma} \tag{7A.3}$$

and

$$\frac{K(\psi_{Sto})}{K(\psi_{St})} = \exp\left(-(z - \frac{z^+}{2})(\overline{\Psi}_{Sto} - \overline{\Psi}_{St})\right) = \overline{\Gamma}^{(1-2z/z^+)} \tag{7A.4}$$

From (7A.2) and (7A.4) the following relation is obtained,

$$\frac{d\overline{\Gamma}}{d\theta} + \overline{\Gamma}^2 \approx \overline{\Gamma}^{(1-2z/z^+)}\left(1 - \delta\int_0^{\sqrt{\theta}}\frac{d}{d\tau}\overline{\Gamma}(\theta - \tau)d\sqrt{\tau}\right) \tag{7A.5}$$

with

$$\delta = \sqrt{\frac{8}{\pi}\frac{c_{el}}{c_o}}\exp\left((z - z^+)\overline{\Psi}_{Sto}\right)\frac{z - z^+/2}{z}. \tag{7A.6}$$

From a comparison of Eq. (7A.6) with (7.51) the most interesting case of strong electrostatic retardation exists at

$$\delta \ll 1. \tag{7A.7}$$

Under the condition (7A.7) the Eq. (7A.5) now reads

$$\frac{d\overline{\Gamma}}{d\theta} + \overline{\Gamma}^2 \approx \overline{\Gamma}^{(1-2z/z^+)}. \tag{7A.8}$$

Eq. (7A.8) suggests an adsorption value at which the contribution of co-ions to the charge of the diffuse layer and to the deceleration of the adsorption process can be neglected. In other words the adsorption is not equal to zero at the very beginning but it is much smaller than the equilibrium adsorption value

$$\bar{\Gamma} \ll 1.$$
(7A.9)

Neglecting the effects of the order of $\bar{\Gamma}$ the following approximate initial condition can be used:

$$\bar{\Gamma}\big|_{\theta=0} = 0$$
(7A.10)

A solution of the problem based on this initial condition will not be exact for the initial period. Nevertheless, its accuracy will increase as θ increases. The solution of Eq.(7A.8) has the form

$$\theta = \int_0^{\bar{\Gamma}} \frac{\xi^{(2z/z^+ - 1)}}{1 - \xi^{(2z/z^+ + 1)}} d\xi.$$
(7A.11)

Eqs (7A.5) and (7A.11) describe the process of adsorption leading to the equilibrium state at infinite time

$$\bar{\Gamma}\big|_{\theta \to \infty} = 1.$$
(7A.12)

If we ignore the state close to equilibrium we obtain

$$\bar{\Gamma}^{(1+2z/z^+)} \ll 1$$
(7A.13)

and we can simplify the integrand in Eq. (7A.11). Integration now results in

$$\theta \approx \int_0^{\bar{\Gamma}} \left(\xi^{(2z/z^+ - 1)} + \xi^{(4z/z^+)} \right) d\xi = \frac{z^+}{2z} \bar{\Gamma}^{(2z/z^+)} + \frac{z^+}{4z + z^+} \bar{\Gamma}^{(4z/z^+ + 1)}$$
(7A.14)

Under the condition Eq. (7A.9) we can omit the second term in (7A.14)

$$\bar{\Gamma}(\theta) = \left(\frac{2z}{z^+} \theta \right)^{(z^+/2z)}.$$
(7A.15)

These simplifications amount to neglecting the second term on the left-hand site of Eq. (7A.5). This means the second term in Eqs (7A.5) and (7A.8) can be omitted under condition (7A.13),

$$\frac{d\overline{\Gamma}^{(2z/z^+)}}{d\theta} + \delta \frac{d}{d\theta} \int_0^{\sqrt{\theta}} \overline{\Gamma}(\theta - \tau) d\sqrt{\tau} \approx 1$$

(7A.16)

Eq. (7A.16) together with the initial condition (7A.10) can easily be integrated and yields

$$\frac{z^+}{2z}\overline{\Gamma}^{(2z/z^+)} + \delta \int_0^{\sqrt{\theta}} \overline{\Gamma}(\theta - \tau) d\sqrt{\tau} \approx \theta$$

(7A.17)

The second term on the left of Eq. (7A.17) is associated with the decrease of surfactant concentration at the outer boundary of the DL. This effect decreases with increasing $\overline{\Gamma}(\theta)$. Thus the applicability of the adsorption kinetics given by Eq. (7A.15) is restricted both at the initial stage and around equilibrium. Nevertheless, the approximation Eq. (7A.15) is a useful solution to the problem of the DL effect on adsorption kinetics because Eq. (7A.8) holds at strong electrostatic retardation.

APPENDIX 8A: SMALL REAR STAGNANT CAP OF BUBBLE AT HIGH REYNOLDS NUMBERS

The relation of $\Gamma(\theta)$ with the integral desorption flux can be calculated if viscous stress distribution is known for the strongly retarded region near the lower pole. It follows from boundary condition (8.143) that

$$\eta \frac{\partial v_\theta}{\partial r}\bigg|_{r=0} = \eta \frac{V}{a_b}\tilde{E} = -\frac{RT}{a_b\phi^*}\frac{\partial \Gamma}{\partial m},$$

(8A.1)

where $\tilde{E}$ is the dimensionless shear stress; $m = (\pi - \theta)/\phi^*$ (Harper, 1973). As distinct from this work, we have obtained results suitable for an arbitrary function $\tilde{E}(m)$. It follows from (8.143) that

$$\Gamma(m) = \frac{\eta V\phi^*}{RT} \int_1^m \tilde{E}\, dm'.$$

(8A.2)

Because of the assumptions made the tangential component of the liquid velocity within the strongly retarded region equals zero. It can be assumed that within a thin diffusion layer the tangential component of velocity is linear in $s = (r - a_b)/(a_b\phi^*)$,

$$v_\theta = V\tilde{E}\phi^\cdot s, \tag{8A.3}$$

and the radial component can be determined from the continuity equation

$$v_r = -\frac{1}{2}V\frac{(\phi^\cdot s)^2}{\sin\theta}\frac{\partial}{\partial\theta}(\tilde{E}\sin\theta). \tag{8A.4}$$

Within the limits of the strongly retarded region, the convective diffusion equation has the form:

$$-s\tilde{E}\frac{\partial c}{\partial m}+\frac{1}{2}\frac{s^2}{m}\frac{\partial}{\partial m}(\tilde{E}m)\frac{\partial c}{\partial s}=\frac{D}{Va_b^2\phi^{\cdot2}}\frac{\partial^2 c}{\partial s^2}. \tag{8A.5}$$

Changing the variables

$$s=\frac{Y}{A\sqrt{\tilde{E}m}};\ X=\int_m^1 \tilde{E}^{1/2}m'^{3/2}dm' \tag{8A.6}$$

with

$$A=\left(\frac{1}{9}\frac{a_b V\phi^{\cdot2}}{D}\right)^{1/3} \tag{8A.7}$$

it simplifies substantially to

$$\frac{\partial c}{\partial X}=\frac{1}{Y}\frac{\partial^2 c}{\partial Y^2} \tag{8A.8}$$

It follows from (8A.6) and (8A.7) that

$$Y^2=A^2\tilde{E}ms^2=\left(\frac{1}{81}\frac{a_b^2V^2\phi^{\cdot4}}{D^2}\right)^{1/3}\tilde{E}ms^2 \tag{8A.9}$$

In our view there was some inaccuracy in a similar formula given by Harper (1973).

The solution of Eq. (8A.8) has the form

$$c=\frac{1}{(1/3)!}\int_0^X dX'\frac{dc}{dX'}\Big|_{Y=0}\int_{Y/(X-X')^{1/3}}^\infty e^{-t'^3}dt' \tag{8A.10}$$

By means of this solution, we can calculate the derivative

$$\left.\frac{\partial c}{\partial Y}\right|_{Y=0} = -\frac{1}{(1/3)!}\int_0^X dX' \left.\frac{dc}{dX'}\right|_{Y=0} \frac{1}{(X-X')^{1/3}}$$

and the integral desorption flux J_d:

$$\frac{J_d}{2\pi a_b^2 D} = \frac{1}{a_b}\phi^\bullet \int_0^1 \left.\frac{\partial c}{\partial s}\right|_{s=0} m\, dm = \frac{1}{a_b}A\phi^\bullet \int_0^{X_m} \left.\frac{\partial c}{\partial Y}\right|_{Y=0} dX =$$

$$= \frac{A}{a_b}\frac{\phi^\bullet}{(1/3)!}\frac{3}{2}\int_0^{X_m} dX' \left.\frac{dc}{dX'}\right|_{Y=0} (X-X')^{2/3} \qquad\qquad\qquad (8A.11)$$

Substituting (8A.2) into (8A.11) the final result is obtained,

$$J_d = \frac{3^{1/3}\pi}{(1/3)!} a_b D\, c \left(\frac{a_b V}{D}\right)^{1/3} \frac{\eta V}{RT\Gamma}\phi^{\bullet 8/3}\, I_1 \qquad\qquad\qquad (8A.12)$$

where

$$I_1 = \int_0^1 dm\, \tilde{E}(m)\left(\int_0^m \tilde{E}^{1/2} m'^{3/2}\, dm'\right)^{2/3} \qquad\qquad\qquad (8A.13)$$

For Re«1, when

$$\tilde{E}(m) = \frac{4}{\pi}\frac{m}{\sqrt{1-m^2}}, \qquad\qquad\qquad (8A.14)$$

the calculated numerical value of integral amounts to $I_1 = 0.515$. Equating the desorption and adsorption fluxes (Eq. (8.142)), we obtain for Re»1

$$\phi^\bullet = 1.29\, Pe^{1/16}\left(\frac{RT\Gamma}{\eta V}\right)^{3/8}. \qquad\qquad\qquad (8A.15)$$

APPENDIX 10A: PROCESSES RESTRICTING WATER PURIFICATION BY MICROFLOTATION, AND PREVENTION OF BUBBLE SURFACE RETARDATION

The surface of a bubble is cleanest at the instant of its formation. Thereafter it becomes covered by adsorbed impurities whose presence at even extremely low bulk concentration suffices to retard the surface of small bubbles. If a large amount of gas (preliminarily cleaned from aerosols) is passed through a small volume of liquid (also preliminarily cleaned from impurities by all available methods) the content of impurities in water should decrease with each new portion of bubbles. One might therefore expect that the surface of each subsequent portion of bubbles should contain less admixture and should be less retarded. It is possible that in this way alone we will succeed in obtaining bar, non-retarded surfaces of small bubbles.

For a successful purification it is necessary to prevent a return of admixtures to purified water once the bubbles have reached the water-air interface and burst. This process has been well studied by Loglio et al. (1984). The rupture of foam films produces aerosol drops enriched with admixtures which contaminate the water again. It is probably impossible completely to eliminate the return of admixtures to the bulk, but it is possible to reduce this disturbing effect. One of the methods consists of blowing away the drops by a tangential air flow. At a sufficiently high speed, the drops will be removed from the air layer adjacent to the water surface in less time than sedimentation takes. The smaller the water surface area the less the necessary tangential gas flow velocity. As it is impossible to purify gas from usual aerosol particles completely, it is also impossible to prevent these aerosol particles reaching the water surface. It is especially difficult to purify gas from submicron particles which deposite on the water surface from air flow through diffusion rather than sedimentation. As the air flow velocity decreases through the water bulk, the thickness of the diffusion layer increases and diffusion transfer of aerosol particles to the water surface should decrease too. Thus the increase of the tangential air velocity prevents liquid droplets to return to the surface and simultaneously enhances the deposition of aerosol paricles.

At first glance it would seem possible to avoid transport of such impurities from the surface into the deeper water layers by using higher water columns. Agitation of water by buoyant bubbles unfortunately prevents this. Because of the bubbles some elementary volumes will sink, others rise. A rising bubble sinks a volume of liquid. This is repeated by any rising bubble so there is a liquid flow downwards and simultaneous transport of impurities.

If there is no correlation of bubble life time at a given space, the motion of liquid initiated by rising bubbles and averaged over the mean bubble life time has a random walk character. Steps of the random walk are large, of the order of the bubble size, so that there can be considerable transport of impurities. Any estimation is complicated because the step and time of individual walks are not constant but vary widely. The element of liquid near the bubble surface moves a large distance upwards for example. Meanwhile a liquid element further from a bubble will move down insignificantly as the bubble passes. The further away the element is from the bubble the smaller this path is.

At high bubbling rates, a more intensive admixture transport mechanism arises opposite to the direction of bubble motion. This is caused by the system of ascending and descending liquid flows (so-called Benard cell, or simply benards). Benards are more or less regular hexagons in the centers of which liquid rises and along its perphery it sinks. It is shown in Section 2.4 of the book by Derjaguin et al. (1986) that the velocity of convective flow can substantially exceed the velocity of buoyant bubbles. So to provide a maximum purity of bubble surfaces, it is advisable not to use large section cells and, probably, to conduct the experiment in vertical long capillaries having cross-sections comparable to the bubble diameter.

If we assume that a number of bubbles rises to the surface along one and the same vertical axis, it is easy to imagine lines of liquid flow representing a periodical structure with a spacing equal to the distance between the centers of the bubbles, in a form of an almost rectangular contour whose vertical dimension exceeds its horizontal dimension by several orders of magnitude.

Liquid moves upwards along these closed flow lines close to the bubble surface at approximately the same velocity and flows slowly down again near the capillary walls. After some time, the concentration of impurities decreases at the bottom because of the transport by rising bubbles, and it increases at the top. When the concentration of impurities decreases at the bottom, their transport with bubbles upwards slows down. When concentration at the top increases, transport of impurities downwards with descending flow of liquid increases. As a result we should expect the establishment of a stationary state at which transport of impurities by bubbles upwards is compensated by the transport of impurities downwards with descending liquid flow.

Because of the difficulties of high purification of water by microflotation, other purification methods are also applied. The greatest successes has been with the technology of high purification of water, a wide field since electronics industry for example consumes a large body of highly desalinated and extremely pure water.

APPENDIX 10B: ROLE OF R.S.C. IN TRANSPORT STAGE AT DIFFERENT PARTICLE ATTACHMENT MECHANISMS

The attachment can happen due to the instability of the wetting film confined between the surfaces of particle and bubble and its break at some critical thickness h_{cr}. If h_{cr} essentially exceeds the effective radius of attraction forces between surfaces, the transport of the particle to distance h_{cr} does not act as attraction forces. This means that the grazing trajectory is not extended to the lower part of bubble surface because gravitation is there to impede sedimentation at $\theta > \pi / 2$. The availability of a rear stagnant cap affects the hydrodynamic field in the neighbourhood of the lower half of bubble surface and has little effect on the leading part of the surface remote from the rear stagnant cap. This means that the rear stagnant cap has no effect on the transport stage of flotation because it is far removed from the grazing trajectory.

A different situation arises when $h_{cr} = 0$ and fixation of particles happens due to surface attraction forces. These forces can exceed the gravitation force so fixation of particles at the lower pole of the bubble becomes possible. The appearance even of a small rear stagnant cap in the neighbourhood of the lower pole results in a change of the course of the grazing trajectory.

Thus, if an experiment shows a sensitivity of the collision efficiency to the appearance of a small rear stagnant cap, it demonstrates that fixation is controlled by the mechanism of surface attraction forces. An opposite experimental result would point to particle fixation according to mechanism controlled by the break of the wetting film. A quantitative description of collision efficiency results in qualitatively different formulas depending on the controlling mechanism, which contain h_{cr} in one case and the Hamaker constant in the other case.

When dimensions of the rear stagnant cap and the bubble are commensurable, the hydrodynamic field close to the upper half of the bubble changes substantially. A rear stagnant cap of substantial size therefore affects the transport stage of flotation at any mechanism of fixation. Under its effect, the tangential component of the liquid velocity must sharply decrease and the normal component increase. This results in a sharp deflection of the liquid flow line normal to the surface at the boundary between rear stagnant cap and free surface (cf. Fig. 10.4). A more detailed analysis shows that this deflection begins earlier. As a result, the angle θ_i decreases (interception angle). This naturally results in a decrease of the integral flow of particles on bubble surface and in a decrease of collision efficiency.

APPENDIX 10C: CHOICE OF HYDRODYNAMIC REGIME UNDER PRODUCTION CONDITIONS. DECIMICRONE-SIZE PARTICLE

Means of control differ substantially depending on whether attachment forces are large or small, i.e. the potential well is not deep or is absent. The modes of control of particle and bubble charges should be selected already at the stage of studying the properties of the flotation system. It is advisable to conduct experiments of flotation recovery of particles by bubbles of small and large size, for example, at $Re \approx 1$ and $Re \approx 400$. Clearly, opposite charges of particle and bubble should be provided and the same dispersion medium should be used contaminated with unknown surfactant, as will be used in the planned technology.

A drift of surfactant to the rear zone can appear for large bubbles with a slightly retarded surface so that their selection is restricted by condition (8.71). In that case it is preferable to change the sign of the particle charge in order to decrease reagent consumption.

If at a given type of particle the recovery decreases with increasing bubble size, it points to small adhesion forces and to a detachment of particles. Especially problematic for such systems is an increase of bubble size. A substantially smaller average sized bubble should for this reason be used. Polydispersity should also be taken into consideration; the fraction of the largest bubbles providing a small collision efficiency will not make an appreciable contribution to flotation recovery. A shift in bubble size distribution towards enlargement caused by the coalescence of bubbles is also possible. Since flotation recovery in such systems takes place only at strong retardation of the bubble surface, application of small amounts of highly surface active reagents is possible in accordance with condition (8.72). Non-reagent non-contact flotation is hampered since at electric repulsion the depth of the potential well, also small at electrostatic attraction, becomes still smaller.

If a noticeable recovery is also observed at large bubble size, the potential well is deep, i.e. an α-film is either absent or very thin.

A wide range of types of flotation is possible at a deep potential well. A non-reagent flotation is also possible, at a high salt content when the bubble potential is small for example. Experiments on the stability of wetting films (cf. Appendix 10.D) at high electrolyte concentration has shown that contact angles grow above centinormal solutions although α-films continue to exist and κ^{-1} and the effective range of electrostatic attraction forces decreases. So when large bubbles and electrostatic attraction appears the possibility of non-reagent flotation should be examined.

Difficulties in realizing the described program may arise from the very beginning: the output of flotated particles is very small despite the provision of electrostatic attraction and the small size of bubbles used. Even at this first stage it is important to evaluate whether the small output of flotated particles is the result of pure fixing or a manifestation of low collision efficiency. The answer is determined by the following quite different courses. Models of flotation kinetics limited only by the transport stage are very useful here (cf. Section 10.1). Sizes of bubbles and particles can be measured easily. The only parameter required for collision efficiency calculations and difficult to measure is the degree of bubble surface retardation. Using small bubbles we can assume that the surface is completely immobilised. If it turns out that the experimentally measured flotation is much less than that calculated by formula (10.57), particle fixing is the limiting stage of microflotation. Thus it is important to improve particle fixing. The reason for a poor fixing at a suppressed electrostatic barrier can be the presence of a potential barrier of nonelectric nature either near the particle surface or at the water-air interface.

This barrier can be caused either by the hydrophilicity of the particles, by the availability of a boundary water layer near the particle surface (cf. Section 10.5) or by a surfactant adsorption layer hydrophilising the particle. Since hydrophilicity of many dispersed systems has been studied in colloid chemistry, independent information can be used to identifying the nature of the barrier at the particle surface. Broad experience in flotation of large particles can be used for selecting surfactants to destruct the boundary layer of particles of a concrete nature.

A quite different technique is necessary when a stabilising adsorption layer on the particle surface is formed. Apparently it is necessary to provide reversibility of surfactant adsorption so that a decrease of surfactant concentration in the bulk results in desorption. This decrease is used in water purification technology based on adsorption methods. Specifically, if a surfactant stabilises the adsorption layers on particles and also adsorbs at the water-air interface, a preliminary flotation of surfactant can decrease their adsorption and thus destabilise the particles. Then microflotation can be applied to extract destabilised particles.

A barrier at the water-air interface can be removed in the same way. Surfactant should be removed from the system by means of preliminary flotation. After the barrier has been removed a fixing of particles takes place on a less-covered bubble surface.

Quite different measures are required when the experimentally determined kinetic constant of microflotation K is close to the calculated one. In this case it is necessary to intensify the transport stage either by arranging preliminary aggregation of particles or by applying two-stage microflotation (Section 10.9).

APPENDIX 10D: NEW DEVELOPMENTS IN SURFACE FORCES

The main characteristic of the stability of bubble-particle aggregates is the contact angle ψ. The equation

$$F_{at} = 2\pi a_p (1 - \cos\psi)$$

$$(10D.1)$$

expresses the attachment force F_{at} as a function of the contact angle. The thermodynamics of surface phenomena connects the value of contact angle with the difference between the free energies of solid-vapour, γ_{sv}, and solid-liquid, γ_{sl}, interfaces. The difference $\gamma_{sl} - \gamma_{sv}$ can be calculated on the basis of the theory of surface forces (Derjaguin et al. 1987).

As a result, the equilibrium contact angle θ_o may be calculated from the known equation of Frumkin-Derjaguin's theory (Frumkin 1938, Derjaguin 1940),

$$\cos\theta_o = 1 - \frac{1}{\gamma} \int_{h_o}^{\infty} \Pi(h)\, dh,$$

$$(10D.2)$$

where h_o is the equilibrium thickness of the wetting film in contact with the bulk meniscus and $\Pi(h)$ is the corresponding disjoining pressure which is equal to the capillary pressure of the meniscus.

The isotherm $\Pi(h)$ can be obtained experimentally or calculated on the basis of the surface forces theory. When using Eq. (10D.2) for wetting films of water or aqueous solutions, it is necessary to take into account at least three components of the disjoining pressure, i.e. the dispersion, Π_m, electrostatic, Π_e, and structural, Π_s, contributions.

First, let us consider the influence of the electrostatic potentials of the bubble ψ_b and particle ψ_p on the contact angle in absence of the structural component of the disjoining pressure. We denote by Φ_1 the larger potential between ψ_b and ψ_p, and by Φ_2 the smaller one.

When the potentials Φ_1 and Φ_2 have the same sign but different orders of magnitude, as is the case for wetting films of water on silica, at some critical thickness the electrostatic forces change from repulsion at $h > h_1$ to attraction at $h < h_1$ (Derjaguin et al. 1987),

$$h_1 = \kappa^{-1} \ln(\Phi_1/\Phi_2)$$

$$(10D.3)$$

Due to the attraction of the film surfaces at $h < h_1$ the wetting water film usually ruptures. Such ruptures have been observed for example for flat water films on a methylated quartz surface (Blake & Kitchener 1972).

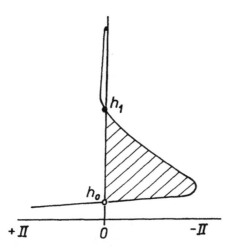

Fig. 10D.1. Schematic representation of the disjoining pressure isotherm of wetting films, $\Pi(h)$ in the case of partial wetting (h is the equilibrium film thickness and h_1 is the film thickness at which the change in sign of the electrostatic forces take place). The hatched area is proportional to the value of the contact angle θ

Fig. 10D.1 shows the isotherm $\Pi(h)$ of an aqueous wetting film for the case where the potentials Φ_1 and Φ_2 are different in magnitude. When the forces of electrical attraction predominate in the region $h < h_1$, there will be a partial wetting. Since the contribution to the integral in Eq. (10D.2) in the region of thick films $h > h_1$ is small, the value of $\cos\theta$ depends mainly on the shaded area in Fig. 10D.1. The greater the area, the smaller is $\cos\theta$ and the higher are the values of the contact angle θ (Derjaguin & Churaev, 1984).

For the region $h < h_1$, the equation for the $\Pi(h)$ isotherm can be written in the form (Churaev 1993)

$$\Pi(h) = \left(A_{130}/6\pi h^3\right) - \left[\varepsilon^2 (\Phi_1 - \Phi_2)^2 / 8\pi h^2\right].$$

(10D.4)

Here the first term expresses the contribution of dispersion forces of repulsion $\Pi_m > 0$, where A_{130} is the composite Hamaker constant. The second term represents the forces of electrostatic attraction $\Pi_e < 0$.

Taking $\Pi = 0$ in Eq. (10D.4), it is possible to derive an expression for the equilibrium film thickness (see Fig. 10D.1)

$$h_o = 4A_{130}/3\varepsilon^2 (\Phi_1 - \Phi_2)^2.$$

(10D.5)

This allows us to derive an analytical expression for $\cos\theta$ by substituting $\Pi(h)$ from Eq. (10D.4) and h_o from Eq. (10D.5) into Eq. (10D.2),

$$\cos\theta = 1 - \left[3\varepsilon^2 \left(\Phi_1 - \Phi_2\right)^4 \big/ 64\pi A_{130}\gamma_{lv}\right].$$ (10D.6)

For differences, $\Delta\Phi = |\Phi_1 - \Phi_2| = 10$ mV, 20 mV, and 30 mV, Eq. (10D.6) gives contact angle values of $\theta = 0.5$, $\theta = 1.5$, $\theta = 3.5°$ and equilibrium film thicknesses $h_o = 11$ nm, $h_o = 2.7$ nm and $h_o = 1.2$ nm, respectively, assuming $A_{130} = 7 \cdot 10^{-21}$ J (Churaev 1984) for quartz.

Attachment forces can stabilize bubble-particle aggregates at these small contact angles if particles are sufficiently small. So it follows significantly that attachment of a small particle is possible even at the same signs of the surface potentials of bubbles and particles, as the necessary condition is only a small difference in their values. However there is a discrepancy between the calculated contact angle values and the measured high values of contact angles for hydrophobic surfaces. This lets Churaev (1993) conclude that the effect of hydrophobic forces has to be taken into account. Such a theory does not exist and experiments have been performed only for aqueous interlayers between two hydrophobic solid surfaces (Israelachvili & Pashley 1984; Claesson & Christenson 1988; Rabinovich & Derjaguin 1988). Churaev proposes that the water-vapour interface may be considered hydrophobic. At least in the transition zone between water and vapour and between water and a hydrophobic surface, the density of water decreases and the dipoles of water molecules are oriented preferentially parallel to the interface (Brodskaja & Rusanov, 1986; Iljin et al., 1991, Derjaguin & Churaev, 1989; Townsen & Rice, 1991). This leads to a sufficient increase of the contact angle values. Botsaris & Glazman (1982) also suggested that structural repulsion can be connected with the hydration effect of the end groups of adsorbed surfactant molecules.

Formation of a dense monolayer (Fig. 3(b)) influences also the dispersion forces, Π_m. At $h < 3$ nm the forces become attractive because the CTAB monolayer screens the film-air interface (Pashley & Israelachvili 1981).

Structural forces for hydrophilic surfaces were included into the theory of stability of colloids and wetting films (Churaev & Derjaguin 1985) earlier than for hydrophobic surfaces. The experimental basis for this generalisation of the DLVO theory was the investigation of polymolecular adsorption of polar liquid vapors on a polished glass surface (Derjaguin & Zorin 1955). A later analysis of these data has demonstrated (Derjaguin & Churaev 1974) that they can be correctly interpreted only when taking into account structural forces. Measurements of polymolecular adsorption of water vapor have shown that the radius of action of structural forces can substantially increase with the surface hydrophilicity.

In case of a hydrophilic quartz surface, the Van der Waals forces are repulsive, but electrostatic forces pass from repulsion to attraction at shorter distances. This results in a rupture of the so-called β-film, and a stabilisation of the α-film occuring at h = 6.4nm which has been interpreted to be the cause of the repulsive structural forces.

Calculations give values of the contact angle of water on quartz of $\psi \sim 4°$ which is close to experimental data. For this calculation based on Eq. (10D.2), the third component, i.e. the repulsive structural force, was introduced into the equation for the disjoining pressure. Thus, the phenomenon of the incomplete wetting of silicate surfaces is connected to the influence of the electrical state of the film-gas interface. A small contact angle is caused by structural repulsive forces. A still higher effect can be provided by recharging the film-gas surface, which is corroborated experimentally (Zorin et al. 1979).

Thus we can conclude that the attachment of a small hydrophilic particle is possible in the presence of a wetting film between particle and bubble. In other words it is possible for hydrophilic or unmodified fine particles to be attached by heteroflocculation. This could be of great value in some practical cases. It also accounts for the low selectivity of fine-grain flotation (Schulze 1984). Recently, the investigation of Derjaguin & Churaev (1989) was reanalysed by Skvarla & Kmet (1993). They received almost the same contact angle value and noted that "in this case the attachment is possible only in the so-called secondary minimum". This remark deserves attention because the analysis is generalised by taking into account the Fowkes approach to interfacial interaction which is based on a splitting of the surface energy of solids and liquids into two components, the so-called apolar and polar interactions (Fowkes 1990; Van Oss et al. 1987). Skvarla & Kmet (1993) have extended the Van Oss surface thermodynamics of interfacial interaction between two identical surfaces to particle-bubble attachment and have derived a formula for the hydrophilic-hydrophobic surface component on the basis of the Lewis acid-base (AB) contribution to the work of adhesion. The agreement of these results with the Derjaguin-Churaev theory of wetting film and contact angle has been verified.

Recently Xu & Yoon (1989, 1990) used the existence of hydrophobic interaction to explain the spontaneous coagulation of hydrophobic particles. Yoon (1991) used a similar approach for weak bubble-particle interactions and emphasises a possible role of hydrophobic interaction in particle-bubble attachment. The investigation underestimates the possible role of the electrostatic interaction stressed by Churaev (1993). Without any ground, Yoon asserts that "bubble-particle adhesion occurs only when the particles are sufficiently hydrophobic".

However, Yoon does not clarify what "sufficiently hydrophobic" means. If this means the presence of an non-zero contact angle, there are no objections to use it in the discussion of the flotation theory by Derjaguin & Dukhin (1961, 1984). In this theory the terms hydrophobicity or hydrophilicity are connected with the presence of hydrophobic or hydrophilic component of the disjoining pressure. Yoon has a reason to emphasise the role of attractive structural forces in flotation, but not to neglect the possibility of microflotation in the presence of repulsive structural force because attachment can be caused by attractive electrostatic forces.

As to small particles, at sufficiently low electrolyte concentration and opposite charges of bubble and particle surfaces, the flotability of hydrophilic quartz particles in the presence of significant cationic surfactant adsorption on the particle surface has been proven by experiment (Bleier et al. 1977). These authors emphasise the peculiar nature of the flotation in their experiments, which was caused by the cationic surfactant adsorption on the bubble surface as distinct from usual practice when flotation is provided by collector adsorption on the hydrophilic mineral particle surface. The authors conclude "the property which determines ultimately whether the bubble and or the particle heterocoalesce is the relative hydrophobicity of the solid surface as described by equations (1a) and (1b)". This statement is quite different from Yoon's above. Bleier et al. (1977) use the term "relative hydrophobicity" to mean something similar to "incomplete wetting". In other words, this term refers to a non-zero value of the contact angle, because their equations (1a) and (1b) "relate the corresponding interfacial free energies to contact angle".

At the transition to high electrolyte concentration ($10^{-2} - 10^{-1}$M), electrostatic interaction is suppressed to some extent and a predomination of the electrostatic attraction force is questionable. It could preserve because the special water structure near a hydrophilic surface is destroyed by increasing salt concentration. More investigation is necessary to clarify which component of the disjoining pressure predominates as both electrostatic and structural components are suppressed to a certain degree at higher salt concentration.

Systematic investigations of Somasundaran & Li (1991, 1992, 1993) proved high adsorbability of OH^- ions and of multivalent inorganic cations on bubble surfaces. Even at high concentration (10^{-1}M) the absolute value of the negative electrokinetic potential of a bubble is sufficiently high (-40 mV) in the alkaline pH region and equals -20 mV at pH=6. The comparison of these results with the old notion of adsorption of inorganic ions at the water-gas interface (Lopis 1971, Randles 1977) shows an essential distinction.

A recharge of the bubble, caused by aluminium adsorption provides an electrokinetic potential value of 15 mV at 0.01M NaCl. These new results show that the conditions for the electrostatic attraction remain even at high salt concentration. The additional condition is not a small Stern potential of the particle surface. It is important to take into account that at high electrolyte concentration the Stern potential exceeds significantly the ζ-potential value. For example, in systematic investigations by Matijevie and co-workers (Colie et al., 1991, Hesleiter et al., 1991) the Stern potential of hematite dispersion exceeds the ζ-potential and chages in the range from 20 mV (pH=6) to 40 mV (pH=4) at an ionic strength $I = 10^{-2}$ M. At pH=4, the Stern potential decreases from 40 mV to 10 mV. These results are in qualitative agreement with measurements of additional surface conductivity by low-frequency dielectric dispersion (Kijlstra et al., 1993) for hematite. The additional (anomalous) surface conductivity is caused by a surface current across the hydrodynamically immobile part of the diffuse double layer where the potential drop $\psi_d - \zeta$ is localized (Dukhin & Semenikhin 1970, Dukhin 1993). Integrated electrosurface investigations, which include measurements of the mobile charge σ_m by means of low-frequency dielectric dispersion (Lyklema et al. 1983) and electrokinetic charge σ_ζ by electrophoresis were applied by many groups in the world for particle surface characterization. In a review Dukhin (1993) summarized these investigations and it appeared that generally σ_m exceeds σ_ζ many times and, consequently, ψ_d essentially exceeds ζ. Taking into account the high sensitivity of contact angle on Stern potential according to Eq. (10A.5), we can conclude that mobile charge measurements of colloidal particles is necessary and the existing practice of ζ-potential measurement is not sufficient for the particle characterisation in flotation systems. Perhaps this conclusion does not relate to bubbles because its surface is molecularly smooth and coincidence of ζ and ψ_d potentials can be assumed.

APPENDIX 10E: MICROFLOTATION OF SUBMICRON, MICRON AND DECIMICRON PARTICLES

10E.1.   MICRON   AND   SUBMICRON   PARTICLES.   AIR-DISSOLVED   FLOTATION   AND MICROFLOTATION

While we satisfy the condition of intensified transport stage, we simultaneously simplify also the analysis of attachment and detachment conditions. As pointed out in Section 10.4, the use of bubbles of regular flotation size results in a very low collision efficiency. The use of bubbles of approximately 20 μm diameter is necessary. This is precisely the reason for the success of electroflotation and air-dissolved flotation (Iunzo & Isao, 1982). As distinct from pneumatic flotation machines producing bubbles of millimeter size, bubbles of smaller size are generated in electroflotation and air-dissolved flotation. There was great interest in air-dissolved flotation in Russia in particular on publication of Klassen's work (1973). Four different schemes were described in books by Matsnev (1976) and Stakhov (1983). A common disadvantage of these schemes is the limit of liquid saturation with gas at the pressures used. To eliminate this disadvantage improved schemes have been developed with additional air ejection into preliminarily saturated water. The spectrum of large bubbles in water expands with additional ejection of air. The size of bubbles increases up to 50 - 1000 μm at flow rates of ejected air of 2.6 wt%. When the flow rate is decreased to 0.57 wt%, the bubble size is reduced to 50-500 μm. A finer fragmentation of bubbles is achieved by additional dispersion using special techniques (cf. Kovalenko et al., 1979). The size of bubbles escaping as a result of electrolysis depends on the way they are generated and amounts to 15-200 μm (cf. Matsnev 1976, Matov 1971). The surface of such bubbles is either completely or at least very strongly retarded. Two factors decrease detachment forces very strongly: a small particle size and a strong (complete) surface retardation. It is likely that the problem of destruction of the bubble-particle aggregate does not arise in flotation of micron particles by decimicron bubbles.

Clearly if the depth of the primary energy minimum and cohesive forces in the aggregate are very small, a detachment can take place. To overcome electrostatic repulsion is a critical task in microflotation systems of such type as well as in the transport stage.

Using the term "microflotation system" attention is paid to the dimension of particles and bubbles as primary importance in microflotation. A microflotation system is characterized first of all by these dimensions. At micron size of particles and a Stokes hydrodynamic regime the hydrodynamic pressing force is very small (cf. Fig. 10.8). Therefore, recharging the bubble surface through adsorption of cationic surfactants is necessary. A strong (complete) retardation requires reagents of very high surface activity.

For submicron particles (for example, at a size of 0.1 - 0.3 μm) even the use of decimicron bubbles does not provide a sufficiently high collision efficiency. Particles of still smaller dimensions are not discussed here since their transport is influenced by Brownian diffusion. These difficulties are overcome by particle aggregation (cf. Section 10.9).

10E.2.   DECIMICRON PARTICLES

Two possibilities are presented here. Bubbles with a completely or sufficiently strongly retarded surface can be used at Reynolds numbers less than 40. The second possibility is to use millimeter-size bubbles with less than completely-retarded surfaces. It follows from the estimate (10.50) that the transport stage proceeds in both cases with approximately equal intensity. Each variant has its advantages and disadvantages.

APPENDIX 10F: FLOTATION WITH CENTIMICRON AND MILLIMETER BUBBLES

10F.1.   THE USE OF CENTIMICRON BUBBLES

Advantages are:

1. Because of strong surface retardation of centimicron bubbles, there is unlikely to be detachment of attached particles.

2. Because of strong retardation of centimicron bubbles, a substantially lower amount of cationic surfactant of high surface activity is needed (cf. Section 10.13).

Disadvantages are:

1. It is difficult to produce centimicron bubbles. Flotation machines generate mainly larger size bubbles, and bubbles of smaller size are obtained in air-dissolved flotation and electroflotation.

2. Because of strong surface retardation and weak hydrodynamic pressing forces cationic surfactants are required.

An evaluation of the possibility of overcoming the electrostatic barrier by hydrodynamic pressing forces in the transient hydrodynamic regime has not been carried out, but it is clear that the situation will be much better at strong surface retardation than in the Stokes regime (cf. Section 10.5.2).

10F.2.   USE OF MILLIMETER BUBBLES

The advantages are the following:

1. It is easy to obtain millimeter bubbles, and the equipment to generate bubbles of this size already exists.

2. There is the possibility of non-reagent flotation. Because of non-retardation of surfaces and a high velocity of bubble rising, the normal component of the hydrodynamic field of bubbles is sufficiently large to provide hydrodynamic pressing forces to ensure the electrostatic barrier is overcome (cf. Fig. 10.8).

The disadvantages is caused by the possibility of detached particles being deposited at the leading bubble surface and displaced along the bubble surface to the zone where the normal velocity component is directed from the bubble surface, and a detachment becomes possible.

APPENDIX 10G: FLOTATION WITH BUBBLES BETWEEN MILLIMETER AND CENTIMICRON

The possibility of detachment of particles is critical when millimeter bubbles are used. If this cannot be prevented, centimicron particles should obviously be used. Millimeter bubbles can be combined with reagents whose adsorption on the particles surface prevents their detachment.

The use of millimeter bubbles is attractive because they are produced by existing technologies and because electrostatic repulsion can be overcome by hydrodynamic pressing forces. The latter advantage get lost if surfactant has to be added to modify the surface of the particle to prevent its detachment.

Attachment of particles on the rear stagnant cap of bubbles of a size between millimeters and centimicrons can combine the advantages of two processes discussed above:

1. It is simple to obtain millimeter-size bubbles.
2. There is not much chance of fixed particles becoming detached because of the strong retardation of the bubble surface in the rear stagnant cap zone.
3. Only small amounts of cationic surfactant are needed. Since attachment takes place on the rear stagnant cap, the drift of adsorbed surfactant from the leading bubble surface and its low stationary values there are not troublesome.

Both the rear stagnant cap at $Re > 40$ and the transport of particles to the rear stagnant cap remain as yet unstudied but this is a purely temporary problem. The predicted advantages point to the necessity of systematic experimental and the verification of theoretical studies. Under industrial conditions the nature of impurities, the degree of retardation of the bubble surface, the reversible or irreversible character of bubble-particle aggregates are unknown. Lack of information about all these properties of a flotation system means the choice of the hydrodynamic and reagent regimes in industrial processes must be made by following the special sequence of investigations discussed in Appendix 10C.

APPENDIX 10H: POSSIBILITY OF MICROBUBBLE CAPTURE FROM BELOW, DYNAMIC ADSORPTION LAYER AND POSSIBILITY OF DECREASE OF SURFACTANT CONSUMPTION

As noted in Sections 10.4. and 10.8., the use of bubble diamters between millimeters and centimicrons can be technologically effective since industrial methods for generating centimicron-size bubbles are less developed.

Here we have to discuss how to fix particles on the rear stagnant cap of a millimeter-size bubble. Sedimentation of particles hampers to some extent the transport of particles from below to the rear stagnant cap against the action of gravity. A more favorable situation exists in two-stage microflotation. Bubbles of decimicron size can be deposited on the rear stagnant cap from below because of their buoyant velocity. They are also transported by convective flow which arises in the vicinity of the rear stagnant cap. Considering only sedimentation to small buoyant bubbles, the lower collision efficiency can be determined from Eq. (10.25). So far there is no theory to describe the buoyance velocity of a bubble with a well-developed rear stagnant cap. The quadratic relation between buoyancy velocity of a bubble and its radius, Eq. (10.6), was obtained by Levich under the assumption of potential velocity conditions, i.e. in absence of any surface retardation. The consideration of surface retardation leads to a change from a quadratic to a linear dependence.

Collision efficiency calculated from Eq. (10.60) is several times smaller than that obtained by Sutherland's formula if a_{b1} amounts to some ten microns and a_{b2} several hundred microns.

The difference is much smaller when the deposition of particles on a bubble from the top can be slown down by the action of inertia forces (cf. Section 10.14).

To apply Eq. (10.25), it is required that the number of particles in the vicinity of the surface differs slightly from that at infinity, which is not proven for rising bubbles at $Re > 40$. The separation of the flow gives rise to two, to some extent autonomous flows with the upper flowing around the upper free surface (primary flow) and the lower flowing around the rear stagnant cap (secondary flow). When the lower flow represents a closed eddy, the number of particles in it can be decreased since microbubbles continuously go away from it to the rear stagnant cap. The supply of microbubbles to this closed eddy from outside can be insufficient and a stationary concentration lower than at infinity results with growing Reynolds numbers. The boundary layer between the upper and the lower flow turbulises and leads to an intensive turbulent exchange and to a compensation of the quantity of particles in the lower curl decreased by deposition on the rear stagnant cap.

A description of the DAL combining the rear stagnant cap and the hydrodynamic vortexes in its neighborhood appears to be necessary, which is a logical conclusion of a large series of investigations described in Chapter 8. It becomes especially vital in connection with the problem of intensification of microflotation and in particular in two-stage microflotation.

APPENDIX 10I: AIR-DISSOLVED FLOTATION AND TWO-STAGE FLOTATION

Particle-bubble aggregates can be formed not only by collisions but also as a result of separation of bubbles from a solution (Klassen, 1973) in a so-called air-dissolved flotation process. Formation of floto-complexes is obtained in several ways: as a result of collision of particles and bubbles (1), by ejection of gas and solution supersaturated with gas (air) to the particle surface (2), and by capture of rising bubbles by particle aggregates (3).

Not only for process (1) but in general the air-dissolved flotation is based on regularities described in preceding sections. The separation of floto-complexes of decimicron size from the liquid is necessary. Air-dissolved flotation can be used as the first stage of two-stage flotation. This two-stage microflotation is now used industrially.

APPENDIX 10J: INDUSTRIAL APPLICATION OF TWO-STAGE MICROFLOTATION

With the use of 12-meter high columns in Mines Gaspe plant in Canada (Coffin, 1982), losses of molybdenum particles with diameters less than 30 μm in the cycle of repurification have been decreased substantially. An increase of recovery in such a high equipment is apparently connected to the release of dissolved air at the surface of particles and with a further capture of activated aggregates by large bubbles, i.e. the coalescence mechanism first proposed by Klassen (1959) acts under these conditions.

A strengthening of the coalescence mechanism of recovery is achieved through saturation of a part of water with air or of the whole pulp, and also by direct ejection. Investigations have shown that the method of presaturation of the whole pulp under a pressure of 343 kPa in a tank and subseqent flotation in a mechanical flotation machine (Shimoiizaka, 1982) is advantegeous.

Column plants (Patent a) in which ejected air is released to the liquid layer in form of microbubbles with a diameter of 17-20 µm and large bubbles with a diameter of 1-2 mm from below through perforated aerators (Nebera, 1982) exhibit high efficiency. The specific volumetric efficiency of such column plant is 100 times higher than that of mechanical ones. The total value of hydrodynamic capture increases, probability by collision and attachment of particles on small bubbles; the velocity of activated rising aggregates is controlled by the velocity of large rising bubbles and strong turbulent flows are absent which benefits the existence of floccules.Thus the rate of inertialess flotation of particles increases 6 to 10 times under flocculation conditions (Samygin et al., 1980).

Application of express-flotation is promising for recovery of small particles (Petrovitz, 1981). The process consists of the following operations. Pulp is saturated with microbubbles 100 µm in diameter in a mechanical flotation machine at intensive agitation of a small amount of air (80 to 100 per m^3 pulp). Saturated pulp is released to a thickener where separation into foam and chamber product takes place. Saturation time is 5 min, flotation time is 1 min. Extraction of finely pulverized iron ore of Olenegorskii deposites makes up 98%. In usual flotation such recovery is attained in 10 min.

Air dispersion to microbubbles can be carried out without application of agitators by surface streams of liquid (Patent b). Such machines are recommended for purification of paper production sewage. Floccules are destroyed by mechanical flotation purification.

APPENDIX 10K: BUBBLE POLYDISPERSITY IN TWO-STAGE FLOTATION

As pointed out in Section 10.8 and shown in Fig. 10.11, a remarkable polydispersity of bubbles is observed in air-dissolved flotation and electroflotation. Its degree is highly dependent on the conditions of electroflotation and microflotation.

If the polydispersity of bubbles is high, the process of particle recovery is performed by a two-stage flotation even without an additional introduction of centimicron bubbles. Small bubbles mostly capture particles and large bubbles mostly coalesce with small bubbles. The capture process is useful and a high rate of the coalescence process can be disturbing. Indeed, it leads to a rapid decrease of concentration of small bubbles and therefore to a decrease of particle capture. A number of conclusions follow.

If the polydispersity of bubbles generated in air-dissolved flotation or electroflotation is high, there is no need for additional introduction of centimicron bubbles. Optimal flow of two-stage flotation corresponds to the maximum attainable degree of monodispersity of bubbles. In this case the ratio between volume fractions of micro- and macrobubbles and collision efficiencies of the processes of particle capture by small bubbles and bubble coagulation must be such that the particle capture process outweighs the process of coalescence.

Since a reduction in the bubble spectrum is connected to additional difficulties and polydispersity of bubbles in electroflotation and air-dissolved flotation cannot be therefore avoided, it is important to know the effect of polydispersity. This effect should not be too high since the process of particle capture should be ahead of the process of coalescence. It should also not be too low so large bubbles can capture small bubbles in a moderate time. Unfortunately, such a scheme which appears convincing at a first glance means a slow rate of coalescence of small bubbles. The coalescence can proceed faster than particle capture, so that the intensification of capture becomes very important. Hence it is necessary to combine introduction of small bubbles with aggregation of particles.

APPENDIX 10L: TWO-STAGE MICROFLOTATION WITH PARTICLE AGGREGATION

Two schemes of combined effects in waste water through introduction of microbubbles and aggregation are known in the literature. Two-stage process flow is characteristic of the first scheme. Aggregation of particles is arranged first and bubble capture follows.

Additional advantages appear with the use of electrocoagulation-flotation enabling both processes to be carried out simultaneously: variation of the dispersed state takes place as a result of their coagulation under an electric current, of ions of dissolving metal of electrodes or of other products of electrochemical reactions in the bulk of electrolyte, and of fixing of electrolytic gas bubbles on the surface of coagulated particles which provides their following flotation (Rogov, 1973). An obligatory condition for the flow of the electrocoagulation-flocculation process is the passage of purified liquid in the interelectrode space during a time required for particle coagulation and formation of floto-complexes.

These examples are presented not only to demonstrate the wide-spread application of microflotation in water purification. It is apparent that optimal technical design (selection of microflotation version and process parameters, such as volume fraction and size of bubbles, hydrodynamic conditions in flotation aggregates, etc.) strongly depends on properties of water which vary over a wide range depending on the plant contaminating the water. Even for one and the same plant waste water varies because of changing conditions of the production process. Optimisation and control of the operation of microflotation plants requires the development of mathematical models of the process. Most difficult is the description of collision efficiencies. Thus, the requirements for a collision efficiency theory increase, most of all due to the need of taking into account adsorption layer dynamics.

APPENDIX 10M: COMMENTS ON THE ROLE OF BOUNDARY LAYER AND CENTRIFUGAL FORCE AS DISCUSSED BY MILEVA (1990)

For the first time Mileva (1990) has considered the effect of a hydrodynamic boundary layer on the elementary act of inertia-free microflotation based on a mobile bubble surface free of an adsorption layer and at high Reynolds numbers. The velocity distribution is a potential one, Eq. (8.117), with an additional contribution of the velocity differential along the cross-section of the boundary layer, Eq. (8.127). The difference between the velocity distribution along the bubble surface and the potential distribution is given by Eq. (8.128). In contrast, Mileva (1990) used the formulas of a related theory by Moore (1963) which are the solution of the same Eq. (8.122) under the same boundary condition (8.118).

Like the potential hydrodynamic field of a bubble, Moore's formulas are only true at high Reynolds numbers (Re > 100) when the effect of the boundary layer on the tangential velocity component is negligible.

Integration of Eq. (10.28) along the cross-section of the hydrodynamic layer allows us to check whether within its limits the radial velocity component is proportional to the tangential derivative of the velocity distribution along the bubble surface, which differs slightly from the potential distribution. The effect of a boundary layer on the normal velocity component and on inertia-free deposition of particles should be therefore very small. The formula for the collision efficiency given by Mileva as an inertia-free approximation is thus $\sqrt{Re}$ times less than the collision efficiency according to Sutherland, which is definitely erroneous.

The error in deriving this formula is that an incomplete expression for the hydrodynamic field of a bubble was substituted into the equation for the liquid stream-line. Of course, it is impossible to obtain a correct result and, in particular, the limiting case of Sutherland's formula by dropping the highest degree term (potential velocity distribution).

10M.1. LIFT FORCES

While the effect of Faxen (1922) is caused by viscous interaction of particles with the spatially inhomogeneous hydrodynamic field, the lift force is an inertia force proportional to the lag velocity, v_L, i.e. to the relative particle velocity. Within a tangential flow in which a velocity gradient v_θ / z exists, a force acts on the particle perpendicularly to the flow direction. The sign of this force depends on v_L: if $v_L < 0$, the lift force is directed towards the bubble, if $v_L > 0$. F_L is a drift force (lift force) directed away from the bubble. According to Saffman (1965) $F_L = 3.2 \, \eta a_p v_L \sqrt{Re}$.

10M.2. REYNOLDS NUMBERS CHARACTERISING THE PARTICLE MOTION RELATIVE TO THE LIQUID

In the case of a retarded bubble surface, the velocity v_L exceeds that for a free surface $\sqrt{Re}$ times since the boundary layer thickness differs slightly and the velocity gradient along the boundary layer is $\sqrt{Re}$ times greater. Thus, at a retarded surface a more substantial F_L can be expected. Mileva (1990) compares F_L and the pressing force (Dukhin & Rylov 1971) for the case of a potential flow and concludes that F_L is negligible. But just from such a comparison it becomes evident that for retarded bubbles the pressing force is substantially less than F_L.

10M.3. LIFT FORCE AND CENTRIFUGAL FORCE

Comparing F_L with different other forces, Mileva (1990) concludes that for non-retarded bubble surfaces F_L considerably exceeds the centrifugal force. Note that the gravitational force exceeds F_L by one order of magnitude. Thus, from comparing Mileva's estimation for the ratio of F_L over the centrifucal force with the ratio of F_L over the gravitational force we can conclude that gravitational forces exceed centrifugal forces by 2 orders of magnitude. The work of Mileva (1990) contains an error in one of the very bulky expressions since the centrifugal force must exceed the gravitational force at bubble radii greater than 200 μm (Eq. 10.79). It is essential that the particle density cancels out in the ratio "centrifugal force/gravitational force". When passing from a non-retarded to a retarded surface, the centrifugal force can strongly decrease while F_L increases, so that it can exceed the centrifugal force. Unfortunately this statement was formulated by Mileva for a non-retarded surface which is absolutely wrong.

APPENDIX 11A: CORRECTION OF THE CALCULATION OF CENTRIFUGAL FORCES AT $St > St_c$

When the Stokes number increases the tangential particle velocity $v_{p\theta}$ deviates more and more from the local liquid velocity v_θ. This velocity was taken into account in the derivation of Eq. (11.36). The decrease in particle velocity is expressed by the multiplier $\frac{2}{3}S$ which leads to a large decrease in the centrifugal forces caused by $\beta \sim \left(\frac{2}{3}S\right)^2$. However, Eq. (11.46) underestimates $v_{p\theta}$.

566

A particle attached to the bubble surface is under the strong action of centrifugal forces which are proportional to $\dfrac{v_\theta^2}{a_b}$. At the same distance such a particle would be under the action of a centrifugal force $\dfrac{2}{9}S^2\dfrac{v_\theta^2}{a_b}$, which is contradictory.

As the distance between bubble and particle becomes small the large viscous stress in the liquid gap prevents the large difference in the velocities of bubble and particle. Thus, the equation for β must be generalised, i.e. by Eq. (11.76).

APPENDIX 11B: ANALYSIS THE THEORY BY SCHULZE AND OTHERS ON ATTACHMENT BY COLLISION

Calculations performed by Schulze et al. (1989) are well known along with the theory described in Sections 11.1 - 11.7. General theoretical prerequisites of these investigations coincide in many respects.

Considering deformation of the surface of a spherical particle, Schulze et al. take into account the formation of a meniscus beyond the plane-parallel spherical film, which was not taken into consideration in Sections 11.1 - 11.2. Moreover, as stated by the authors, the impact of a particle on a bubble surface generates a wave propagating along the bubble surface. We agree that a term reflecting the energy loss connected with this wave must be introduced into the energy balance at the basis of the theory. Its neglecting in the theory of Schulze et al. (1989) means neglect of the energy dissipation connected with drainage.

The theory of Schulze et al. has been developed to interpret experiments by Stechemesser et al. (1980), in which the impact of spherical particles against the water-air interface was initiated by the effect of gravity. In this case the velocity of the particle at the beginning of the impact was given as the boundary condition. This velocity is unknown under flotation conditions. A method of its calculation is developed in Section 11.2, which is based on the consideration of particle energy loss during the process of a rising bubble approach to the surface. Thus, the theory of Rulyov & Dukhin (1986) is developed with a view to the hydrodynamic conditions of inertia particle-bubble interactions under flotation conditions, in contrast to the theory of Schulze et al. (1989). The theory by Rulyov & Dukhin (1986) considers sub-processes of this interaction and their interrelation. The drainage of liquid from the water interlayer between particle and bubble, described in Section 11.2, is caused at any instant of time by the inertial particle-bubble interaction. The inverse effect of drainage on inertia interaction is also taken into account since drainage is accompanied by viscous dissipation of energy (Section 10.2). In the theory of Schulze et al. (1989) the single processes taking place in a bubble-particle system and the water film in between are presented artificially as two independent sub-processes: particle oscillation on the bubble surface and drainage. Clearly, neglecting the interaction of these sub-processes results in distortion of the physical picture of the phenomenon and thus in substantial quantitative errors.

Describing inertial interaction, Schulze et al. (1989) do not take into account energy dissipation caused by drainage. Thus, damped oscillations of particles are substituted by continuous oscillations. Section 11.2 showed that the energy dissipation in the film results in a damping of oscillations. The accuracy of the description of drainage duration the so-called induction time τ_i is decreased to a still greater extent (Schulze et al., 1989). When the bubble surface deformation grows under the effect of a particle collision, the radius of the film between them also grows (cf. Fig. 11.1b) and its thickness reduces. This means that the drainage ratio decreases very strongly and the drainage (induction time τ_i) can be calculated only through the integration of an equation of the type of Eq. (11.29). This equation was not available and, therefore, the strong time dependence of the drainage rate was neglected by Schulze et al. Thus they have to restrict themselves by the estimation of τ_i by Reynolds formula for hydrodynamic resistance. The hydrodynamic resistance which strongly varies with time is replaced by a value at the last instant when film thinning has reached the critical value h_{cr}.

568

LIST OF SYMBOLS

A	area of the interface [cm²]
A_H	Hamaker constant
a_b	bubble radius
a_L	constant of the Langmuir isotherm [mol/cm²]
a_F	constant of the Frumkin isotherm [mol/cm²]
a_p	spherical particle radius
a'	interaction constant in the Frumkin isotherm [mN/m]
C	dimensionless surfactant concentration [-]
c	bulk concentration [mol/cm³]
c_o	equilibrium bulk concentration [mol/cm³]
c_{el}	electrolyte concentration [mol/cm³]
D	diffusion coefficient [cm²/s]
D_{eff}	effective diffusion coefficient [cm²/s]
D_{el}	diffusion coefficient of electrolyte ions[cm²/s]
E	electric field
E	collision efficiency
E_a	attachment efficiency
E_0	collision efficiency caused by LRHI
E_o	thermodynamic surface dilational modulus [mN/m]
E_c	capture efficiency
$E_c = E' + E''$	complex surface elasticity
E_{dif}	electric field of the diffusion potential
E'	real part of the complex modulus
E''	imaginary part of the complex modulus
e	charge of an electron [1.6 10⁻¹⁹ coul]
f	frequency of oscillations [Hz]
$f(x)$	$= e F / kT$, dimensionless potential distribution
F	Faraday constant [9.652 10⁴ coul/g]
F	Fourier-Laplace operator
F^{-1}	inverse Fourier-Laplace operator
G	interfacial Gibbs free energy
g	acceleration constant [cm/s²]
$H = h / a_p$	dimensionless thickness
H	enthalpy
ΔH	amplitude of oscillations [cm]
h	depth of a bulk phase; film or layer thickness [cm]
h_{cr}	critical thickness

$i = \sqrt{-1}$	imaginary unit
I	current density
$j^{\pm}$	ion stream
K	distribution coefficient
K_H	Henry constant $= \Gamma_o/c_o$ [cm]
$K(\phi)$	coefficient of electrostatic deceleration
k	Boltzmann's constant [1.38 10^{-23} J/mol/K]
k_{ad}	rate constant of adsorption [cm/s]
k_{des}	rate constant of desorption [1/s]
k_f	rate constant of micelle formation [1/s]
k_d	rate constant of micelle dissolution [1/s]
L	length of a trough [cm]
N	Avogadro number
n	number of molecules
P	pressure [N/m²]
Pe	Pecklet number
P_γ	capillary pressure [mN/m²]
q(t)	flux of molecules towards the interface
R	gas law constant [8.314 g cm²/(s² mol K)]
Re	Reynolds number
R_m	reflection of a monolay-covered interface
R_w	reflection of a monolay-free interface
r	number of components
r_o	radius of curvature of a bubble or a drop at its apex [cm]
S	entropy
St	Stokes number
T	absolute temperature [K]
t	time [s]
u	flow velocity [cm/s]
V	volume [cm³]
V_{dif}	diffusion potential difference
V_o	volume of a measuring chamber [cm³]
ΔV	Volta potential difference [V]
v	buoyancy bubble velocity
W	interaction energy
$W^{\pm}$	adsorption energy of anions or cations
X	dimensionless coordinate in x direction
x	direction normal to the interface [cm]

Y	dimensionless coordinate in y direction
y	direction tangential to the interface [cm]
z	radial coordinate in a spherical system
$z^{\pm}$	valency of ions
z_s	valency of surfactant ions
α	inclination angle [deg]
α_i	coefficients of a series or a polynomial
β	damping coefficient
β_s	damping coefficient of shearing
χ	coefficient of retardation
χ_b	coefficient of retardation for a bubble surface
Δ	thickness of adsorption layer [cm]
δ_D	thickness of the diffusion layer [cm]
δ_G	hydrodynamic thickness [cm]
ε	dielectric constant
$\Phi(\omega)$	system function of the oscillating bubble method
ϕ	dimensionless potential
φ	loss angle
$\bar{\Gamma} = \Gamma(t)/\Gamma_o$	dimensionless adsorption
Γ_d	adsorption at time t=0 [mol/cm²]
Γ_i	adsorption of component i [mol/cm²]
Γ_o	equilibrium surface concentration [mol/cm²]
Γ_∞	maximum surface concentration [mol/cm²]
γ	interfacial tension [mN/m]
γ_o	interfacial tension of the pure solvents [mN/m]
η_α	bulk viscosity
η_d	surface dilational viscosity
η_s	surface shear viscosity
κ	complex wave number
κ	thickness of the electric double layer [cm⁻¹]
κ	line tension
λ	wavelength [cm]
μ	chemical potential
μ_i	chemical potential of component i
$\nu = \eta/\rho$	dynamic viscosity
$\Pi = \gamma_o - \gamma$	surface pressure
Π	disjoining pressure

Θ	dimensionless time
θ	angle coordinate in a spherical system
ρ	density [g/cm³]
ρ_ψ	volume density of charge
σ	charge
τ	thickness of the interfacial region [cm]
τ_D	relaxation time of the diffuse part of the electric double layer [s⁻¹]
τ_{ad}	characteristic time of the adsorption process
$\psi(x)$	potential distribution in the diffusion layer of the electric double layer
$\overline{\psi}(x) = \dfrac{e\psi(x)}{kT}$	dimensionless potential
ψ_{St}	potential of the Stern layer
ψ_{Sto}	equilibrium Stern potential
$\overline{\psi}_{St} = \dfrac{e\psi_{St}(x)}{kT}$	dimensionless Stern potential
$\overline{\psi}_{Sto} = \dfrac{e\psi_{Sto}(x)}{kT}$	dimensionless equilibrium Stern potential
Ω	area disturbance (d ln A/dt)
ω	angular frequency [rad⁻¹]
ω_o	relaxation frequency [rad⁻¹]
ξ	displacement of a surface element
ψ	electric potential
Ψ	angle characterising the rear stagnant cup

Indices

α	bulk phase α
a	attachment
ad	adsorption
β	bulk phase β
b	bubble
B	boundary
c	capture
cr	critical
des	desorption
diff	diffuse part of the electric double layer
dyn	under dynamic conditions
P	potential flow
p	particle
R	cation

RY	anion-cation complex
S	Stokes flow
s	surface
st	static
St	Stern
x	normal to the interface
y	tangential to the interface
Y	anion
±	anion, cation

Abbreviations

ACS	after collision stage
BCS	before collision stage
BSA	bovine serum albumin
BZR	Belousov-Zhabotinski reaction
CMC	critical concentration of micellisation
CTAB	cetyl trimethyl ammonium bromide
DAL	dynamic adsorption layer
DDR	Derjaguin Dukhin Rulyov model
DL	electric double layer
DLVO	Derjaguin-Landau-Verwey-Overbeek theory
FRAP	fluorescence recovery after photobleaching
f.s.f.z.	front stagnation free zone
HA	human serum albumin
HHF	Hogg-Healy-Fuerstenau approximation
IUPAC	International Union of Pure and Applied Chemistry
LRHI	long-range hydrodynamic interaction
r.s.c.	rear stagnant cap
RSP	rear stagnant point
SDS	sodium dodecyl sulphate
SGC	Stern-Gouy-Chapman theory
SRHI	short-range hydrodynamic interaction
t.p.c.	three phase contact
2D	two dimensional
3D	three dimensional

Subject Index

574

Bakker's equation, 41
Bancroft's rule, 21
Belousov-Zhabotinski reaction, 510
bilayers, 9
biological membranes, 9
blow up effect, 155
Bogolubow equation, 259
Boltzmann distribution, 59
Boltzmann law, 239
boundary layer, hydrodynamic, 301
boundary layer, of diffusion, 301
Boussinesq equation, 93
Boussinesq's theory, surface viscosity in, 296
bovine serum albumin, 188
BRIJ58, 227
BSA, 228
BSA adsorption kinetics, 191
bubble coalescence, 389
bubble dead time, 535
bubble formation frequency, 158
bubble fractionation, 313
bubble surface, deformation of, 342
bubble surface, mobile, 270
bubble surface, molecular contaminants, 364
bubble surface, particle reflection, 436
bubble surface, residual mobility of, 356
bubble time, 121
bubble times, definition of characteristic, 159
bubble velocity, 389
bubble, dead time, 158
bubble, physico-chemical hydrodynamics of, 278
bubble, surface lifetime, 158
bubble-particle aggregate, 342
bubbles, volume of detaching, 535
bubbling , steady-state, 390
buoyant bubble, rear pole, 269
buoyant bubble, rear stagnant region of, 308
buoyant bubble, 243
    rising time, 315
    hydrodynamic field of, 269

Burgers model, 79
capillary length, effect on bubble dead time, 535
capillary waves, generation of, 207
capture effectiveness, 440
casein, 188; 189
cetyltrimethyl ammonium bromide, 230
chaotic turbulence-like behaviour, 512
characteristic frequency, 210; 214
characteristic length, 315
characteristic time,
    of adsorption , 205
    of bubble surface restoration, 324
    of ionic adsorption, 251
charge determining ions, 55
charged interface, 52
charged surfactants, 31
Chebyshev approximation, 516
co-surfactant, 24
coating process, 128, 202
coion concentration in the DL, 239
collapse pressure of a monolayer, 540
collision efficiency, 345; 389; 440
collocation method, 109
    solution for diffusion model, 524
complex elasticity modulus, 209
computer simulation, 33
concentration distribution, 327
concentration gradients in surface films, 513
contact angle measurements
    liquid films, 500
    dynamic contact angle, 442
convective diffusion, 242
    equation, 275
    adsorption under, 255
    non-stationary state, 269
    steady-state, 269
correction factors of Wilkinson, 525
Coulomb interaction, 53
coupling
    of hydrodynamic with surface forces, 1
    of transport of momentum and mass, 319
critical film thickness, 350
critical micelle concentration, 23; 124

Printed and bound by CPI Group (UK) Ltd, Croydon, CR0 4YY

03/10/2024

01040330-0011